T0181179

Ways of Knowing in HCI

Judith S. Olson • Wendy A. Kellogg
Editors

Ways of Knowing in HCI

Editors
Judith S. Olson
Donald Bren Professor of Information and Computer Sciences
University of California, Irvine
Irvine, CA, USA

Wendy A. Kellogg
IBM T.J. Watson Research Center
Yorktown Heights, NY, USA

ISBN 978-1-4939-4405-7 ISBN 978-1-4939-0378-8 (eBook)
DOI 10.1007/978-1-4939-0378-8
Springer New York Heidelberg Dordrecht London

Printed on acid-free paper

Springer is part of Springer Science+Business Media (www.springer.com)

We dedicate this book to Gary Olson and John Thomas, who are not only colleagues who know a lot about many of these methods but also partners and supporters in the academic life. We appreciate your input, patience, and encouragement.

Special Dedication

During the last year before publication, one of our valued colleagues and authors, John Reidl, continued to work, laugh, and be friends while enduring the scourge of melanoma. We lost John, Date is July 15, 2013. He will be sorely missed.

Contents

Prologue

The field of HCI grew from the field of human factors applied to computing, with strong roots in cognitive psychology. The seminal book by Card, Moran, and Newell, *The Psychology of Human Computer Interaction*, (1983) gathered what was known from cognitive psychology to apply to the design of computer interfaces—aspects of motor movement with Fitts' law, perception with gestalt laws, the differences between recall and recognition guiding the preference of menus over command lines, etc. Studies involving experiments and surveys dominated.

But the field has grown since then. The roots from psychology are still relevant: aspects of perception and motor movement and memory still guide the rationale behind designs of mobile computing, Google Glass, and embedded computing. But the focus is wider. We study not only the design of the interface but also the setting in which computing is embedded, the needs of people in various contexts, and the activities they engage in while using various forms of computing. In 1987, Lucy Suchman introduced ethnographic methods of knowing how people navigate their lives in her seminal book *Plans and Situated Action*. Later came the advent of design as a way of knowing in which designers would push the boundaries of what is introduced in the world in order to find out more about the world. All along technology researchers were building amazing new capabilities to meet people's needs or making it easier for other developers to build things by giving them toolkits.

With the variety of research methods came challenges to the field. Reviewers as well as readers asked, "What counts as good research? How do we know whether the advance is done well? Can we trust the findings? Do we know more now than we did?"

This book grew from this challenge. From discussions among the attendees of the 2010 Human Computer Interaction Consortium (HCIC), we decided to hold a number of tutorials on the variety of methods that researchers use in HCIC, not with the goal of becoming expert in each, but to be tutored to a level where, in reading a paper, we could tell whether the method was done well and what the contribution was. We wanted to become able reviewers of a variety of methods well beyond those we were formally trained in. The success of the tutorials, run for 2 years because the first year was so well received, generated the idea of making this knowledge, these sensibilities, more widely available. Thus came this book: *Ways of Knowing in HCI*.

The chapters in this book are remarkably diverse. There are chapters on ethnography, grounded theory method, and action research. Three chapters focus on system building: technical research, building an experimental online community, and field deployments. Two chapters focus on design research, one contrasting design research with science and one explicating what is involved in research through design. There are two chapters covering experiments and surveys, with an additional chapter showing how crowdsourcing can help both. Three chapters address new sources of digital data: sensory systems, eye tracking, and log analysis. Following these are three newer analysis techniques in HCI: retrospective analysis, agent-based modeling, and social network analysis. Because many of these methods extend to the world of online activity, there are new ethical challenges, described and discussed in a new chapter on ethics.

Though this collection represents a remarkably broad set of methods, or as we prefer, "ways of knowing," it is not complete. We have no explicit chapter on how to conduct and analyze interviews from the field; ethnography has more stringent requirements on the researcher in that the interviewer/observer is a player in the activities and attempts to understand the experience of the people, not just their activities. For those interested in learning more about interviewing we point them to the Sage publication, "Doing Interviews," by Steinar Kvale.

Similarly, we do not have a chapter on Contextual Inquiry, the method of examining a complex situation in order to generate ideas on how to make the situation better in one way or the other. We point readers to Beyer and Holtzblatt (both books).

We also are lacking in a chapter building on the roots of cognitive modeling, the work that grew direction from *The Psychology of Human Computer Interaction*, used in designing interactions, especially for people doing a task all-day-every-day in operations. And, there are ways of knowing over longer periods of time under the rubric of historical research, recently exemplified in Edwards on global warming, and maybe Bowker and Starr on medical categorization.

In spite of these omissions, we believe this collection of chapters to be highly useful, and, for some, enlightening *in the whole*. We have had the privilege of reading them all, and in close proximity, giving us a perspective on how the methods compare, how they might be used in conjunction, etc. We recommend such a reading, and suggest that in reading in the whole that the following aspects of the methods be called out:

- What is the situation in which the data are collected?
- What do the data consist of?
- What kind(s) of analyses are performed on the data to generate "knowing?"
- What kinds of questions can this method answer (and what not)?

In the epilogue, we will attempt to point out some comparisons on these and other relevant dimensions, helping the readers to see a bigger picture of the methods in our field, and where these might be going in the future.

The 34 authors of these chapters were asked to provide not a tutorial, per se, but advice on what is entailed in doing this kind of research, what cautions to attend to, and what kinds of things to report in publications so reviewers as well as readers could judge the work for trustworthiness and value. All authors were asked to cover the following topics, not as a template, but to provide some consistent coverage from chapter to chapter:

- A short description of the essence of the method
- Its history or intellectual tradition, and evolution
- What questions the method can and cannot answer
- How to do it: What constitutes good work

One of the most valuable resources in this book is the reference list. To make it a bit more useful yet, we asked authors to indicate which references would help a reader become more expert in the method, if such references exist, and to indicate somewhere in the paper some examples of where this method was used well in HCI. In addition, we asked authors to say something about what attracted them to this particular approach to knowing in HCI; a bit of history of them doing this kind of work.

Irvine, CA, USA Judith S. Olson
Yorktown Heights, NY, USA Wendy A. Kellogg

Reading and Interpreting Ethnography

Paul Dourish

Ethnography Simply Defined

In the context of this volume, it is appropriate to begin with a simple definition. Ethnography is an approach to understanding cultural life that is founded not on witnessing but on participation, with the goal of understanding not simply what people are doing, but how they experience what they do. This idea has many significant consequences, and this chapter attempts to make them clear.

Introduction

Although ethnographic methods are still regarded, to an extent, as new aspects of HCI research practice, they have been part of HCI research almost since its inception, and certainly since the early 1980s, about the same time as the CHI conference was founded. What, then, accounts for this sense of novelty and the mystery that goes along with it? One reason is that ethnographic methods have generally been associated with what we might call nontraditional settings in relation to HCI's cognitive science roots, emerging at first in organizational studies of collaborative work (in the domain of CSCW), subsequently applied in studies of alternative interaction patterns in ubiquitous computing, and later still associated with domains such as domestic life, experience design, and cultural analysis that have been more recent arrivals on the scene. Another is that ethnographic methods are often associated with forms of analysis and theorizing of human action—ethnomethodology stands out as an example here—that are themselves alien to HCI's intellectual traditions

P. Dourish (✉)
Donald Bren School of Information and Computer Sciences, University of California Irvine, 5086 Donald Bren Hall, Irvine, CA 92797-3440, USA
e-mail: jpd@uci.edu

J.S. Olson and W.A. Kellogg (eds.), *Ways of Knowing in HCI*,
DOI 10.1007/978-1-4939-0378-8_1,

and which have not always been clearly explained. Indeed, debates within the field have often been founded on these sorts of confusions, so that in the internecine battles amongst social theorists, ethnographic methods suffer collateral damage (e.g., Crabtree, Rodden, Tolmie, & Button, 2009). Finally, in a discipline that has often proceeded with something of a mix-and-match approach, liberally and creatively borrowing ideas and elements from different places, ethnography has often been seen instrumentally as a way of understanding important aspects of technological practice while its own epistemological commitments have remained somewhat murky.

The focus of this chapter is on this last consideration—examining foundational commitments associated with the main stream of ethnographic work as borrowed from anthropology and, to an extent, from sociology. So, this chapter does not set out to instruct the reader on conducting ethnographic research. In such a small space, any account would inevitably be misleadingly partial, and besides, several excellent overviews are already available (see the Recommended Reading section at the end of the chapter.) Besides, not everyone in HCI who encounters ethnographic work wants to carry it out. My goal here then is somewhat different, and, I hope, more broadly useful. It is to explain how to read, interpret, and understand ethnographic work. That is, the focus here is on what ethnography does and how it does it, so as to provide those who read, review, and consume ethnographic research with a sound basis for understanding what it sets out to do and how it achieves its ends. The approach that I will take here is largely historical, or at least uses a historical frame as a way of contextualizing contemporary ethnographic work. By explaining something of where ethnography began and what issues it responded to, and by then tracing some of the debates and intellectual currents that have shaped different periods of ethnographic research, I hope to be able to provide some insight into the rationales of ethnographic practice. Arguably, this is no less fraught with peril than the tutorial approach, and no less subject to partiality and revisionism; hopefully, though, the omissions will perhaps be less consequential and the benefits more widely felt. The approach that I take here is one that is shaped in particular by recent interest in what has been labeled "third wave" (Bødker, 2006) or "third paradigm" (Harrison, Sengers, & Tatar, 2011) HCI—an approach that focuses on technology not so much in utilitarian terms but more in experiential and ones. The epistemological challenges of third paradigm HCI warrant a reassessment of ethnographic methods, ethnographic theorizing, and ethnographic data in HCI research, one that attempts to recover ethnography's context.

Perspectives

When teaching ethnography, I often begin with two remarks about ethnographic practice from well-known anthropologists, both of which emphasize the questions of engagement and emergence in ethnographic work.

The first is from Marilyn Strathern (2003), who comments that ethnography is "the deliberate attempt to generate more data than the investigator is aware of at the time of collection." Two aspects of this comment are particularly significant in terms of ethnography as a means of knowing within HCI. One, to which we will return later, is the idea that ethnographic data is generated rather than simply amassed; that data is the result of an ethnographer's participation in a site rather than simply a feature or aspect of the site that the ethnographer harvests while hanging around. The second and more immediately relevant consideration, though, is the fundamental notion expressed here. How is it that more data can be generated than the ethnographer is aware of? From the perspective of traditional forms of HCI analysis, this seems nonsensical; the idea that data is not simply what is recorded in notebooks, gathered in spreadsheets, or captured on tape or digital materials is already a move beyond the cycle of define–measure–record–analyze–report. It speaks instead to a process of unpredictability, of interpretation and reinterpretation, and of ongoing reflection; it speaks also to a provisional and open-ended process in which (again in Strathern's words) "rather than devising research protocols that will purify the data in advance of analysis, the anthropologist embarks on a participatory exercise which yields materials for which analytic protocols are often devised after the fact." Ethnography, then, is data production rather than data gathering, in the sense that an it is only the ethnographer's presence in the field and engagement with the site—through action and interaction—that produces the data that is then the basis of analysis.

The second remark is by Sherry Ortner (2006), who describes ethnography as "the attempt to understand another life world using the self—or as much of it as possible—as the instrument of knowing." There are several important considerations to take from this felicitous phrase.

The first is the emphasis on the *life world* as the central topic into which ethnographic work inquires. This implies a holistic concern with forms of being and experience, a perspective that often seems to be at odds with a more circumscribed, task-oriented perspective at work in HCI studies, in which we might be more interested in smaller fragments of experience—writing documents, videoconferencing with the grandkids, going to the bank, sharing photographs, or navigating urban space, for example. Indeed, this holistic perspective is frequently a source of tension in multidisciplinary HCI teams, for example on the occasions where ethnographic research frames going to the bank in terms of the broader embedding of people in the logic of finance capital or attempts to understand video conferencing in terms of the responsibilities of kinship.

The second consideration is the focus on the self. What does it mean to suggest that the self is an instrument of knowing? It requires us to imagine that the process of ethnographic fieldwork—going places to see what happens—is not merely a question of traveling to the places where things happen in order to witness them but is more about the insertion of the ethnographer into the scene. That is, if we think about ethnography's primary method as participant-observation, then it directs our attention towards the importance of participation not just as a natural and unavoidable consequence of going somewhere, but as the fundamental point.

This, in turn, suggests that question that often arises in interdisciplinary investigations—"doesn't the ethnographer alter things by being there?"—is ill-founded on the face of it. That is, the ethnographer absolutely alters things by being there, in exactly the same way as every other participant to the scene alters things by being there; indeed, there is "no there" without the participation of whatever motley band of people produce any particular occasion, from a cocktail party to a dissertation defense. The ethnographer is just another party to the scene.

The third is the important elaboration of this form of participation suggested by the phrase "as much of it as possible." This formulation underscores that there are no aspects of that participation that are not germane. It is not simply what the ethnographer might see or hear, but also, for example, what the ethnographer might feel; that is, the ethnographer's discomforts, disquiets, joys, and anticipations are as much ethnographic data as the statements of others to the extent that they reveal something of how a setting is organized (whether it is organized to produce the same forms of emotional response in its subjects, for example, or whether there are aspects of one's participation in a setting that serve to mitigate or defuse these kinds of responses, or whether again these are perhaps the point in the first place).

Ortner's pithy description of ethnographic method cuts straight to the heart of the matter, then, in terms of the kinds of participation that are fundamental to the production of ethnographic accounts. We will be able to understand this better, though, if we can place it in some sort of context. The history summarily sketched in the pages that follow will attempt to do just that.

1910s: Origins

The history of ethnography begins in anthropology, although anthropology itself does not begin with ethnography. The systematic study of culture is a discipline that arose in consequence of European exploration and particularly of colonial expansion, which created a context of cultural encounter to which anthropology was an academic response. Early anthropology, though, was often something of an armchair discipline, conducted in the libraries and museums of colonial metropoles like London and Paris, where artifacts, reports, and materials from around the world were collected, collated, and compared. Even when anthropologists ventured out to the places inhabited by the people they studied, they typically did so as members of larger expeditions—military, scientific, and exploratory—and conducted their work from the safety of the stockade and the shaded comfort of the verandah.

The traditional (although partial) history of the development of the ethnographic method begins with a Polish scholar, Bronislaw Malinowski, who worked in England for most of his professional life. Studying at the London School of Economics in 1914, Malinowski joined an expedition to Papua, lead by one of his advisors, Charles Seligman. Shortly after the expedition set out, the First World War began, and Malinowski, a subject of the Austro-Hungarian and therefore an enemy of the Allies, found himself stranded in British Australia on arrival.

An agreement was worked out whereby Malinowski would spend his time on the Trobriand Islands (now part of Papua New Guinea.) Almost by accident, then, Malinowski found himself conducting a style of research that became known as ethnographic; living daily life along with the Trobrianders, participating alongside them in the attempt, as he put it, to "grasp the native's point of view." By living with and living like a group, he argued, one might begin to apprehend the world from their perspective and be in a position to document not only what they do but also something of what they experience in the doing. It is this shift to the topic of experience, and the concomitant methods of observation in and through participation in daily life, with its implications too of long-term immersive engagement, that fundamentally characterized the Malinowskian ethnographic shift.

On returning to England after the War, Malinowski took up a faculty position at the LSE, and published a series of books on the Trobriand Islands that also set out his distinctive form of inquiry through participation and immersion (Malinowski, 1922, 1929, 1935). From his position at the LSE, he became a leader in the British social anthropology community, while ethnographic participant-observation became the dominant, even defining, method of anthropological inquiry.[1]

1920s and Onwards: Spreading Out

Beginning in the 1920s, then, and proceeding for several decades, we see a gradual diffusion and evolution of ethnographic practice. What began as a means to understand the ways of the Trobriand Islanders, their religion, trading practices, and experience of everyday life became the method of inquiry that anthropologists applied all over the world—in Australia, in South America, in Africa, in Asia, in Melanesia, or wherever they traveled. They brought with them (and then brought home with them again) an evolving toolbox of practices of participant observation.

Ethnography of necessity looked slightly different every time and on every occasion, although the ethnographic anthropology of this period by and large evidenced some commonalities. It focused on cultural life, which had suggested particular concerns—language, religion, art, leadership, conflict, birth, death, ritual, and the stuff of life. It focused largely on distinct groups—this people or that, the Nuer or Zande or Arrente—in geographically bounded locations—the Rift Valley, the Simpson Desert, Highland Burma, Mato Grosso—and attempted to understand them as independent and individuable social wholes. Ethnographic inquiry was also often paired with particular forms of social analysis, especially the functionalism of which Malinowski had been a champion, which attempted to understand the interrelated and mutually supportive roles of different elements of social life and society.

[1] It should be noted that this is a very European history. Many of the same considerations animated the approximately contemporaneous work of Franz Boas in the USA, although their context was quite different.

Examples of ethnographic studies from this period include those by Radcliffe-Brown (1922), Firth (1929), and Evans-Prichard (1937).

During the period too, though, interest in ethnography also spread into related domains. In particular, a group at the University of Chicago recognized the opportunity to use the participant-observation methods developed in anthropology as a tool for sociological investigations of urban life. The so-called Chicago School (more accurately, Chicago Schools) sociologists used ethnography's approach to the examination of cultural practice to inquire into the experience of urban subcultures—taxi drivers, hobos, medical students, drug users, school teachers, gamblers, jazz musicians, numbers runners, and more. The immersive ethnographic approach, qualitative analysis, and a focus on experience, meaning, and interpretation (framed, in something of a post hoc rationalization, as symbolic interactionism) became a characteristic of a form of sociological inquiry that took its lead not just methodologically but also, to an extent, conceptually, from anthropological practice.

1960s: Structuralism

With the usual provisos, we might broadly characterize the 1960s in terms of the rise of structuralist anthropology with its impacts on ethnographic practice. Structuralist anthropology is often associated most particularly with the work of Claude Levi-Strauss, who drew on other currents in intellectual life of the 1950s and 1960s to fashion a novel approach to the interpretation of cultural settings and mythology (e.g., Levi-Strauss, 1958, 1962).

Levi-Strauss's analysis was deeply structuralist. Structuralism is a broad approach to understanding human phenomena that has its origins in linguistics, and in particular the approach developed by Ferdinand Saussure. Saussure was concerned with semiotics—how language comes to carry meaning. His observation was that the elements of language that carry meaning—words and letters—are essentially arbitrary. Unlike a picture of a dog, which bears some visual relationship to the animal that it depicts, the word "dog" has no inherent relationship to that animal. In that sense, it is entirely arbitrary. The meanings of words, then, are not based on any relationship between those words and the objects or phenomena that they denote. Instead, Saussure argued, we can find the source of the meaningfulness of words within the linguistic system itself. Meaning arises through patterns of difference. So, the meaningfulness of the term "dog" arises in the relationship of that word to other words—"cat," "lion," "bitch," "mutt," "hound," "puppy," "follow," "chase," "blackguard," and so on. What conveys meaning is the pattern of differences.

Saussure's structuralist semiotics is a foundation for Levi-Strauss' analysis of culture and myth systems. What matters in mythology, Levi-Strauss argues, are the arrangements of things and the distinctions that are drawn. When we combine individual myths to understand them as systems, patterns of distinction and relationships between categories emerge, and it is these patterns that matter. For instance, in his classic examination of the Oedipus myth, Levi-Strauss examines the structural relationships amongst elements (including actions, actors, and features of the settings in which they interact) in order to highlight the binary oppositions at the heart of the myth (e.g., between heroic prowess on the one hand, and lameness and debility on the other). As the analysis proceeds, the details of the story fall away to reveal the structures that animate and it. On the basis of this analysis, he argues that the central topic of the myth is the contrast between two notions of the origins of humans (born of the earth or born of people).

The structuralist approach has at least two consequences for ethnographic analysis that concern us here. The first is that it turns the object of ethnographic analysis from the event to the system of events, or from the experience to the system of meaning within which that experience is embedded, because it is that system of differences that makes particular events, actions, experiences and moments meaningful. These broader structures may be both synchronic and diachronic, and so we may need to look at the evolution of patterns over time and at particular ethnographic moments as instances of broader patterns of possibility. The second and broader consideration is the way it more explicitly focuses ethnographic attention on the decoding of patterns of meaning and the symbolic nature of culture and paves the way for further examinations of cultural life (and ethnography itself) as an interpretive process.

1970s: The Hermeneutic Turn

Just as the structuralist anthropology of the 1960s was a response to (and an example of) broader intellectual trends, so too in the 1970s did a progressive turn towards hermeneutics and textuality reflect broader currents. Clifford Geertz (1973), one of the most prominent anthropologists of his generation (and others), signals this turn explicitly in his landmark text *The Interpretation of Culture*:

> Man is an animal suspended in webs of significance he himself has spun. I take culture to be those webs, and the analysis of it therefore not an experimental science in search of law but an interpretive one in search of meaning. Geertz, 1973: 5

The hermeneutic turn, then, is one that places interpretation at its core, in at least two ways—first, it focuses on the work of the ethnographer as essentially interpretive, and second, it draws attention to the interpretive practices that participants themselves are engaged in as they go about everyday life. That is, if culture is a text to be read and interpreted, then that is simply what people are doing themselves. This hermeneutic or textual turn was by no means particular to anthropology. However, it has some particular consequences for ethnography.

First—as explicitly signaled by Geertz above—it reconfigures our expectations of what the ethnographer is doing—from providing an explanation to offering an interpretation. An interpretation illuminates, for sure, and it unpacks and accounts for actions in the world, but it is contestable and provisional. This is at best unsettling as goal for academic (or "social science") inquiry.

Second—and following on from the first—an even more unsettling consideration arises when we recognize that this interpretive stance is also here posited as the stance of cultural participants towards the occasions in which they find themselves, meaning that their own accounts—their own understandings—are themselves equally contestable and provisional. Taken to its conclusion, then, this turn suggests that there is no underlying "fact of the matter" as to the organization of a sociocultural setting; there is merely what people do, and what they understand, and how they act on the basis of those understandings, and on and on again.

Third—and this is a matter that will be of more concern shortly—if the ethnographer and the participants are both interpreters of the settings in which they find themselves, then what kind of relationship is postulated amongst them? Remember here that the essential feature of ethnographic inquiry, after all, is that it is grounded in participation, always with the proviso of course that that participation is limited, circumscribed and partial. This unsettling hermeneutic shift suggests first that the participation of "participants" is itself limited, circumscribed and partial, and in turn suggests that distinctions between ethnographers and other participants may simply be matters of degree. (This is not to mention the problem of how the ethnographer or analyst is an interpreter of his or her own setting—a question of reflexivity that is foundational to ethnomethodology and its position on the epistemological status of sociological theory.)

These perspectives are not simply unsettling but destabilizing within a positivist tradition, a topic to which we will return when exploring further the relationship between ethnographic work and contemporary HCI.

First, though, we should ask what Geertz suggests, in this interpretive vein, provides ethnography with the means to make progress and offer up its interpretations. His answer lies in thick description, a term he borrows from Gilbert Ryle. The essence of thick description is the multiple levels of understanding that it captures—different frames of interpretation, layers of meaning, contradictions and elaborations woven together. The goal of an ethnographic description, then, is not merely to set down on the page what happens in front of the eyes, but to do so in a way that allows for multiple, repeated, indefinite processes of interpretation; the goal is to open up, not to close down, the play of meaning. Geertz is trying in this description then to resituate ethnographic reports within an interpretive frame.

One critical aspect of this turn towards significance and interpretation is a transformation in the topic of culture itself, from what we might call a "taxonomic" view to a "generative" view (Dourish & Bell, 2011).

The taxonomic view of culture is one that attempts to differentiate one cultural practice from another and to be able to set out a framework of cultural classification by which we could, for example, discuss the differences between Chinese culture and German culture, or between Latin culture and Scandinavian culture. From this perspective, different groups have different cultural practices and understandings that can be analyzed in terms of their similarities and differences to build up larger pictures of the operation of broader cultural complexes. Ethnography, in this view, documents particular cultures, supporting a broader analysis of the cultural patterns that their behaviors exhibit. The focus here, then, is on difference and distinction, and the operation of culture as a categorization device—a way of distinguishing between and then relating different cultural groups.

The taxonomic view of culture is one that had operated since Malinowski or before. However, this view throws up a range of conceptual and methodological problems. For example, when our notion of culture is geographically bound, how do find the "central" considerations, and how do we handle borders and boundaries? Where do we draw the boundaries of different cultural groups? How, for that matter, might we handle the problem of the broad traffic in culture associated with the movements of goods, media, capital, and people? As a dual-national and a Scot living in America, how should I be categorized, for example? In turn, this causes us to stumble on the problems of the relationship of individuals to broader cultural groups identified in the taxonomic view.

In contrast to the taxonomic argument that culture exists and we all live within it, the generative view of culture argues that culture is produced as a continual, ongoing process of interpretation. We do not so much live inside of a culture as participate in one, or more usually in many. Culture as Geertz lays out is a system of meaning and meaning-making. The domain of the cultural, then, is the domain of the more-or-less collectively symbolic, and culture operates through processes of interpretation that reflect the multiple embeddings of people, so that college professor, researcher, computer scientist, and white middle-class male are every bit as much cultural categories as Scot, European, or American. The generative view of culture loosens the ties that bind culture to place, while at the same time accommodating considerably more diversity and turning our attention to the processes of culture rather than reifying it as an object.

1980s: Reflexivity

While the hermeneutic turn of the 1970s reflected an early encounter between anthropology's concern with culture and that arising out of contemporary literary and cultural theory, this wave broke with considerably more force during the 1980s, with, arguably, considerably greater significance not just for anthropological theorizing but also for the practice of ethnographic work. Most particularly, and for the purposes of this rough-and-ready historical account, these related to the question of ethnographic reflexivity and the roles of both ethnographers and participants in the ethnographic enterprise.

For the editors and authors of *Writing Culture* (Clifford & Marcus, 1986), one of the landmark texts of this turn, the primary focus is the production of ethnographic texts and the understanding of ethnography as a writing practice—not just the ethno- but the ethno-*graphy*. What does it mean to write about another? What is the role and the status of the author, as someone who creates and crafts a narrative, selects and shapes the data to be presented, who presents an account in which others are actors but the ethnographer's name is the one that appears on the cover? Think for example of the mode of presentation of traditional ethnography—"The Nuer trade in cattle,"

"The Zande consult the poison oracle for important decisions," "The Yonglu believe that their land was created by ancestral beings"—and notice, first, the definitiveness of the sentences, second, the eternal ethnographic present as the tense in which these observations are offered, and, third, the disappearance of the ethnographer as author of these statements. If we believe that it might matter whether the ethnographer arrived at the head of a column of colonial soldiers, whether the ethnographer was informed about local practice on a two-week visit or a year-long stay, whether the ethnographer's ethnicity, language, gender, religion, attitude, experience, political support, perceived interests, suspected allegiances, or history of engagement might make a difference to what is said, what is done, and what is learned, it certainly is not on display in these classical texts.

As in earlier discussions, we see here too a response within ethnographic practice to broader cultural and intellectual considerations. Questions of power, situatedness and subject position, for example, also animated feminist debate—although feminist anthropologists noted with disappointment that the authors collected in *Writing Culture* are almost entirely white males (Behar & Gordon, 1996)—as well as in postcolonial studies (which, of course, set an important context for any kind of self-reflection on the part of anthropology as a discipline.) When placed in this context, then, we can see the impact of this reasoning on three levels—political, conceptual, and methodological. On the political level, it addresses the question of the power relations of ethnographic work and the nature of the ethnographic program as a whole, including its emancipatory potential, the questions of voice and witness, and the questions of the groups on whom the ethnographic gaze might fall in the first place (Nader, 1972). On a conceptual level, it focuses attention on the question of classificatory schemes, the models of narrative, and the sources of epistemological authority within anthropological and social science practice. On a methodological level, it speaks to the importance of subject position as both a tool and a topic of ethnographic work, and hence to the significance of accounting for it and being able to find such an account within ethnographic projects, as well as the potential need for a reformulation of the conditions of participation and partnership. Self-consciousness and self-awareness become important tools of the job, and at the same time we are forced to confront the question of whether the people whom we have already stopped calling "subjects" and started calling "participants" might better be labeled "collaborators" (cf. Hayes, this volume).

In *Anthropology as Cultural Critique*, Marcus and Fischer (1986) note that one aspect of subject position in the production of ethnographic texts is the figuring of a culture for a specific audience. That is, although ethnography is often characterized as a process of "going there" (wherever "there" might be) we need to recognize that it also depends on "coming back again," and the question of just how and just where one comes back, and what, on the basis of one's trip one feels one now has the warrant to say, matters greatly. Anthropology, they observe, is generally in the business not merely of reporting on "them" but on reporting, at least implicitly, on the relationship between us and them, and so, through the encounter with an ethnographic other, of reflecting upon, de-familiarizing, and critiquing the institutions and structures of (generally) the West. In their attempt to draw attention to the implicit function of subject position in the crafting of ethnographic texts, Marcus and Fischer identify cultural critique as an element of the anthropological program and, in line with the considerations at the heart of *Writing Culture*, elaborate what consequences this might have for a reflexive human science.

1990s: Globalization and Multisitedness

If the developments that significantly affected ethnographic practice in the 1970s and 1980s were those of an evolving academic discourse and a re-theorization of human sciences, then the developments that significantly affected ethnographic practice in the 1990s were less those of the academy and more those of political and economic reality. Certainly, the theoretical arguments recounted above conspired to threaten easy categorizations of peoples and cultures, naive separations between "us" and "them," and the idea of a world of distinct, geographically bounded cultural groups. In the 1990s, these concerns became more prominent within ethnographic circles, compounded by a range of factors, including the increasing reach of electronic and digital media, an intensification in multinational commercial practice, the neoliberal reach of corporate considerations into the functioning of the nation-state, and the increasing significance of transnational governance.

Globalization is by no means a new phenomenon, but the 1990s saw a recognition of its contemporary intensification and the increasing importance of transnational or supranational agencies and organizations—the United Nations, the International Monetary Funds, the World Intellectual Property Organization, the General Agreement on Tariffs and Trade (which later gave rise to the World Trade Organization), and more—on the conditions of daily life all over the world. What sense could it make, in this context, to conduct ethnography as if its topics are entirely bounded by a specific place? What influence might the boundaries between sites have, and how might we go about studying phenomena that inherently escape the boundedness of particular geographical locales. People, objects, practices, customs, media, and ideas certainly occur in particular places, but they do not do so in isolation.

In the mid-1990s, Marcus explicitly articulated this as his call for "multi-sited ethnography" (Marcus, 1995). Multi-sited ethnography is not explicitly a comparative project; the goal of the incorporation of multiple sites is not to line them up next to each other and see what differs. Nor is it an attempt to achieve some kind of statistical validity by leaning towards the quantitative and amassing large data sets. Rather, it reflects a recognition that the objects of ethnographic inquiry inevitably escape the bounds of particular sites, and that following objects, ideas, and practices as they travel amongst different sites is both a valuable and a necessary part of contemporary ethnographic practice. Similarly, it argues that we need to proceed from a recognition that those self-same objects, ideas, and practices do, already, travel, and that therefore as part of understanding them we need to figure them in their trajectories. Miller's studies of topics such as the Internet (Miller & Slater, 2000) or commercial practice (Miller, 1997) may be grounded particularly in Trinidad, for example, but they highlight the "local"-ness of engagements with these phenomena. If the Internet is a technology that allows people to "be Trini," then it does so in ways that reflect the relationship between images of Trini-ness and of its alternatives. Similarly, Lindtner's study of Chinese players of World of Warcraft shows them encountering the game as intrinsically Chinese, in, for instance, what is perceived as its dependence upon systems of mutual support and reciprocal obligation, or *guanxi* (Lindtner, Mainwaring, Dourish, & Wang, 2009). This way of thinking about transnational patterns of cultural practice reflects, again, a generative rather than taxonomic view of culture.

In this context, the traditional "field" of ethnographic fieldwork begins to dissolve (Gupta & Ferguson, 1997), its boundaries irredeemably porous. The field becomes less of a site to which an ethnographer might travel than a phenomenon that an ethnographer might seek to identify and explain; that is, the question for the ethnographer might be how a particular complex or assemblage of ideas, concerns, people, practices, and objects cohere and condense for some group of participants as a stable, identifiable, and operable whole in the midst of a maelstrom.

Ethnography and Contemporary HCI

This historical backdrop may provide some context that helps us understand the encounter between ethnography and HCI. Several concerns stand out, including the production of ethnographic data through participation and engagement, the concern with subjectivity and reflexivity as components of the research method, the skepticism towards the boundedness of sites, and the interpretive stance on the part of both researchers and participants. Each of these, of course, is a significant departure from traditional HCI approaches, not simply in terms of techniques (that is, as a matter of methods) but in terms of the fundamental epistemological stance towards investigation and knowledge production (that is, as a matter of methodology). It is precisely these sorts of concerns on which communication around ethnographic work often falters in HCI contexts. In light of the historical account, then, let's try to explore some common topics of discussion and debate.

Ethnography and Generalization

One of the most frequent sources of confusion or frustration around ethnographic data is the question of generalization. Ethnography revels in particulars, and seeks to explain actual human occasions and circumstances; it is deeply situated in particular settings and contexts. Traditional HCI, and in particular design-oriented HCI, seeks generalized understandings and abstract models that apply across a wide range of settings.

First, we should distinguish between generalization and abstraction. Generalization concerns making statements that have import beyond the specific circumstances from which they are generated. Abstraction concerns the creation of new entities that operate on a conceptual plane rather than a plane of actualities and that have generalized reach through the removal of specifics and particulars.

Making this distinction allows us to make two important observations concerning the generalizability of ethnographic work in comparison to other types of investigations.

The first is that it allows us to observe that the nature of generalization in, say, survey work is a particular sort. Survey data can have statistical power, which it achieves through abstracting away particulars, reducing people and issues to parameter sets. The question, of course, is the meaningfulness of this when applied to any particular case. Ethnographers argue that the details matter, and so they resist the forms of abstraction upon which much scientific generalization relies.

The second observation that follows from this distinction is that there might be other forms of generalization that do not depend upon abstraction. Essentially, ethnographic work often generalizes, but it does so through juxtaposition—contradistinction, comparison, sequentiality, referentiality, resonance, and other ways of patterning across multiple observations. This form of ethnographic juxtaposition does not in itself truck in abstractions but it extends itself beyond the circumstances of specific observation. It does not imagine specific observations to be particularlized instances of abstract entities, but understands them to be things-in-themselves that can be related to other things-in-themselves in a range of ways without the mediation of abstractions as formal entities.

The level of ethnographic generalization then is often the corpus, rather than the specific study; the body of detailed observational material and analysis that is built up across a broad historical literature.[2] This in turn also helps us to understand the problems of seeking generalizations from singular studies, singular papers, and singular investigations rather than thinking about the ways that one might read a single study against or alongside one or more others in order to examine the resonances amongst them.

Ethnography and Theory

This in turn leads us to think about the relationship between ethnography and theory. To the extent that ethnography is often thought of as a data collection technique, or even as a method to be applied, then it might seem at first blush to be independent of and devoid of theory (at least from the perspective of those areas of HCI that feel that a theory is something you do to your data or evaluate with your data after you gather it).

[2] This is an activity not for an individual but for a discipline, although articles in places like the *Annual Review of Anthropology* clearly provide some insight. More broadly, this approach signals the way that literature reviews do more than simply demonstrate that things have been read.

However, as the foregoing should make clear, ethnography always and inevitably theorizes its subjects (including the ethnographer), and the debates that have shaped ethnographic practice are debates about exactly this process. Ethnographers coming to HCI have not always been as clear as they might have been about ethnography's theoretical and conceptual claims, with the unfortunate consequence that these sometimes are not distinguished as clearly as they should be. The result is that conceptual claims are read as empirical, empirical ones are read as conceptual, and the entire enterprise is seen as somehow just about saving people the cost of a plane flight to find out what happens somewhere.

Ethnography in HCI has most commonly been associated with one particular analytic position, ethnomethodology. Ethnography may or may not be ethnomethodological, and ethnomethodology may or may not be ethnographic, although in the HCI research record we have plenty of examples of research that is both (e.g., O'Brien, Rodden, Rouncefield, & Hughes, 1999; Swan, Taylor, & Harper, 2008; Tolmie, Pycock, Diggins, MacLean, & Karsenty, 2002). As I have outlined here, ethnography advocates an approach to understanding social phenomena through participation. Ethnomethodology, on the other hand, is a particular analytic position on the organization of social action and in turn on the role of analysis and theorization within sociology (Garfinkel, 1996). One can adopt an ethnomethodological stance towards one's ethnographic work, but ethnography and ethnomethodology remain quite distinct.

Within HCI, though, they are routinely confused, perhaps for historical reasons. Given that several of the earliest practitioners of ethnography within CSCW and HCI were ethnomethodologists, ethnomethodology essentially "came along for free" in HCI's turn towards ethnographic method, and so it may not be surprising that confusion about the relationship between the two might arise. Since both were unfamiliar to early readers of work in this area, the boundaries were not intelligible when they were practiced together. For instance, Suchman's classic studies of plans (Suchman, 1987) is ethnomethodological but not ethnographic, while her studies of the work of accountants (1983) and airport operations staff (1993) are both ethnographic and ethnomethodological. These confusions, however, have persisted. More recently, some seem to have quite pointedly refused to take opportunities to clarify the distinction—in an impassioned argument for ethnomethodological work, Crabtree et al. (2009) manage not to mention ethnomethodology directly at all, instead pitching their argument in terms of "new ethnography" (by which they refer to ethnography in the anthropological tradition) and "traditional ethnography" (by which they generally mean ethnomethodological research, not all of which is ethnographic even in the examples the paper cites.) HCI researchers can be forgiven for being confused.

More broadly, the extent to which different pieces of ethnographic work take on board or respond to, for instance, the post-structuralist concerns of the 1970s or the reflexive considerations of the 1980s, will vary; by these degrees do different theoretical positions become articulated in and through ethnographic work. (In light of these developments, though, we should be in no doubt that the absence of any account of subject position, the suggestion of geographical and historical boundedness, or the construction of ethnographic facts as somehow unproblematically "out there" are themselves theoretical statements of considerable heft). Similarly, as outlined above, the forms of juxtaposition and discursive embedding within ethnographic work set out a conceptual position and frame any piece of work as making contributions within a theoretical tradition.

Ethnography and Design

How then should we understand the role of ethnography within a design process? There is no single answer, just as there is no canonical ethnographic project nor a canonical design project. Certainly, the idea that, on the basis of understandings produced ethnographically, we might be able to formulate design requirements is one useful relationship. However, as I have argued elsewhere (Dourish, 2006), it is not the only one, and to imagine it so is to misunderstand ethnographic practice. Quite apart from the narrow conception of people as "users" (Dourish & Mainwaring, 2012; Satchell & Dourish, 2009), examining ethnographic accounts purely in terms of their statements about potential design interventions focuses on the empirical and ignores the conceptual.

Ethnographic work at the conceptual level may work best not by providing answers but by raising questions, challenging perceived understandings, giving silenced perspectives voice, and creating new conceptual understandings. That is, it may be destabilizing rather than instrumental, defamiliarizing topics, sites, and settings understood complacently (Bell, Blythe, & Sengers, 2005; Marcus & Fischer, 1986). However, this is not to say that raising questions is not usefully engaged with the design concerns of some in HCI; conceptual reformulation is itself a basis for design thinking. Arguably, indeed, the notion that what ethnography should provide is implications for design similarly misconstrues the design process. In particular, recent years have seen HCI engage more broadly with the design community and so broaden a former focus on design as a process of product engineering to a more holistic form of practice which is, itself, conceptual and research-oriented (Zimmerman, Stolterman, & Forlizzi, 2010). So, for instance, where Crabtree and colleagues (2009) concern themselves with the "sorts of ethnography most useful for designers," they do so largely in terms of engineering design practice in search of requirements, rather than critical designers engaged in design-oriented analysis (e.g., Dunne & Raby, 2001) or what Cross (2007) has called "designerly ways of knowing."

Ethnography and Cultural Analysis

Broadly, we might associate ethnography with a shift in attention in HCI towards cultural analysis, by which I mean not a reductive, psychometric account of cross-cultural differences but rather a form of humanistically inspired analysis of cultural practice. Scholars working within HCI have increasingly recognized the relevance of the humanities for their work, and that interactive systems in contemporary society should be understood not simply as instrumental tools to be evaluated for their efficiency but as cultural objects to be understood in terms of the forms of expression and engagement that they engender. This position basically argues that if you restrict your vocabulary to bandwidth, storage, and encoding technologies, it's difficult to capture the essence of YouTube, and that menu layouts have little to do with people's attitudes towards Facebook. Ethnographic investigation implies more than simply a different way of getting at data, or a way of getting at it in a different setting ("in the wild" rather than "in the lab") but also signals, in this context, a shift in the objects or concerns of inquiry that asks what cultural work digital media and interactive systems do, how they fit into broader patterns of practice and how the two coevolve. This is not simply, then, using the tools of anthropology to study interactive systems; it is also studying interactive systems anthropologically as sites of social and cultural production, in the sense of the generative (rather than taxonomic) reading of culture. What emerges is a new disciplinary hybrid, and so the epistemological foundations shift. This implies then that ethnography is not simply a tool to be picked up in order to better carry out the same old job; the job changes, its demands and requirements change, the qualifications to undertake the work change, and our expectations of what we're doing change too. Or so, at least, we should hope.

Asking Questions of Ethnography

This chapter is written with the expectation that many more people may come across ethnographic work in HCI, may read it, review it, or attempt to employ it than will ever actually attempt to conduct it. It is for this reason that it has taken as its topic not how to do ethnography, but rather what ethnography tries to do, and why, through a discussion of the historical debates and currents that have shaped contemporary ethnographic practice. In HCI, as in many other disciplines, ethnography has become a technique that many use, often in different ways. The historical account given here, rough and ready as it is, provides some tools for assessing that work and for understanding how it should be read. In light of this, it should be clear that there are some good questions that one might choose to ask of ethnographic work, and some less good ones.

Among the good questions to ask might be "What are this work's empirical claims?" and "What are this work's conceptual claims?" with an emphasis on the fact that these are two different questions. That is, ethnographies make both empirical and conceptual claims, and they should be distinguished from each other. Ethnography has often been thought of in HCI as a purely empirical activity, a way of uncovering facts about places and people. However, this is at best a partial view and often a deeply problematic one if one is unable to recognize conceptual claims as being just that. (Hopefully, in light of the preceding pages, we know better now than to say "uncovering" and might perhaps say "generating" instead.)

"What was the context of production?" How was this work produced, and in what ways? What, in particular, is the foundation for the kinds of participation that the work discloses? Indeed, is this participation even made clear? Many ethnographic texts in HCI resemble anthropological ethnography of the 1950s or before, couched in authoritative claims of the lives of others with little, if any, recognition of the person of the ethnographer as a party to the production of ethnographic data. Such an account supports the position that I have tried to steer readers away from here that ethnographic data is simply lying around on the ground waiting for the ethnographer to pick it up and bring it home. If we accept a view of ethnographic material as the product of occasions of participative engagement, then we surely need to be able to inquire into the nature of that engagement. Or, thinking of it another way, the question the ethnographer asks of events and utterances is, what makes just this statement or action make sense in context? So similarly, we as readers should be able to ask the same question of ethnographic texts, and so need some account of this context in order to proceed.

How does this contribute to the corpus? If the broad ethnographic corpus is the site of engagement and generalization, then how should particular texts be read against, alongside, or in response to others? Reading ethnographic material purely as a cataloging of observations garnered in some particular place or time renders its conceptual contributions largely invisible. At the same time, in the design context, it rules as largely irrelevant any work that arises at a time, in a place, with a group, or organized around a topic not immediately germane to the domain of application. On the other hand, when read as a corpus contribution, and as something that not only supplements but also comments upon an existing corpus of materials, ethnographic research has the potential for much greater impact and significance.

If some questions are good ones to ask, others are less so, although they do arise frequently, not least perhaps due to the epistemological mismatch between different disciplinary perspectives. What are some of these?

"Is this a representative sample?" Ethnographers certainly use the term "sampling" but since they do not seek to make statistical statements about the settings under investigation, issues of representativeness do not arise in the way in which they do in quantitative work. The concern for the ethnographer is to understand and account for what arises in the data. Statements made by participants, events seen to play out, and so on are not necessarily taken as evidence of anything more than the possibility of exactly these occurrences; specifically, they are not generally taken in and of themselves as exemplars of putative more abstract phenomena. Quite apart from the question of what "the average American," "the average HCI researcher," "the average New Yorker," "the average banker" or "the average southern Californian adolescent" might be as anything other than an academically convenient statistical fiction, ethnographic work does not seek to operate in those terms; it seeks to interpret and account for things that actually happened. This is not to say that ethnography does not seek to make broader statements based on repeated observation (and ethnographers most certainly count things—see, for example, the charts and even graphs in Becker and colleagues' classic *Boys in White* (Becker, Geer, Hughes, & Strauss, 1961)). However, the point is rather that questions of representativeness are not immediately germane because ethnographic data does not "stand for" a broader statistical phenomenon in the ways in which survey data or other quantitative approaches might attempt to do.

Methodologically, in fact, it can be of particular value to seek out the unusual. It is frequently observed that the most valuable informants are often people whose status is somewhat marginal or peripheral (since they have a useful insider/outsider perspective on the situation). Similarly, we might deliberately choose to look for and talk with people whose position on a phenomenon is unusual because of the precise nature of their unusual relationship. In a study of public transit in London, for instance, we found it fruitful to talk to people who, for example, refused to use the Underground system precisely because of the kind of perspective that that might give on the questions of the public transit system as an aspect of everyday life.

"How can you tell if what people told you is right?" This question arises from time to time and signals something of a misapprehension about the nature of ethnographic interviews. In general, when we ask questions in an ethnographic context, plugging a gap in our knowledge is only one aspect of what is being done; another is learning about the answer. A statement, utterance or action is taken, ethnographically, as documentary evidence of its own production; that is, the interesting thing isn't necessarily what was said, but that it could be said, that it was a sensible thing to be said by just that person in just those circumstances and in just that way. The question to ask, analytically, isn't "do I believe this to be true?" or "is this person lying to me?" but "what warrants that answer?" In other words, what is it about the relations that obtain in the moment between the ethnographer and participant that make the participant's answer a sensible one for the participant to give? What allows this to be an answer that is appropriate? What does the answer reveal about the organization or meaningfulness of the topic? A lie is revealing; it suggests that there is something worth lying about, and the choice of lie matters. So too do circumlocutions, partial answers, and so on. More importantly, it is not a question of dividing the world into true statements and false ones; all statements and all actions at all times are produced to meet the immediate circumstances of the moment, whether those circumstances are a wedding, drinks with friends, an intimate moment, an encounter with authority, a lecture, or an interview with a nosy social scientist.

"Didn't you affect things by being there?" My usual answer to this is, "I should hope so"; if I am being less flippant I might add, "in exactly the same way as every other person who was there changed things by being there." That is, the scenes into which ethnographers inquire are themselves ever changing and dynamic, and there is no simple fact of the matter as to what happens independently of the particular set of people who are parties to the scene and participants within it. The ethnographer is one of those, as are others, each engaged in the production of social life as a lived and enacted accomplishment. Certainly, it would be different if the ethnographer had not been there, just as it would have been different if a slightly different cast of characters had turned up. What's important of course is to make this clear. In two pieces that appear in the same volume, for example, Yvonne Rogers (1997) and

Blomberg, Suchman, & Trigg (1997) thoughtfully reflect on cases where their own presence as ethnographers precipitated disruptions that were methodologically, analytically, and politically significant.

"What should I build now that I know this?" Much research in HCI is concerned with technology design (not all, by any means, but a good deal.) So, the question of "what to build" is one that preoccupies many researchers and practitioners. I have it listed here under "less good questions" not because it is not, in itself, a sensible question but rather because it is a less good question to ask *of an ethnographic text*. As elaborated and exemplified elsewhere (Dourish, 2007), ethnographic research may inspire design practice, but the value that it offers is in an encounter with design rather than in its own terms. The implications for design, that is, lie not within the ethnographic text itself but rather in the way in which it reframes the contexts and questions of design. Again, if we think of the corpus as the site of ethnographic generalization, then we may see too the need to move to a different level in order to engage more fruitfully with design.

Recommended Reading

There are any number of basic how-to books that will provide you with an overview of the ethnographic method and hard-won lessons from the field. Examples include Agar's *The Professional Stranger* agar (1996), Fetterman's *Ethnography* Fetterman (1998), Snow, Lofland, Lofland and Anderson's *Analyzing Social Settings* Lofland et al. (2006), and DeWalt and DeWalt's *Participant Observation* DeWalt & DeWalt (2002). Different people have their favorites amongst these for different reasons, although they broadly cover the same ground. In my classes, I like to use Emerson, Fretz, and Shaw's *Writing Ethnographic Fieldnotes* Emerson et al. (1995); despite the title, its focus is considerably broader than fieldnotes, but it does take an approach based on the generation and analysis of texts, which I find very useful.

Spradley's *The Ethnographic Interview* Spradley (1979) and Weiss' *Learning From Strangers* Weiss (1994) are particularly good on interview techniques (the latter features useful transcripts annotated with notes on strategies, tactics and occasional blunders.) Sarah Pink's *Doing Visual Ethnography* Pink (2001) explores the use of visual materials as tools in ethnographic research.

Howard Becker's books *Tricks of the Trade* becker (1998) and *Telling About Society* becker (2007) are both filled with insight and advice for conducting and writing about ethnographic research, but in doing so they provide too considerable background that unpacks the nature of qualitative research and its documents.

Moore's *Visions of Culture* Moore (1997), while not focused on ethnographic research in particular, provides overview sketches of the theoretical positions of a wide range of anthropologists and social scientists, which can be helpful in recognizing a range of alternative positions that ethnographic material might take.

Geertz's landmark text *The Interpretation of Culture* paints a vivid and detailed picture of a program of interpretive, semiotic anthropology, illustrated with ethnographic essays of his own including his classic study of the Balinese cockfight.

Clifford and Marcus' collection *Writing Culture* explores the question of how ethnographic texts work; its publication was something of a watershed moment in ethnographic methodology. Geertz's *Works and Lives* Geertz (1988) and Van Maanen's *Tales of the Field* Van maanen (2011) both reflect on the production of ethnographic texts too, although in different ways—in Geertz's case, approached more as literary criticism, and in Van Maanen's, as something of a manual for practitioners.

As with several other disciplines, the Annual Reviews series includes a volume—*Annual Reviews in Anthropology*—which publishes extensive reviews in particular areas of work—the anthropology of money, the anthropology of time, the anthropology of visual practice, and so on—that collect, organize, and interpret extensive sets of research reports. If ethnographic generalization lies in juxtaposition as much as in abstraction, this is a good place to see it on display.

The traditional ethnographic form is a monograph, and so HCI, with its emphasis on highly abbreviated texts, rarely provides the space needed for thick description and conceptual development. Nonetheless, we do find within HCI examples of ethnographic material that is both analytically and empirically rich—some representative examples include Bowers, Button, and Sharrock's (1995) studies of the print shop floor (which is both ethnographic and ethnomethodological), Jenna Burrell's (2012) investigations of Ghanaian Internet cafes, Mainwaring, Chang, and Anderson's (2004) studies of people's relationships to infrastructure, and Vertesi's (2012) study of human–robot interaction and embodied interaction in planetary exploration. Increasingly too, though, we find ethnographic work of considerable interest to HCI arising outside of HCI itself, including studies of design practice (e.g., Loukissas, 2012), of mobile communications (e.g., Horst & Miller, 2006), of the production of games and virtual worlds (e.g., Malaby, 2009), and of gambling machines (Schüll, 2012).

Exercises

1. What would the challenges be in pooling related ethnographies to create a portfolio from which to make generalizations? What aspects of Grounded Theory Method would apply to the generating of these generalizations?
2. Describe the positive aspects of the researcher being a participant in the experience?

Acknowledgements The first version of the tutorial on which this document is based was developed and taught along with Ken Anderson. Wendy Kellogg and Judy Olson provided me with the opportunity to present it to an engaged audience at the HCIC meeting, and then another at UCI, both of whom helped me shape up the presentation considerably. I have written some about these topics before with Genevieve Bell, and taught some of them alongside Martha Feldman and Cal Morrill, all of whom have taught me a great deal, and shown me how much more I still have to learn. Work towards this chapter has been supported by the National Science Foundation under awards 0917401, 0968616, 1025761, and 1042678 and by the Intel Science and Technology Center for Social Computing.

References

Agar, M. (1996). *The professional stranger* (2nd ed.). San Diego, CA: Academic Press.

Becker, H. (1998). *Tricks of the trade: How to think about your research while you're doing it.* Chicago, IL: University of Chicago Press.

Becker, H. (2007). *Telling about society*. Chicago, IL: University of Chicago Press.

Becker, H., Geer, B., Hughes, E., & Strauss, A. (1961). *Boys in White: Student culture in medical school*. Chicago, IL: University of Chicago Press.

Behar, R., & Gordon, D. (1996). *Women writing culture*. Berkeley, CA: University of California Press.

Bell, G., Blythe, M., & Sengers, P. (2005). Making by making strange: Defamiliarization and the design of domestic technology. *ACM Transactions on Computer–Human Interaction, 12*(2), 149–173.

Blomberg, J., Suchman, L., & Trigg, T. (1997). Reflections on a work-oriented design project. In G. C. Bowker, S. L. Starr, W. Turner, & L. Gasser (Eds.), *Social science, technological systems, and cooperative work: Beyond the great divide*. Mahwah, NJ: Lawrence Erlbaum Associates.

Bødker, S. (2006). When second wave HCI meets third wave challenges. *Proceedings of the Nordic Conference on Human–Computer Interaction NordiCHI 2006* (pp. 1–8).

Bowers, J., Button, G., & Sharrock, W. (1995). Workflow from within and without. *Proceedings of the European Conference on Computer-Supported Cooperative Work ECSCW'95, Stockholm, Sweden* (pp. 51–66).

Burrell, J. (2012). *Invisible users: Youth in the Internet cafés of urban Ghana*. Cambridge, MA: MIT Press.

Clifford, J., & Marcus, G. (1986). *Writing culture: The politics and poetics of ethnography*. Berkeley, CA: University of California Press.

Crabtree, A., Rodden, T., Tolmie, P., & Button, G. (2009). Ethnography considered harmful. *Proceedings of the ACM Conference on Human Factors in Computing Systems CHI 2009* (pp. 879–888).

Cross, N. (2007). *Designerly ways of knowing*. Basel, Switzerland: Birkhäuser Varlag AG.

DeWalt, K., & DeWalt, B. (2002). *Participant observation: A guide for fieldworkers*. Walnut Creek, CA: AltaMira.

Dourish, P. (2006). Implications for design. *Proceedings of the ACM Conference on Human Factors in Computing Systems CHI 2006, Montreal, Canada* (pp. 541–550).

Dourish, P. (2007). Responsibilities and implications: Further thoughts on ethnography and design. *Proceedings of the ACM Conference on Designing for the User Experience DUX 2007, Chicago, IL.*

Dourish, P., & Bell, G. (2011). *Divining a digital future: Mess and mythology in ubiquitous computing*. Cambridge, MA: MIT Press.

Dourish, P. & Mainwaring, S. (2012). Ubicomp's colonial impulse. *Proceedings of the ACM Conference on Ubiquitous Computing Ubicomp 2012, Pittsburgh, PA.*

Dunne, A., & Raby, F. (2001). *Design Noir: The secret life of electronic objects*. Basel: Birkhäuser.

Emerson, R., Fretz, R., & Shaw, L. (1995). *Writing ethnographic fieldnotes*. Chicago, IL: Univesrity of Chicago Press.

Evans-Prichard, E. E. (1937). *Witchcraft, Oracles, and Magic amongst the Azande*. Oxford: Oxford University Press.

Fetterman, D. (1998). *Ethnography: Step by step* (2nd ed.). Thousand Oaks, CA: Sage.

Firth, R. (1929). *Primitive economics of the New Zealand Māori*. London: Routledge.

Garfinkel, H. (1996). Ethnomethodology's program. *Social Psychology Quarterly, 59*(1), 5–21.

Geertz, C. (1973). *The interpretation of cultures*. New York: Basic Books.

Geertz, C. (1988). *Works and lives: The anthropologist as author*. Stanford, CA: Stanford University Press.

Gupta, A., & Ferguson, J. (1997). *Anthropological locations: Boundaries and grounds of a field science*. Berkeley, CA: University of California Press.

Harrison, S., Sengers, P., & Tatar, D. (2011). Making epistemological trouble: Third paradigm HCI as successor science. *Interacting with Computers, 23*(5), 385–392.

Horst, H., & Miller, D. (2006). *The cell phone: An anthropology of communication*. Oxford: Berg.

Levi-Strauss, C. (1958). *Structural anthropology (anthropologie structurale)*. Paris: Plon.

Levi-Strauss, C. (1962). *The savage mind (La Pensee Sauvage)*. London: Weidenfeld and Nicolson.

Lindtner, S., Mainwaring, S., Dourish, P., & Wang, Y. (2009). Situating productive play: Online gaming practices and Guanxi in China. *Proceedings of the IFIP Conference on Human–Computer Interaction INTERACT 2009 (Stockholm, Sweden), Lecture Notes in Computer Science LNCS* (Vol. 5736, pp. 328–341).

Lofland, J., Snow, D., Anderson, D., & Lofland, L. (2006). *Analyzing social settings: A guide to qualitative observation and analysis* (4th ed.). Belmont, CA: Wadsworth/Thompson.

Loukissas, Y. (2012). *Co-designers: Cultures of computer simulation in architecture*. Cambridge, MA: MIT Press.

Mainwaring, S., Chang, M., & Anderson, K. (2004). Infrastructures and their discontents: Implications for Ubicomp. *Proceedings of the International Conference on Ubiquitous Computing Ubicomp 2004* (*LNCS* 3205, pp. 418–432).

Malaby, T. (2009). *Making virtual worlds: Linden lab and second life*. Ithaca: Cornell University Press.

Malinowski, B. (1922). *Argonauts of the Western Pacific*. London: Routledge.

Malinowski, B. (1929). *The sexual life of savages in North-West Melanesia*. London: Routledge.

Malinowski, B. (1935). *Coral gardens and their magic*. London: Routledge.

Marcus, G. E. (1995). Ethnography in/of the World System: The emergence of multi-sited ethnography. *Annual Review of Anthropology, 24*(1), 95–117.

Marcus, G., & Fischer, M. (1986). *Anthropology as cultural critique: An experimental moment in the human sciences*. Chicago, IL: University of Chicago Press.

Miller, D. (1997). *Capitalism: An ethnographic approach*. Oxford: Berg.

Miller, D., & Slater, D. (2000). *The Internet: An ethnographic approach*. Oxford: Berg.

Moore, J. (1997). *Visions of culture: An introduction to anthropological theories and theorists*. Walnut Creek, CA: AltaMira.

Nader, L. (1972). Up the anthropologist: Perspectives gained from studying up. In D. Hymes (Ed.), *Reinventing anthropology* (pp. 284–311). New York: Pantheon.

O'Brien, J., Rodden, T., Rouncefield, M., & Hughes, J. (1999). At home with the technology: An ethnographic study of a set-top box trial. *ACM Transactions on Computer–Human Interaction, 6*(3), 282–308.

Ortner, S. (2006). *Anthropology and social theory: Culture, power, and the acting subject*. Durham, NC: Duke University Press.

Pink, S. (2001). *Doing visual ethnography*. Thousand Oaks, CA: Sage.

Radcliffe-Brown, A. R. (1922). *The Andaman Islanders*. Cambridge: Cambridge University Press.

Rogers, Y. (1997). Reconfiguring the social scientist: Shifting from telling designers what to do to getting more involved. In G. C. Bowker, S. L. Star, W. Turner, & L. Gasser (Eds.), *Social science, technological systems, and cooperative work: Beyond the great divide*. Mahwah, NJ: Lawrence Erlbaum Associates.

Satchell, C. & Dourish, P. (2009). Beyond the user: Use and non-use in HCI. *Proceedings of the Australasian Conference on Computer–Human Interaction OzCHI 2009, Melbourne, Australia* (pp. 9–16).

Schüll, N. (2012). *Addiction by design: Machine gambling in Las Vegas*. Cambridge, MA: MIT Press.

Spradley, J. (1979). *The ethnographic interview*. Wadsworth/Thompson. Belmont, CA.

Strathern, M. (2003). *Commons and Borderlands: Working Papers on Interdisciplinary, Accountability and the Flow of Knowledge*. Sean Kingston.

Suchman, L. (1983). Office procedure as practical action: Models of work and system design. *ACM Transactions on Information Systems, 1*(4), 320–328.

Suchman, L. (1987). *Plans and Situated Actions: The Problem of Human-Machine Communication*. Cambridge, UK: Cambridge University Press.

Suchman, L. (1993). Technologies of accountability: On lizards and airplanes. In G. Button (Ed.), *Technology in working order* (pp. 113–126). London: Routledge.

Swan, L., Taylor, A., & Harper, R. (2008). Making place for clutter and other ideas of home. *ACM Transactions on Computer–Human Interaction, 15*(2), 1–24.

Tolmie, P., Pycock, J., Diggins, T., MacLean. A., & Karsenty, A. (2002). Unremarkable computing. *Proceedings of the ACM Conference on Human Factors in Computing Systems CHI 2002* (pp. 399–406).

Van Maanen, J. (2011). *Tales of the field: On writing ethnography* (2nd ed.). Chicago, IL: University of Chicago Press.

Vertesi, J. (2012). Seeing like a rover: Visualization, embodiment and interaction on the mars exploration rover mission. *Social Studies of Science, 42*(3), 393–414.

Weiss, R. (1994). *Learning from strangers: The art and method of qualitative interview studies*. New York, NY: The Free Press.

Zimmerman, J., Stolterman, E., & Forlizzi, J. (2010). An analysis and critique of research through design: Towards a formalization of a research agenda. *Proceedings of the ACM Conference on Designing Interactive Systems DIS 2010, Aarhus, Denmark* (pp. 310–319).

Curiosity, Creativity, and Surprise as Analytic Tools: Grounded Theory Method

Michael Muller

In memory of Susan Leigh Star (1954–2010), whose insights and humanity helped many of us to find our ways, individually and collectively.

Introduction: Why Use Grounded Theory Method?

Grounded theory method (GTM) is increasingly used in HCI and CSCW research (Fig. 1). GTM offers a rigorous way to explore a domain, with an emphasis on discovering new insights, testing those insights, and building partial understandings into a broader theory of the domain. The strength of the method—as a full method—is the ability to make sense of diverse phenomena, to construct an account of those phenomena that is strongly based in the data ("grounded" in the data), to develop that account through an iterative and principled series of challenges and modifications, and to communicate the end result to others in a way that is convincing and valuable to their own research and understanding. GTM is particularly appropriate for making sense of a domain without a dominant theory. It is *not* concerned with testing existing theories. Rather, GTM is concerned with the *creation* of theory, and with the rigorous and even ruthless examination of that new theory.

Grounded Theory Method is exactly that—a *method*, or rather, a family of methods (Babchuk, 2010)—for the development of theory. GTM makes explicit use of the capabilities that nearly all human share, to be curious about the world, to understand the world, and to communicate that understanding to others. GTM adds to these lay human capabilities a rigorous, scientific set of ways of inquiring, ways of thinking, and ways of knowing that can add power and explanatory strength to HCI and CSCW research.

M. Muller (✉)
IBM Research, Cambridge, MA, USA
e-mail: michale_muller@us.ibm

J.S. Olson and W.A. Kellogg (eds.), *Ways of Knowing in HCI*,
DOI 10.1007/978-1-4939-0378-8_2,

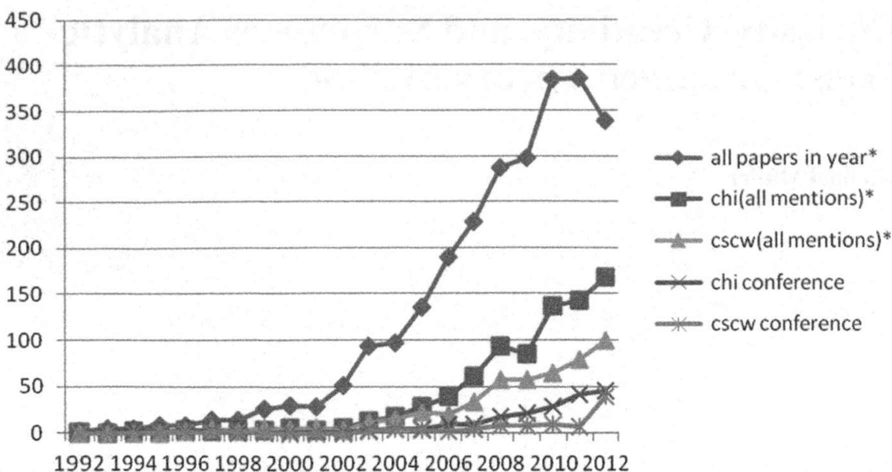

Fig. 1 Papers in the ACM Digital Library that mention "grounded theory." The line labeled "all" is for all references to "grounded theory" in each year. The line labeled "chi" shows papers that mentioned both "grounded theory" and "chi." The line labeled "cscw" shows papers that mentioned both "grounded theory" and "cscw." The remaining two lines are grounded theory papers in the CHI or CSCW *Conference Proceedings* (including *Extended Abstracts*). Figures for 2012 (*asterisk*) are estimates, based on entries from January to September

GTM has been used to study diverse phenomena that pertain or contribute to HCI and CSCW studies. Matavire and Brown (2008) surveyed the use of GTM in information systems research; (Riitta, Urquhart, & Iivari, 2009; Riitta & Newman, 2011) used GTM to understand information systems project management (*see* also Seidel & Recker, 2009, for a grounded theory study of business process management). Adolph, Hall and Kruchten (2008) applied GTM to understand software development. More specifically, Hoda (2011) used GTM to develop an account of agile software teams. In a contrastive pairing, Macrì, Tagliaventi and Bertolotti (2002) conducted a grounded theory study of resistance to change in organizations, and Pauleen and Yoong (2004) studied innovation in organizations. Locke (2001) focused on management studies.

Within HCI and CSCW, various forms of GTM have been used to study phenomena such as boundary objects and infrastructures (1985, 1999, 2002; Star & Griesemer, 1989), appropriation (Kim & Lee, 2012), decision-making (Lopes, 2010), personas (Faily & Flechals, 2011), HCI education (Cennamo et al., 2011), social media (Blythe & Cairns, 2009; Thom-Santelli, Muller, & Millen, 2008), and the use of classifications in organizations (Bowker & Star, 1999). Among domains that can be addressed via information and computing technologies, GTM was used in studies of diverse populations ranging from homeless people (Eyrich-Garg, 2011) to seniors (Sayago & Blat, 2009; Vines et al., 2012) to parents (Rode, 2009) to families in various configurations (Odom, Zimmerman, & Forlizzi, 2010; Yardi & Bruckman, 2012) to the founders of ventures (Ambos and Birkinshaw (2010). GTM has also been invoked in studies that focused primarily on technologies (Chetty et al., 2011; Faste & Lin, 2012; Kim, Hong, & Magerko, 2010) and on the social attributes of technologies (Kjeldskov & Paay, 2005; Lewis & Lewis, 2012; Mathiasen & Bødker, 2011; Paay, Kjeldskov, Howard, & Dave, 2009; Rode, 2009; Wyche, Smyth, Chetty, Aoki, & Grinter, 2010; Yardi & Bruckman, 2012).

However, the development of GTM has been complex and even schismatic. After the initial *Discovery of Grounded Theory* (Glaser & Strauss, 1967), two major orientations to grounded theory diverged from one another (Babchuk, 2010; Kelle, 2005, 2007), followed by a "second generation" of grounded theorists who creatively extended and recombined one or both of the major orientations (Morse et al., 2009) and further offshoots as well (Matavire & Brown, 2008), described in more detail later. Also, the application of GTM in HCI and CSCW has been uneven (see Furniss, Blandford, & Curson, 2011, for a recent discussion). Some researchers adopt the concept of grounded theory as a full methodology (e.g., Star, 1999, 2007). Other researchers make selective use of a subset of GT practices (e.g., Paay et al., 2009; Thom-Santelli et al., 2008). Yet other researchers invoke GTM as a kind of signal to indicate an extended qualitative data analysis. Taken together, these problems have led to a blurring of the definition and the practices of GTM in HCI and CSCW research. It is difficult to know what a reference to "grounded theory" means in CSCW and HCI, and it is correspondingly difficult to assess the quality and rigor of grounded theory reports.

This chapter attempts to address some of these problems. Because the theme of this volume is "ways of knowing," I use the grounded theory approach of *abductive inference* as a core distinguishing contribution of GTM to HCI and CSCW, and as the central organizing principle of this chapter. As with many papers on GTM, my excerpting from the literature is necessarily personal; I provide citations to different perspectives as well.

Grounded Theory Method as a Way of Knowing

Ground Theory Method is concerned with *knowing* as a human endeavor, using the unique capabilities of humans as active inquirers who construct their interpretations of the world and its phenomena (Charmaz, 2006; Gasson, 2003; Lincoln & Guba, 2000). In this way, GTM differs from many conventional "objective" approaches to HCI, which often define their methods as a series of procedural steps that should result in a replicable outcome regardless of the identity of the researcher(s) involved (e.g., Popper, 1968). Grounded theory recognizes that human researchers are curious and active agents, who are constantly thinking about their research questions, and who can make, modify and strengthen their research questions as they learn more. The procedural steps of conventional approaches are replaced with a different logic of inquiry derived generally from the philosophy of pragmatism (Peirce, 1903), with its own standards of rigor.

Conventional approaches advise a linear sequence of actions in which the researcher (1) defines a theoretical question, (2) collects data, (3) analyzes the data, and (4) interprets the analysis to answer the theoretical question. Grounded theory makes a virtue of our human inclination to ask "what's going on here" long before we have completed our data collection (Charmaz, 2006; Gasson, 2003). Instead of waiting to theorize until all the data are collected, GTM provides ways of thinking that depend crucially on the iterative development of interpretation and theory, using principles of *constant comparison* of data-with-data, and data-with-theory (Charmaz, 2006; Corbin & Strauss, 2008; Glaser & Strauss, 1967; Kelle, 2007; Urquhart & Fernández, 2006). Data collection is guided by the iteratively developing theory, usually in ways that involve challenging that theory through additional data samples that are chosen to test the theory at its weakest points (e.g., Awbrey & Awbrey, 1995). For example, we might ask, "is this finding universal, or does it occur only among a subset of the population?" or, using a more targeted strategy, "what *other* situations are crucially different, such that we should we *not* be able to replicate this finding in those situations?" A theory that survives this process is likely to be broad and robust, and is therefore likely to provide explanatory value and power to the researcher and the field.

Abductive Inference and Surprise

According to many GTM researchers, the core concept of GTM is a way of reasoning that is distinct among most other methods in HCI and CSCW. *Abductive inference* is a "logic of discovery" (Paavola, 2012) concerned with finding new interpretations (theories) for data that do not fit old ideas (Reichertz, 2007; Shannak & Aldhmour, 2009). As such, it is neither inductive nor deductive, although some theorists claim that it incorporates both of these inferential operations (e.g., Haig, 1995). The logic of abduction is to find a surprising phenomenon, and then to try to explain it. Haig (2005) describes the process as follows:

> [S]ome observations (phenomena) are encountered which are surprising because they do not follow from any accepted hypothesis; we come to notice that those observations (phenomena) would follow as a matter of course from the truth of a new hypothesis in conjunction with accepted auxiliary claims; we therefore conclude that the new hypothesis is plausible and thus deserves to be seriously entertained and further investigated. (Parentheses in original)

The new idea is a "hypothesis on probation" (Gold, Walton, Cureton, & Anderson, 2011), and must be rigorously tested. GTM provides disciplined ways of "managing" one or more "hypotheses on probation," and of testing them in ways that make the hypothesis stronger, more internally consistent, and broadly applicable.

Most grounded theorists trace the concept of abduction to Pierce's philosophy of pragmatism (Peirce, 1903): "Deduction proves that something must be; Induction shows that something actually is operative; Abduction… suggests that something may be."[1] But how can we move from the tentative position of "may be" to a stance of greater confidence? Quoting from Peirce, Reichertz (2010) summarizes: "One may [achieve] a discovery of this sort as a result of an intellectual process and, if this

[1] For more discussion of pragmatism, see Hayes' chapter on Action Research in this volume.

happens, it takes place 'like lightning,' and the thought process 'is very little hampered by logical rules.'" While intriguing, Peirce's theorizing would seem to make for poor science. The *method* aspects of Grounded Theory Method are designed to resolve these problems in detail.

What Grounded Theory Is and Is Not

Grounded theory is not a theory!—at least, not in the conventional sense of theory, such as Activity Theory (Nardi, 1996) or Structuration Theory (Orlikowski, 1992). Grounded theory is a family of methods (Babchuk, 2010)—hence, the more accurate term of Grounded Theory Method (Charmaz, 2006). The methods are used to *construct* theories of particular phenomena or domains that are "grounded" in the data. In this way, GTM puts its emphasis on data, and on thinking about the data. The methods of GTM help researchers to describe data, to build increasingly powerful abstractions based on the data, and to collect additional data that can provide the most effective tests of those abstractions.

Grounded Theory as Method

History and Sources of Grounded Theory

Grounded Theory began as a "discovery" of two sociologists (Glaser & Strauss, 1967) who had enjoyed a fruitful collaboration (Glaser & Strauss, 1965, 1968; Strauss & Glaser, 1970), but who eventually disagreed with one another, sometimes profoundly (Corbin & Strauss, 2008; Glaser, 1978, 1992; Strauss, 1993). The core of their shared insights was a rejection of the positivist sociology that was dominant in the US in the 1960s (Star, 2007) and the development of an approach that emphasized the gradual development of new theories based on continual reference ("constant comparison") to data. They rejected the conventional approaches that begin with a theory, collect data in a uniform manner, and then test that theory. Instead, they pioneered methods for making sense of data through iterative coding and theorizing, in which theory guided codes and codes guided theory, and in which the theory was understood to be under constant development.[2] A direct consequence of the focus of theory and ongoing development was the requirement to reshape the inquiry based on the developing theory (see "theoretical sampling," below).

The disagreement between Glaser and Strauss has been discussed by many grounded theory researchers (Bryant & Charmaz, 2007; Charmaz, 2006, 2008; Locke, 2001; Morse et al., 2009), including an HCI-oriented account (Muller & Kogan, 2012).

[2] This approach is similar to HCI ideas of iterative design, and the quick, in-process evaluations of designs through formative evaluation (Nielsen, 1992). GTM adds methodological rigor and the coordinated development of both data and theory.

Strauss focused on a set of methods for conducting grounded theory research. Consistent with the themes of ongoing development and discovery, Strauss made significant modifications to his treatment, sometimes discarding entire "paradigms" in favor of more open procedures (Corbin & Strauss, 2008). Glaser disagreed with many of the specific methodologies, which he considered to be "forcing" the data into preexisting structures (e.g., Glaser, 1992), with a potential loss of the ability ("sensitivity") to discern and create new theories (e.g., Glaser, 1978). Students of the two founders developed their own practices and their own philosophical orientations. Today, grounded theory spans multiple positions, from quasi-positivist (e.g., Corbin & Strauss, 2008) to constructivist (e.g., Charmaz, 2006) to explicitly postmodern (e.g., Clarke, 2005). In what follows, I focus more on the Strauss and Charmaz approaches, because they offer relatively clear guidance for HCI and CSCW. I encourage interested readers to consult many of the other sources, because of the strongly personal and personalized nature of much of grounded theory methods. GTM methods are *ways of knowing*; each GTM practitioner will need to make choices about the best (sub) methods through which she or he perceives and knows.

Major Resources for Grounded Theory

As mentioned above, the founding text of grounded theory was the book about its "discovery" by Glaser and Strauss (1967). Strauss's work proceeded through a methodological evolution, sometimes informally referred to as "the cookbook"; the most recent version appeared as Corbin and Strauss (2008). Glaser published a series of theoretical evolutions, with a diminished focus on methods; Glaser (1998) is a good summary.

Students of the founders developed their own approaches. One group of students described themselves as "the second generation," and published a summary of their approaches in Morse et al. (2009). Several of them also published influential versions of grounded theory research methods, such as the constructivist methodology of Charmaz (2006), the postmodern and cartographic approach of Clarke (2005), and the more pragmatic, business-applied version of Locke (2001). Like any field of committed scholars, grounded theory has needed its own handbook to pursue diverse specialized topics. An influential handbook has appeared in the Sage series by Bryant and Charmaz (2007).

Grounded Theory Practices

The Abstraction of the New: Codes, Coding, and Categories

Grounded theory begins not with theory, but with data. Data are connected to thinking, and to theorizing, through a formal vocabulary known as *codes* (Holton, 2007), as shown by lozenge shapes in the left side of Fig. 2. Star (2007) wrote, "A code sets

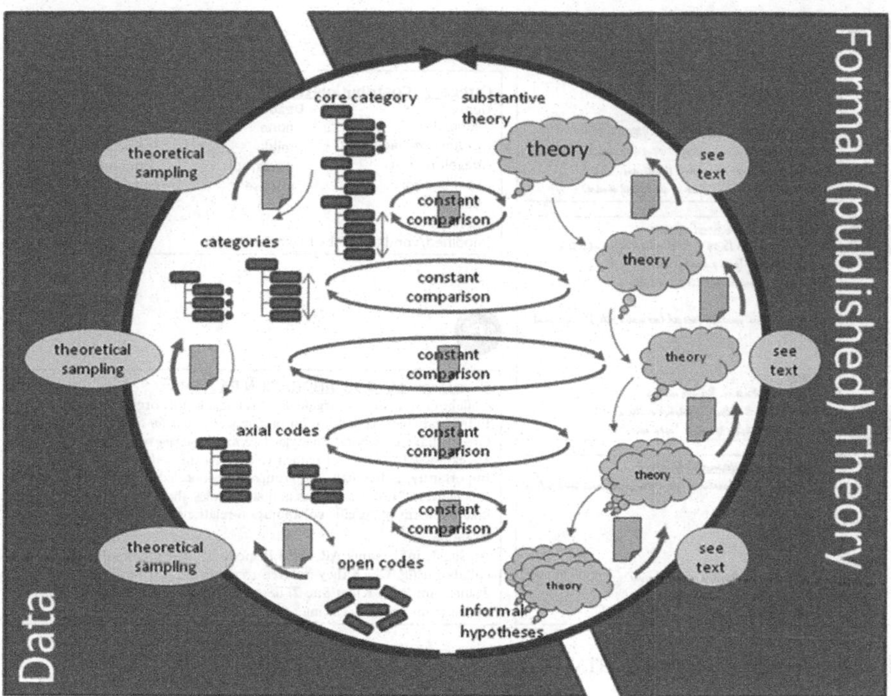

Fig. 2 A sketch of the major components of GTM practices

up a relationship with your data, and with your respondents [Codes are] a matter of both attachment and separation [...] Codes allow us to know about the field we study, and yet carry the abstraction of the new." Writing descriptions that are both accurately detailed and powerfully abstract is challenging. A code is a descriptor of some aspect of a particular situation (a site, informant or group of informants, episode, conversational turn, action, etc.). When codes are reused across more diverse situations, they gain explanatory power. Each situation becomes a test of the power of the codes to explain an increasing rich set of data. Codes are initially descriptive and tied to particular aspects of the data. Over time, the researcher(s) develop more abstract codes, which become one instantiation of the developing theory as shown by the thought-bubbles on the right side of Fig. 2. GTM provides guidance about how this happens, how to assess the resulting set of codes (see "Research Quality and Rigor," below), and how to record the emerging theory through informal documents called "memos" (the paper icons in the central column of Fig. 2).[3] Several influential accounts of GTM converge on a four-level schema to help to meet the challenge of how to get started in coding (Charmaz, 2006; Corbin & Strauss, 2008;

[3]Note that there is controversy among GTM researchers about the appropriate time to consult Formal Theory (i.e., the research literature). See "Creativity and Imagination," below.

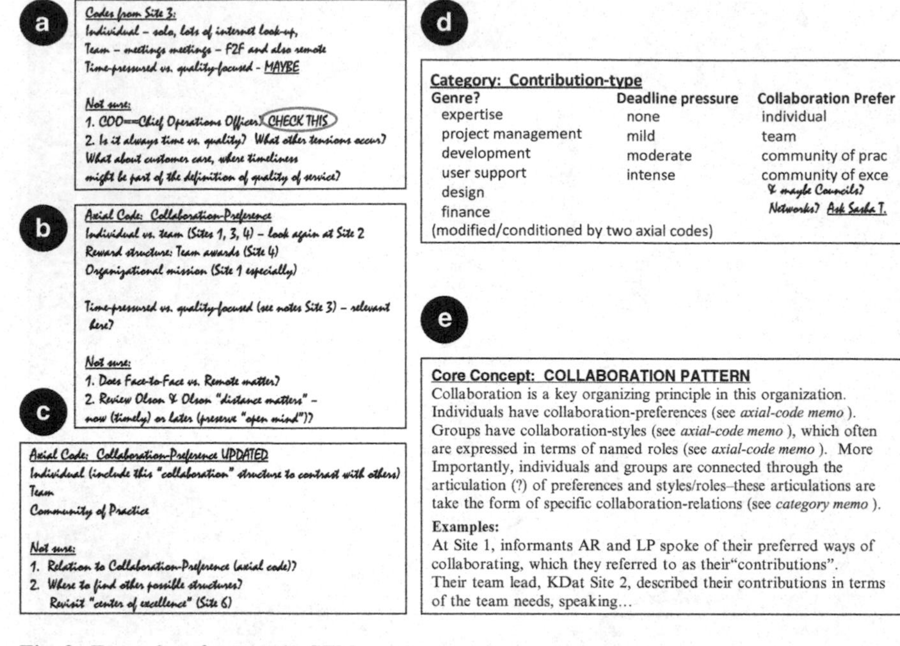

Fig. 3 Examples of memos in GTM

Dick, 2005; Star, 2007): open coding, axial coding, selective coding, and the designation of the core concept.

Open coding is the phrase used for the initial description of a situation. An open code is a kind of label placed on a phenomenon. Open codes are "open" in the sense that they are "open minded" (not governed by formal prior knowledge) and also in the sense that they are relatively unconstrained.

Suppose that we want to understand work practices in organizations. We might begin by interviewing people about their work. In this particular case, a "situation" is a person, and we are coding attributes of the person's job, tasks, and responsibilities.[4] Possible codes might include "individual" or "team," or "time-pressured," or "quality-focused." Some codes turn out to be useful in more than one situation, while others, which are mentioned by only a single informant, turn out to have little generality. Codes should be recorded in brief, informal researchers "memos" (see below). At this stage, the memo would be likely to contain a list of codes, the informal rules or heuristics for applying those codes to the data, and the beginnings of a list of reasons to doubt that the codes are complete descriptions of the data (Fig. 3a).

Axial coding is the first of several practices to organize the open codes into broader abstractions or more generalized meanings—a continuing integration of one's understandings, moving from *describing* to *knowing*. Axial codes are

[4] In other cases, the "situation" could be a group, or an organization, or a document, or a conversation.

collections of related open codes. It is tempting to say that axial codes are built up from the component open codes in a bottom-up way, and that's partially true. For example, the open codes of "individual" and "team" may suggest an axial code of "collaboration-preference." That might seem like a nice research outcome, and we should record it in another informal memo, describing the axial code and its component open codes—perhaps with back-references to the memo(s) created in the preceding paragraph (Fig. 3b).

The concept of collaboration-preference might provide a good basis for writing a paper. However, an axial code may also be used to interrogate the open codes, leading to more data collection and more open coding. Suppose, in the example of the axial code for collaboration-preference, we might have heard references to other configurations of work, such as communities of practice, or less structured networks. If we have already accepted the axial code as "individual-or-group," then these other configurations of work would come as a surprise. If we were guided by a hypothesis-testing approach, we might try to force a community of practice into the "team" category. Because we are using GTM, we can instead interrogate our initial theory and its axial code of collaboration-preference, to see how it can be expanded and strengthened based on the tentative evidence of communities of practice.

This way of thinking may lead to a search beyond the current sample of informants, for people who work in those other configurations, such as communities of practice (see "theoretical sampling," below). If we find people who work in those configurations, then the axial code must be broadened, and has thus become stronger and more generalizable: The axial code now organizes more cases, and (crucially) it *sets each case in relation to other cases along a common frame of reference*—that is, each case has a unitary description (the open code), and those unitary descriptions make more sense because they can be thought about in relation to other unitary descriptions (other open codes). The axial code sets these open codes into that relationship, which should be recorded in another informal memo. Like the preceding memo, this new document could be quite short, describing the axial codes, their constituent open codes, and the emergent concepts that are related to this new cluster of labels (Fig. 3c).

Categories begin to emerge as we focus our attention and insight upon certain axial codes. A category is a well-understood set of attributes of known relation to one another. A simple example might be "contribution type" (as a component of collaboration-preference). Continued interviews may show that the informants typically make contributions such as "expertise," "project management," "development," or "user support." If we become convinced that these four types of contributions are sufficient to describe all (or most) cases, then these terms become properties of the category of contribution-type. Another example might be "deadline-pressure," which might be summarized as "none," "mild," "moderate," or "intense." As these clarifications occur, they too should be recorded in another informal memo. This memo might be longer than the previous ones, because it would detail the category and the several axial codes that contribute to it (Fig. 3d).

More radically, we might recall that some informants seemed to refer to different kinds of roles, with different collaboration attributes, in different working groups. Further interviews confirm that this phenomenon is widespread. However, the pattern of *multiple* collaboration attributes for the same person, could not occur if each person had a single, *personal* collaboration-preference. This is a key moment in abductive inference, because we have to think of a new informal theory to make sense of this insight. Is the collaboration-preference really an attribute of the person? This thinking suggests additional questions, and those questions might lead us both to find new informants, and also to return to previous informants to get answers to those additional questions. In some cases, we could return to a set of interview transcripts, or documents, and use our new understanding to ask new questions of these "old" data. As we find that some people participate in *different* collaboration-patterns, then the attribute of collaboration-preference has moved away from the *person* and has become instead a characterization of each collaborative *group*, such as the collaboration personas of Matthews, Whittaker, Moran, and Yuen (2011). It might be appropriate to rename the axial code at this point, to make its group-basis clearer—perhaps "collaboration-style." The evolving theory has become much stronger, because we have a new understanding of *what entity* is properly described by the collaboration-style. Another memo is needed to record this new understanding. As in the preceding paragraph, we may find that the memos are getting longer, comprising lists of open and axial codes, but also greater depth and integration of the emergent theory.

The collaboration attribute now appears to be a defining aspect of each group. That's an interesting new theory, but we need to test it further. In GTM, we usually test a theory at its weakest point. We might therefore ask if all of the members of each group have the *same* kind of relationship to the group. And indeed, we learn that some people serve as core members of a group, while other people serve as more peripheral members (e.g., subject-matter experts, who are called upon from time to time for specific types of expertise)—another surprise. On this basis, the "collaboration-style" theory appears to be insufficient, because it proposed that the group had a single collaboration-pattern. How can the theory be broadened and deepened, to accommodate these new insights from the data?

We could hypothesize another kind of theoretical "relocation" of the characteristic of collaboration-style. First, we thought that collaboration-preference was a characteristic of a *person*. Then we thought that collaboration-style was a characteristic of a *group*. Through a series of surprises, we realized that neither of those theories was capable of describing the richness of the data. Now we hypothesize that the collaboration attribute is a characteristic not of a person nor of a group, but rather of the *relationship* of a person to a group. Perhaps now we should use the phrase "collaboration-relation," and we should document this subtle but important distinction in another memo. This new memo describes not only the new configuration of codes, but also the theoretical concepts that led to that reconfiguration (Fig. 3e). The developing theory has changed again, and has become more powerful, and capable of describing a broader set of phenomena. Further interviews and observations present no further surprises: The theory appears to explain all of the data, and this phase of theory development is complete.

Additional work could be done to expand the theory beyond this situation or to test the theory in more detail. For example, are there certain *types of groups* that have a set of characteristic collaboration-relations (Matthews et al., 2011) that link people to each group? Or it might be useful to determine if certain job titles have a set of characteristic collaboration-styles that link people in that job title to other people (through groups). And it might be helpful to see if certain people tend to have a single, predominant collaboration-relation with their groups.

A further test of the theory could be done via a social network analysis, and some of the hypotheses could have been evaluated through such a network analysis (see Chapter on Social Network Analysis in this volume). Alternatively, we could have been using a statistical summary of individuals and their group memberships all along, to help us find appropriate next people for interviews (e.g., as we did, in a more primitive fashion, in Muller, Millen, & Feinberg, 2009; Thom-Santelli et al., 2008). This is to say, while GTM is most commonly used for qualitative data, it can also be used for a quantitative exploration, and both qualitative and quantitative methods may be used together.

The *core concept* emerges through this kind of intense comparison of data to data, and data to emerging theory (some grounded theorists make reference to *selective coding*, which is approximately the choice of the core concept). Could it be that we are thinking about a complex set of inter-related axial codes? We are currently thinking about collaboration-relations as describing the links between people and groups. But we earlier thought about collaboration-preferences of individuals, and *perhaps that concept is still useful* to us. Also, we earlier thought about the collaboration-styles of groups, and *perhaps that concept is also useful* to us. The general concept of collaboration-pattern appears to apply, in *different but related ways*, to persons, groups, and the relationships among them. This three-way analysis of collaboration-pattern is becoming a powerful and generalizable theory. At this point, we can retrieve the two memos describing collaboration-preference and collaboration-style, and combine them with the more recent memo on collaboration-relations. With those source materials in hand, we can write a longer, more integrative memo about the core concept of collaboration-patterns, making use of each of the three preceding memos. This new memo is likely to be the basis of the results and/or discussion section of our report of this research. We should record other ideas in other memos, and save them for later. The core concept that we have chosen now will be the basis for one report of the work. We may want to revisit the data and our memos later, for additional insights, and perhaps additional papers.

Substantive Theory

Glaser (1978) proposed the heuristic question, "What is this data a study of?" (Charmaz, 2006, might rephrase this as "what story do I want to tell about the data?"). In this example, the answer is becoming:

> The data are a study about a broad concept of collaboration-patterns, which are manifested in individuals as a subset of attributes that we've called collaboration-preferences, in groups as a related subset of attributes that we've called collaboration-styles, and in connections between groups and individuals as collaboration-relations.

This has become a powerful theory *based in the data*, and we may now be ready to begin to write a report of what we have concluded. The report will be centered on the *core concept* of collaboration-pattern, and will make use of the *categories* of collaboration-preference, collaboration-style, and collaboration-relation. Each of these categories has multiple *axial codes* which organize the original *open-coded* data. Our intense thinking, sampling, and theorizing about the core concept has resulted in what grounded theorists call a *substantive theory*—that is, a well-developed, well-integrated set of internally consistent concepts that provide a thorough description of the data. The work is not over. The next step, in beginning to write the report, is to relate this substantive theory to previously published or "formal" theories in the research literature (see "Case Studies of Grounded Theory Method in HCI and CSCW," below).

From the perspective of this book, we have used the powerful methods of grounded theory to shape our *knowing* about this domain, through a disciplined series of movements up and down a scale of abstraction. Initially, all we knew were the data. Keeping an open mind, we looked for regularities in the data (repeating phenomena, repeating patterns), and we began to hypothesize how those phenomena and relations could be related to one another. We tried various informal theories, and for each theory we immediately returned to the data, asking more questions, *testing* the theory to see if it was an adequate description. The goal of GTM at this point is to find out *what's wrong with the developing theory*, so that we can replace weak parts with stronger conceptions. Our testing led to these kinds of desirable failures, and ultimately to a much stronger, much more generalized theory. Now we *know* more about our domain, and we know it because we based and tested each of our theoretical developments on the data. The theory is *grounded* in the data.

Grounded theory researchers would describe this journey in different ways. Glaser held that the theory *emerged* from the data (1992), and that a principal task of the research is to cultivate sufficient *theoretical sensitivity* to be able to discern the theory in the data (1978). Corbin and Strauss also focused on finding patterns that were present in the data, using well-defined procedures and coding practices to find the right data, and to describe the phenomena in those data (2008). In retrospect, both of these approaches seem to reflect the objectivism of the times. *Knowing* takes place through discovery—grounded in the data.

By contrast, Charmaz (2006) and Clarke (2005) emphasize the researcher as an active interpreter of the description and the developing theory. In their postmodern approach, theory is constructed (not discovered), and the researcher is accountable both for the theory that she/he creates, and for the path through which she/he arrived at that theory (Charmaz, 2006, 2008; see Dourish's chapter on Ethnography in this volume for a similar movement toward accountability in ethnography). Clarke particularizes the role and responsibility of the researcher, asking *whose voice is not being heard* (*and why*)? *Whose silence is significant* (*and why*)? From the perspective of this volume, *knowing* in these postmodern accounts of GTM is an active process of construction, and takes place through cognitive and/or social acts of interpretation, conceptualization, hypothesis-creation and testing, and construction of theory—grounded in the data.

Creativity and Imagination: Memos

GTM describes a series of rigorous steps through which theory development occurs incrementally. In that spirit, most grounded theory researchers advocate an iterative series of documents (memos) that record the development of our understandings, including descriptions of codes and their meanings, thoughts about what might be going on, descriptions of how data fit (or do *not* fit) the developing theory, strategies for new samples, and so on. Corbin and Strauss write, "[Memos] force the analyst to work with ideas instead of just raw data. Also, they enable analysts to use creativity and imagination, often stimulating new insights into data." Charmaz (2006) agrees: "Memo-writing constitutes a crucial method in grounded theory because it prompts you to analyze your data and codes early in the research process […] [N]ote where you are on firm ground, and where you are making conjectures. Then go back to the field to check your conjectures." Memo-writing is an essential component of the *knowing* that occurs during GTM: "[M]emos… grow in complexity, density, clarity, and accuracy as the research progresses… They… are just as important to the research process as data gathering itself" (Corbin & Strauss, 2008).

Advice about the practices of memo-writing practices varies widely. At perhaps one extreme of brevity is Dick (2005), who recommends that a grounded theory researcher carry file cards in a pocket, so that she/he can record one of *several* memos on *each* file card. Corbin and Strauss (2008) provide examples of memos that range from a single paragraph to a page or more. Charmaz's examples include single paragraphs and well-structured essays, the latter including headers and subheaders within a single memo (2006). As theory-development progresses, memos may take on greater structure, such as the essays in Charmaz's account (2006), causal diagrams (e.g., Corbin & Strauss, 2008), formal tables that lay out each category with its component codes (Muller & Kogan, 2012), and a cartographic technique called *situational maps* (Clarke, 2005), as shown in Fig. 4. Each researcher, and each research team, will probably need to experiment to find the form or forms that suit their work.

The important point is that memo-writing is a way for the researcher to construct her/his knowledge, and to put that evidence of *knowing* into a concrete form. Activity theorists might say that, through memos, the act of knowing is externalized or crystallized (e.g., Nardi, 1996). To coin a phrase, memos are a crucial step in *making the knowing known*—to oneself and others. Memos help us to remember old ideas that we thought were not relevant (as in the examples about collaboration-preference and collaboration-style, above). Memos are the expression of theory, and guide data collection, as well as being useful in writing reports of a GTM research project.

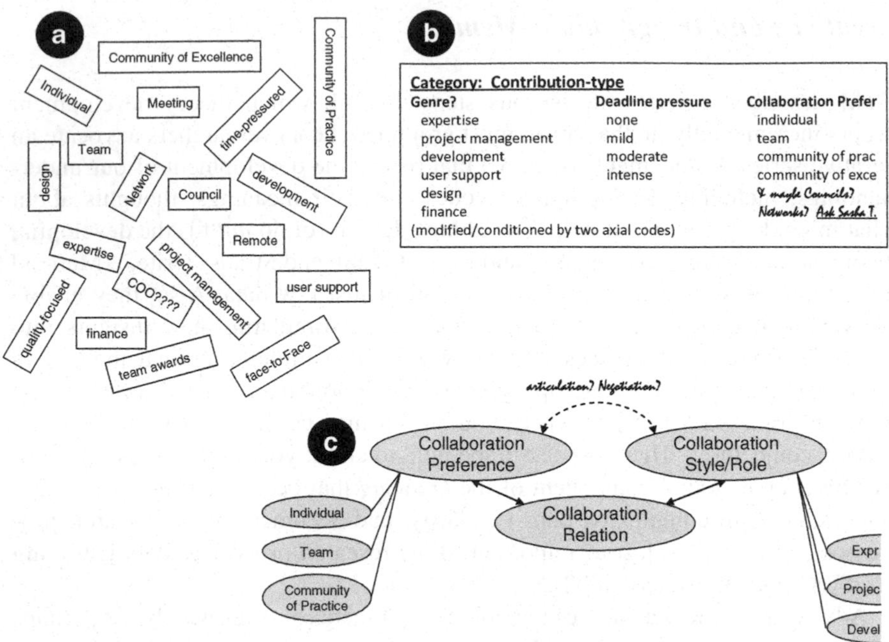

Fig. 4 The scenario in this chapter, represented according to Clarke's situational maps. (**a**) "Messy" situation map. (**b**) Ordered situational map. (**c**) Relationship map

Surprise as a Cognitive Tool: Theoretical Sampling and Constant Comparison

A core cognitive strategy of GTM is to make human capacities, such as curiosity and sensemaking, into tools of inquiry throughout the research process. Surprise is one of those tools. In the example of collaboration-patterns (above), we were repeatedly surprised to find data that did not fit the current state of our theory, and we ended the data collection when there were "no further surprises." In accord with abductive inference, each surprise led to new hypotheses ("How could this be? What would have to be true, for this new information to make sense?"). We then sought new data, to test each new hypothesis, and to strengthen and broaden the theory accordingly.

In GTM, this overall strategy is called *theoretical sampling* (Corbin & Strauss, 2008; Glaser, 1978), a rigorous form of abductive reasoning that is "strategic, specific, and systematic" (Charmaz, 2006), exactly because it is guided by the questions needed to strengthen the developing theory. We gather the new data to test the hypothesis. The data inform the hypothesis, leading to stronger hypotheses which in turn guide further data collection: "Theoretical sampling tells where to go"

(Charmaz, 2006), and memos record our progress. Theoretical sampling is one of the major strategies within the overall GTM concept of *constant comparison* of data with data, and of data with theory (Glaser & Strauss, 1967).

But if data lead to new hypotheses, and new hypotheses lead to more data, and we need to make informal documentation of each new understanding through memos, then how will we ever stop? This is where surprise again becomes an important cognitive tool. Grounded theory researchers often write about the need to *saturate one's categories* (Glaser & Strauss, 1967) or to achieve *theoretical saturation* (Gasson, 2003). A coding category is considered "saturated" when all the available data are explained by the codes in that category. There is no further surprise. Similarly, theoretical saturation is reached when all of the categories appear to be adequate to explain all of the data. Phrased in this way, the concept seems very abstract. Stern (2007) concretized it as follows, describing her study of family violence: "I realized that I had reached the point of saturation when the [informant] was telling me how when he was a small child he stood witness as his mother shot his father dead, *and I was bored*. I made all the right noises… but I knew that my data collection for that study had come to an end." (italics in the original).

Summary and Recapitulation

In the section on abductive interference, I reviewed Peirce's philosophy of abductive inference (1903), and showed that it depends crucially on (a) recognizing when one is surprised, and (b) searching for an alternative explanation. Peirce's account of how that alternative explanation is found—"like lightning"—was unsatisfactory for scientific work. I then promised that GTM would provide a principled way of moving from lightning to careful thought and deep involvement in data. The long, imaginary research story about collaboration-patterns showed key aspects of abductive inference in GTM, in the form of interleaved and interdependent practices of data collection, coding, theorizing, and documenting. Surprise played a crucial role—in concert with the principle of constant comparison of data-with-data, and data-with-theory—to show where the developing theory failed to describe the available data. We then used theoretical sampling, allowing the problems with our theory to help us choose the next people to interview (or, more generally, the next data to collect). Theoretical sampling is the rigorous GTM response to the problem of Peirce's lightning, replacing mysterious intuitions with disciplined guidance toward collecting the best data to lead toward a productive new understanding. To borrow a turn of phrase from Stern (2007), we continued until we were bored—that is, until there were no more surprises when we compared data-with-data, and data-with-theory. Surprise told us where to go next. Lack of surprise told us we were done.

Different Forms of Grounded Theory Method in HCI and CSCW

In HCI and CSCW, GTM has developed in several distinct ways. One important distinction is in the use of the research literature. Glaser and Strauss (1967) seemed to advocate that the researcher should approach the data as a kind of tabula rasa (blank slate), and should therefore avoid reading the formal or published research literature, to keep her/his mind free of bias. Subsequent researchers noted that both Glaser and Strauss had already read hundreds of books and papers about theory, and that they already had this knowledge somewhere in the background of their thinking (Morse et al., 2009). Glaser remained adamant on this point, insisting that theory emerged from an immersion in the data (e.g., Glaser, 1992). Dey (1999) phrased the objection to Glaser's position succinctly as "there is a difference between an open mind and an empty head. To analyse data, we need to use accumulated knowledge, not dispense with it" (see also Bryant & Charmaz, 2007; Funder, 2005; Kelle, 2005). Corbin and Strauss (2008) cautiously suggested that the research literature can be considered another form of data, and can be used in that way (e.g., through constant comparison) as part of a grounded theory investigation. Opinion continues to vary across a wide range of positions.

It is unlikely that an HCI or CSCW project could find successful publication if it did not include a detailed literature review. Indeed, as Urquhart and Fernández (2006), most graduate students who undertake a grounded theory study must first pass their qualifying examinations, in which they are expected to demonstrate deep engagement with the research literature. If GTM is to serve as a way of knowing, then the knowledge that it produces should be placed in relation to other knowledge. For these reasons, I believe that GTM in HCI and CSCW will probably be closer to the position of Corbin and Strauss (2008); and Bryant & Charmaz, 2007; Dey, 1999; Funder, 2005; Kelle, 2005).

Three Usage Patterns of Grounded Theory in HCI and CSCW

GTM has been invoked in three different ways in HCI and CSCW. Two of the three usage patterns appear to have a valuable place in HCI and CSCW research. In my opinion, the third way is more problematic from a GTM perspective.

Using GTM to Structure Data Collection and Analysis

The first type of invocation of grounded theory is a series of variations on the practices sketched in this chapter—i.e., iterative episodes of data collection and theorizing, guided by theoretical sampling, and the use of constant comparison as a way to think about and develop theory during ongoing data collection.

Susan Leigh Star is perhaps one of the best known grounded theory researchers in HCI and CSCW. She used grounded theory as an organizing method in a life's work that spanned the use of concepts and artifacts (boundary objects and infrastructures, Star, 1999, 2002; Star & Griesemer, 1989), the implications of classifications for organizations and inquiries (Bowker & Star, 1999), and the sources of uncertainty in nineteenth century science (Star, 1985). In Star's research, grounded theory became a powerful way of knowing which informed highly influential theorizing.

Using GTM to Analyze a Completed Dataset

The second type of invocation of grounded theory applies deep and iterative coding to a complete set of data that have already been collected, gradually building theory from the data, often through explicit use of concepts of open coding, axial coding, categories, and core concepts. While this process involves constant comparison, the application of theoretical sampling is more subtle. If the dataset must be treated "as is" (i.e., no further data can be collected), then how can the researcher use the developing theory to guide further data collection? One answer that can occur in large datasets is that the developing theory suggests different ways of sorting and excerpting from the data. In this way, the researcher finds new insights and new concepts through a process that is very similar to theoretical sampling.

An example of this approach appeared in a well-regarded paper by Wyche and Grinter (2009) about religion in the home. Wyche and Grinter conducted interviews in 20 home settings. They ended their data collection when they reached saturation (i.e., no further surprises). This appears to have been the *beginning* of their grounded theory analysis: They describe an enormous dataset of interviews, photographs, and field notes, and make explicit reference to the constant comparative method for deep and iterative coding, in conjunction with reading the research literature. Their analysis is fascinating, and has been cited as an example of excellent and influential research, with implications for theory as well as design.

Paay et al. (2009) conducted a similar post-data-collection grounded theory analysis of hybridized digital-social-material urban environments, which was explicitly guided by theoretical concepts from the research literature. In addition to a very detailed discussion of open and axial coding, they used an affinity-diagramming method from Beyer and Holtzblatt (1998) that is similar in some ways to Clarke's cartographic techniques (2005). Outcomes included a process model for this complex design domain, as well as qualitative critiques of design prototypes.

Using GTM to Signal a Deep and Iterative Coding Approach

The third type of invocation of ground theory is, to me, more problematic. Some researchers make a general reference to grounded theory as a kind of signal that they coded their data carefully. However, they give no details of their coding strategies or outcomes, and it is difficult to find any convincing evidence that they built theory from the data. In some cases, they appear to have begun their study with very specific questions, and then collected data to answer those questions. It might make more sense for these papers to make reference to more general guidance in coding data (e.g., Dey, 1993; Lincoln & Guba, 1995; Miles & Huberman, 1994). As with much of grounded theory work, this point is probably controversial. My purpose in this chapter is not to criticize authors of good work over a difference in nomenclature, so I will not name specific examples of this kind of invocation. However, from the perspective of GTM, a lack of detail about the process makes it impossible to take up the work into the corpus; it is in this sense that this use of "grounded theory" as a description of method is problematic.

Research Quality and Rigor

The preceding section suggests some indicators of quality and rigor in grounded theory research when applied to HCI and CSCW research. However, it is important to note that issues of quality remain unresolved within the broader community of grounded theory researchers, with diverse views from many researchers (Adolph et al., 2008; Corbin & Strauss, 2008; Charmaz, 2006, 2008; Gasson, 2003; Locke, 2001; Matavire & Brown, 2008), dating back, in part, to the earlier split between Glaser and Strauss (Kelle, 2007; Morse et al., 2009).

Glaser and Strauss (1967) proposed some very general qualities for evaluation of grounded theory outcomes, focusing on four terms:

- Fit: How well does the theoretical description describe the data?
- Relevance: Does the description appear to answer important questions? (See Hayes, "Knowing by Doing: Action Research as an Approach to HCI" this volume, for a related perspective).
- Work (ability): Do the components of the theoretical description lead to useful predictions?
- Modifiability: Is the theory presented in a way that will encourage other researchers to use it, test it, and change it over time?

Charmaz (2006) proposed a similar set of criteria: credibility (overlapping with fit), resonance (overlapping with relevance), originality (overlapping with some aspects of relevance and work), and usefulness (overlapping with some aspects of work and modifiability). However, most of this advice remains very general, and it is difficult translate the generalities into criteria for review of grounded theory work.

Gasson (2003) argues that grounded theory research has been difficult to evaluate (or defend) because of the default assumptions about what makes "good" research (e.g., hypothesis-testing, confirmatory/disconfirmatory expectations—see Popper, 1968). In the general context of information systems research, she calls for researchers to move from a positivist stance of "objective" facts, to an interpretivist stance that each researcher reports her/his findings as honestly as possible, for comparison with the interpretations of other researchers. Here is a partial summary of her proposed movement from positivist to interpretivism criteria in evaluating research:

From positivist	To interpretivist
Objectivity	Confirmability (emphasis is placed on informants, not researchers)
Reliability	Dependability/auditability (clear path to conclusions)
Internal validity	Internal consistency (related to GTM concept of saturation—i.e., all the components of the theory work together; there are no more surprises)
External validity	Transferability (generalizability). Cooney (2011) recommends that external "experts" be requested to render judgments of validity, as well

Even these broad criteria may be problematic. For example, Gasson proposes to test via confirmability with informants, as do Cooney (2011) and Hall and Callery (2001); similar proposals have been made for collaborative ethnography (Lassiter, 2005) and action research (Hayes, "Knowing by Doing: Action Research as an Approach to HCI" this volume). However, Elliott and Lazenbatt (2005) argue that grounded theory uniquely *combines* perspectives of many informants (constant comparison of data-with-data), and also *abstracts* a more formalized theoretical description from their combined accounts (constant comparison of data-with-theory—see also Star, 2007). In this way, grounded theory method may produce a theoretical description (i.e., the core concept and its elaboration) that presents perspectives that would be rejected by some of the informants.

Within HCI and CSCW, applying these changes in criteria may take some time, and some further development to meet our own diverse subfields' requirements. During this period, it may be useful to follow the advices of Charmaz (2006), Hall and Callery (2001), and Locke (2001). They recommend making the research process transparent to the reader, so that the reader can make her or his own assessment of the quality of methods followed and their results (Locke, 2001).

For HCI and CSCW, a citation to Glaser and Strauss (1967) provides only a general orientation to the "family of methods" that collectively describe (but do not yet define) grounded theory (Babchuk, 2010; Bryant & Charmaz, 2007; Morse et al., 2009). A more useful citation would provide a later reference, preferably after the split between Glaser and Strauss, and preferably to a methodologist who provides specific guidance—e.g., Charmaz (2006), Corbin and Strauss (2008, or the previous editions of their procedural guide), Clarke (2005), Glaser (1992, 1998), or Locke (2001). It would be useful to know which specific coding practices were used in the analysis, and it may also be useful to see a brief recapitulation of the axial coding, leading to the core concept. It would also be useful to know how the research literature was used—e.g., as a source of candidate axial codes, or as a follow-on after the analysis was largely completed. The works that I cited in the preceding section ("Case Studies") provide this kind of methodological detail, and are strengthened by it.

Conclusion

Charmaz (2006), Gasson (2003), Hall and Callery (2001), and Locke (2001) recommend that grounded theory reports be reflective on their own process, and provide transparency into that process. This advice is, of course, exactly what is needed for GTM to work as a way of knowing. The researcher needs cognitive and methodological tools to be assured of the quality of her/his own knowing, and the reader needs strong visibility into the research methods to be convinced that she/he, too, wants to share in that knowledge.

In this chapter, I have provided an inevitably personal account of grounded theory. My account focused on the virtues of human curiosity, creativity, and surprise as cognitive tools for scientific rigor. I began with Peirce's analysis of abductive inference, and went on to detail some of the rich and powerful methods that grounded theory researchers have developed to turn Peirce's insight into scientific method. People think about what they are learning *while* they are learning, and GTM turns that tendency into a scientific strength through methodological underpinnings of disciplined coding practices, guided by principles of constant comparison and theoretical sampling. The goal is to remain faithful to the data, and to draw conclusions that are firmly grounded in the data. People (not procedures or methods) construct meaning and knowledge, and GTM can help them to do that, and to share their new knowledge credibly with one another.

Exercises

1. What modifications to GTM would need to be made if the researcher has an inkling of what theory might be relevant to their observations?
2. How well does GTM accommodate a team of researchers? Where would they work independently and where collaboratively?

References[5]

Adolph, S., Hall, W., & Kruchten, P. (2008). A methodological leg to stand on: Lessons learned using grounded theory to study software development. *Proceedings of Conference on the Center for Advanced Studies on Collaborative Research 2008*. Toronto, Ontario, Canada

Ambos, T. C., & Birkinshaw, J. (2010). How do new ventures evolve? An inductive study of archetype changes in science-based ventures. *Organization Science, 21*(6), 1125–1140.

Awbrey, J., & Awbrey, S. (1995). Interpretation as action: The risk of inquiry. *Inquiry Critical Thinking Across the Disciplines, 15*, 40–52.

Babchuk, W. A. (2010). *Grounded theory as a "Family of Methods": A genealogical analysis to guide research.* US-China Education Review, 8(9). http://www.adulterc.org/Proceedings/2010/proceedings/babchuk.pdf

[5] All URLs were accessed successfully during July 2012.

Beyer, H., & Holtzblatt, K. (1998). *Contextual design: Defining customer-centered systems*. San Francisco, CA, USA: Morgan-Kaufmann.

Blythe, M., & Cairns, P. (2009). Critical methods and user generated content: The iPhone on YouTube. *Proceedings of the CHI 2009*, pp. 1467–1476

Bowker, G. C., & Star, S. L. (1999). *Sorting things out: Classification and its consequences*. Cambridge, MA, USA: MIT Press.

Bryant, A., & Charmaz, K. (Eds.). (2007). *The Sage handbook of grounded theory*. Thousand Oaks, CA, USA: Sage.

Cennamo, K., Douglass, S.A., Vernon, M., Brandt, C., Scott, B., Reimer, Y., et al. (2011). Promoting creativity in the computer science design studio. *Proceedings of the SIGCSE 2011*, pp. 649–654

Charmaz, K. (2006). *Constructing grounded theory: A practical guide through qualitative analysis*. London: Sage.

Charmaz, K. (2008). Grounded theory in the 21st century: Applications for advancing social justice. In N. K. Denzin & Y. S. Lincoln (Eds.), *Strategies of qualitative research* (3rd ed., pp. 203–241). Thousand Oaks, CA, USA: Sage.

Chetty, M., Hashim, D., Baird, A., Ofoha, U., Sumner, B., & Grinter, R.E. (2011). Why is my internet slow? Making network speeds visible. *Proceedings of the CHI 2011*, pp. 1889–1898

Clarke, A. E. (2005). *Situational analysis: Grounded theory after the postmodern turn*. Thousand Oaks, CA, USA: Sage.

Cooney, A. (2011). Rigour and grounded theory. *Nurse Researcher, 18*(4), 17–22.

Corbin, J., & Strauss, A. L. (2008). *Basics of qualitative research 3e*. Thousand Oaks, CA, USA: Sage.

Dey, I. (1993). *Qualitative data analysis: A user-friendly guide for social scientists*. London, UK: Routledge.

Dey, I. (1999). *Grounding grounded theory: Guidelines for qualitative inquiry*. San Diego, CA, USA: Academic.

Dick, B. (2005). *Grounded theory* (paper 59). Previously *Grounded theory: A thumbnail sketch*. http://www.aral.com.au/DLitt/DLitt_P59ground.pdf

Elliott, N., & Lazenbatt, A. (2005). How to recognise a "quality" grounded theory research study. *Australian Journal of Nursing, 22*(3), 48–52.

Eyrich-Garg, K. M. (2011). Sheltered in cyberspace? Computer use among the unsheltered "street" homeless. *Computers in Human Behavior, 27*(1), 296–303.

Faily, S., & Flechals, I. (2011). Persona cases: A technique for grounding personas. *Proceedings of the CHI 2011*, pp. 2267–2270

Faste, H., & Lin, H. (2012). The untapped potential of digital mind maps. *Proceedings of the CHI 2012*, pp. 1017–1026

Funder, M. (2005). Bias, intimacy, and power in qualitative fieldwork strategies. *Journal of Transdisciplinary Environmental Studes, 4*(1), 1–9.

Furniss, D., Blandford, A., & Curson, P. (2011). Confessions from a grounded theory phd: Experiences and lessons learnt. *Proceedings of the CHI 2011*, pp. 113–122

Gasson, S. (2003). Rigor in grounded theory research: An interpretive perspective on generating theory from qualitative field studies. In M. Whitman & A. Woszczynski (Eds.), *Handbook for information systems research* (pp. 79–102). Hershey, PA, US: Idea Group.

Glaser, B. G. (1978). *Theoretical sensitivity*. Mill Valley, CA, USA: Sociology Press.

Glaser, B. G. (1992). *Basics of grounded theory analysis: Emergence vs forcing*. Mill Valley, CA: Sociology Press.

Glaser, B. G. (1998). *Doing grounded theory: Issues and discussions*. Mill Valley, CA, USA: Sociology Press.

Glaser, B. G., & Strauss, A. L. (1965). *Awareness of dying*. Chicago, IL, USA: Aldine.

Glaser, B. G., & Strauss, A. L. (1967). *The discovery of grounded theory*. Chicago, IL, USA: Aldine [[[e.g., 'The published word is not the final one, but only a pause in the never-ending process of generating theory']]].

Glaser, B. G., & Strauss, A. L. (1968). *A time for dying*. Chicago, IL, USA: Aldine.
Gold, J., Walton, J., Cureton, P., & Anderson, L. (2011). Theorising and practitioners in HRD. *European Journal of Training and Development, 35*(3), 230–246.
Haig, B. D. (1995). Grounded theory as scientific method. In *Philosophy of education yearbook 1995*.
Haig, B.D. (2005). Grounded theory as scientific method. *Philosophy of Education Yearbook 2005*. Philosophy of Education Society.
Hall, W., & Callery, P. (2001). Enhancing the rigor of grounded theory: Incorporating reflexivity and relationality. *Qualitative Health Research, 11*(2), 257–272.
Hoda, R. (2011). *Self-organizing agile teams: A grounded theory*. Wellington, NZ: Victoria University of Wellington.
Holton, J. (2007). The coding process and its challenges. In A. Bryant & K. Charmaz (Eds.), *The Sage handbook of grounded theory*. Los Angeles: Sage [[[coding practices + Charmaz + Corbin]]].
Kelle, U. (2005). "Emergence" vs. "Forcing" of Empirical Data? A crucial problem of "Grounded Theory" reconsidered [52 paragraphs]. *Forum Qualitative Sozialforschung/Forum: Qualitative Social Research, 6*(2), Art. 27, http://nbn-resolving.de/urn:nbn:de:0114-fqs0502275
Kelle, U. (2007). The development of categories: Different approaches in grounded theory. In A. Bryant & K. Charmaz (Eds.), *The Sage handbook of grounded theory* (pp. 191–213). Thousand Oaks, CA, USA: Sage.
Kim, H., & Lee, W. (2012). Framing creative users for describing cases of appropriation (poster). *Proceedings of the CSCW 2012*, pp. 135–138
Kim, T., Hong, H., & Magerko, B. (2010). Design requirements for ambient display that supports sustainable lifestyle. *Proceeding of the DIS 2010*, pp. 103–112
Kjeldskov, J., & Paay, J. (2005). Just-for-us: A context-aware mobile information system facilitating sociality. *Proceedings of the MobileCHI 2005*, pp. 23–30
Lassiter, L. E. (2005). *The Chicago guide to collaborative ethnography*. Chicago, IL, USA: University of Chicago Press.
Lewis, S., & Lewis, D.A. (2012). Examining technology that supports community policing. *Proceedings of the CHI 2012*, pp. 1371–1380
Lincoln, Y. S., & Guba, E. G. (1995). *Naturalistic inquiry*. London, UK: Sage.
Lincoln, Y. S., & Guba, E. G. (2000). Paradigmatic controversies, contradictions, and emerging confluences. In N. K. Denzin & Y. S. Lincoln (Eds.), *The handbook of qualitative research* (pp. 163–188). Beverly Hills, CA: Sage.
Locke, K. (2001). *Grounded theory in management research*. Thousand Oaks, CA, USA: Sage.
Lopes, E. (2010). *A grounded theory of decision-making under uncertainty and complexity*. Saarbrücken: VDM Verlag.
Macrì, D. M., Tagliaventi, M. R., & Bertolotti, F. (2002). A grounded theory for resistance to change in a small organization. *Journal of Organisational Change Management, 15*(3), 292–310.
Matavire, R. & Brown, I. 2008. Investigating the use of "Grounded Theory" in information systems research. *Proceedings of the SAICSIT 2008*, pp. 139–147
Mathiasen, N.R., & Bødker, S. (2011). Experiencing security in interaction design. *Proceedings of the CHI 2011*, pp. 2325–2334
Matthews, T., Whittaker, S., Moran, T., & Yuen, S. (2011). Collaboration personas: A new approach to designing workplace collaboration. *Proceedings of the CHI 2011*, pp. 2247–2256
Miles, M. B., & Huberman, A. M. (1994). *Qualitative data analysis: An expanded sourcebook* (2nd ed.). Thousand Oaks: Sage Publications.
Morse, J. M., Stern, P. N., Corbin, J., Bowers, B., Charmaz, K., & Clarke, A. E. (2009). *Developing grounded theory: The second generation*. Walnut Creek, CA, USA: Left Coast Press.
Muller, M., & Kogan, S. (2012). Ground theory method in HCI. In J. Jacko (Ed.), *Human computer interaction handbook*. Florence, KY, USA: CRC Press.
Muller, M., Millen, D.R., & Feinberg, J. (2009). Information curators in an enterprise file-sharing service. *Proceedings of the ECSCW 2009*. Vienna, Austria: Springer

Nardi, B. (Ed.). (1996). *Context and consciousness: Activity theory and human-computer interaction*. Cambridge, MA, USA: MIT Press.

Nielsen, J. (1992). The usability engineering lifecycle. *IEEE Computer, 25*(3), 12–22.

Odom, W., Zimmerman, J., & Forlizzi, J. (2010). Designing for dynamic family structures: Divorced families and interactive systems. *Proceedings of the DIS 2010*, pp. 151–160

Orlikowski, W. J. (1992). The duality of technology: Rethinking the concept of technology in organizations. *Organization Science, 3*(3), 398–427.

Paay, J., Kjeldskov, J., Howard, S., & Dave, B. (2009). Out on the town: A socio-physical approach to the design of a context-aware urban guide. *ACM Transactions on Computer-Human Interaction, 16*(2), 1–34. article 7.

Pauleen, D. J., & Yoong, P. (2004). Studying human-centered IT innovation using a grounded action learning approach. *The Qualitative Report, 9*(1), 137–160.

Paavola, S. (2012). *On the origin of ideas: An abductivist approach to discovery*. University of Helsinki, Saabrücken, DE: Lap Lambert Academic Publishing. Excerpts available at http://helsinki.academia.edu/SamiPaavola/Books/1585338/On_the_Origin_of_Ideas._An_Abductivist_Approach_to_Discovery._Revised_and_enlarged_edition. Original PhD thesis (2006) available at http://ethesis.helsinki.fi/julkaisut/hum/filos/vk/paavola/

Peirce, C.S. (1903). Harvard lectures on pragmatism. *Collected Papers 5.1*, pp. 71–172

Popper, K. (1968). *The logic of scientific discovery* (2nd ed.). New York, NY: Harper Torchbook [[[falsifiability]]].

Reichertz, J. (2007). Abduction: The logic of discovery in grounded theory. In A. Bryant & K. Charmaz (Eds.), *The Sage handbook of grounded theory* (pp. 214–228). London: Sage.

Reichertz, J. (2010). Abduction: The logic of discovery of grounded theory. *Forum Qualitative Social Research, 11*(1), 16. Art 13.

Riitta, H., Urquhart, C., & Iivari, N. (2009). "Who's in charge, and whose rules are followed…?" Power in an inter-organisational IS project. *Proceedings of the ECIS 2009*, pp. 943–956

Riitta, H., & Newman, M. (2011). The complex nature of emotions in an inter-organisational information system project. *Proceedings of the ECIS 2011*, pp. 943–956

Rode, J. (2009). Digital parenting: Designing for children's safety. *Proceedings of the BCS HCI 2009*, pp. 244–251

Sayago, S., & Blat, J. (2009). *About the relevance of accessibility barriers in the everyday interactions of older people with th web*. Proc. W4A2009-Technical, 104–113

Seidel, S., & Recker, J. (2009). *Using grounded theory for studying business process management phenomena*. Proc. ECIS 2009, 490–501

Shannak, R. O., & Aldhmour, F. M. (2009). Grounded theory as methodology for theory generation in information systems research. *European Journal of Economics, Finance, and Administrative Services, 15*.

Star, S. L. (2007). Living grounded theory. In A. Bryant & K. Charmaz (Eds.), *The Sage handbook of grounded theory*. Thousand Oaks, CA, USA: Sage.

Star, S.L. (2002). Got infrastructure? How standards, categories, and other aspects of infrastructure influence communication. 2nd Social Study of IT Workshop at the LSE ICT and Globalization. http://csrc.lse.ac.uk/events/ssit2/LeighStar.pdf .

Star, S. L. (1999). The ethnography of infrastructure. *American Behavioral Scientist, 43*(3), 377–392.

Star, S. L. (1985). Scientific work and uncertainty. *Social Studies of Science, 15*(3), 391–427.

Star, S. L., & Griesemer, J. R. (1989). Institutional ecology, "translations" and boundary objects: Amateurs and professionals in Berkeley's Museum of Vertebrate Zoology, 1907–39. *Social Studies of Science, 19*(3), 387–420.

Stern, P. N. (2007). Properties for growing grounded theory. In A. Bryant & K. Charmaz (Eds.), *The Sage handbook of grounded theory*. Thousand Oaks, CA, USA: Sage.

Strauss, A. L. (1993). *Continual permutations of action*. New York, NY, USA: Aldine.

Strauss, A. L., & Glaser, B. G. (1970). *Anguish*. Mill Valley, CA, USA: Sociology Press.

Thom-Santelli, J., Muller, M.J., & Millen, D.R. (2008) Social tagging roles: Publishers, evangelists, leaders. *Proceedings of the CHI 2008*

Urquhart, C., & Fernández, W.D. (2006). Grounded theory method: The researcher as blank slate and other myths. *Proceedings of the ICIS 2006*, pp. 457–464

Vines, J., Blythe, M., Lindsay, S., Dunphy, P., Monk, A., & Olivier, P. (2012). Questionable concepts: Critique as a resource for designing with eighty-somethings. *Proceedings of the CHI 2012*, pp. 1169–1178

Wyche, S.P., & Grinter, R.E. (2009). Extraordinary computing: Religion as a lens for reconsidering the home. *Proceedings of the CHI 2009*, pp. 749–758

Wyche, S.P., Smyth, T.N., Chetty, M., Aoki, P.M., & Grinter, R.E. (2010). Deliberate interactions: Characterizing technology use in Nairobi, Kenya. *Proceedings of the CHI 2010*, pp. 2593–2602

Yardi, S., & Bruckman, A. (2012). Income, race, and class: Exploiting socioeconomic differences in family technology use. *Proceedings of the CHI 20120*, pp. 3041–3050

Knowing by Doing: Action Research as an Approach to HCI

Gillian R. Hayes

What Is Action Research?

Action research (AR) is an approach to research that involves engaging with a community to address some problem or challenge and through this problem solving to develop scholarly knowledge. AR is method agnostic, which is to say action researchers make use of a large variety of qualitative and quantitative methods to understand the change they are undertaking in communities. In HCI, AR often also uses design, development, and deployment of technologies as methods of knowing and of enacting change. The cornerstone of AR is that these two cannot be disentangled: the doing and the knowing, the intervention and the learning.

AR is explicitly democratic, collaborative, and interdisciplinary. The focus when conducting AR is to create research efforts "with" people experiencing real problems in their everyday lives not "for," "about," or "focused on" them. Thus, AR research focuses on highly contextualized, localized solutions with a greater emphasis on transferability than generalizability. That is to say, the knowledge generated in an AR project should be contextualized enough to enable someone else to use this information to create their own change—which may or may not be similar—in another environment—which again may or may not be similar.

AR offers a systematic collaborative approach to conducting research in HCI that satisfies both the need for scientific rigor and promotion of sustainable social change and has been taken up by a variety of researchers in HCI (e.g., Foth & Axup, 2006; Palen, 2010) and information systems (e.g., Baskerville & Pries-Heje, 1999) research. AR "aims to contribute both to the practical concerns of people" in problematic situations and to the academic goals of science "by joint collaboration within a mutually acceptable ethical framework" (Rapoport, 1970, p. 499). AR includes "systemic inquiry that is collective, collaborative, self-reflective,

G.R. Hayes (✉)
Donald Bren School of Information and Computer Sciences, University of California Irvine,
5072 Donald Bren Hall, Irvine, CA 92797–3440, USA
e-mail: gillian.hayes@uci.edu

J.S. Olson and W.A. Kellogg (eds.), *Ways of Knowing in HCI*,
DOI 10.1007/978-1-4939-0378-8_3, © Springer Science+Business Media New York 2014

critical and undertaken by participants in the inquiry" (McCutcheon & Jung, 1990, p. 148). Procedurally, AR is "comparative research on the conditions and effects of various forms of social action, and research leading to social action" that uses "a spiral of steps, each of which is composed of a circle of planning, action, and fact-finding about the result of the action" (Lewin, 1946, 1948). AR necessitates that researchers become "facilitators" of the intervention and research process, enabling collaborators from the community to step up into the researcher role alongside the rest of the team. Researchers in this model must become conscious of their own positions and allow for the prioritization of different values than they might typically. This transformation from leader, with expertise that is prioritized above participant knowledge, to coach, who draws out participant ideas and places them centrally within the project, allows space for all viewpoints. This approach privileges the local knowledge contributed by community insiders as much as the academic knowledge contributed by community outsiders.

History and Intellectual Tradition of AR

Although there is some debate about exactly when and how AR emerged (Masters, 1995), most scholars credit Kurt Lewin, a psychologist who escaped Nazi Germany for the USA in the 1930s, with first defining a theory of action research in 1944 while a professor at MIT. He published "Action research and minority problems" (Lewin, 1946) shortly thereafter, creating the first published piece of scholarship explicitly describing AR. By arguing that knowledge could best be constructed by real-world tests and that "nothing is as practical as a good theory" Lewin began to make AR and intervene in research settings acceptable as a means for scholarly inquiry. To make this kind of progress, however, Lewin relied on an emergent scientific culture led by the pragmatists, perhaps most notably John Dewey and William James, that saw science as relevant and available to everyone, not just the ivory tower elites interested in "esoteric knowledge" (Dewey, 1976; Greenwood & Levin, 2007; James, 1948). Particularly relevant to the ideas that would form the basis for AR, Dewey saw the process of generating knowledge as the product of cycles of action and reflection (Dewey, 1991/1927). He advanced the idea that thought and action cannot be separated, a cornerstone of Lewin's approach to research and of AR more generally.[1]

What Can Action Research Do for You?

There are numerous resources on AR, including books that will walk you through the history or the application of AR as well as works critically reconstructing AR and its tenants. In one such book, "The Action Research Reader," Grundy breaks

[1] Interested readers are referred to the chapter on Grounded Theory Method in this same volume, which also engages the concerns and ideas of the pragmatists.

AR projects into a taxonomy that includes technical, practical, and emancipatory (Grundy, 1988). Likewise, McKernan describes three views on problem solving prevalent in AR: scientific-technical, practical-deliberative, and critical-emancipatory (McKernan, 1991).

The first type (technical or scientific-technical) is traditionally most relevant to the natural and computational sciences in which truth and reality are generally thought to be knowable and measurable, and knowledge produced may be predictive and generalizable. In this case, the facilitator engages with collaborators to test an intervention based on a predefined scientific theory. This intervention is designed to create some change in the setting, which can include new practices and approaches, different power structures or group dynamics, altered patterns of action, or simply the incorporation of a new piece of technology into daily practice. This approach may then result in change less likely to be carried on by the community partners after completion of the project, depending on how "bought in" to this theory and intervention the community collaborators are (or can be given the resources they have available). Indeed, in my own research, when I have used this approach to AR, I am left with results that are very comfortable to an HCI audience and which could be useful in creating long-term sustained change, but the specific projects themselves did not succeed in making those changes. For example, in one school-based research effort, I developed a system that fit well with established educational theory around behavior management for children with severe disabilities. However, in practice, the teachers and administrators did not have the resources available to them to continue to use the system after the end of the research project. This system has since experienced commercial success with other schools, in which resources are not as constrained and practices are more closely aligned with those theoretically recommended.

The other two approaches (practical-deliberative and critical-emancipatory), which are more familiar to a humanities or a critical theory research team, focus more on unknowable, social realities with research problems that are constantly evolving and defined in the situation by a variety of stakeholders with dynamic and mixed values. The relationships in these models are based on a move towards mutual understanding and shared solution development as opposed to a model in which the researchers, while working in democratic partnership with community members, are still ultimately interested in technical design, validation, and refinement. These latter AR approaches tend to produce sustainable change more reliably but may not produce innovative solutions that would warrant additional interest from the computing research community. Both approaches rely heavily on interpretivist data analysis and the development of shared understanding among all participants. The primary difference in these two approaches lies in the degree to which the research facilitators seek to identify problems in collaboration with partner participants. Practical AR is largely about understanding local practices and solving locally identified problems, whereas emancipatory AR promotes a kind of consciousness raising and criticality that seeks to empower partners to identify and rise up against problems they may not have identified initially on their own.

AR is essentially method agnostic. Ultimately, researchers interested in AR must decide what they hope to understand: an underlying technical reality that will produce generalizable results (*technical*), a local problem and its (potentially technological) solutions (*practical*), or how to change practices towards those that enhance or produce equity amongst underrepresented and mistreated communities (*emancipatory*). AR can support researchers in any of these goals, but the approaches may be different depending upon what is in focus. Regardless of what type of AR one undertakes, there are common underlying tenets, as described in the following section.

Doing Action Research (and Doing It Well)

Good AR is fundamentally empirical and cyclical, which is to say the actions undertaken are responsive to emergent evidence. This responsiveness is required of research settings in which the goal is to achieve both intervention and understanding. Furthermore, this understanding must unpack both the setting itself and the outcome of the intervention—whether successful or not. Thus, the research questions and methods must continually evolve alongside the context of the setting, which allows researchers to capitalize on the knowledge developed in earlier stages of the project with the involvement and engagement of those most affected by the intervention. Additionally, good AR must be critical, which of course is easier in a cyclic process, in which action always follows planning and is followed by reflection and review. Schon (1983) references this kind of criticality as "reflection in action," a process by which the research team unpack both the outcomes of the intervention(s) and the means by which they were accomplished interdependently. Given the limited separation of research and practice in AR, this kind of reflection must consider not only the specific research questions initially posed and those that have evolved from the work but also questions of practice. The research team then must ask the following: What happened? Did the intervention work (as planned)? What do we know about the site, our theories, and the empirical data that can explain why or why not? Now what?

The emphasis on incorporating multiple stakeholder viewpoints[2] alongside literature reviews and empirical evidence can enable researchers to engage more critically with the field site, as described here. This must include critical reflection on the interests and values of the community. For example, as noted in other chapters, particularly those focused on qualitative research methods, researchers engaged deeply in field sites must recognize their own taken-for-granted positions and beliefs. The same is true for AR. One cannot go into an AR project with a mind completely clear of our own cultural and personal beliefs. Instead, AR requires us to uncover our own prejudices alongside those of the field site. Thus, good action

[2] Interested readers should also explore value-sensitive design, values in design, and participatory design as design-oriented approaches that focus on multiple-stakeholder viewpoints.

researchers use a multitude of methods to gather evidence about complex situations and varied viewpoints while critiquing their own practices and knowledge production. AR then requires careful discrimination among the data, summaries of those data, and interpretations or judgment based on the data and theory. The inherent flexibility of AR allows these researchers to balance critical reflection and scientific rigor, as defined by an eye towards trustworthiness (Lincoln & Guba, 1985). Likewise, by examining transferability rather than generalizability, researchers can ensure that even in the face of multifaceted and complicated projects and field sites enough information about the projects is documented to allow other researchers to take up the results.

Trustworthiness stems from four distinct but related concepts: credibility, transferability, dependability, and confirmability (Lincoln & Guba, 1985; Stringer, 2007). The notion of trustworthiness as a measure of scientific rigor can be—and often is—applied to other related approaches to research (e.g., ethnographic methods, collaborative inquiry). AR is particularly well suited to address issues of credibility and integrity of studies. First, the prolonged engagement common to AR projects ensures that the kinds of deep-seated emotional responses or hidden tacit knowledge that are nearly impossible to retrieve in a single interview or focus group will emerge. Second, AR projects typically include persistent and explicit observation over this extended period of engagement enabling researchers to gather data directly in the field while it is happening as well as from informant accounts. Furthermore, both in interviews and observations, AR places an emphasis on participant language and perspectives as opposed to the layering of scientific language from the literature on participant concepts. To this end, Stringer advocates the use of the *verbatim principle*, in which researchers use terms and concepts "drawn from the words of the participants themselves" to "minimize the propensity to conceptualize events through their own interpretive lenses" [Stringer, 2007, p. 99]. Third, AR ensures credibility of data through the inclusion of multiple perspectives which can allow conflict, disagreement, and therefore data triangulation to occur (Lincoln & Guba, 1985) followed by member checking—in which informants verify data collected about them—and debriefing—in which participants are encouraged to voice concerns and comment on the science itself. Furthermore, through an emphasis on standpoint analysis, by which researchers are encouraged to understand and to describe both their own perspectives and those of the participants with whom they are working (Denzin, 1997; Smith, 1989; Stringer, 2007), AR reminds us that no singular account with one voice can describe the myriad complex viewpoints in any research setting. Finally, the credibility and validity of AR knowledge are measured to a large degree by the "workability" of solutions—that is, their ability to address real problems in the lives of the participants (Greenwood & Levin, 2007, p. 63). The workability requirement of solutions enforces the tight link between theory and practice by ensuring that theoretical knowledge generated in and from the field is returned to the field in the form of some sort of action that can be evaluated.

AR intentionally de-emphasizes the notion that research results can or should be made generalizable to some larger population beyond the one present. Researchers engaged directly and closely with communities, as in AR, recognize the inherent contextualization and localization of any developed solution. Thus, the goal is instead transferability. To accomplish this goal, data must be collected, analyzed, and described as transparently as possible (dependability). Furthermore, enough evidence must be presented to confirm that the events transpired as described (confirmability).

This transparency in the development of solutions, collection of data about them, and analysis in results enables other researchers—or community members and other stakeholders in related situations—to trust the results enough to examine what is similar and what is different about their setting in an attempt to replicate parts of the solution while changing others. Thus, AR does not say that no solution can ever be successful outside of the local context for which it was developed. Instead, AR provides a rigorous framework for generating and sharing sufficient knowledge about a solution that it may potentially be transferred to other contexts.

AR shares many methods and issues familiar to HCI researchers: working with community partners, engaging in fieldwork, and designing and developing solutions iteratively. However, an AR approach alters these processes in significant ways. First, the researcher in an AR project takes on the role of a "friendly outsider" (Greenwood & Levin, 2007, p. 124–128). The researcher as friendly outsider is an approach that explicitly rejects the idea that researchers should distance themselves from the "subjects" of their research in the name of objectivity. Instead, AR requires researchers to become "coaches" who are skilled at opening up lines of communication and facilitating research activities *with* community partners rather than designing and implementing research *about* them. Likewise, the research facilitator co-designs interventions and change *with* community partners, not *for* them. In this model, researchers may support community collaborators in critical thinking and academic reasoning, but this view privileges local knowledge as being as important as scientific or scholarly knowledge. Thus, all involved are co-investigators of, co-participants in, and co-subjects of both the change and evaluation activities of the project. Importantly, as Light et al. note, finding and working with community partners are not as simple as identifying someone in need (or someone representative of a group in need) and placing them in the collaborative relationship of the research team. Rather, there is a process by which these individuals are made participants—and a parallel process by which the researchers are also made participants—resulting in the entire team being together rather than from the university or from the community (Light, Egglestone, Wakeford & Rogers, 2011), a process that bears some similarity to the notion of collaborative ethnography (Lassiter, 2005). In this section, I describe some of the considerations and procedures relevant to taking an AR approach in HCI with examples from my own work when appropriate.

Establishing a Relationship with a Community Partner

The first step in many scientific research projects is to formulate a problem statement or collection of research questions. In AR, these research questions should be developed collaboratively in partnership with members of the community you wish to engage and thus tend to be inherently interdisciplinary in nature. Thus, the first step in an AR project is often to engage with a community partner. Community partners can be people with whom one has a long-standing relationship or they may emerge once a researcher has decided to address a particular problem or a set of problems. For example, one action researcher might choose to work with the school where he or she teaches or his or her daughter attends, while another may hold a workshop for local teachers and attempt to identify someone sympathetic to the research problem being addressed. Likewise, community partners may be recruited by or may recruit researchers. For example, one researcher may be called upon by a nonprofit familiar with their work and interested in what technology can do for their organization while others may have to call a set of nonprofits working in their areas of interest.

Regardless of the means by which community partners are identified, it is incumbent upon the research team in an AR project to grow those relationships and establish trust among all parties before work can begin. Typical relationship-building approaches can include researchers presenting some of their work to the potential partners, partners presenting some of their challenges to the research team as well as any ideas they have of how they might work together, and less formal approaches like just "hanging out" together. Even after an initial relationship is established, it can take a long time to develop into a workable partnership for an AR project. Signs that the relationship is established to that point include indicators that all team members trust each other, they have a shared commitment to working together, and there is general amiable casual communication.

Research Questions and Problem Statements

Once a relationship has been established, the AR project team—including researchers and community partners—can begin to develop shared research questions and problem statements. AR inherently includes the development of some action, often a technological intervention in HCI research. Before such an intervention can be designed, vision and operational statements should be crafted collaboratively [Stringer, p. 151]. Vision statements enable the AR team to work together to decide what the issues are and to develop methods for accounting for all of the concerns of the varied people involved. They provide the means by which all voices are heard and all concerns are included and often include a list of goals or a "vision" for the outcomes of the project. Vision statements often arise from substantial fieldwork, surveys, focus groups, and interviews, activities that are described in other chapters of this book and in other reference materials.

As one example, in working with an afterschool program that supported teaching children about technology in inner-city Atlanta, I struggled to craft a vision with the local leaders of the program for successful change in their efforts.

The dominant issue in our struggles was whether the program, which appeared successful in the literature and thus was being replicated in Atlanta, would in fact translate from the program in which it originated in a larger city in the Northeast United States. I was inclined to follow the literature and thus viewed our project as focusing on developing action to support getting their program "back on track" with the national efforts. The local leaders, however, believed that the processes and ideas that originated elsewhere would not work for their population. Thus, I then spent several months conducting fieldwork to understand the nuances of their population and the implementation of the program at their site before we could begin to craft a collaborative vision statement. By working together over these months, we were ultimately able to articulate multiple research questions and a general direction that incorporated portions of each of our original ideas and some that emerged during our time working together. These research questions were both substantially more relevant to the real issues at hand and more credible in terms of developing knowledge due to their connection to both the literature and the local context. Context and community are thorny words in any research, but in HCI, they can become even more so. When considering information and communication technologies, knowledge is no longer strictly place or infrastructure based. They can include people, structures, technologies, localities, and virtual spaces. Of course, not every community collaborator is interested in traditional academic research questions, regardless of discipline. Thus, in AR, the notion of a "research question" must be broader than those that can be published and include questions about process and outcomes that are important to the community partners who are interested in quality improvement and assessing the impact of an intervention on their sites.

Operational statements follow from vision statements and specifically detail how all of the individuals involved will work together to ensure that the vision statements can be met [Stringer, p. 151]. As such, operational statements operationalize the vision and often include phrasing such as "the [organization] will enact its vision through" followed by a list of detailed changes that will be made. Operational statements can be hard to craft and even harder to support and commit to completing. Thus, the action researcher, as research facilitator, must work to support participants in communicating with one another, compromising, and prioritizing some activities over others. In terms of HCI projects, these activities can also include prioritizing some features and functionality in technological artifacts over others. Again, it is important to recognize here that the researchers have some expert knowledge (e.g., what can be done technologically, what timeline and funding resources constrain the project) as well as the ability to see things as outsiders. However, local knowledge is also vitally important and should be treated as expert knowledge in its own right. Thus, these decisions should be made collaboratively as part of a negotiation between all of the stakeholders and participants in the project. Addressing these issues early in the project can enhance the commitment of all members of the team to ensuring that both the intervention and the research are completed successfully as well as enable the airing of any potential concerns before they grow into substantial problems.

Action and Intervention

The action in AR can include any of a variety of social and technological changes within the larger sociotechnical context in which the AR project is situated. Adjustments must be made to both the technological and the organizational systems at the same time. This "joint optimization" accounts for the necessary training required to "operate in [the new] technical environment" and the necessary design required given the particular behaviors or features of the organization into which the technology will be deployed (Greenwood & Levin, 2007). Technological and organizational designs are therefore inseparable. Furthermore, as with the research questions and vision and operational statements, design of these sociotechnical interventions must be conducted collaboratively with the community partners. This kind of engagement is related to but distinct from that traditionally advocated in participatory design (PD) (e.g., Greenbaum & Kyng, 1992; Muller, 2007; Schuler & Namioka, 1993). Both PD and AR stem from the notion that change should be designed and implemented democratically and inclusively (Foth & Axup, 2006). However, the scope of PD is typically more limited to the design of solutions, whereas the scope of AR includes the notion of *learning through action*.

Although this kind of reflection is important to design, and in particular the PD process, it is not the same as the construction of scholarly knowledge through action required of AR. This kind of learning stems from the extensive co-construction of knowledge before, during, and after the implementation of any change—technological or otherwise. This broad scope ensures that the problems as well as the solutions are collaboratively developed and articulated. Furthermore, the emphasis on research over design in AR drives home the idea that the end goal of AR is not *the best* solution to a problem but rather greater understanding of the setting through engagement in change and production of *potentially better* solutions iteratively and over long periods of engagement. For example, in a 5-year ongoing project with a public school in Southern California, we have been working towards an understanding of the role of digital tools in providing visual support for students. Over time, these tools have taken a variety of forms, and as the teaching practices and available hardware have both changed, so too has our software. Being unafraid of using something that is not "done" has enabled us to make positive change in the classroom activities and to unpack interesting research questions about the design of these artifacts and their use in schools. Recognition that the ultimate goal of AR is to learn through doing can free the designers and researchers in the project from what Stolterman refers to as "design paralysis" that can occur through "endless opportunities" in a messy design space (Stolterman, 2008). AR teams create interventions after thoughtful consideration. However, an attitude that focuses on the outcome of learning something, regardless of the "success" of the design or the intervention, can free up the team to attempt interventions that may be risky or underdetermined.

Evaluation

Proponents of AR frequently note that evaluation is neither a natural nor a neutral act. Evaluation as a process begs the following questions: Who evaluates? What gets evaluated? What power structures and decision processes led to this evaluation strategy? Thus, evaluation in AR, just like problem definition and intervention design, is recognized to be a value-laden enterprise. AR projects seek to ask and answer questions of interest to the research community as well as those that are of interest to the community partners. Furthermore, AR seeks to "define outcomes in ends that are acceptable to stakeholders, rather than those whose degree of success may be measured against some set of fixed criteria" [Stringer, p. 141]. In this model, evaluation is carried out as a joint construction among all the participants. Stakeholder groups are encouraged to air all of their concerns, review data that has been collected about and for the project, resolve any issues they can, and prioritize a list of unresolved items ("future work," in HCI parlance) (Guba & Lincoln, 1989). Both scholarly and practical questions around the change must be addressed. Because action researchers often engage deeply with a field site prior to any change occurring, some traditional measures of change can often be deployed (e.g., surveys, observational measures). Ideally, the change is sustained, but the use of the technology might not be, leaving room for a pre–during–post-intervention study design. For example, in one project focused on teaching social skills to students in elementary school, we removed the technology after an intervention period, but the adjustments to student behavior remained. This research finding had the positive practical impact of allowing the teachers to create a curriculum model that includes a brief but intense intervention each year that has lasting effects on the social behavior of the students throughout the year.

These methods inevitably lead to disagreement in some projects. Furthermore, the academic pressures of publishing—and the position of the research facilitators as people who know what is of interest to the academic community—can privilege some portions of the evaluation activities over others. Academic researchers are skilled in arguing their points, have deeper knowledge of the research literature than community partners, and carry with them innate status. Thus, they must be careful of "model monopoly" (Braten, 1973), in which the professional researcher dominates the conversation with their own models of the community partners and the situation. This kind of dominance ultimately enables the professional researcher to thereby dominate the plan of action. It is important during evaluation as much as at any point in an AR project to remember that the researcher should act as facilitator for a team, not leader of a project, and ensure that all of the perspectives are represented in the plans for evaluation and analysis.

A compromise on the means for evaluation to ensure that all perspectives are represented is core to the AR approach, even when it means substantial additional work on the part of the research team. One example of such compromise occurred in my work with a special education school over a 2-year period. The research questions we initially developed as a team focused on whether teachers would be able to collect the data required for a particular school practice more efficiently and with less burden using the technological intervention we had designed. As it became clear that the teachers would easily be able to conduct these practices using the technology, the teachers and other school-based professionals began to iterate on the goals of the project, noting that the quality—beyond efficiency—of teacher practice might be changed using the tools we had provided. There were also questions about the quality of teacher-based assessments when compared with professional experimental assessments regardless of whether the teachers were using our tool. This quality was best measured by gathering substantially more data and analyzing these data in a way that would produce rigorous results that could be included in year-end reports about each child as well as each teacher's progress but would be of little to no interest to the HCI research community. Because our first duty in AR is always to our community partners, we included these issues in our evaluation and analysis. The additional data not only addressed the questions raised by our community partners, but it also enabled the co-construction of new knowledge that was unexpected by both the community partners and the research facilitators but emerged through the partnership. These results, though not directly relevant to HCI researchers, were of much interest to the community partners, to special education researchers, and to our interdisciplinary team. Ultimately, their inclusion strengthened the work and led to further publications outside the HCI domain. Of course, respect for all the viewpoints in an AR project could mean not collecting data that the researchers themselves want. For example, collecting the data might be too invasive or too cumbersome for the community partners, particularly when for legal, access, or ethical issues these data must be collected by the partners. In these cases, a compromise would have to be created that respects the viewpoints of the entire team.

Disseminating Knowledge and Documenting Progress

The full inclusion of community partners in AR projects does not end with the implementation of the research or with the analysis of results. Rather, AR explicitly requires writing with engaged partners. The written material generated from these collaborative activities can come in three forms: reports written for the local group only, scholarly works written for the research community most closely aligned with the community partners, and scholarly works for the research facilitator's research community.

Reports generated for the local group should have a written component, both to serve as a formal record of the project and to ensure the specificity of language and reflection by all participants. However, they may also be accompanied by presentations or even dramatic plays and other performances. For example, in a project in Southern California, we recently created a video report to show to busy members of a local school board who were unable to spend more than a few minutes discussing any particular project or issue at their meetings.

These reports can serve multiple purposes in an AR project. First, and most importantly, the activity of creating a report itself makes space for an explicit time during which the entire research team comes together and reflects on the actions they have taken. By doing so in writing or another presentation medium, team members must carefully articulate their responses and the results of this reflection to one another and potentially to the outside world. Second, these reports often serve to update local sponsors and gatekeepers (e.g., a local school board or hospital administration) on the project's progress, the research outcomes, and the results of the action in terms these stakeholders use and find important. Third, community partners are often accountable to outside organizations, such as funding organizations. Reports written in lay terms for a local audience can often be appropriated by the community partners in their communication activities with these external bodies. For example, when conducting a research project focused on a technology-oriented curriculum for adolescent girls during summer camp, we worked with a local branch of a major national girls organization. Our community partners used our local report, which included a video, to present the results of the camp to both the national board of their organization and local donors. We have since used the created video in fund-raising and recruiting efforts at our university, an unexpected benefit of the creation of this video report.

Scholarly writing and academic papers may be more familiar to researchers than the kinds of local reports described above. However, scholarly works—particularly in computer science, information science, and HCI—are almost certainly more familiar to the research facilitators than to the community partners in an AR project. Many community partners may never have published in an academic venue, and if they have, the publications may not have been in the disciplinary style or the venues of the professional researchers. Thus, researchers must attend carefully to ensuring empowerment to influence the scholarly production for all members of the team. Specifically, teams should work to ensure that alternate ways of contributing to the scholarly publication are available for those not as comfortable with this format of reporting. Additionally, scholarly publications should be submitted to places that can help the careers of both the research facilitators and the community partners when possible. For example, top-tier conference publications are often the primary goal for HCI researchers (e.g., CHI, CSCW). However, the computer science tradition of low acceptance rates and high prestige being afforded to these venues does not translate well into many other disciplines. Thus, decisions about publication venues should be made collaboratively when possible. Furthermore, an appropriate amount of time must be built into the writing plan to ensure for translation of language among different communities and inclusion of everyone's input. When writing a paper for a computing venue, for example, the HCI research facilitators may need to take extra time to explain the venue, the types of papers, and the questions of interest in this community to the research partners. Often, it would be simpler and more expedient to skip these steps, writing the reports within the academic portion of the research team and then asking for feedback on a nearly completed draft from the community research partners. However, to meet the goals of a truly collaborative AR experience, the entire team should be included from the beginning to the end when possible, and a variety of reporting mechanisms should be employed.

Moments of Celebration

Getting results of an AR project published is certainly a cause for celebration, and presenting the results whether at a local event or a national conference provides the team with a defined moment of celebration. However, in AR projects, because there is no clearly defined ending point in most cases, it is also important to recognize intermediate moments of celebration throughout the project.

In one research project in a school, teachers were asked to perform a set of activities with two children in each of their classes. They worked with me as well as with my community research partner in the schools to complete the tasks in their classrooms over the course of approximately 3–5 weeks per child. Once everything for an individual child was completed, we brought the teacher a gift bag filled with things she needed for her classroom: hand sanitizer, snacks, school supplies, and so on. Each time they would receive their gifts, the teachers called over their aides and sometimes the students as well to publicly open the gift bags and join us in thanking the entire classroom and celebrating the completion of one portion of the research effort. These kinds of public displays of celebration can be much more effective in building good will and compensating research participants for involvement than simple cash payments.

In this same project, we also celebrated at bigger milestones. Once all four teachers involved had completed their work with two children each, the first phase of our project was completed. We took advantage of the ending of the school year for these teachers, which coincided with their completion of this first phase, to throw a party at my house. At this party, all of the researchers on the academic side who had helped in building the system we were testing, transcribing interviews, and performing other activities were present along with the teachers, school administrators, aides, and other team members from the schools. Many of the people present were meeting each other for the first time, with only a few of us having been heavily involved across sites. The team should be emphasized during these moments of celebration, not the individuals. So, at this party, I gave everyone a present from both the academic research team and the community research team and thanked them collectively and very briefly.

AR requires sustained long-term engagement with research sites and community partners. Although the exact time frame depends largely on the composition of the team and the work involved, this kind of relationship and effort can be exhausting to all involved. I have had sites begin to fall apart within months of engagement whereas others are still wonderful collaborative relationships years later. There are even examples in the AR literature, outside of anything involving technology, that last decades. As milestones are met and the iterative cycle of the project continues, it can be easy to lose some of the drive and focus that began the project in the first place. Thus, using moments of celebration to demark beginnings of new phases and endings of old ones can serve to build more collaborative teams as well as to reinvigorate everyone involved.

Leaving the Site

Although AR projects tend not to begin with defined ending points in mind, invariably the realities of the academic process and the constraints of the community partners' lives necessitate that the research facilitators leave the site. This time can be a painful one for all involved. In the worst cases, the team wishes to keep working together, but a change at the field site has eliminated the project, the academic team has lost funding, or some other problem has befallen the project. However, more frequently, members of the team have begun to recognize that the time for the collaborative part of the project may be ending. Faculty members and community partners move jobs, students complete their degrees, and researchers may be interested in exploring different research questions that may or may not build on the work done at the current site. Furthermore, successful AR projects result in sustainable, dependable change, which can be less interesting from a research standpoint than the implementation of novel solutions and the study of changes immediately following. Thus, action researchers must be prepared to leave the sites and the people with whom they have become intimately intertwined, and their community research collaborators must also be prepared for this inevitability.

In AR, the goal is ultimately to create sustainable change. That is to say, once the research facilitators leave, the community partners should be able to maintain the positive changes that have been made. In many AR projects, the changes made are based in the creation of new policies or the changing of the old, the development of new programs, restructuring of staff roles, and so on. In HCI, however, AR project changes often include the deployment of novel technologies. In these cases, one of the challenges to leaving the AR site is ensuring that the technologies can be left behind and when left behind can be maintained. It is neither in the best interests of the academic researchers—who have limited resources and other commitments—nor the community partners—who should be made to feel in power and in control of their own projects, particularly after the facilitators leave—for the technological infrastructure to continue to be maintained by the academic partners.

In some AR projects with which I have been engaged, such as those at hospitals and medical centers, IT support is already available within the organization. These individuals can be trained to maintain the equipment brought into the research site by the AR project. Of course, the request for this additional work on the part of the IT organization should be managed carefully as all relationships and new activities should be in an AR project. If possible, it may even be useful to include them on the project team from the beginning.

As an example, in one project I developed a simple mobile phone application to help medical clinicians implement a change in the way they monitored compliance with a home-based intervention. The IT support person who worked with this medical team primarily focused on more traditional enterprise issues (e.g., ensuring that the videoconferencing system was working before meetings, troubleshooting e-mail, and setting up servers). As part of the project, however, I had meetings with him to discuss his ideas for the phone application. He requested some changes be made to the back end of the system so that he could more easily manage it, which I was able to make. After a few weeks of use, he no longer needed my help and had begun managing all parts of the system along with a nurse he had pulled into the process simply because she liked technology and wanted to learn more. Although my involvement in the project lasted for many months after this transition, when I did eventually leave the team, they had already become self-sustaining.

In other organizations, however, this solution may not be viable. For example, in many schools, although IT support personnel are available, they are usually already spread too thin and cannot easily take on additional responsibilities. In such a situation, a member of the original community partner research team or a research participant at the field site might take on the role of champion for the project and volunteer to maintain the technologies moving forward. This situation can offer a solution to the issue of sustainable change but should be managed carefully, because the change in role for this individual can effect a change in status or power dynamics within the team. Such was the case in a school-based effort in which two teachers wanted to continue to use the system we had developed after we left the research site. One had been enthusiastic from the beginning and, though she had no formal training, had a particular aptitude for handling computing systems. The other had originally been wary of the system and only engaged with it positively towards the end of my involvement in the site. Ultimately, we chose to leave the equipment in the hands of the teacher who had always demonstrated enthusiasm and aptitude. This decision strained their relationship, which was already tenuous for other reasons, and my relationship with the teacher who had not been chosen. Had we had the resources available, it would have been a better choice to provide them both with equipment and instruction for long-term maintenance.

Some Examples of AR in HCI Research

An early example of AR in HCI-related research—in this case, information systems—can be found in Ned Kock's AR study of communication media and group work (Kock, 1998). In this work, the researchers partnered with university-based process improvement groups to understand how groups might begin to adopt a new communication medium voluntarily, even as they perceived it as highly limited. Just as action researchers have come to use more and more ICT in their solutions, so too have ICT researchers begun to seriously engage AR in their work. The results of this confluence of activities are present in a variety of venues, including venues that focus on these approaches, such as the *Journal of Community Informatics* and *Action Research* (from Sage Journals). A recent special issue of Community Informatics on "Research in Action" includes multiple examples of high-quality AR projects that use ICT in their solutions or have access to and education about AR as their focus (Allen & Foth, 2011). For example, Carroll and colleagues describe their efforts to develop a community network over several years, including their interest in and approaches to enhancing "end-user participation in the design of information technology" (Carroll et al., 2011). Other venues, that are not AR specific, have also found engagement with AR to be useful and meaningful. For example, in 2004, MIS Quarterly devoted a special issue to AR. In this work, a variety of approaches were demonstrated, all resulting in high-quality research findings. For example, Kohli and Kettinger described a project focused on working with hospital management and physicians to add digital resources and tools to help manage complex hospital information (Kohli & Kettinger, 2004).

Closing Thoughts and My Own AR Story

My career as an academic and as a researcher has been heavily influenced by being a child of academics. My parents, both educational psychologists by training, took different career paths, but both consistently tackled projects that were personally meaningful, democratically constructed, and in all the best ways quite practical. My father has written extensively on this topic (see, e.g., Blackman, Hayes, Reeves & Paisley, 2002; Hayes, Paisley, Phelps, Pearson & Salter, 1997; Paisley, Bailey, Hayes, McMahon & Grimmett, 2010; Paisley, Hayes & Bailey, 1999), all publications I neglected to read until after my formal introduction to action research outside my family influence.

I first formally learned about AR in May of 2005 at the Public Responsibility in Medicine and Research (PRIM&R) conference, a meeting meant to orient new members and provide continuing education for staffs and senior members of Institutional Review Boards (IRB). At the time, I had just joined the IRB at Georgia Tech as a student member, and I was in the middle of my dissertation work, which involved participatory research with educators of children with autism. I attended the AR session not because I was interested in AR per se but rather because the advertised talks seemed like they included research that I found intellectually fascinating and relevant to society's problems: needle exchange programs and transformation of school curricula for inner-city students. After introducing the research projects, the group began a somewhat heated discussion about how to ensure that the federal definition of research—which notably requires an attempt at "generalizable knowledge"—included AR. The intense discussion about the ethics of AR, how to write and talk about local solutions in a scholarly manner, and challenges for AR participants were quite useful in framing my dissertation work and sparked my interest in exploring the various ways an AR approach can be helpful in research projects.

My work at the time would best be described as a mix of technical and practical-deliberative AR, using McKernan's framework. As a student hoping to defend a successful dissertation, I was inclined to present the work as measurable and knowable, and the process of preparing a thesis proposal meant that much of the problem had been defined in advance. Schools are delicate places though, and working in them requires a lot of compromise, collaboration, and democratically determined research questions and approaches. Through my years of working with—and in some cases for—the teachers who were participating in my studies, I developed a variety of new interests and problem statements, defined in the situation by the stakeholders and community partners who cared most. Substantial time in the field also taught me what many researchers know instinctively: that reality is messy, constructed, and complex. AR handles this kind of mess quite well by acknowledging it and incorporating the knowledge to be gleaned from the mess into the scholarship of the research program. Furthermore, an approach within AR that allows for the idea that some results may be predictive while others cannot be enables researchers to produce knowledge about particular situations while informing others about what solutions might work in other situations, a result that is both scholarly and practical at once. This kind of transferability does not speak well to an idea of generalizability at the level of the individual AR project. However, as theories are produced and lessons learned from these efforts, the corpus of work in the field—alongside other research projects, whether they take an AR approach or not—enables a kind of generalized thinking in the form of new theoretical models or common frameworks for the design of solutions.

This chapter serves as an introduction to action research within the framework of "ways of knowing" for human–computer interaction researchers. My hope is that it will be useful to those people, who like me are focused on attempting to create real solutions to real problems and want to include those most affected by those problems in the design of the solutions. The approaches outlined here echo those in the chapters on design (Research Through Design), ethnography (Reading and Interpreting Ethnography), and field deployments (Field Deployments: Knowing from Using in Context). Furthermore, action researchers can take advantage—in cooperation with their community partners—of a variety of the specific methods outlined in chapters here and in other research method publications. The pragmatic nature of AR does not require adherence to specific methods but is instead a way of knowing that reflects an agreement of sorts that we are all in this together—researchers, designers, community partners, and participants—and together we can develop solutions to sticky problems and through these solutions learn about our world.

Additional Reading for Gaining Expertise in Action Research and Related Areas

Core Action Research Readings

- Chevalier, J.M. and Buckles, D.J. 2013. *Participatory Action Research: Theory and Methods*. Routledge.
- Greenwood, D.J. and Levin, M. 2007. *Introduction to Action Research 2e*. Sage Publications.
- Herr, K.G. and Anderson, G.L. 2005. *The Action Research Dissertation: A Guide for Students and Faculty*. Sage Publications.

- McIntyre, A. 2007. *Participatory Action Research.* Sage Publications.
- McNiff, J. and Whitehead, J. 2006. *All you need to know about Action Research.* Sage Publications.
- Reason, P. and Bradbury-Huang, H. (Eds.) 2007. *Handbook of Action Research: Participative Inquiry and Practice.* Sage Publications.
- Stringer, E.T. 2007. *Action Research.* Sage Publications.

Reflective, Collaborative, and Critical Inquiry

- Alvesson, M. and Sköldberg, K. 2000. *Reflexive methodology: new vistas for qualitative research.* Sage Publications.
- Beebe, J. 2001. *Rapid appraisal process: an introduction.* Alta Mira Press.
- Malhotra Bentz, V. and Shapiro, J.J. 1998. *Mindful inquiry in social research.* Sage Publications.
- Bray, J., Lee, J., Smith, L., and Yorks, L. 2000. *Collaborative inquiry in practice: action, reflection, and making meaning.* Sage Publications.
- Carr, W. and Kemmis, S. 1986. *Becoming critical: education knowledge and action research.* The Falmer Press.
- Van de Ven, A.W. 2007. *Engaged Scholarship: A Guide for Organizational and Social Research.* Oxford University Press.

Collaborative Design and Information Systems Research

- Checkland, P. 1981. *Systems thinking, systems practice.* Wiley.
- Checkland, P. and Holwell, S. 1997. *Information, systems, and information systems: making sense of the field.* Wiley.
- Schuler, D. and Namioka, A. (Eds.) 1993. *Participatory Design: Principles and Practices.* CRC/Lawrence Erlbaum Associates.
- Moore Trauth, E. (Ed.) *Qualitative research in IS: issues and trends.* Idea Group Publishing.

Exercises

1. Compare and contrast action research with ethnography?
2. What are the negative aspects of having the participants be co-researchers in this endeavor?
3. What are the dangers when the project ends? How can those dangers be mitigated?

References

Allen, M., & Foth, M. (2011). Research in action for community informatics: A matter of conversation. *The Journal of Community Informatics. 7*(3).

Baskerville, R., & Pries-Heje, J. (1999). Grounded action research: A method for understanding IT in practice. *Accounting, Management and Information Technology, 9*, 1–23.

Blackman, L., Hayes, R. L., Reeves, P., & Paisley, P. (2002). Building a bridge: Counselor educator–school counselor collaboration. *Counselor Education and Supervision, 41*, 243–255.

Braten, S. (1973). Model monopoly and communication: Systems theoretical notes on democratization. *Acta Sociologica, 16*(2), 98–107.

Carroll, J. M., Horning, M. A., Hoffman, B., Ganoe, C. H., Robinson, H. R., & Rosson, M. B. (2011). Visions, participation and engagement in new community information infrastructures. *The Journal of Community Informatics. 7*(3).

Denzin, N. K. (1997). *Interpretive ethnography*. Thousand Oaks, CA: Sage.

Dewey, J. (1976). Essays on logical theory. 1902–1903. In A. Boydston (Ed.), Carbondale: Southern Illinois University Press.

Dewey J. (1991/1927). *The public and its problems*. Athens: Ohio University Press.

Foth, M., & Axup, J. (2006). Participatory design and action research: Identical twins or synergetic pair? In *Proceedings of the Participatory Design Conference* (pp. 93–96).

Greenbaum, J., & Kyng, M. (Eds.). (1992). *Design at work: Cooperative design of computer systems*. Hillsdale, NJ: L. Erlbaum Assoc. Inc.

Greenwood, D. J., & Levin, M. (2007). *Introduction to action research 2e*. Thousand Oaks, CA: Sage.

Grundy, S. (1988). Three modes of action research. In S. Kemmis & R. McTaggert (Eds.), *The action research reader* (3rd ed.). Geelong: Deakin University Press.

Guba, E. G., & Lincoln, Y. S. (1989). Judging the quality of fourth generation evaluation. In E. G. Guba & Y. Lincoln (Eds.), *Fourth generation evaluation* (pp. 228–251). Newbury Park, CA: Sage.

Hayes, R. L., Paisley, P. O., Phelps, R. E., Pearson, G., & Salter, R. (1997). Integrating theory and practice: Counselor educator – school counselor collaborative. *Professional School Counseling, 1*, 9–12.

James, W. (1948). *Essays in pragmatism*. New York: Hafner.

Kock, N. (1998). Can communication medium limitations foster better group outcomes? An action research study. *Information and Management, 34*(5), 295–305.

Kohli, R., & Kettinger, W. J. (2004). Informating the clan: Controlling physicians costs and outcomes. *MIS Quarterly, 28*(3), 363–394.

Lassiter, L. E. (2005). *Chicago guide to collaborative ethnography*. Chicago, IL: University of Chicago Press.

Lewin, K. (1946). Action research and minority problems. *Journal of Social Issues, 2*(4), 34–46.

Lewin, K. (1948). *Resolving social conflicts*. New York: Harper.

Light, A., Egglestone, P., Wakeford, T., & Rogers, J. (2011). Participant making: Bridging the gulf between community knowledge and academic research. *The Journal of Community Informatics. 7*(3).

Lincoln, Y. S., & Guba, E. G. (1985). *Naturalistic inquiry*. Beverly Hills, CA: Sage.

Masters, J. (1995). The history of action research. In I. Hughes (Ed.), *Action research electronic reader* The University of Sydney, Retrieved from March 3, 2012, from http://www.behs.cchs.usyd.edu.au/arow/Reader/rmasters.htm.

McCutcheon, G., & Jung, B. (1990). Alternative perspectives on action research. *Theory into Practice 24*(3): Summer.

McKernan, J. (1991). *Curriculum action research. A handbook of methods and resources for the reflective practitioner*. London: Kogan Page.

Muller, M. J. (2007). Participatory design: The third space in HCI (revised). In J. Jacko & A. Sears (Eds.), *Handbook of HCI 2e*. Mahway, NJ, USA: Erlbaum.

Paisley, P. O., Bailey, D. F., Hayes, R. L., McMahon, H. G., & Grimmett, M. A. (2010). Using a cohort model for school counselor preparation to enhance commitment to social justice. *Journal for Specialists in Group Work, 35*(3), 262–270. Reprinted in A. A. Singh & C. F. Salazar (Eds.), *Social justice in group work: Practical interventions for change*. London: Routledge/Taylor and Francis.

Paisley, P. O., Hayes, R. L., & Bailey, D. F. (1999). School counselor practice and preparation: A local partnership for change. *Georgia School Counselors Association Journal, 1*(6), 52–57.

Palen, L. (2010). Better ODDS than "Snowballs in Hell?" – or – What might action research do for HCI?. Presented at *Human-Computer Interaction Consortium*. Fraser, CO, USA, Retrieved February 27, 2010.

Rapoport, R. (1970). Three dilemmas in action research. *Stronger Families Learning Exchange Bulletin, 23*(6), 499–513.

Schon, D. A. (1983). *The reflective practitioner: How professionals think in action*. New York: Basic Books.

Schuler, D., & Namioka, A. (1993). *Participatory design: Principles and practices*. Hillsdale, NJ: Erlbaum.

Smith, D. E. (1989). Sociological theory: Methods of writing patriarchy. In R. A. Wallace (Ed.), *Feminism and sociological theory* (pp. 34–64). Newbury Park, CA: Sage.

Stolterman, E. (2008). The nature of design practice and implications for interaction design research. *International Journal of Design, 2*(1), 55–65.

Stringer, E. T. (2007). *Action research*. Newbury Park, CA: Sage.

Concepts, Values, and Methods for Technical Human–Computer Interaction Research

Scott E. Hudson and Jennifer Mankoff

This chapter seeks to illuminate the core values driving technical research in human–computer interaction (HCI) and use these as a guide to understanding how it is typically carried out and why these approaches are appropriate to the work. HCI overall seeks to both understand and improve how humans interact with technology. Technical HCI focuses on the technology and improvement aspects of this task—it seeks to use technology to solve human problems and improve the world. To accomplish this, the fundamental activity of technical HCI is one of *invention*—we seek to use technology to *expand what can be done* or to *find how best to do things* that can already be done. Inventing new solutions to human problems, increasing the potential capabilities of advanced technologies, and (in a spiral fashion) enabling others to invent new solutions and/or apply advanced technical capabilities are all central to technical HCI. The ability to create new things, to mold technology (and the world), and to enhance what people (or technology) can do drives our fascination with technical work; hence, the core value at the heart of technical HCI is invention.

One way of understanding the work of technical HCI research is by contrasting it with other types of HCI research. In an interdisciplinary setting such as HCI, we often shift between disciplines that have stable and functional but potentially contradictory world views. In doing so, we are confronted with the need to select and use (or at least appreciate, understand, and evaluate) a wide range of methods and with them a wide range of expectations and values. For example, different disciplines, such as social and cognitive psychology, design, and computer science, have evolved their own methods, value systems, and expectations about what constitutes appropriate and impactful work. Because they work well for individual disciplines, these expectations and values are often left somewhat unexamined within the discipline itself. For a researcher to work effectively within a discipline, it is critical to know and heed these expectations and values (and hence be able to distinguish and

S.E. Hudson (✉) • J. Mankoff
Human–Computer Interaction Institute, School of Computer Science, Carnegie Mellon University, 5000 Forbes Ave., Pittsburgh, PA 15213, USA
e-mail: scott.hudson@cs.cmu.edu; jmankoff@cs.cmu.edu

J.S. Olson and W.A. Kellogg (eds.), *Ways of Knowing in HCI*,
DOI 10.1007/978-1-4939-0378-8_4,

produce good vs. not-so-good work). But in turn it is often less critical to fully understand the variety of perspectives held in other disciplines. However, in an interdisciplinary setting like HCI, examining *why* particular methods are suited to particular kinds of work is important. While invention is not unique to technical HCI (this is also a clear component of design-oriented HCI; see Chaps. Science and Design and Research Through Design in HCI), this distinction does separate it from parts of HCI that aim to describe or understand the world through, for example, new discoveries about how the world works, critical theory, or models of human behavior. Thus, as we lay out the expectations, values, and approaches inherent in technical HCI, we will use as a touchstone the contrast between its main activity of *invention* with the focus on *discovery* that typifies some approaches to HCI research.

Another way of understanding technical HCI research is by contrasting it with other types of technical work that is not research. For our purposes, *research* can be seen as having *the creation of reusable knowledge* at its core. More specifically *technical HCI research* emphasizes knowledge about how to create something (invention) but also knowledge that might be reused to assist in the creation of a whole class of similar things or even multiple types of different things. For example, several decades ago considerable research effort went into developing ways to allow graphical user interface (GUI) layout to be specified graphically, including the first modern "interface builder" (Hullot, 1986). In contrast, *development* has at its core the creation of a particular thing—something we might often consider a *product*. Development generally requires creation of knowledge, but there is no particular need that this knowledge be reusable. So for example, numerous similar "interface builder" tools are now available in various development environments. Each of these required substantial effort to create and perfect. But only a small part of those efforts have produced reusable concepts.

The distinction between research and development leads to differences in approach among those who practice purer versions of each. However, there is no clear dividing line between them. For example, development of nearly any useful artifact can provide knowledge about how to (or perhaps how not to) build another similar artifact. Further, as will be considered later in this chapter, good evaluation of inventive research almost always mandates some development work—the building of some, or all, of the thing invented. In the end, this means that research and development activities are often intertwined and can be difficult to cleanly separate. Thus, in the second half of this chapter we describe the work of invention in HCI, focusing on types of impact, the essential role of development in validating any invention (through a proof-of-concept implementation), and other forms of validation.

Einstein Versus Edison: Invention as the Basis of Technical HCI Work

Activities of *invention* at their core seek to bring useful new things into the world. This nearly always requires knowing facts about the world and may entail pursuit of new discoveries if the necessary facts are not known or not known well enough. But the heart of invention is *changing how the world works through innovation and creation.*

This is the core purpose and the typical passion of those who undertake activities of invention. In contrast, activities of *discovery* at their core seek to *develop new understandings of the world.* To the extent that inventions play a role in these activities, they are in the service of discovery.

Albert Einstein (1879-1955)
Thomas Edison (1847-1931)

Photos are in the public domain

Note that we might have used the terms *science* and *engineering* here, rather than discovery and invention. In our view, discovery and invention are at once more descriptive and more neutral terms. Both activities are critically important to the success of HCI, but there is a discernible bias, at least in academic circles, towards science and away from engineering. We can see this by noting that we often hear phrases such as "Where's the science in this?" and "That's just engineering," but we pretty much never hear "Where's the engineering in this?" or "That's just science." In fact "science" is often misused as a synonym for "rigorous" or just "good work" irrespective of whether the work is actually scientific in nature. On the other hand, both discovery and invention can confer great benefits to society, and as such both have been honored. We can see this by noting that exemplars such as Einstein and Edison are both held in high regard in many societies.

There are many similarities in the work of discovery and invention but also some key differences. These have to do with the underlying values and goals of each type of work, specifically what constitutes how work in the field moves forward and what constitutes a trustworthy and valuable result.

Differences in How Fields Move Forward

Activities of discovery can have a variety of aims, including generating rich, empirically based descriptions, and creating new theoretical understandings (see Chaps. Reading and Interpreting Ethnography and Grounded Theory Method, this volume). Once articulated, theories typically form *framing truths* that establish a context for the work. The work of discovery often proceeds by elaborating and refining these framing truths to progress towards improved understandings. An initial theory that explains the most easily observable facts may be refined to explain more phenomena or to be more predictive. This progression requires developing and testing competing ideas (which might both be consistent with a framing theory). For example, both Newtonian and Einsteinian notions of gravity explain everyday objects falling to earth, and even the motion of planets, quite well. Only when we consider finer and more difficult-to-observe phenomena does one clearly improve on the other. As another example, the speed and accuracy of directed reaching movements are well described in one dimension by Fitts' law (Fitts, 1954). However, this theory has various limits (for example, when applied to 2D targets of arbitrary shape). Newer theories, for example those based on the microstructure of movements (see Meyer, 1990), provide a more detailed account of the same phenomena and allow us to overcome some of these limitations (see for example, Grossman & Balakrishnan, 2005b; Grossman, Kong, & Balakrishnan, 2007).

In contrast, activities of invention almost always progress towards the creation of new or better things but not necessarily through refinement. Normally we invent by combining a set of things we already understand how to create into larger, more complex, or more capable things that did not previously exist. The early phonograph for example made use of existing concepts such as a mechanism for rotary motion at a carefully controlled rate and the use of a horn shape for directing sound and combined these with a new method for recording and reproducing small vibrations with a needle in a trace scored in a tinfoil sheet. Similarly, in an HCI context the first graphical interfaces (Sutherland, 1963) were created using existing input and display devices (a light pen, buttons, rotary input knobs, and a random dot CRT) along with new concepts expressed in software to create (among other pioneering advances) the ability of users to manipulate objects displayed graphically by pointing at them. In both inventions each of the detailed precursors was combined to create a much more complex and functional whole based on these smaller and simpler parts. In some cases, activities of invention may start with a larger truth (about something that *should* be possible), but the detailed process of invention still typically depends on the combining of smaller or simpler parts into a larger and more complex whole. Hence, in contrast to discovery, as we progress in invention we are not necessarily refining a framing truth. In fact, our understanding can sometimes actually decrease because we are creating more complex things that are less well understood than the simpler things they are made of. However, the things created are more *capable*—they do more or better things in the world—and this is the core of inventive progress.

Differences in What Makes a Result Valuable and Trustworthy

In discovery work, the properties of valuable and trustworthy results are intertwined. Core values in discovery work include increasing understanding (e.g., of new phenomena) or understanding in more powerful ways (e.g., more profoundly or in some cases predictively). But the desire to *know* and have confidence in results makes the details and reliability of the methods used to reach a result of central importance (what Gaver calls "epistemological accountability" in Chap. Science and Design). In some sense, the methods used to obtain a result are part of the result. The assertion of an understanding about the world cannot stand on its own; it is necessary to know about the method (or in some perspectives, the person; see Chap. Reading and Interpreting Ethnography).

The need for high confidence in results drives the familiar tactic of isolating and testing a small number of variables—often just one or two—in an attempt to separate their effects from other confounds. This tactic achieves increased trustworthiness at the cost of focusing on less complex circumstances. As a result, a study that tests a theory in a specific context may only be able to make claims about a narrow slice of reality. This can make it hard to generalize to more complex, real-world settings without replicating the study in many different but similar settings to be sure that the underlying theory is robust across changing circumstances. To be sure, some forms of discovery grapple more directly with complexities of the real world (see many chapters herein), but confidence in the results, building consensus, and causal attribution can be more difficult.

Invention, in contrast, privileges the value of creating something that has the potential for practical impact. To improve practicality, inventions are most valued if they work within the full complexity of the world around us. In fact, in many cases, if we limit the scope of work to very controlled situations (e.g., with only one or two degrees of freedom), it can easily *destroy* the value of the work. Often we start with specifics and use them to create something that has multiple uses. Indeed to the extent it is possible to apply the result (the invention) in multiple domains it may be considered more valuable.

For invention, the goodness of a result is a property of the concept invented. The properties of the thing invented generally stand alone and can be understood and evaluated independently of the particular methods used in the inventive process. It might be that the inventors came up with their result by means of an arduous process of testing many alternatives, or it might be that the concept came to them in a dream the night before. However, if both paths lead to the same invention, it is equally good. The trustworthiness of an inventive result depends on an examination of the thing that was invented (almost always through consideration of an implementation of it).

The Work of Invention in Technical HCI

We have shown that invention can be seen as an activity that creates artifacts that can solve problems in the world and that the things that make a result trustworthy and valuable differ between activities of invention and discovery. In this section we explore the process of invention, focusing on key aspects of technical HCI research.

Our focus in this section is not on the creative process per say, but rather on the directions from which one might approach invention.

We begin by reviewing the types of contributions typically found in technical HCI research (direct creation and enabling research). Next we review approaches to concept creation, followed by proof-of-concept implementations, the core form of validation for invention. This form of validation is a crucial and inseparable part of the process of concept creation. However, while building takes on a central role, additional validations may help to show the specific impacts of different types of technical contributions. We then present a review of these types of secondary validations. Thus we might well break up the work of inventive research into three parts rather than two: concept creation, proof-of-concept implementation, and (additional) validation.

Types of Contributions

The contributions that can be made by inventive HCI research can come in a number of forms. Many of them might be summed up at the highest level as supporting the invention of things that meet human needs. This can in turn be separated into at least two overall categories: *direct creation* of things meeting human needs and development of things that *enable* further invention.

Direct creation is most straightforward. This might involve creation of something that improves some aspect of a long-standing goal such as supporting collaborative work at a distance (Engelbart & English, 1968; Ishii, Kobayashi, & Arita, 1994) or selecting items on a screen more quickly (Sutherland, 1963; Grossman & Balakrishnan, 2005a); that introduces a new capability such as interacting with wall displays that are larger than the reach of a person's arms (Khan et al., 2004; Shoemaker, Tang, & Booth, 2007); or that brings a capability to a new user population such as photography by the blind (Jayant, Ji, White, & Bigham, 2011).

Enabling research on the other hand is more indirect. It has as a goal not directly addressing an end-user need, but rather to enable others to address a need by making it possible, easier, or less expensive for future inventive work to do so. Enabling research can also come in a number of forms. These include development of tools, systems, and basic capabilities.

Tools generally seek to make it much easier to create a certain class of things. Tools normally do not directly meet end-user needs. Instead, they act indirectly by enabling developers to quickly and easily meet end-user needs or to construct complex and functional artifacts. For example, through extensive UI tools research in the 1980s (such as Buxton, Lamb, Sherman, & Smith, 1983, Cardelli, 1988), specifying the appearance and basic functioning of a GUI is now a simple enough matter that it can often be done by those with only minimal programming ability. Tools also often bring a benefit of making it practical to create a broader set of things. For example, subArctic (Hudson, Mankoff, & Smith, 2005) and Amulet (Myers et al., 1997) are GUI toolkits that provide high-level abstractions that make it much easier to build interactive systems. Tools may not provide any new capabilities at all, but instead make existing capabilities much more accessible or useful for developers (see threshold and ceiling effects, below).

Systems bring together a set of capabilities into a single working whole—often providing abstractions that make these capabilities more useful, more manageable, and/or easier to deploy or reuse. For example, the input handling abstractions in the Garnet toolkit (Myer, 1990) made use of finite state machines for controlling interaction as many systems do—something already widely used. However, it provided a new highly parameterized abstraction of that concept which made it much easier for developers to use. Systems also sometimes bring together a disparate set of capabilities that has not been combined before or combine capabilities in new ways that make them more useful. As an example, every major operating system today includes a subsystem specifically for handling overlapping windows, which provides basic input and output capability on a single set of devices that can be shared by many programs.

Basic capabilities: Another enabling contribution is an advance on a specific and difficult problem that is holding up progress in a problem domain. The advance made may be very narrow but have value in the breadth of the things it enables. By creating new or improved algorithms, new circuits, or new sensors, we can enable a range of new inventions. Examples of HCI-relevant basic capacities that have been introduced, e.g., to modern operating systems include input device drivers, event modeling (providing an abstraction that describes user input in a device-independent fashion), and graphics systems (which provide an abstraction for displaying images on a screen; typically one that can be transparently translated into a range of fast graphics hardware). In another example, algorithms for face recognition and tracking that were able to operate at frame rate (Viola & Jones, 2001) enabled a range of new capabilities such as digital cameras that automatically focus on faces, thus producing better photography by average consumers with no additional effort on their part.

Finally, it is important to note that enabling research also often takes the form of *importing* and *adapting* advances made in other technical areas and putting them to use for new purposes. In some respects this might not be considered invention per se. However, it surely must be considered a research advance, as in the modern world substantial progress is made in exactly this fashion—an idea or a concept originally created in one research domain is first imported, and then typically adapted, for use in others. For example, finite-state automata are now heavily used in implementing interaction techniques. This concept was first introduced for HCI use by Newman (1968). However, Newman clearly did not invent finite-state automata (they were originally devised to model neuronal activity (McCulloch & Pitts, 1943) and subsequently used in many other ways). Nonetheless, the idea has been of great benefit in user interface implementation and has since been built on and improved upon numerous times (Wasserman, 1985; Jacob, 1986; Appert & Beaudouin-Lafon, 2008; Schwarz, Mankoff, & Hudson, 2011). As such this importing and adaptation of a powerful technique can have great value and so must be considered a contribution in its own right.

Approaches to Concept Creation

It is extremely difficult to put one's finger on the best approach to inventing a new concept. However, there are some strategies that have been shown to be productive in our experience. One of the most frequent outcomes of inventive work in HCI is to devise a new way to bridge between technical capabilities and human needs. This simple framing points the way to some of the most common strategies for developing technical contributions. A researcher can start from an observed human need and seek to find a technical approach that can make a positive impact on the need. This approach often leads one to specialize in one or more application areas, learning more and more about the details of human needs in that area. For example, systems supporting special-needs populations such as elder care (see for example Mynatt, Essa, & Rogers, 2000; Mynatt, Rowan, Craighill, & Jacobs, 2001) have often taken this approach. A researcher may do discovery-based work to better understand these needs (and human properties that impact them) and then seek (mostly existing) technological capabilities that might be used to meet these needs.

Within this general framework, one can also work from the technology side: a researcher may specialize in one or more areas of useful or promising technology—learning a substantial amount about how they work (and/or where their weaknesses lie), and extending and improving them, and then seeking to find existing human needs that the technology might have a positive impact on. For example Shwetak Patel and his colleagues have produced several related types of sensors that work by observing changes in the noise found on household power lines (see Patel, Robertson, Kientz, Reynolds, & Abowd, 2007; Cohn, Morris, Patel, & Tan, 2011; Gupta, Chen, Reynolds, & Patel, 2011). This work was undertaken not because of a human need but because of a new technological opportunity that the researchers have considerable expertise with (the ability to rapidly analyze and classify minute variations in "noise" as an intentional signal). Initially, the research was used to sense the location of people within the home, but the researchers also developed the capability to sense appliance use and then simple gestures. These potentially very useful sensing capabilities could be installed simply by plugging a device in (as opposed to hiring an electrician). Thus, as it happened, the resulting product was able to meet several human needs, once it was packaged in an easily deployable box and tied to applications of interest.

This type of technology-first approach has developed a bad reputation within the HCI research community. Historically, researchers coming from technological disciplines have not always matched their emphasis on progress in the technology with careful attention to true human needs. However, if inventions are in the end really valued in proportion to their positive effect on human needs, then it does not fundamentally matter whether a technology-driven or a needs-driven approach was driving the effort to meet those needs. Not only that, technology is currently changing very quickly, while human needs are changing relatively slowly. Indeed, technology is becoming pervasive so rapidly that it is beginning to drive change in human needs. Also, invention that focuses ahead of the technology curve is more likely to be relevant in the 5–10-year horizon that matters in research. These factors combine to make technology-first invention an effective way to build bridges between technology and human needs.

Of course, in practice good researchers often do not limit themselves to either pure needs-first or technology-first approaches. Instead a common approach is to study (or simply stay informed about) the properties of people and the progress in meeting needs within a few application areas and at the same time carefully track progress in a range of potentially useful technologies, searching for new things that might meet outstanding needs. This points to another important property of inventive work—that progress is very often made not by conceiving of entirely new things but instead by *recognizing* that innovations might be used in additional ways and adapting or combining them to meet existing needs. While we often think of invention at its heart as the conception of new things, in fact it much more often involves recognition of new possibilities within already invented things or enabled by new combinations of things (followed in many cases by some adaptation). For example, low-cost MEMS-based accelerometers were originally marketed in large part to support the deployment of airbags in automobiles. But once these devices became available, they were adapted for HCI use. First they were used for exploring the use of tilt as a general form of input (Harrison, Fishkin, Gujar, Mochon, & Want, 1998). This in turn was adapted in additional research on the use of sensors in mobile devices to support landscape/portrait display orientation switching (Hinckley, Pierce, Sinclair, & Horvitz, 2000), which was in turn adopted with small modifications in most current smartphone and tablet interfaces.

In addition to bridging between technology and needs, another typical strategy for making progress is to seek out particular roadblocks to advancement and focus specifically on those. This strategy typically involves carefully tracking progress in some application or technological area, analyzing what the roadblocks to progress or limitations of current solutions are, and then producing concepts targeted specifically at these roadblocks. This approach can often be more indirect—it does not seek to directly impact a human need but instead enables something else that (eventually) will. For example, the authors' joint work on tools and techniques for dealing with uncertainty (Mankoff, Hudson, & Abowd, 2000a, 2000b; Schwarz, Hudson, & Mankoff, 2010b) arose in part from the difficulty of building a specific recognition-based interface to address the need of people with certain disabilities to use something other than the keyboard and mouse for computer input. Tools are a common outcome of this paradigm.

Validation Through Building of Proof-of-Concept Implementations

When we consider validation of an invented concept there are many criteria with which we might judge it. However, most fundamental is the question of "does it work?". A concept can have many good properties, but unless and until it can be realized in a form that actually functions, none of those properties matter very much. Further, experience with invented concepts shows that many ideas that seem excellent at the early point we might call *on paper* fail in the details that they must confront during implementation. That is, there are one or more seemingly small or hidden details that end up becoming a major obstacle to practical implementation of the concept. Most small details are relatively unimportant. However, some details can end up critically important, and experience has clearly shown that it is very difficult to segregate the critical from the trivial details in advance. This difficulty leads to the most fundamental of validation approaches for inventive work: *proof-of-concept implementation*. Because of the difficulty of uncovering critical details, experienced inventors do not put much credence in an idea until it has been at least partly implemented; in short: *you do not believe it until it has been built*.

The centrality of proof-of-concept implementations as a validation mechanism is so strong that the evolved value system gives *building* a central role. Even a really strong user study or other empirical evaluation cannot improve a mediocre concept (or tell us how good an invention it is). In contrast, a proof-of-concept implementation is a critical form of validation because an invented concept is not normally trusted to be more than mediocre without an implementation.

While the creation of concepts is arguably the most important aspect of invention, proof-of-concept implementations typically consume the most time and effort in inventive work. Building things is typically hard, so hard that it is often impractical to build a complete implementation of a candidate concept. This should not be surprising since it is not uncommon to spend millions of dollars and years of time on the development of a significant real-world product. However, it makes little sense to expend the resources necessary to create a complete implementation of a concept before much is known about how well, or even whether, it might work. Hence, in research most proof-of-concept implementations are compromises that implement some of the critical aspects of an idea but do not necessarily consider all the different factors that must be addressed for a full complete product. Such a compromise seeks to maximize the knowledge gained while working within appropriate constraints on the resources required for building.

Questions Proof-of-Concept Implementations Answer

Proof-of-concept implementations normally seek to elicit particular types of knowledge. This knowledge most often starts with some variation on the basic question of "does it work?". However, we often end up asking "does it work well enough?". How we choose to define "well enough" in turn has a strong impact on the type and extent of implementation we undertake. Sometimes we are looking for evidence indicating that the concept offers some advantage over existing solutions to the same problem. For example there were a number of promising input devices for pointing at displays devised before the mouse (English, Engelbart, & Berman, 1967), but the mouse was found to be a particularly good pointing device compared to its competitors (Card, English, & Burr, 1978). Sometimes, particularly when creating a completely new capability or overcoming a critical stumbling block, we are only looking for evidence that the concept works at a minimal level (but perhaps shows promise to be improved). An example is our exploration of the value of cords as an input device (Schwarz, Harrison, Hudson, & Mankoff, 2010a). Sometimes we require information about accuracy, accessibility, or effectiveness of the technical concepts with respect to end users of some type, in which case a certain level of robustness may be required.

The question of "does it work (well enough)?" is also complicated by the fact that the inventions most valued are often those that are most robust to the widely varying conditions of the real world. Similarly, for tools, we ask which ones enable the widest range of other things to be created, potentially even unanticipated ones. So the question almost always also starts to shift into one of "in what circumstances does it work?". Finally, even when a system does not work well, we may still learn something useful if there is enough promise that the concept might be made to work and we uncover information about what problems need to be overcome.

Overall, the knowledge we seek to elicit through an implementation tends to be rich and varied. Correspondingly, as described in the next section, the types of implementation approaches seen in typical practice also tend to take on a wide variety of forms and approaches (and none really dominates). There are many different implementation platforms that may be used, ranging from scripting or prototyping platforms not normally suitable for production use to "industrial strength" platforms of the same type that might be used for a final implementation. Similarly, implementations may consider only a very narrow range of function—only that which is new or what is strictly necessary to demonstrate the concept alone—or may include a richer set of functions necessary to make use of it in more realistic settings. In the end, to be sufficient, a proof-of-concept implementation needs to be complete enough to answer both the basic questions of "does it work (well enough, etc.)?" and any set of additional questions that we might wish to ask in an extended evaluation.

Types of Proof-of-Concept Implementations

Many proof-of-concept implementations take a form that can best be described as a *demonstration*. To succeed, that demonstration must illustrate the worth of the invention and in many cases motivate why it should be considered a success. Demonstrations fall along a rough scale of completeness or robustness. As used in the HCI research community, the presentation form of a demonstration is an indirect measure of its robustness, ordered below from the least to the most robust:

- Description in prose
- Presentation through photos (or screen dumps) showing the invention working
- Video showing the invention in use
- Live demonstration by the inventors
- Testing of properties with users
- Deployment to others to use independently

Presentation type works as a rough surrogate indicator because as we progress along this scale, more and more robustness or completeness is required to adequately present it (in part because the circumstances become less and less controlled or more open and arbitrary).

While higher levels of robustness or completeness clearly provide improved evidence about the quality of the invention, progression along this scale also involves dramatically increased levels of effort and resources. For example, deployment for widespread use can require a level of completeness nearly identical to a full product. (see Chap. Field Deployments: Knowing from Using in Context, this volume.) This often brings with it a need for development efforts that touch on many things not particularly relevant to evaluating the invention in question. Yet this extremely high level of effort may provide only a small increment in additional knowledge. In fact in the worst case, a high-end demonstration involving something like a deployment can even introduce enough confounds unrelated to the core invention that it actually obscures our understanding of it. For example, a deployment may fare very poorly with end users, but this might be due to factors completely unrelated to the worth of the core invention.

For example, suppose we have invented a way to help people who are deaf to find out about the content of ambient sounds in their environment (e.g., Matthews, Fong, Ho-Ching, & Mankoff, 2006). This piece of work, originally completed in 2004, depended on a human to transcribe audio that was shipped to them at the request of a participant who pressed a "What happened?" button on their mobile phone. At the time, technologies that would make this easy to implement today were not available: smartphones were just beginning to be available (but Android and the iPhone were not), Mechanical Turk was less than a year old, speech recognition could only function in constrained environments, and non-speech audio was not easily recognized. Our "deployment" lasted only a few weeks and required of users that they deal with cellular network wait times of up to 9 h and depend on a single human transcriber who was only available for a limited set of hours each day. From a technical perspective, *all of these barriers were peripheral to the invention itself.*

Our validation consisted of our proof-of-concept implementation and was (in this case) enhanced by some data on places and ways in which the technology was used by users who were willing to put up with the other difficulties. At the time, nothing similar existed, so the appropriate goal for the work was to answer the question "can we do this at all?". Our study also answered some questions about "what sounds need to be recognized to automate this?" (such as emotion, non-speech audio) and in the process answered some questions about "where might people use this?" though the last contribution was not strictly necessary for the work to make a technical contribution. In the six years since the work was published all but one (the recognition of non-speech audio) have been "solved." Thus, similar work done more recently has pushed much further on raising the ceiling for what can be done. An example is VizWiz (Bigham et al., 2010) that introduced a new way to use crowd workers to increase the speed of real-time image interpretation for the blind, and Legion Scribe (Lasecki, Miller, Kushalnagar, & Bigham, 2013), which made further advances to enable real-time captioning of videos.

However, from a technical HCI perspective, the value of the invention was clear (and publishable) irrespective of these difficulties.

As a result, it is critical to find an appropriate trade-off between robustness and completeness compared to the cost and effort necessary to create such an implementation. If we were to insist that each invention has the most robust implementation before we could trust its worth enough to build on it, progress in the field would be dramatically reduced—we would spend our time creating many fewer things and so decrease our ability to learn from, and build on, the previous efforts.

Alternatives to Proof-of-Concept Implementations

Although proof-of-concept implementations at some level are considered necessary as a basic validation, there are times when they are either not appropriate or not possible. For example, one less common way to make a contribution is to categorize or organize prior work in an area in a way that places it in a much more understandable light. This includes for example creating a useful taxonomy for a body of work, such as the design space of input devices put forth by Card and Mackinlay (1990). While this does not involve the creation of any new invention per se, it requires the creation of a conceptual framework of new organizing principles. Such a framework may highlight properties that have not been combined or identify areas that have not been explored. For example, our review of approaches to handling uncertainty in user input (such as touch screen input or gestural input) breaks uncertainty down into target uncertainty (where did the user click or what did he or she intend to interact with), recognition uncertainty (what interaction type is indicated) and segmentation uncertainty (where did an input begin and end) (Mankoff, Hudson, & Abowd, 2000a, 2000b). By viewing related work through the lens of different types of uncertainty, we can see that very few if any researchers have addressed segmentation uncertainty in the same depth that other forms of uncertainty have been addressed. Observations such as these can point to areas that are "ripe" for new work and thus make it easier to invent new things.

Another occasion when proof-of-concept implementations are less viable is when a concept requires something beyond the current state of the art to realize. While we might consider such concepts impractical and discard them, they can be very valuable contributions. For example, imagine an application that requires two problems to be solved (such as more accurate eye tracking in real-world contexts and more robust registration of the user's head position with the world). It may be possible to make progress in one area (more robust registration, say) while waiting for progress in the other. Similarly, we may want to demonstrate the high value in terms of unrealized applications of a currently unsolved problem as motivation for others to direct their attention and resources to solving it. Because of the value of being able to consider concepts seemingly beyond the present capability, the community has developed several approaches to learning about the properties of these concepts. These include *buying a time machine*, *Wizard of Oz* approaches, and *simulation*.

Buying a Time Machine

One approach to working beyond the state of the art is what is sometimes called *buying a time machine*. This approach involves spending a comparatively large sum of money or other resources—a sum too large to be justified for a real product of the same type—to get access now to technology that we can expect to be much more affordable and/or practical in the future. For example, we might be able to explore the capabilities of a future home vacuum-cleaning robot with very sophisticated vision processing by implementing the vision processing on a rented high-end supercomputer that communicates with the robot wirelessly. It is not currently practical to put a supercomputer in a vacuum cleaner, but the exponential growth of computing power described by Moore's law makes it reasonable to assume that the equivalent computing power will be available in a single-chip computer in the future.

Unfortunately, in the area of general-purpose computing, it is harder to *buy a time machine* today than it has been in the past. For example, in the middle of 1980s technical HCI researchers could employ what were then high-end workstations that performed 10 or even 100 times faster than typical consumer products of the era. This allowed them to explore the properties of systems that would not be widely practical for consumers for another 5–10 years. However, because of changes in the market for personal computers, it is not that easy to leap ahead of the "average" system today. On the other hand, advanced systems today are incredibly capable and diverse in comparison to past systems. Additionally, today's researchers may exploit graphic processing units (GPUs), create custom electronic circuits, or use (currently) more expensive fabrication techniques such as 3D printing to explore concepts. Each of these technologies allows us to make use of technologies that will likely be more practical and ubiquitous in the future but also currently comes at a cost in terms of requiring specialized skills or approaches.

Wizard of Oz Prototyping

Wizard of Oz prototyping involves simulating advanced capabilities by means of a hidden human who performs actions that a future system might be able to provide autonomously. This method was originally developed to explore user interface aspects of natural language understanding systems that could not yet be built in order to inform how such a system should be structured (Kelley, 1983, 1984). The Wizard of Oz approach clearly has some substantial advantages, both for exploring currently unattainable capabilities and simply for more rapidly and inexpensively simulating attainable ones. However, care must be taken to limit the wizard to an appropriate set of actions and to understand the effects that differences such as slower response times might have.

Simulation

A final way in which we might explore concepts that are impractical or impossible to build is to make use of simulation. This can take the form of simulating some or all of a system or of providing simulated rather than actual input data. A related set of techniques has recently emerged in the form of *crowdsourcing* (see Chap. Crowdsourcing in HCI Research, this volume), wherein large numbers of human workers recruited by services such as Amazon's Mechanical Turk can provide forms of *human computation* (simulating what otherwise might be computed by a machine). Interestingly, recent research shows that it may be possible not only to temporarily substitute human computation for future parts of a system but also to consider using crowdsourcing techniques as a part of a deployed system (Bernstein et al., 2010; Bernstein, Brandt, Miller, & Karger, 2011).

Secondary Forms of Validation

Beyond the central questions surrounding "(In what circumstances) does it work (well enough)?" there are a wide range of other criteria by which we can validate invention in HCI. These follow a set of properties that the community often sees as valuable.

Validations of Inventions Providing Direct Value for Human Needs

For inventions that are providing a direct contribution, we value creating an artifact that meets a stated human need. These needs are often met by creating a new capability or by speeding or otherwise improving a current capability. Perhaps the most common evaluation methods we see employed to demonstrate this are usability tests, human and machine performance tests, and what we will call expert judgment and the prima facie case. Although these are not universally appropriate, they are the most common in the literature.

Usability Tests

Because of the current and historical importance of usability and related properties as a central factor in the *practice* of HCI, usability tests of various sorts have been very widely used in HCI work and are the most recognizable of evaluation methods across the field. In fact the authors have frequently heard the assertion among students and other beginning HCI researchers that "you can't get a paper into CHI[1] without a user test!"

[1] *The ACM SIGCHI Conference on Human Factors in Computing Systems*, which is the largest HCI conference and seen by many as the most prestigious publication venue for HCI work.

This assertion is demonstrably false. An invention must be validated, but validation can take many forms. Even if a usability test shows that an invention is easy to use, it may not be very impactful. Its ability to be modified, extended, or applied to a different purpose may be much more important than its usability. Additionally, while user-centered methods may help with iterative design of a product, for the actual act of inventing—the conception of a new thing—usability tests offer relatively little assistance. However, usability testing (and other user-centered methods) does represent a bias of the community at large, particularly when results are going to be presented to, or evaluated by, a wide audience within our diverse field. This is likely true because they are one of the few evaluation methods with which every HCI researcher is sure to be familiar with.

On the other hand, usability tests are clearly appropriate when they match the properties of a research advance. Any research that puts forward an artifact or a system intended to provide improvements in usability, user experience, etc. clearly needs to present evidence that this is the case. There are a range of widely employed methods for doing this. Not all inventive research seeks to improve on user-centered properties. Indeed, it is critically important that we do not push for all or even most inventive research to aim mainly at these goals. If we were to do that, the field would suffer substantially because in early stages of work on a new type of artifact we must often first get past the questions such as "can we do this at all?" and "what capabilities are most important?" before considering whether something is useful/usable/desirable/etc.

To get to this:

Or this

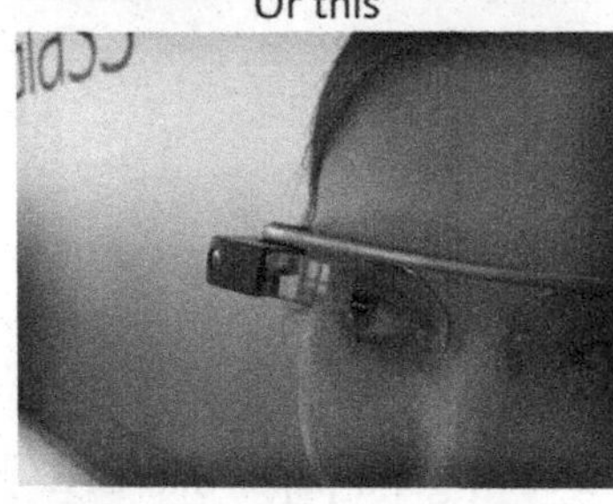

We must often pass through something like this:

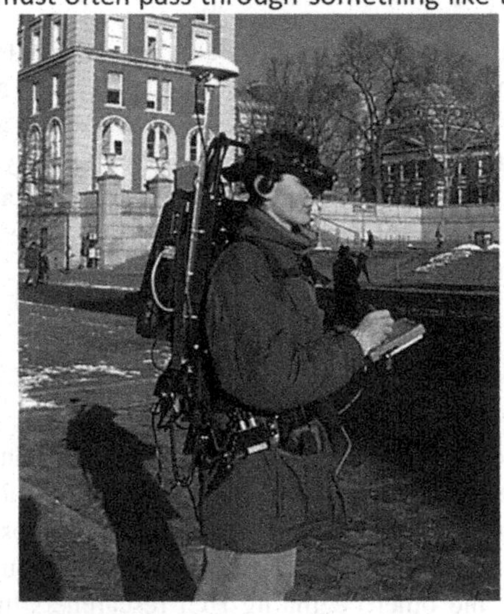

Photo in the right is copyright ©1997 by Steven Feiner (used with permission). Photos in the left are (*top*) "New York Times on iPhone 3GS" by Robert Scoble, http://www.flickr.com/photos/scobleizer/4697192856, and (*bottom*) "Details of Google Glass" by Antonio Zugaldia, http://www.flickr.com/photos/azugaldia/7457645618/, both published under a Creative Commons Attribution 2.0 Generic License

In short, as illustrated in the figure above, it is often necessary to pass through decidedly non-usable stages to create the technology necessary to make something that in the end delivers a great user experience.

Human Performance Tests

Another very widely used class of evaluation methods involves measuring the performance of typical users on some set of tasks. These tests are most applicable when goals for results revolve around a small set of well-defined tasks. Work in interaction techniques is one of the few areas where this type of validation is consistently appropriate. Because some interactive tasks recur frequently, this is also one of the few areas where at least some consistent and reusable measures have emerged. In particular, measurement of pointing performance within a Fitts' law framework (e.g., determining Fitts' law coefficients for devices and interaction techniques) is common because pointing and selection tasks are fundamental to many interactive systems (MacKenzie, 1992; Wobbrock, Cutrell, Harada, & MacKenzie, 2008). Similarly measures of efficiency in text entry such as keystrokes per character (Mackenzie, 2002) have become well developed because text entry is a common task that has received considerable inventive attention.

One danger in using this kind of evaluation is that human performance tests are easiest to apply to narrow and well-defined tasks and generally seen as most valid when they are carefully controlled. Unfortunately, this leads away from the values of wide applicability of results (e.g., an invention useful for a wide range of tasks) and so can be in conflict with other properties of interest for inventive HCI research. Instead of looking for statistically significant improvements, it is important to focus on practical significance (effect size), and unfortunately there are no simple or widely accepted criteria for that. So while human performance tests are widely accepted and understood by the community, without care they can be much less useful than their popularity might indicate. (See Chapter on Experimental Research in HCI, this volume.)

Machine Performance Tests

Tests can also be done to measure the performance of an artifact or an algorithm rather than the person who uses it. These can be very practical in providing information about the technical performance of a result such as expected speed, storage usage, and power consumption. These measures resemble the validation measures commonly used in other domains such as systems research in computer science. It is often considered valid to *simulate* use across a range of conditions to generate such measures. Although this may be indirect and lack real-world validity, such tests of technical performance can in turn point to likely effects on end users such as expected response times or battery life of a device. Similarly, tests could indicate properties such as "*runs at frame rate*[2]" that may indicate that the part of the system being tested is unlikely to be a bottleneck in overall performance, thus telling the researcher that it may be appropriate to turn to improving other parts of the system in the future.

Expert Judgment and the Prima Facie Case

Properties such as innovation and inspiration are of substantial value for many research results. Opening new areas others had not considered before and providing a motivated basis for others to build within them are central to progress within the community. However, these factors are extremely hard if not impossible to measure in any standardized way. For these important but more nebulous properties we most typically must rely on what amounts to expert opinion—whether the result impresses other researchers experienced in the area. This is often done with demonstrations and/or scenarios that are intended to present a prima facie case for innovation and/or inspiration. In essence these are intended to elicit a reaction of "Wow, that's cool!" from experts who know the area well and can informally compare it to the state of the art. Such a reaction is a rapid and informal but an experienced-based assessment that the work has important properties such as advancing the state of the art, opening up new possibilities, or taking a fresh approach to an established problem. For example, inventions may open a new area that had not been conceived of before (such as inspiring large numbers of people to do small bits of useful work by playing a game, see von Ahn & Dabbish, 2008) or take a substantially different approach to a problem that many others have worked on (such as recognizing activities in a home by listening to water pipes and electrical noise in the basement (Fogarty, Au, & Hudson, 2006; Patel et al., 2007) or identifying people based on recognizing their shoes, see Augsten et al., 2010).

Clearly this type of validation has problems. It is very dependent on the subjective opinion of experts (most notably reviewers of papers seeking to publish the results) and as such is not very reliable or repeatable. Applying validations of this form to activities of discovery would normally be unacceptable. But in activities of invention where we usually must deal with the uncontrolled complexity of the world, and often seek the widest circumstances for applicability, we are almost never able to know everything we need to know with certainty. As a result follow-on work tends not to make strong assumptions about the applicability of past validation to current circumstances. This means that the uncertainty associated with this type of validation can be more acceptable and less damaging if it turns out to be wrong.

Validation of this form is seen fairly widely in practice—things are valued based on informal assessment of their level of innovation and inspiration by experts, in colloquial terms things treated as having value in part because "they seem cool" to

---

[2] This is the rate of display refresh (which is typically 50 or 60 times per second in order to avoid perceived flicker). This rate is of particular interest because even if internal updates to visual material occur faster than this, they will still never be presented to the user any faster than this.

those with experience in similar work. However, the uncertain properties of this approach make reliance on this type of validation alone a rather risky and unpredictable approach, both for the inventor seeking acceptance of a single invention and the field in making progress overall. To overcome this, most inventions that are validated in this way often seek to provide additional forms of validation (starting with proof-of-concept implementations).

Validations of Tools That Have Indirect Impact on Human Needs

We now consider validation methods for our second set of contributions: those that provide indirect value—that contribute to something that enables or promotes an eventual practical impact rather than providing it directly. For these properties, a rather different set of approaches to validation are appropriate.

One of the most important forms of validation for enabling tools is the use of examples of things that can be built with the tool that demonstrate certain desirable properties of the tool. These can include demonstrations of lower threshold, higher ceiling, breadth of coverage of a desirable design space, increased automation, and good abstractions or extensibility, discussed in more detail below. For inventions involving base capabilities (which are often aimed at overcoming specific roadblocks or limitations of prior work) machine performance tests and in some cases illustration of a prima facie case may be useful.

Threshold, Ceiling, and Breadth of Coverage

A primary example of how inventions help researchers make useful things is *improvements in threshold or ceiling effects* (Myers, Hudson, & Pausch, 2000). (Threshold effects relate to the ease with which simple things can be accomplished and/or novice users can get started, whereas ceiling effects are related to the limitations of a tool or a system for creation of complex or unexpected things.) Validating a low threshold for a tool is often done with a demonstration where the inventor illustrates that something, which in other tools requires considerable work, can be created in their tool easily. For example, the inventor may demonstrate how something can be built in a small number of steps or using a small amount of specification code. Validating a high ceiling is most typically done via a demonstration wherein the inventor shows that one or more sophisticated or complex things—often things that are out of the practical reach of other tools—can be created with their tool. Unfortunately, low threshold tools often tend to impose a low ceiling, and high ceiling tools often come with a high threshold. Consequently, finding ways to ensure both low threshold and high ceiling in one tool is highly valued. Illustration of breadth of coverage is often provided by demonstrating a *spread of examples*—that is, a set of examples that are very different and that span a large(r) space within the set of possible results.

These types of validation all involve creating examples with the tool. Note that the validation is about creation of the examples, but the full properties of the resulting examples are usually not the central issue. So validations that address the properties of the examples themselves are generally not appropriate. For example, performing a usability test on an example built with a tool would likely tell us almost nothing about the tool—many different things might be built with any good tool, and the usability of those things is at least as much a reflection of the designer using tool as it is a property of the tool. Instead the simplicity of creation (for threshold), the power or complexity (for ceiling), or the variety (for breadth of coverage) of the examples is what is critical.

As with other sorts of inventions, machine performance tests may be valuable for enabling technologies. For example, in the case of increased automation it can be appropriate to use performance tests to show that the results are comparable to what is created by previous non-automated methods. Similarly, it may be valuable to demonstrate that the abstractions employed work as the use of the tool scales up. This can be proven in part using simulation, but description and logic may also play a role.

Presentation of Good Abstractions

Like the other validations appropriate for tools and systems, a typical validation for good abstractions is through a set of illustrative examples. To illustrate extensibility, these examples are often similar to breadth of coverage examples, in that illustrating a spread of applicability is useful. For illustrating improved understanding, or ease of application, sets of examples are often similar to those used to illustrate improvement in floor or ceiling effects. While at times this is validated by having developers actually use a toolkit and exploring the details of what they built (see below) this is in many cases a prohibitively expensive way to validate, and it is often considered sufficient to describe abstractions and clearly contrast them with prior alternatives.

Usability for Developers

In some cases, usability tests may be carried out with enabling tools. However these tests need to focus on the *developers* who may be using the tool to create applications, not on the *end users* of the applications created. The number of confounds affecting our ability to evaluate whether a tool engenders usable applications from an end-user perspective is enormous, and the usability of applications is often not the primary value of the tool and should not be the central focus of validation efforts.

Some evaluation of developers working with tools has focused on what abstractions they make use of. When a tool is sufficiently far along to have a large developer community, it can also be interesting to look at metrics such as what types of applications were built with the tool and how the tool was extended. This begins to resemble studies of programmers, programming, and open-source communities. However the cost of bringing a tool this far along may be prohibitive especially when compared to the benefits for invention. Further, because of the high number of confounding factors that may be outside the scope of the tool advance being presented, this type of validation can actually be quite "noisy." In particular, it is very difficult to separate the effects arising from extraneous usability issues in tool interfaces being compared from those related to the core concepts of the tools.

Summary

At this point we must step back and note that the primary form of evaluation for enabling technologies is to build key parts of the technology (proof-of-concept creation). As outlined above, after this primary step it is typical to consider additional validation that highlights the specific goals of the work, that is, to describe the abstractions it employs clearly or to build examples that demonstrate the capabilities of the technology. While there are some secondary evaluations that involve (end) user studies, these are rarely employed.

Summary and Conclusion

In this chapter we have considered the nature of technical work in HCI. To do this we have first situated the work in a broad framework that contrasts its inventive character with one of the other dominant bodies of activities within HCI: those of discovery. This high-level characterization of the work is useful because it allows us to see fundamental differences in the nature of the two kinds of work. These in turn lead to very different values and methods that have evolved to suit each type of work. For example, we conclude that the specifics of methods used in activities of discovery are extremely important—so much so that results are not really understandable in isolation from the methods used to reach them, and so they really become part of the results themselves. In contrast, for activities of invention, the use of one method versus another is much more fluid and less fundamental. Instead, the application of the invention, as demonstrated through a proof-of-concept implementation of the thing invented, is a crucial component of the result.

Using this overall conceptual framework we then consider inventive HCI work itself. We characterize two broad categories of contributions: direct and indirect—where direct contributions directly contribute to meeting some human need, while indirect contributions serve as enablers for later work that meets some human need.

We then go on to characterize the tasks of inventive work in HCI. These tasks include concept creation and validation of concepts. However, we note that one form of validation—the building of proof-of-concept implementations—is more fundamental than other forms. Because it addresses the basic issue of "does it work?"

a proof-of-concept implementation represents a prerequisite for other validation of the work. Because of its special nature it is the normal practice in technical HCI to give proof-of-concept implementations separate and stronger consideration than other forms of validation. As a result, we conclude that technical HCI work should be considered in three parts: concept creation, validation through proof-of-concept implementation, and other validation. The creation of a proof-of-concept implementation (which may need to be quite complex in some cases, as with a toolkit) is a key point of difference with other forms of HCI: Technical HCI is about making things that work, and the work of technical HCI is not done until the validation inherent in an implementation (at a minimum) is complete.

We explore each of these three parts separately. There are few specific methods that one can expect to provide consistently positive outcomes for concept generation. However, we do consider several general strategies for going about the work. These include needs-first and technology-first approaches. We also point to some advantages for technology-first approaches, even though they have developed a somewhat tarnished reputation within the HCI research community. We then consider validation through proof-of-concept implementations by looking at why they are so critical and central. We elucidate the questions that they can address and highlight the diminishing returns inherent in making a prototype complete and robust.

Finally, we consider a range of different forms of secondary validation that can be useful. We characterize a range of different measures we might be interested in and then consider an equally wide range of techniques that can be applied to provide information in those areas. We emphasize again that we must consider a trade-off between the level of knowledge to be gained and the costs of these evaluations and point to places where our community has not always succeeded in choosing the best evaluation methods.

It is typical that technical researchers learn these methods and ideas through osmosis—few courses teach approaches to validating technical work or concept creation in the way that study design and analysis are taught, for example. Instead, technical education programs tend to give researchers the necessary knowledge base from which to invent (how to program, how to use machine learning, how to build circuits, and so on) and hope that with that knowledge, the examples of those who came before (and the guidance of mentors), and a good dose of creativity the novice research will create good results. This chapter has set out to rectify some of those gaps by putting common practice, and the rationale behind it, into words.

Exercises

1. Compare and contrasts technical HCI research with research through design.
2. Where do the ideas come from for technical HCI research? What is the problem that researchers are solving?

References

Appert, C., & Beaudouin-Lafon, M. (2008). SwingStates: adding state machines to Java and the Swing toolkit. *Software: Practice and Experience, 38*(11), 1149–1182.

Augsten, T., Kaefer, K., Meusel, R., Fetzer, C., Kanitz, D., Stoff, T., et al. (2010). Multitoe: high-precision interaction with back-projected floors based on high-resolution multi-touch input. *Proceedings of the 23nd Annual ACM Symposium on User Interface Software and Technology (UIST' 10)*, (pp. 209–218).

Bernstein, M. S., Little, G., Miller, R. C., Hartmann, B., Ackerman, M. S., Karger, D. R., et al. (2010). Soylent: a word processor with a crowd inside. *Proceedings of the 23nd Annual ACM Symposium on User Interface Software and Technology (UIST' 10)*, (pp. 313–322).

Bernstein, M. S., Brandt, J., Miller, R. C., & Karger, D. R. (2011). Crowds in two seconds: enabling real-time crowd-powered interfaces. *Proceedings of the 24th Annual ACM Symposium on User Interface Software and Technology (UIST' 11)*, (pp. 33–42).

Bigham, J. P., Jayant, C., Ji, H., Little, G., Miller, A., Miller, R. C., et al. (2010) VizWiz: nearly real-time answers to visual questions. *UIST 2010* (pp. 333–342).

Buxton, W., Lamb, M. R., Sherman, D., & Smith, K. C. (1983). Towards a comprehensive user interface management system. *SIGGRAPH Computer Graphics, 17*(3), 35–42.

Card, S. K., English, W. K., & Burr, B. J. (1978). Evaluation of mouse, rate-controlled isometric joystick, step keys, and text keys for text selection on a CRT. *Ergonomics, 21*, 601–613.

Card, S. K., Mackinlay, J. D., & Robertson, G. G. (1990). The design space of input devices. *Proceedings of the SIGCHI Conference on Human Factors in Computing Systems (CHI' 90)*, (pp.117–124).

Cardelli, L. (1988). Building user interfaces by direct manipulation. *Proceedings of the 1st Annual ACM SIGGRAPH Symposium on User Interface Software (UIST' 88)*, (pp. 152–166).

Jayant, C., Ji, H., White, S., & Bigham, J. P. (2011). Supporting blind photography. In *The proceedings of the 13th international ACM SIGACCESS conference on computers and accessibility* (pp. 203–210). New York, NY: ACM.

Cohn, G., Morris, D., Patel, S. N., & Tan, D. S. (2011). Your noise is my command: sensing gestures using the body as an antenna. *Proceedings of the SIGCHI Conference on Human Factors in Computing Systems (CHI '11)*, (pp. 791–800).

Engelbart, C., & English, W. K. (1968). *AFIPS Conference Proceedings of the 1968 Fall Joint Computer Conference,* San Francisco, CA, December 1968, (Vol. 33, pp. 395–410)

English, W. K., Engelbart, D. C., & Berman, M. L. (1967). Display-Selection Techniques for Text Manipulation. *IEEE Transactions on Human Factors in Electronics, HFE-8*(1), 5–15.

Feiner, S., MacIntyre, B., Hollerer, T., & Webster, A. (1997). A touring machine: Prototyping 3D mobile augmented reality systems for exploring the urban environment. *Proceedings of the 1st IEEE International Symposium on Wearable Computers (ISWC' 97)*, (pp. 74–81).

Fogarty, J., Au, C., & Hudson, S.E. (2006). Sensing from the basement: A feasibility study of unobtrusive and low-cost home activity recognition. *Proceedings of the 19th Annual ACM Symposium on User Interface Software and Technology (UIST' 06)*, (pp. 91–100).

Fitts, P. M. (1954). The information capacity of the human motor system in controlling the amplitude of movement. *Journal of Experimental Psychology, 47*, 381–391. Reprinted in Journal of Experimental Psychology: General, 1992, 121(3), 262–269.

Grossman, T., & Balakrishnan, R. (2005a). The bubble cursor: enhancing target acquisition by dynamic resizing of the cursor's activation area. *Proceedings of the SIGCHI Conference on Human Factors in Computing Systems (CHI' 05)*, (pp. 281–290).

Grossman, T., & Balakrishnan, R. (2005b). A probabilistic approach to modeling two-dimensional pointing. *ACM Transactions on Computer–Human Interaction, 12*(3), 435–459.

Grossman, T., Kong, T., & Balakrishnan, R. (2007). Modeling pointing at targets of arbitrary shapes. *Proceedings of the SIGCHI Conference on Human Factors in Computing Systems (CHI' 07)*, (pp.463–472).

Gupta, S., Chen, K. Y., Reynolds, M. S., & Patel, S. N. (2011). LightWave: Using compact fluorescent lights as sensors. *Proceedings of the 13th International Conference on Ubiquitous Computing (UbiComp' 11)*, (pp. 65–74).

Harrison, B. L., Fishkin, K. P., Gujar, A., Mochon, C., & Want, R. (1998). Squeeze me, hold me, tilt me! An exploration of manipulative user interfaces. *Proceedings of the SIGCHI Conference on Human Factors in Computing Systems (CHI' 98)*, (pp. 17–24).

Hinckley, K., Pierce, J., Sinclair, M., & Horvitz, E. (2000). Sensing techniques for mobile interaction. *Proceedings of the 13th Annual ACM Symposium on User Interface Software and Technology (UIST' 00)*, (pp. 91–100).

Hudson, S. E., Mankoff, J., & Smith, I. (2005). Extensible input handling in the subArctic toolkit. *Proceedings of the SIGCHI Conference on Human Factors in Computing Systems (CHI' 05)*, (pp. 381–390).

Hullot, J. M. (1986). SOS Interface. *Proceedings of the 3rd Workshop on Object Oriented Programming, Paris, France*, Jan. 1986.

Ishii, H., Kobayashi, M., & Arita, K. (1994). Iterative design of seamless collaboration media. *Communications of the ACM, 37*(8), 83–97.

Jacob, R. J. K. (1986). A specification language for direct-manipulation user interfaces. *ACM Transactions on Graphics, 5*(4), 283–317.

Kelley, J. F. (1983). *Natural language and computers: Six empirical steps for writing an easy-to-use computer application*. Doctoral dissertation, The Johns Hopkins University, Maryland.

Kelley, J. F. (1984). An iterative design methodology for user-friendly natural language office information applications. *ACM Transactions on Information Systems, 2*(1), 26–41.

Khan, A., Fitzmaurice, G., Almeida, D., Burtnyk, N., & Kurtenbach, G. (2004). A remote control interface for large displays. *Proceedings of the 17th Annual ACM Symposium on User Interface Software and Technology (UIST' 04)*, (pp. 127–136).

Lasecki, W. S., Miller, C. D., Kushalnagar, R. S., Bigham, J. P. (2013). *Legion scribe: real-time captioning by the non-experts. W4A 2013* (pp. 22).

MacKenzie, I. S. (1992). Fitts' law as a research and design tool in human-computer interaction. *Human–Computer Interaction, 7*, 91–139.

MacKenzie, I. S. (2002). KSPC (keystrokes per character) as a characteristic of text entry techniques. In Fabio Paternò (Ed.), *Proceedings of the 4th international symposium on mobile human–computer interaction (Mobile HCI' 02)*, (pp. 195–210).

Mankoff, J., Hudson, S. E., Abowd, G. D. (2000a). Providing integrated toolkit-level support for ambiguity in recognition-based interfaces. *Proceedings of CHI 2000*, (pp. 368–375).

Mankoff, J., Hudson, S. E., & Abowd, G. D. (2000b). Interaction techniques for ambiguity resolution in recognition-based interfaces. *Proceedings of UIST 2000*, (pp. 11–20).

Matthews, T., Fong, J., Ho-Ching, F. W.-L., & Mankoff, J. (2006). Evaluating visualizations of non-speech sounds for the deaf. *Behavior and Information Technology, 25*(4), 333–351.

McCulloch, W., & Pitts, W. (1943). A logical calculus of the ideas immanent in nervous activity. *Bulletin of Mathematical Biophysics, 5*(4), 115–133.

Meyer, D. E., Smith, K. J. E., Kornblum, S., Abrams, R. A., & Wright, C. E. (1990). Speed-accuracy tradeoffs in aimed movements: Toward a theory of rapid voluntary action. In M. Jeanerod (Ed.), *Attention and performance XIII* (pp. 173–226). Erlbaum: Hillsdale, NJ.

Myer, B. A. (1990). A new model for handling input. *ACM Transactions on Information Systems, 8*(3), 289–320.

Myers, B. A., McDaniel, R. G., Miller, R. C., Ferrency, A. S., Faulring, A., Kyle, B. D., et al. (1997). The Amulet environment: New models for effective user interface software development. *IEEE Transactions on Software Engineering, 23*(6), 347–365.

Myers, B., Hudson, S. E., & Pausch, R. (2000). Past, present, and future of user interface software tools. *ACM Transactions on Computer-Human Interaction, 7*(1), 3–28.

Mynatt, E.D., Essa, I., & Rogers, W. (2000). Increasing the opportunities for aging in place. *Proceedings on the 2000 Conference on Universal Usability (CUU' 00)* (pp. 65–71).

Mynatt, E. D., Rowan, J., Craighill, S., & Jacobs, A. (2001). Digital family portraits: supporting peace of mind for extended family members. *Proceedings of the SIGCHI Conference on Human Factors in Computing Systems (CHI' 01)*, (pp. 333–340).

Newman, W.M. (1968). A system for interactive graphical programming. *Proceedings of the April 30–May 2, 1968, Spring Joint Computer Conference (AFIPS' 68 (Spring))* (pp. 47–54).
Patel, S. N., Robertson, T., Kientz, J. A., Reynolds, M. S., & Abowd, G. D. (2007). At the flick of a switch: Detecting and classifying unique electrical events on the residential power line. *Proceedings of the 9th International Conference on Ubiquitous Computing (Ubicomp' 07)*, (pp. 271–288).
Schwarz, J., Harrison, C., Hudson, S., & Mankoff, J. (2010a). Cord input: An intuitive, high-accuracy, multi-degree-of-freedom input method for mobile devices. *Proceedings of CHI' 10*, (pp. 1657–1660).
Schwarz, J., Hudson, S., & Mankoff, J. (2010b) A robust and flexible framework for handling inputs with uncertainty. *Proceedings of UIST' 10*, (pp. 47–56).
Schwarz, J., Mankoff, J., & Hudson, S. (2011). Monte Carlo methods for managing interactive state, action and feedback under uncertainty. *Proceedings of the 24th annual ACM symposium on user interface software and technology (UIST' 11)*, (pp. 235–244). New York, NY: ACM.
Shoemaker, G., Tang, A., & Booth, K. S. (2007). Shadow reaching: a new perspective on interaction for large displays. *Proceedings of the 20th Annual ACM Symposium on User Interface Software and Technology (UIST' 07)*, (pp. 53–56).
Sutherland, I. E. (1963). SketchPad: A man–machine graphical communication system. *AFIPS Conference Proceedings, 23*, 323–328.
Viola, P., & Jones, M. (2001). Rapid object detection using a boosted cascade of simple features. *Proceedings of the IEEE Conference on Computer Vision and Pattern Recognition, 1*, 511–518.
von Ahn, L., & Dabbish, L. (2008). Designing games with a purpose. *Communications of the ACM, 51*(8), 58–67.
Wasserman, A. I. (1985). Extending state transition diagrams for the specification of human-computer interaction. *IEEE Transactions on Software Engineering, 11*(8), 699–713.
Wobbrock, J. O, Cutrell, E., Harada, S., & MacKenzie, I. S. (2008). An error model for pointing based on Fitts' law. *Proceedings of the SIGCHI Conference on Human Factors in Computing Systems (CHI' 08)*, (pp. 1613–1622).

Schwarz, J., Hudson, S., Mankoff, J., & Wilson, A. D. (2010). A framework for robust and flexible handling of inputs with uncertainty. *UIST '10* (pp. 47–56).
Schwarz, J., Mankoff, J., & Hudson, S. E. (2015). An architecture for generating interactive feedback in probabilistic user interfaces. In *Proceedings of the 33rd Annual ACM Conference on Human Factors in Computing Systems (CHI '15)* (pp. 2545–2554). New York, NY: ACM.
[illegible]

Viola, P., & Jones, M. (2001). Rapid object detection using a boosted cascade of simple features. In *Proceedings of the 2001 IEEE Computer Society Conference on Computer Vision and Pattern Recognition (CVPR 2001)* (pp. I-511–I-518).

Wobbrock, J. O., & Kientz, J. A. (2016). Research contributions in human–computer interaction. *ACM Interactions, 23*(3), 38–44.
Wobbrock, J. O., Findlater, L., Gergle, D., & Higgins, J. J. (2011). The aligned rank transform for nonparametric factorial analyses using only ANOVA procedures. In *Proceedings of the SIGCHI Conference on Human Factors in Computing Systems (CHI '11)* (pp. 143–146).

Study, Build, Repeat: Using Online Communities as a Research Platform

Loren Terveen, John Riedl, Joseph A. Konstan, and Cliff Lampe

Introduction: Using Online Communities as a Research Platform

We do research on social computing and online communities. We begin from the premise that a deep understanding of online social interaction and adequate evaluation of new social interaction algorithms and interfaces requires *access to real online communities*. But what does "access" consist of? By reflecting on our own and other's work, we can identify four levels of access, each of which enables additional research methods:

1 *Access to usage data*: This enables behavioral analysis, modeling, simulation, and evaluation of algorithms.
2 *Access to users*: This enables random assignment experiments, surveys, and interviews.
3 *Access to APIs/plug-ins*[1]: This enables the empirical evaluation of new social interaction algorithms and user interfaces, as long as they can be implemented within the available APIs; systematic methods of subject recruitment may or may not be possible.

[1] API stands for Application Programming Interface, a published protocol that defines a set of functionality that a software component makes available to programmers who may want to use that component. A plug-in is a piece of software that is added to a larger software application to extend or customize its functionality.

L. Terveen (✉) • J.A. Konstan
Department of Computer Science and Engineering, The University of Minnesota, Minneapolis, MN 55455, USA
e-mail: terveen@cs.umn.edu

J. Riedl (deceased)

C. Lampe
School of Information, University of Michigan, 105 S. State Street, Ann Arbor, MI 48109, USA

J.S. Olson and W.A. Kellogg (eds.), *Ways of Knowing in HCI*,
DOI 10.1007/978-1-4939-0378-8_5, © Springer Science+Business Media New York 2014

4 *Access to software infrastructure*: This allows for the introduction of arbitrary new features, full logging of behavioral data, systematic recruitment of subjects, and random assignment experiments.

In general, as one ascends the levels, more powerful methods can be deployed to answer research questions in more authentic contexts. However, the costs and risks also increase: the *costs* of assembling a team with the diverse skills required to design, build, and maintain online community software and the *risk* that the effort will not pay off: if a community system does not attract users, it does not enable interesting research.

In the rest of this chapter, we draw heavily on our personal experience to elaborate on and exemplify this approach.

History and Evolution

We have been developing our approach since the mid-1990s. We describe a few key theories, projects, and technological developments that were important intellectual and practical influences on our approach.

Artifacts as psychological theories. We found Carroll and colleagues' (Carroll & Campbell, 1989; Carroll & Kellogg, 1989) notion of "artifacts as psychological theories" conceptually inspiring, with its argument that designed artifacts embody claims about user behavior, that it is instructive to make these claims explicit, and that evaluating the use of an artifact also amounts to evaluating these behavioral claims. As we elaborate below, the features we include in our systems often have associated psychological claims, and we sometimes design features explicitly guided by psychological theory.

Project Athena and Andrew were 1980s' projects to deploy networked workstations throughout the MIT and CMU campuses (respectively) to improve the quality of education and support research. Creating these networks required many design choices on issues that were then at the frontier of distributed computing and personal computing, such as reliability, security, scalability, interoperability, distributed file systems, name services, window managers, and user interfaces. By deploying these systems for real use, the designers were able to evaluate how well their design choices fared in practice. Given our computer science background, we found these examples of large-scale system building and deployment *done by researchers* (at least in part) f*or research purposes* inspiring. However, since we have a quite different research emphasis—focusing on different research questions and using different methods—our work looks quite different. Notably, we focus on social interaction among users, we use psychological theory to guide our designs, and we use controlled field experiments to evaluate our designs.

The Web: do it yourself. For the authors, the advent of the World Wide Web was a direct gateway to developing our research approach. If you had an idea for a new interactive system, the Web was an environment where you could implement the idea and reach a potentially unlimited audience. The problem of information overload was attracting a lot of attention in the mid-1990s, and Riedl and Konstan (already at the University of Minnesota) and Terveen (then at AT&T Labs, now also at the University of Minnesota) explored their ideas in the emerging area of recommender systems as a means to address this problem. Riedl, Paul Resnick, and colleagues developed GroupLens, a system for collaborative filtering of Usenet news (Konstan et al., 1997; Resnick, Iacovou, Suchak, Bergstrom, & Riedl, 1994), and Konstan and Riedl followed this up by creating the MovieLens (Herlocker, Konstan, Borchers, & Riedl, 1999) movie recommendation site (more on MovieLens below). Terveen, Will HIll, and colleagues created PHOAKS (Hill & Terveen, 1996; Terveen, Hill, Amento, McDonald, & Creter, 1997), which used data mining to extract recommended web pages from Usenet newsgroups. The PHOAKS Web site contained "Top 10" lists of recommended web pages mined from thousands of newsgroups and attracted thousands of daily visitors during the late 1990s. These early efforts gave us our first taste of online community-centered research. We had built Web sites that attracted users because of the utility they found there, not because we recruited them to evaluate our ideas. Yet this authentic usage enabled us to evaluate the algorithms and user interfaces we created as part of our research program.

We next describe our approach in more detail. We first discuss the type of research questions and methods it enables. We then give an in-depth portrait of the approach by describing important online community sites that we use as vehicles for our research.

What Questions Is This Approach Suitable for Answering?

Since we are describing not a single research method, but rather a general approach to doing research, there is a very broad variety of questions that can be answered. Therefore, it is more appropriate to consider *methods* that fit best with the approach, *skills* required to do follow the approach, and *benefits*, *challenges*, and *risks of* doing so. We organize this discussion around the four levels of access to online communities we introduce in the "Introduction." Note that each access level enables all the methods listed at all "lower" levels as well as the methods listed at the level itself (Table 1).

The primary benefit of this approach is that it enables good science. At all levels, it enables testing ideas and hypotheses with authentic data such as actual behavioral data and user responses based on their participation in an online community. Even better, at the fourth (and possibly third) level, we can perform *field experiments*. Field experiments are studies done in a natural setting where an intervention is made, e.g., a new system feature is introduced, and its effects are studied. According to McGrath's (1984) taxonomy of research strategies for studying groups, field experiments maximize *realism* of the setting or the context while still affording some *experimental control*. A further benefit of this approach is that once you have

Table 1 Levels of access to online communities, with enabled methods

Access to	Enabled methods
1. Usage data	Behavioral analysis, including statistical analysis and simulation Longitudinal data enables analysis of behavior change over time Development and testing of algorithms
2. Users	Random assignment experiments, surveys, and interviews
3. APIs/plug-ins	Empirical evaluation of new social interaction algorithms and interfaces and psychological theories as long as they can be implemented within a published API
4. Software infrastructure	Empirical evaluation of arbitrary new social interaction algorithms and interfaces and psychological theories. Novel data logging. Random assignment experiments

put in the effort to develop a system (and *if* you are able to attract a user community), there is both the possibility and a natural inclination to do a sequence of studies that build upon earlier results. This too makes for good science.

This approach also creates the opportunity for productive collaborations: if you analyze data from an existing community, the owners of that community will be interested in the results, and if you build a system that attracts a real user community, organizations with an interest in the community's topic will be interested in pursuing collaborative projects. Our experience with communities such as Wikipedia and Cyclopath (details below) illustrate this point.

However, there also are significant challenges and risks to the approach we advocate, including the following:

- A research team needs expertise in a wide variety of disciplines and skills, including social science theory and methods, user interface design, algorithm development, and software engineering. This requires either individuals who have the time and capability to master these skills or a larger team of interdisciplinary specialists. Either approach raises challenges in an academic setting; for example, GroupLens students typically take courses ranging from highly technical computer science topics (e.g., data mining, machine learning) to advanced statistical methods to design. This can increase the time a student is taking classes, and not every student is capable of mastering such diverse skills.
- The system development and maintenance resources needed can be considerable. It is not enough just to build software; the software must be reliable and robust to support the needs of a user community. This requires at least some adherence to production software engineering practices, e.g., the use of version control software, code reviews, and project management and scheduling. Many researchers have neither training nor skills with these tools and practices, and in some cases, such resources simply may not be available. If they are available, they represent a significant investment. For example, our group at the University of Minnesota has supported a full-time project software engineer for over 10 years, as well as a dedicated Cyclopath software engineer for 3 years, with cumulative costs of over $1 million.

- Research goals and the needs of the system and community members must be balanced. There are many potential trade-offs here.
  - Sometimes features must be introduced to keep a system attractive to users, even if there is no direct research benefit.
  - Sometimes a system must be redesigned and re-implemented simply to keep up with changing expectations for Web applications. For example, MovieLens had to be redesigned completely about 10 years ago to include Web 2.0 interactive features; however, given how long it has been since its last update, the MovieLens experience again is dated, and we are discussing whether the effort required to bring it up to date is worth the significant design and development costs this would entail.
  - Significant time may have to be spent working with collaborative partners and user groups; for example, our team members have spent considerable time with partners from the Wikimedia Foundation, the Everything2 community, and various Minnesota transportation agencies to define problems of mutual interest and define ways to address these problems that can produce both research results and practical benefits and that follow the ethical standards of all parties involved.
  - If a site does attract a user community, it becomes difficult (and perhaps unethical) for researchers to abandon it if their interests change or their resources become depleted.
  - Since in many cases graduate student researchers do a large part of the development work, research productivity measured in papers produced is almost necessarily lower. However, we believe that there is a corresponding advantage: the papers that are produced can answer questions in ways that otherwise would be impossible. The detailed discussion of our research sites and studies below is intended to support this claim.
- Finally, the major risk is that if the system you create fails to attract or retain sufficient users, all your effort may be wasted. While failure is in principle a good teacher, many of these types of failures are rather boring: you did not pick a problem that people really cared about, your system was too slow and did not offer sufficient basic features, etc.

We next use work we have done on a variety of online community platforms to describe our approach in detail.

How to Follow This Approach/What Constitutes Good Work

Facebook

We began studying Facebook in 2005, shortly after the site was introduced to the majority of universities (Lampe, Ellison, & Steinfield, 2006). Early on, we had permission from Facebook to "scrape" data from the site using automated scripts, enabling us to conduct a study that compared behaviors captured in user profiles (like listing friends and interests) with site perceptions collected through user surveys (Lampe, Ellison, & Steinfield, 2007). Other work we did was based on surveys of college-aged users in the university system and was focused on the social capital outcomes of Facebook use in that population. We found that social capital, or the extent to which people perceived they had access to novel resources from the people in their networks, was associated with higher levels of Facebook use. That finding was confirmed in a study that looked at change in this population over time (Steinfield, Ellison, & Lampe, 2008) and has been confirmed by other researchers (Burke, Kraut, & Marlow, 2011; Burke, Marlow, & Lento, 2010; Valenzuela, Park, & Kee, 2009). This research has consistently found that people who use Facebook perceive themselves as having more access to resources from their social networks, particularly benefits from weaker ties that have often been associated with bridging social capital (Burt, 1992). This form of social capital has often been associated with novel information and expanded worldviews. Put more directly, this research has shown that people are using sites like Facebook to nurture their social networks and access the resources from them, using the features of the site to more efficiently manage large, distributed networks.

At the same time as we have examined the role of Facebook in people's daily lives, we have continued to explore the relationships between the psychosocial characteristics of users, how they use Facebook, and the outcomes of that use. For example, we used survey research to study people's different motivations for using Facebook and how those people used different tools to satisfy those motivations (Smock, Ellison, Lampe, & Wohn, 2011). We found that people motivated for social interaction (as opposed to entertainment or self-presentation) were more likely to use direct messaging features. In addition, in following up on work about the relationship between Facebook and bridging social capital, we found that it was not the total number of friends in a person's Facebook network that was associated with social capital but rather the number of "actual" friends they felt were part of their articulated Facebook network (Ellison, Steinfield, & Lampe, 2011). This work has also been expanded to show that it is not simply the existence of connections that matter, but how users "groom" those connections (Ellison, Vitak, Gray, & Lampe, 2011) through actions like responding to comments, "Liking" posts, and sending birthday greetings.

The overall pattern in these studies of Facebook use highlights the complex interplay between personal characteristics of users, the types of tasks they are bringing to the system, and the behaviors they engage in as they interact with their networks.

In terms of our hierarchy of access levels, our early work was at Level 1, as we did have access to actual Facebook usage data. However, our later work instead relied on surveys of Facebook users. It is ambiguous, however, whether to consider this work being at Level 2, as we could not recruit users from within Facebook itself, but rather only through external means such as posting messages to University e-mail lists. This puts limits on research; it is impossible to accurately represent the population of Facebook users without being able to access a random sample of those users. After we began our work, Facebook created a public API (see http://developers.facebook.com/) that allowed anyone to create new add-on Facebook applications; for example, popular games like FarmVille and Words With Friends were built on this platform. Further, researchers have used the Facebook API to build Facebook apps to explore ideas such as patterns of collaboration around shared video watching (Weisz, 2010) and commitment in online groups (Dabbish, Farzan, Kraut, & Postmes, 2012). However, Facebook apps do not change the core Facebook experience, nor can they form the basis of true random assignment experiments. Most works that have used interviews, surveys, or experiments of Facebook users have used some other sampling frame, often drawn from registrar lists or convenience samples of university students.

Of course, researchers at companies such as Facebook and Google (including student interns) typically have access to their products' software infrastructure, so they are not subject to these limits. For example, Bakshy, Rosenn, Marlow, and Adamic (2012), working for the Facebook Data Science team, conducted an experiment to investigate how information embedded as links to news articles was diffused through the network of Facebook users. They could experimentally manipulate what users saw in their Newsfeed and use system logs to measure network tie differences between conditions. Google researchers studied how Google+ users constructed user groups on the site using a combination of server-level data, surveys, and interviews (Kairam, Brzozowski, Huffaker, & Chi, 2012). Studies by researchers at these companies can offer interesting results but are necessarily limited in reproducibility due to a variety of legal and ethical hurdles that make sharing data between industry and academia complicated. Recently, Facebook has been establishing processes to enable partnerships with researchers based in academic settings, negotiating the legal and technical needs of these collaborations. These research partnerships could help provide Level 3 access to this important source of data.

Wikipedia

We began doing research on Wikipedia in 2006, leading to a paper that studied two research topics: what types of editors produced the value of Wikipedia articles, and what is the impact of damage[2] on Wikipedia articles (Priedhorsky et al., 2007). This work was an early example of the now common research genre of "download and analyze the Wikipedia dump" (Level 1 access). However, there was one important addition: we also obtained data (including some provided by the Wikimedia Foundation) that let us estimate article views. View data gave us a way to formalize the notions of article value and damage. Intuitively, it is more valuable to contribute content to articles that are viewed more, and damage to articles that are viewed more

[2] We defined "damage" to an article through a combination of algorithmic detection and manual coding to evaluate the accuracy of our algorithm.

is more harmful. With our formal definitions in hand, we found that a very small minority of active editors contributed a large proportion of the value of Wikipedia articles, that their domination was increasing over time, and that the probability of an article being viewed while damaged was small but increasing (over the time period we analyzed).

We have continued to do research involving analysis of Wikipedia data, studying topics such as the following:

How editors change their behavior as they gain experience (Panciera, Halfaker, & Terveen, 2009), how diversity of experience and interests affects editing success and member retention in WikiProjects (Chen, Ren, & Riedl, 2010), the effect of reverting edits on the quality and quantity of work and on editor retention (Halfaker, Kittur, & Riedl, 2011), and the gender gap in editing[3] and its effects on Wikipedia content (Lam et al., 2011).

As with Facebook, there is a sense in which anyone does have access to Wikipedia users (where "users" here means editors). Various pages serve as public forums, and communication can be directed to individual users by editing their "user talk" pages; thus, in principle, a researcher could recruit subjects simply by inserting invitations to participate on their user talk pages. However, these techniques do not enable experimental control: crucially, there is no accepted way to randomly assign editors to experimental groups. Moreover, Wikipedia editors long have had a strong resistance to being treated as "experimental subjects." We learned about these problems through bitter experience. Our first attempt to run an experiment with Wikipedia editors was with SuggestBot, our article recommendation tool (Cosley, Frankowski, Terveen, & Riedl, 2007). In the initial version of our experiment, SuggestBot automatically inserted recommendations of articles to edit on randomly selected editors' user talk pages. However, this went against Wikipedia norms that participation in Wikipedia is "opt in," and the reaction from editors was very negative. We therefore changed our model so that editors had to explicitly request recommendations from SuggestBot. In a subsequent project, we attempted to recruit editors for interviews but once again fell afoul of Wikipedia norms. This time one of our team members was accused of violating Wikipedia policies, there was a proposal to ban this person from Wikipedia, and we had to abandon the study. The root cause of these reactions was that Wikipedia editors were extremely averse to being treated as "guinea pigs," and more generally, they objected to people using Wikipedia for any purpose other than building an encyclopedia. Thus, at this point in its development, Wikipedia did not support Level 2 access as we define it.

External researchers cannot introduce new features directly into Wikipedia (Level 4 access). Thus, we implemented SuggestBot as a purely external service running on our own servers, which Wikipedia editors could opt in to; if they do, it computes recommended articles for these users to edit and inserts them on their talk pages. However, note that this does not change the Wikipedia user experience per se. Subsequently, Wikipedia did provide mechanisms for developers to implement changes to the Wikipedia user experience: user scripts (http://en.wikipedia.org/

[3] As detailed in the paper, our analysis relied on Wikipedia editors' self-reported gender.

wiki/Wikipedia:WikiProject_User_scripts); this enables Level 3 access. Users must download and install these scripts themselves if they want the modified user experience. We used this mechanism to implement NICE (http://en.wikipedia.org/wiki/User:EpochFail/NICE), which embodies ideas about how reverts (undoing edits) can be made in a way that is less likely to de-motivate and drive away reverted editors, especially newcomers (Halfaker, Song, Stuart, Kittur, & Riedl, 2011b). NICE is implemented as a Wikipedia user script, which anyone can download and install to change their Wikipedia editing experience. While this approach does let us test new software features for Wikipedia editors "in the wild," it still has a number of undesirable features, including selection bias (as noted above) and a software distribution problem. If we want to make changes, users will have to explicitly download a new version or else we will have multiple perhaps inconsistent versions running.

To address these specific problems, and more generally to enable responsible scientific experiments to be done in Wikipedia, members of our team (Riedl, along with one of our current graduate students, Aaron Halfaker) joined and became active participants in the Wikimedia Foundation Research Committee. The goal of this committee is "to help organize policies, practices and priorities around Wikimedia-related research" (http://meta.wikimedia.org/wiki/Research:Committee). In particular, they are in the process of defining acceptable protocols for recruiting subjects; more generally, they will review planned research projects to make sure that they are compatible with Wikipedia's goals and community norms.

Transition: building our own communities. Now that we have seen both the power and limits of doing research on third-party sites, we turn to sites we currently maintain to illustrate the additional types of research that Level 4 access enables. At Minnesota, we have created a number of online communities to serve as research sites for our studies in social computing. Some have failed to attract a lasting community (e.g., CHIplace; Kapoor, Konstan, & Terveen, 2005), and some have become useful for their intended user group but have not led to significant amounts of research (e.g., EthicShare.org). Two that have succeeded are *MovieLens* and *Cyclopath*. At Michigan State (and now Michigan), Lampe took responsibility for the already existing site *Everything2*, a user-generated encyclopedia formed in 1999, 2 years before Wikipedia. We discuss the three sites by examining a number of studies we have conducted with each.

MovieLens

Origin. In the mid-1990s, DEC Research ran a movie recommendation Web site called EachMovie. While the site was popular and well received, in 1997 DEC Research decided to take it down and solicited researchers who might be interested in the dataset or the site. The GroupLens Research group volunteered to take ownership of EachMovie; while legal issues prevented a handover of the site itself, DEC did make an anonymized dataset available for download. With this dataset as a basis, MovieLens was born.

Early algorithmic research. The initial use we made of MovieLens as a research platform was to explore the performance of different recommender system algorithms (Herlocker et al., 1999; Sarwar, Karypis, Konstan, & Riedl, 2000). This work is interestingly different from our later work in several key respects:

- The focus was on algorithms rather than interaction techniques.
- Social science theory was not used to inform the research.
- We primarily used MovieLens usage data (Level 1), and the experiments we did were not field experiments (deployed in the site) but rather separate "online laboratory experiments" conducted with MovieLens users (who volunteered, and whose profiles were often not even used in the experiment). In this case, MovieLens was a source of research data and subjects but not yet a living laboratory.

Turning toward people, looking to theory. However, we soon began to move up the access level hierarchy. We did this because we wanted to evaluate our algorithms in actual usage and because we expanded our interests to include user interfaces for recommended systems. Three studies used a combination of field experiments and surveys to evaluate: algorithms and interfaces to explain why an item was recommended (Herlocker et al., 1999); algorithms to select initial sets of movies for users to rate (Rashid et al., 2002); and user interfaces to present initial sets of movies for users to rate (McNee, Lam, Konstan, & Riedl, 2003).[4]

These three studies aimed to solve recommender *system* problems: which items to present to users and how to help users evaluate recommended items. However, at about the same time, we began to incorporate another perspective into our work: the use of social science theory to guide the design of our experiments and software features.

A notable early example of this work is presented in "Is Seeing Believing?" (Cosley, Lam, Albert, Konstan, & Riedl, 2003). This work used the psychological literature on conformity (Asch, 1951) to frame research questions concerning user rating behavior and rating displays in recommender systems. Most generally, there was a concern that the standard recommender system practice of displaying, for a movie that the user had not yet rated, the rating that the system predicted the user would give the movie could bias the user to rate according to the prediction. Specific research questions that were studied included the following:

- Are users consistent in their ratings of items?
- Do different rating scales affect user ratings?
- If the system displays deliberately inaccurate predicted ratings, will user's actual ratings follow these inaccurate predictions?
- Will users notice when predicted ratings are manipulated?

[4] When we began doing these live studies, we realized that we had to obtain Institutional Review Board approval, which we did and which has become routine across all our communities and experiments. Note that our "terms of use" say that we have the right to log and analyze behavioral data for research purposes; we also guarantee that we will not disclose any personal or identifying data in our published research. However, we do obtain IRB approval when we do surveys and interviews or introduce new features explicitly to evaluate for research purposes.

The study had both practical results and theoretically interesting implications. First, we modified the MovieLens rating scale based on the findings. Second, while users were influenced by the predicted ratings they were shown, they seemed to sense when these ratings were inaccurate and to become less satisfied with the system.

From a methodological point, it is worth noting that while an experiment was done with MovieLens users, it was explicitly presented to users *as an experiment* rather than involving authentic ongoing use.

Theory-guided design. Our interests continued to evolve to include (in addition to algorithms and user interfaces) social community aspects of MovieLens such as how to foster explicit interaction between users and how to motivate users to participate in the community. Thus, the GroupLens team began collaborating with HCI researchers trained in the social sciences, notably Robert Kraut and Sara Kiesler of CMU and Paul Resnick and Yan Chen from the University of Michigan. Through these collaborations social science theory came to play a central role in our research. We used theory to *guide our designs*, with the goal to create new features that would achieve a desired effect, such as attracting more ratings for movies that had not received many. An additional benefit was that this enabled us to test theories that had been developed for face-to-face interaction in the new context of online interaction to see how they generalized. We and our collaborators used theories including the collective effort model (Cosley, Frankowski, Kiesler, Terveen, & Riedl, 2005; Cosley, Frankowski, Terveen, & Riedl, 2006; Karau & Williams, 1993; Ling et al., 2005), goal setting (Ling et al., 2005; Locke & Latham, 2002), social comparison theory (Chen, Ren, & Riedl, 2010; Suls, Martin, & Wheeler, 2002), and common identity and common bond theories of group attachment (Prentice, Miller, & Lightdale, 1994; Ren, Kraut, & Kiesler, 2007; Ren et al., 2012). One productive line of work within this approach is *intelligent task routing*, which extends recommender algorithms to suggest tasks for users in open content systems. This is very useful, as open content systems often suffer from problems of *under-contribution*. We began this work in MovieLens (Cosley et al., 2006) but subsequently applied it to Wikipedia (Cosley et al., 2007) and Cyclopath (Priedhorsky, Masli, & Terveen, 2010).

We also have collaborated with Mark Snyder to apply his research on volunteerism (e.g., Clary et al., 1998; Omoto & Snyder, 1995; Snyder & Omoto, 2008) to study motivation for participation in online communities, using MovieLens as a research site. We surveyed thousands of new MovieLens users over a 5-month period, using several standard instruments to assess their motivations for trying the site, and then correlated their motivations with subsequent behavior on the site. As with Lampe and colleagues' study of Facebook (Smock et al., 2011), we found that people who had different motivations for joining the community behaved differently: for example, people with more socially oriented motives engaged in more basic MovieLens behaviors (like rating movies) and connected more with other users (through the MovieLens Q&A forum).[5] Notice that we were able to correlate attitudes and personality characteristics with behaviors only because we had both Level 1 (usage data) and Level 2 (users; experimental control) access to MovieLens.

Cyclopath

Cyclopath was created by Priedhorsky, former GroupLens PhD student, and Terveen. Cyclopath is an interactive bicycle routing site and geographic wiki. Users can get personalized bike-friendly routes. They can edit the transportation map itself, monitor the changes of others, and revert them if necessary. Cyclopath has been available to cyclists in the Minneapolis/St. Paul metropolitan area since August 2008. As of Spring 2012, there were over 2,500 registered users, users have entered about 80,000 ratings and made over 10,000 edits to the map, and each day during riding season, several dozen registered users and a hundred or more anonymous users visit the site and request more than 150 routes (Fig. 1).

Like MovieLens, Cyclopath was a "target of opportunity"; where MovieLens was created with EachMovie data, Priedhorsky was motivated to create Cyclopath because he was an avid cyclist, and he had strong personal knowledge of the limits of existing methods for cyclists obtain and share routing knowledge. Of course, his intuition also was that other cyclists would find such a system useful, and obviously, a basic tenet of HCI is that taking *only* your own preferences in account when designing a system may well result in a system that is of interest only to yourself. Further, we did preliminary empirical work to verify our general design concepts as well as specific design ideas (Priedhorsky, Jordan, & Terveen, 2007). Also like MovieLens, Cyclopath has proved to be a productive research platform for us. However, there are a number of significant differences in the two platforms, some intrinsic to the technology and domain, and some historical, due to when they were developed. First, Cyclopath was created after GroupLens had 10 years' experience running MovieLens as a research platform and had begun research on Wikipedia. Thus, we were able to build on and generalize results and methods from these other platforms. Second, Cyclopath has served as a significant vehicle for collaboration between GroupLens and a number of local government agencies and nonprofits that focus on bicycling. This has created diverse opportunities as well as challenges.

We elaborate on both of these themes next.

---

[5] Many of our effect sizes were small, although still significant. Note that we achieved these results with *thousands* of users. This illustrates that size does matter: the number of users in a community limits the number and types of experiments it can support. For example, as of this writing, we typically can get 50–80 experiments for Cyclopath experiments, while we can get an order of magnitude more subjects in MovieLens. Nonetheless, we sometimes have to schedule several MovieLens experiments in sequence because there are not enough users (or at least not of the desired type, say new users) for both experiments to run in parallel.

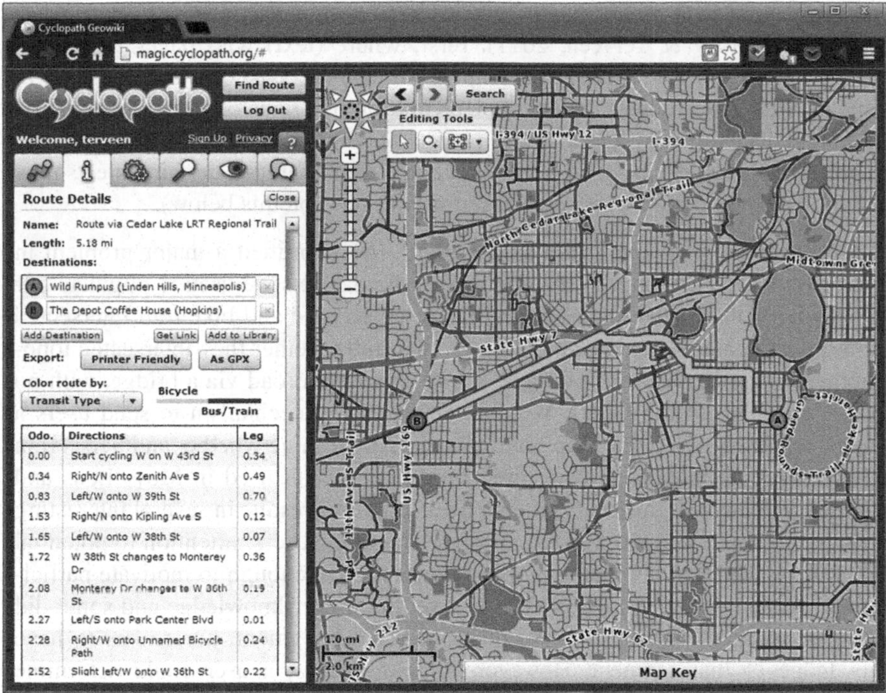

Fig. 1 The Cyclopath bicycle routing Web site, showing a bicycle route computed in response to a user request

Cyclopath: Generalizing Previous Research

Personalized route finding. From the beginning, we wanted to apply our long-standing expertise on recommender algorithms to the route finding problem. We wanted Cyclopath to compute routes *personalized* to the preferences of the requesting user (Priedhorsky & Terveen, 2008). Thus, users are able to enter their personal bikeability ratings for road and trail segments. However, as of this writing, the Cyclopath ratings database is very sparse, an order of magnitude sparser than MovieLens; thus, traditional collaborative filtering recommender algorithms are not practical. On the other hand, we tried machine learning techniques that considered the features (such as speed limit, auto traffic, and lane width) of the segments rated by users to develop predictors of users' bikeability preferences; these predictors were very accurate and practical (Priedhorsky, Pitchford, Sen, & Terveen, 2012).

A geographic wiki. We needed to create analogues of the essential wiki mechanisms, porting them from a text to geographic context. Thus, we developed geographic editing tools, went from watch lists to watch regions, and designed an interactive geographic "diff" visualization. We also were forced to modify the traditional wiki data model (notably, as exemplified in Wikipedia) in two major ways (Priedhorsky & Terveen, 2011). First, where (text) wiki pages have been treated as independent entities, geographic objects are linked. This forced us to come up with a new definition and implementation of a revision as an operation on the entire database, not just a single object. Second, many applications of geographic wikis require fine-grained access control: for example, certain objects may be edited only by certain users or types of users (more on this below).

Theory-based design: intelligent task routing. We identified a major problem in Cyclopath where user input was required: in the datasets we used to populate our map initially, there were thousands of cases where road and bike trail intersected geometrically, but we had no automated way to tell whether they intersected topographically (rather than, for example, a trail crossing a road via a bridge, with no access from one to the other). We thus developed a mechanism to send users a request about an area of the map, asking them to determine whether an intersection existed. This mechanism was inspired by those we had used in MovieLens and Wikipedia. However, this study extended our previous results in several interesting ways. First, we developed a visual interface that drew users' attention to potential intersection, and this interface seemed to be attractive enough to motivate participation. Second, we found that some tasks required user knowledge and some did not. For example, to rate the bikeability of a road segment, a user has to have knowledge of that segment. However, in many cases, a user could determine whether an intersection existed just by zooming in on the map and looking at the aerial photo. This has obvious implications for routing algorithms: some tasks may require users with specific knowledge, while others only require users who are motivated to perform them.

Theory-based design: User life cycles. When we analyzed Wikipedia data to investigate whether and how editors changed over time (Panciera et al., 2009), we found little evidence for development over time. However, limits of the Wikipedia data available for analysis raised several issues concerning our conclusions. In particular, we wondered whether Wikipedia editors might have learned by doing anonymous editing before creating accounts; we also wondered how viewing behavior might have influenced their development as editors. Since we have access to *all* Cyclopath data, we were able to study these issues in this context. In particular, for a number of editors, we were able to associate at least some of the actions they took before creating an account (and while logged out) with the actions they took after creating an account (and while logged in). Our results were analogous to our Wikipedia results: again, we saw little evidence for "becoming" at least in terms of quantity of editing (Panciera, Priedhorsky, Erickson, & Terveen, 2010). However, in subsequent work, we looked at the *type* of editing users did, and here we did observe some transitions over time (Masli, Priedhorsky, & Terveen, 2011), for example, beginning by adding textual annotations to road segment and transitioning to adding new road and trail segments and linking them into the rest of the map.

Cyclopath: Catalyzing Collaboration

Cycloplath has attracted significant attention and support from several Minnesota local government agencies and nonprofits. This has led to projects to add functionality to support analysis by bicycle transportation planners and to extend Cyclopath to cover the entire state of Minnesota. These projects have led to significant technical and conceptual developments, including the following:

- Extending the wiki data model to allow fine-grained access control (Priedhorsky & Terveen, 2011): Transportation planners consider this necessary in order to retain the strengths of open content while still allowing certain information to be treated as authoritative.
- A "what if" analysis feature that uses the library of all route requests ever issued by Cyclopath users: This enables transportation planners to determine where new bicycle facilities are most needed, estimate the impact of a new facility, and get quick and focused feedback from the bicycling public.

In other collaborative projects, we extended Cyclopath to do *multimodal* (bike + public transit) routing, which required changes to our basic routing algorithm, and are extending Cyclopath to cover the entire state of Minnesota. Both projects were funded by Minnesota state and local government agencies.

Everything2

Chi (2009) defined three ways to create what he called "living laboratories." One involves building one's own sites, but another was to "adopt" an existing system and study it in the field. Several years ago, Lampe "adopted" the already existing site *Everything2*, a user-generated encyclopedia formed in 1999, 2 years before Wikipedia. Everything2 was formed by the same group that established the news and discussion site Slashdot but struggled for commercial success after the first dot-com bubble. In exchange for hosting services, the site has had an agreement with Lampe for the past several years to participate in research in many ways, including providing server logs, access to users for interviews and surveys, and rights to add new features to the site for field experiments or Level 3 access in our framework above. The agreement between Lampe and the owners of Everything2 exchanged this level of access for research purposes in exchange for hosting services at the university.

Although Everything2 never achieved the widespread use of Wikipedia, it has an active user population of several thousand users and receives over 300,000 unique visits per month. This activity provided ample behavioral data to examine but also enabled a stable population from which to draw survey samples. This has allowed the Michigan team to study motivations of both anonymous and registered users of the site (Lampe, Wash, Velasquez, & Ozkaya, 2010), finding that both registered and anonymous users had heterogeneous motivations for using the site and that motivations like being entertained were not associated with contributing to the site in the same way as motivations like providing information. We also looked at how habit interacts with those motivations as a predictor of online community participation (Wohn, Velasquez, Bjornrud, & Lampe, 2012), finding that habits are better predictors for less cognitively involved tasks like voting and tagging while being less associated with more involved tasks like contributing articles. Our team also researched which types of initial use and feedback are associated with long-term participation (Sarkar, Wohn, Lampe, & DeMaagd, 2012), finding that those users who had their first two articles deleted were very unlikely to participate in the site again.

Adopting a site in this fashion can have many benefits for both the researcher and the site. For sites that are active, but perhaps not commercially viable on their own, the arrangement provides the community with stability and some measure of security. For the researcher, it can provide access to a community successful enough to research without the difficulty of trying to create a self-sustaining, viable online community. For example, Lampe tried to make several online communities related to public sector interests, none of which achieved the critical mass necessary to conduct the type of research being described here (Lampe & Roth, 2012). Adopting an active online community with a sustainable user population helps to short-circuit some of the major risks and costs of building one's own community.

However, there are some problems with the adoption path, too. For example, with Everything2 the site ownership changed, and the original arrangement had to be renegotiated. The research location also changed, requiring yet more renegotiation. In addition, some users of the site did not appreciate the research agreement and either left the site to avoid "being mice in a maze" or demanded more active agency in the type of research being conducted (similar to Wikipedia editors' reactions described above; for a more general treatment of this topic, see Bruckman's chapter on "Research and Ethics in HCI"). This regular interaction with the community is an additional cost for the management of the research project. Also, just because site owners gave permission to interview and survey members of the community, it did not guarantee that those users would respond to our requests for data.

Sidebar: How Many Users?

We sometimes are asked how many members and how much activity a community must have before it serves as a viable vehicle for research. The answer is that it depends. It depends significantly on your research questions, methods, participation rate of community members, and (if appropriate) effect sizes. If one uses qualitative methods, the ability to interview even ten or so people may suffice. On the other hand, we often assign users to different experimental conditions and then do quantitative analysis of user behaviors, some of which might be rare. In such cases, hundreds of users may be required. MovieLens and Everything2 both enable this.

For example, in the work reported by Fuglestad et al. (2012), nearly 4,000 MovieLens users filled out at least part of a survey. On the other hand, in Cyclopath we can obtain responses from at most several hundred users, but 50–70 is more typical. Oddly enough, it can be difficult to obtain sufficient users for our Wikipedia research as well, since, as we mentioned above, Wikipedia editors have to take some explicit action to opt in to a study. In Everything2, even though there are several hundred active users, we have found that only 150–200 users will respond to surveys during a study period.

Related Work

Since our approach is not a well-defined standard method (yet), we found it appropriate to illustrate with examples from our own research. However, other researchers in social computing and other areas of human–computer interaction have built systems as vehicles for their research and have at least sought to obtain an authentic user community for their systems. While space does not allow a detailed treatment, we wanted to direct the reader to some other noteworthy examples of researchers who have taken similar approaches to ours:

- *Alice* (Pausch et al., 1995) is a 3D programming environment for creating simple animated games or stories. It is a teaching tool, designed to make the concepts of programming accessible to all, including children. Alice has been used in introductory programming classes, and there has been extensive evaluation of its effectiveness as a teaching tool (Cooper, Dann, & Pausch, 2003; Moskal, Lurie, & Cooper, 2004). This led, for example, to the development of Storytelling Alice, which is tailored specifically for middle school children, particularly girls (Kelleher, Pausch, & Kiesler, 2007).
- *Beehive* (since renamed *SocialBlue*) is a social networking and content sharing system created by IBM researchers and deployed to IBM employees worldwide. It has been used as a platform to research issues such as participation incentives and friend recommendation algorithms (Chen, Geyer, Dugan, Muller, & Guy, 2009; Daly, Geyer, & Millen, 2010; DiMicco et al., 2008; Farzan et al., 2008; Steinfield, DiMicco, Ellison, & Lampe, 2009).
- The *International Children's Digital Library* is a Web site that makes children's books from many languages and cultures available online for free. It was created by researchers at the University of Maryland. It has served as a platform for research on topics such as how children search for information online, effective search interfaces for children, design studies, and crowdsourced translation. See http://en.childrenslibrary.org/about/research/papers.shtml (retrieved April 9, 2012) for a lengthy list of references.
- Von Ahn created the *ESP Game* and followed it up with other "games with a purpose." These systems were used by hundreds of thousands of people on the Web, pioneered the area of human computation, and were evaluated in a

number of studies, including investigations of the effectiveness of human computation and how to organize human computation (von Ahn, 2006; von Ahn & Dabbish, 2008).

- PARC researchers created *Mr. Taggy* (http://mrtaggy.com/) to explore tag-based web search and the *WikiDashboard* (http://wikidashboard.appspot.com/; Suh, Chi, Kittur, & Pendleton, 2008) to investigate how people make sense of Wikipedia.
- Dan Cosley and his students at Cornell created *Pensieve* as a tool to support and investigate the process of reminiscing (http://pensieve.cornellhci.org/; Peesapati et al., 2010).
- Eric Gilbert created *We Meddle* to evaluate models of predicting tie strength from social media interaction (Gilbert, 2012).
- Brent Hecht and the CollabLab at Northwestern University have created *Omnipedia*, a tool that allows a user to search for the same word or term across multiple different language versions of Wikipedia to see the prevalence of that entry in the different language versions. This is an example of a tool that adds value to an online community by providing a layer of analysis that increases the opportunities to participate across groups (http://omnipedia.northwestern.edu/; Bao et al. 2012).

Summary and Future Directions

We outlined our approach to doing research on online communities, defined four levels of access researchers can have to a community, and gave a number of in-depth examples of research done at the various levels. We specifically sought to illustrate the *limits* of conducting research where one does not have full access to a community and the *benefits*—but also *risks* and *costs*—of building and maintaining one's own community as a research platform. Most notably, our communities have enabled us to carry out numerous studies where we introduced new—and often theory-based—algorithms and user interfaces and where we were able to evaluate their effects on users' actual behavior as well as users' subjective reactions.

To elaborate on the final point, we are interested in ways to make the benefits of full access to an online community—crucially, the ability to introduce new software features and to conduct random assignment field experiments—widely available to the research community. There are several existing (or emerging) routes to this already as well as possible new directions.

First, there are no (or very little) technical barriers to sharing datasets. This means that groups that maintain online communities can choose to produce (suitably anonymized) datasets available for other researchers to use. Indeed, our group at Minnesota makes several MovieLens datasets available, and these datasets have been used in over 300 published papers. It would be helpful if more large-scale communities follow the lead of Wikipedia and make datasets available for analysis.

Second, researchers should work with commercial sites to try to increase the access researchers have to these sites while ensuring that the values of the community and the desires of its members are respected. The work of Riedl and Halfaker on the Wikimedia Foundation Research Committee is a model here; the results will give all researchers the chance to do controlled experiments and test interventions at a large scale.

Third, we encourage researchers who do maintain successful online communities to make it possible for other researchers to run experiments in their communities. One requirement would be to define APIs to let others write programs designed to run on a site. Another would be to create some sort of management structure to approve proposed experiments, e.g., to be sure that they do not use too many resources or abuse user expectations. The GroupLens research group at the University of Minnesota has developed a proposal to turn MovieLens into this sort of open laboratory, but the development and administrative costs are nontrivial, so dedicated funding to enable this is required.

Exercises

1. Online communities pose special problems to both technical research and field deployments. What are those problems, and what do these researchers have to overcome to be successful?
2. Summarize the various ways of motivating people to contribute to a community and their pros and cons.

Acknowledgments This work has been supported by the National Science Foundation under grants IIS 08-08692, IIS 10-17697, IIS 09-68483, IIS 08-12148, and IIS 09-64695.

References

Asch, S. E. (1951). Effects of group pressure upon the modification and distortion of judgments. *Groups, Leadership, and Men, 27*(3), 177–190.

Bakshy, E., Rosenn, I., Marlow, C., & Adamic, L. A. (2012). The role of social networks in information diffusion. In International World Wide Web Conference (WWW) '12, Rio de Janeiro, Brazil.

Bao, P., Hecht, B., Carton, S., Quaderi, M., Horn, M., & Gergle, D. (2012). Omnipedia: bridging the wikipedia language gap. In *Proceedings of the 2012 ACM annual conference on Human Factors in Computing Systems* (pp. 1075–1084). ACM.

Burke, M., Kraut, R., & Marlow, C. (2011). Social capital on facebook: Differentiating uses and users. In *Proceedings of the 2011 Annual Conference on Human Factors in Computing Systems* (pp. 571–580). New York, NY: ACM Press.

Burke, M., Marlow, C., & Lento, T. (2010). Social network activity and social well-being. In *Proceedings of the 28th International Conference on Human Factors in Computing Systems* (pp. 1909–1912). New York, NY: ACM Press.

Burt, R. S. (1992). *Structural holes*. Cambridge, MA: Harvard University Press.

Carroll, J. M., & Campbell, R. L. (1989). Artifacts as psychological theories: The case of human-computer interaction. *Behaviour and Information Technology, 8*(4), 247–256.

Carroll, J. M., & Kellogg, W. (1989). Artifact as theory-nexus: Hermeneutics meets theory-based design. In *Proceedings of Human Factors in Computing Systems* (pp. 7–14). New York, NY: ACM Press.

Chen, J., Geyer, W., Dugan, D., Muller, M., & Guy, I. (2009). Make new friends, but keep the old: Recommending people on social networking sites. In *Proceedings of the 27th International Conference on Human Factors in Computing Systems (CHI '09)* (pp. 201–210). New York, NY: ACM Press.

Chen, J., Ren, Y., & Riedl, J. (2010). The effects of diversity on group productivity and member withdrawal in online volunteer groups. In *Proceedings of the 28th International Conference on Human Factors in Computing Systems* (pp. 821–830). New York, NY: ACM Press.

Chi, E. H. (2009). A position paper on 'living laboratories': Rethinking ecological designs and experimentation in human-computer interaction. In *Presented at the Proceedings of the 13th International Conference on Human-Computer Interaction. Part I: New Trends, San Diego, CA*.

Clary, E. G., Snyder, M., Ridge, R. D., Copeland, J., Stukas, A. A., Haugen, J., et al. (1998). Understanding and assessing the motivations of volunteers: A functional approach. *Journal of Personality and Social Psychology, 74*(6), 1516–1530.

Cooper, S., Dann, W., & Pausch, R. (2003). Teaching objects-first in introductory computer science. In *Proceedings of the 34th SIGCSE Technical Symposium on Computer Science Education (SIGCSE '03)* (pp. 191–195). New York, NY: ACM Press.

Cosley, D., Frankowski, D., Kiesler, S., Terveen, L., & Riedl, J. (2005). How oversight improves member-maintained communities. In *Proceedings of the SIGCHI Conference on Human Factors in Computing Systems (CHI '05)* (pp. 11–20). New York, NY: ACM Press.

Cosley, D., Frankowski, D., Terveen, L., & Riedl, J. (2006). Using intelligent task routing and contribution review to help communities build artifacts of lasting value. In *Proceedings of the ACM SIGCHI Conference on Human Factors in Computing Systems* (pp. 1037–1046). New York, NY: ACM Press.

Cosley, D., Frankowski, D., Terveen, L., & Riedl, J. (2007). SuggestBot: Using intelligent task routing to help people find work in Wikipedia. In *Proceedings of the International Conference on Intelligent User Interfaces* (pp. 32–41). New York, NY: ACM Press.

Cosley, D., Lam, S. K., Albert, I., Konstan, J. A, & Riedl, J. (2003). Is seeing believing?: How recommender system interfaces affect users' opinions. In *Proceedings of the SIGCHI Conference on Human Factors in Computing Systems (CHI '03)* (pp. 585–592). New York, NY: ACM Press.

Dabbish, L., Farzan, R., Kraut, R., & Postmes, T. (2012). Fresh faces in the crowd: Turnover, identity, and commitment in online groups. In *Proceedings of the ACM 2012 Conference on Computer Supported Cooperative Work* (pp. 245–248). New York, NY: ACM Press.

Daly, E. M., Geyer, W., & Millen, D. R. (2010). The network effects of recommending social connections. In *Proceedings of the Fourth ACM Conference on Recommender Systems (RecSys '10)* (pp. 301–304). New York, NY: ACM Press.

DiMicco, J., Millen, D. R., Geyer, W., Duganm, C., Brownholtz, B., & Muller, M. (2008). Motivations for social networking at work. In *Proceedings of the 2008 ACM Conference on Computer Supported Cooperative Work (CSCW '08)* (pp. 711–720). New York, NY: ACM Press.

Ellison, N. B., Steinfield, C., & Lampe, C. (2011). Connection strategies: Social capital implications of facebook-enabled communication practices. *New Media & Society, 13*, 873–892.

Ellison, N., Vitak, J., Gray, R., & Lampe, C. (2011). Cultivating social resources on facebook: Signals of relational investment and their role in social capital processes. In *Proceedings of the Sixth International AAAI Conference on Weblogs and Social Media* (pp. 330–337). Menlo Park, CA: AAAI Press.

Farzan, R., DiMicco, J. M., Millen, D. R., Dugan, C., Geyer, W., & Brownholtz, E. A. (2008). Results from deploying a participation incentive mechanism within the enterprise. In *Proceedings of the Twenty-Sixth Annual SIGCHI Conference on Human Factors in Computing Systems (CHI '08)* (pp. 563–572). New York, NY: ACM Press.

Fuglestad, P. T., Dwyer, P. C., Filson Moses, J., Kim, J. S., Mannino, C. A., Terveen, L., et al. (2012). What makes users rate (share, tag, edit…)? Predicting patterns of participation in online communities. In *Proceedings of the ACM 2012 Conference on Computer Supported Cooperative Work* (pp. 969–978). New York, NY: ACM Press.

Gilbert, E. (2012). Predicting tie strength in a new medium. In *Proceedings of the ACM 2012 Conference on Computer Supported Cooperative Work (CSCW '12)* (pp. 1047–1056). New York, NY: ACM Press.

Halfaker, A., Kittur, A., & Riedl, J. (2011). Don't bite the newbies: How reverts affect the quantity and quality of Wikipedia work. In *Proceedings of the 7th International Symposium on Wikis and Open Collaboration* (pp. 163–172). New York, NY: ACM Press.

Halfaker, A., Song, B., Stuart, D. A., Kittur, A., & Riedl, J. (2011). NICE: Social translucence through UI intervention. In *Proceedings of the 7th International Symposium on Wikis and Open Collaboration* (pp. 101–104). New York, NY: ACM Press.

Herlocker, J. L., Konstan, J. A., Borchers, A., & Riedl, J. (1999). An algorithmic framework for performing collaborative filtering. In *Proceedings of the 22nd Annual International ACM SIGIR Conference on Research and Development in Information Retrieval (SIGIR '99)* (pp. 230–237). New York, NY: ACM Press.

Hill, W. C., & Terveen, L. G. (1996) Using frequency-of-mention in public conversations for social filtering. In *Proceedings of the ACM 1996 Conference on Computer Supported Cooperative Work (CSCW '96)* (pp. 106–112). New York, NY: ACM Press.

Kairam, S., Brzozowski, M. J., Huffaker, D., & Chi, E. H. (2012) Talking in circles: Selective sharing in Google+. In *Conference on Human Factors in Computing Systems (CHI), Austin, TX.*

Kapoor, N., Konstan, J. A., & Terveen, L. G. (2005). How peer photos influence member participation in online communities. In *Proceedings of the SIGCHI Conference on Human Factors in Computing Systems (CHI '05)* (pp. 1525–1528). New York, NY: ACM Press.

Karau, S. J., & Williams, K. D. (1993). Social loafing: A meta-analytic review and theoretical integration. *Journal of Personality and Social Psychology, 65*(4), 681–706.

Kelleher, C., Pausch, R., & Kiesler, S. (2007). Storytelling alice motivates middle school girls to learn computer programming. In *Proceedings of the SIGCHI Conference on Human Factors in Computing Systems (CHI '07)* (pp. 1455–1467). New York, NY: ACM Press.

Konstan, J. A., Miller, B., Maltz, D., Herlocker, J., Gordon, L., & Riedl, J. (1997). GroupLens: Applying collaborative filtering to usenet news. *Communications of the ACM, 40*(3), 77–87.

Lam, S. K., Uduwage, A., Dong, Z., Sen, S., Musicant, D. R., Terveen, L., et al. (2011). WP:Clubhouse? An exploration of Wikipedia's gender imbalance. In *Proceedings of the 7th International Symposium on Wikis and Open Collaboration* (pp. 1–10). New York, NY: ACM Press.

Lampe, C., Ellison, N., & Steinfield, C. (2006). A face(book) in the crowd: Social searching vs. social browsing. In *Proceedings of the 2006 20th Anniversary Conference on Computer Supported Cooperative Work* (pp. 167–170). New York, NY: ACM Press.

Lampe, C., Ellison, N., & Steinfield, C. (2007). Profile elements as signals in an online social network. In *Proceedings of the SIGCHI Conference on Human Factors in Computing Systems (CHI '07)* (pp. 435–444).New York, NY: ACM Press.

Lampe, C., & Roth, B. (2012). *Implementing social media in public sector organizations*. Paper presented at the iConference '12, Toronto.

Lampe, C., Wash, R., Velasquez, A., & Ozkaya, E. (2010). Motivations to participate in online communities. In *Proceedings of the 28th International Conference on Human Factors in Computing Systems* (pp: 1927–1936). New York, NY: ACM Press.

Ling, K., Beenen, G., Ludford, P., Wang, X., Chang, K., Li, X., et al. (2005). Using social psychology to motivate contributions to online communities. *Journal of Computer-Mediated Communication, 10*(4), article 10.

Locke, E. A., & Latham, G. P. (2002). Building a practically useful theory of goal setting and task motivation: A 35 year odyssey. *American Psychologist, 57*(9), 705–717.

Masli, M., Priedhorsky, R., & Terveen, L. (2011). Task specialization in social production communities: The case of geographic volunteer work. In *Proceedings of the 4th International AAAI Conference on Weblogs and Social Media (ICWSM 2011)* (pp. 217–224). Palo Alto, CA: AAAI Press.

McGrath, J. E. (1984). *Groups: Interaction and performance*. Inglewood, NJ: Prentice Hall, Inc.

McNee, S. M., Lam, S. K., Konstan, J. A., & Riedl, J. (2003). Interfaces for eliciting new user preferences in recommender systems. In: *Proceedings of the 9th International Conference on User Modeling* (pp. 178–188). Berlin, Heidelberg: Springer-Verlag.

Moskal, B., Lurie, D., & Cooper, S. (2004). Evaluating the effectiveness of a new instructional approach. *SIGCSE Bulletin, 36*(1), 75–79.

Omoto, A. M., & Snyder, M. (1995). Sustained helping without obligation: Motivation, longevity of service, and perceived attitude change among AIDS volunteers. *Journal of Personality and Social Psychology, 68*(4), 671–686.

Panciera, K., Halfaker, A., & Terveen, L. (2009) Wikipedians are born, not made: A study of power editors on Wikipedia. In *Proceedings of the ACM 2009 International Conference on Group Work* (pp. 51–60). New York, NY: ACM Press.

Panciera, K., Priedhorsky, R., Erickson, T., & Terveen, L. (2010). Lurking? Cyclopaths?: A quantitative lifecycle analysis of user behavior in a geowiki. In *Proceedings of the 28th International Conference on Human Factors in Computing Systems (CHI '10)* (pp. 1917–1926). New York, NY: ACM Press.

Pausch, R., Burnette, T., Capeheart, A. C., Conway, M., Cosgrove, D., DeLine, R., et al. (1995). Alice: Rapid prototyping system for virtual reality. *IEEE Computer Graphics and Applications, 15*(3), 8–11.

Peesapati, S. T., Schwanda, V., Schultz, J., Lepage, M., Jeong, S. Y., & Cosley, D. (2010). Pensieve: Supporting everyday reminiscence. In *Proceedings of the 28th International Conference on Human Factors in Computing Systems (CHI '10)* (pp. 2027–2036). New York, NY: ACM Press.

Prentice, D. A., Miller, D. T., & Lightdale, J. R. (1994). Asymmetries in attachments to groups and to their members: Distinguishing between common identity and common-bond groups. *Personality and Social Psychology Bulletin, 20*(5), 484–493.

Priedhorsky, R., Chen, J., Lam, S. K., Panciera, K., Terveen, L., & Riedl, J. (2007). Creating, destroying, and restoring value in Wikipedia. In *Proceedings of the 2007 International ACM Conference on Conference on Supporting Group Work* (pp. 259–268). New York, NY: ACM Press.

Priedhorsky, R., Jordan, B., & Terveen, L. (2007). How a personalized geowiki can help bicyclists share information more effectively. In *Proceedings of the 2007 International Symposium on Wikis (WikiSym '07)* (pp. 93–98). New York, NY: ACM Press.

Priedhorsky, R., Masli, M., & Terveen, L. (2010). Eliciting and focusing geographic volunteer work. In *Proceedings of the 2010 ACM Conference on Computer Supported Cooperative Work (CSCW '10)* (pp. 61–70). New York, NY: ACM Press.

Priedhorsky, R., Pitchford, D., Sen, S., & Terveen, L. (2012). Recommending routes in the context of bicycling: Algorithms, evaluation, and the value of personalization. In *Proceedings of the ACM 2012 Conference on Computer Supported Cooperative Work (CSCW '12)* (pp. 979–988). New York, NY: ACM Press.

Priedhorsky, R., & Terveen, L. (2008). The computational geowiki: What, why, and how. In *Proceedings of the 2008 ACM Conference on Computer Supported Cooperative Work (CSCW '08)* (pp. 267–276). New York, NY: ACM Press.

Priedhorsky R., & Terveen, L. (2011). Wiki grows up: Arbitrary data models, access control, and beyond. In *Proceedings of the 7th International Symposium on Wikis and Open Collaboration (WikiSym '11)* (pp. 63–71). New York, NY: ACM Press.

Rashid, A. M., Albert, I., Cosley, D., Lam, S. K., McNee, S., Konstan, J. A., et al. (2002). Getting to know you: Learning new user preferences in recommender systems. In *Proceedings of the*

2002 International Conference on Intelligent User Interfaces (IUI2002) (pp. 127–134). New York, NY: ACM Press.

Ren, Y., Harper, F. M., Drenner, S., Terveen, L., Kiesler, S., Riedl, J., & Kraut, R. E. (2012). Building member attachment in online communities: Applying theories of group identity and interpersonal bonds. *Mis Quarterly, 36*(3)

Ren, Y., Kraut, R., & Kiesler, S. (2007). Applying common identity and bond theory to design of online communities. *Organization Studies, 28*(3), 377–408.

Resnick, P., Iacovou, N., Suchak, M., Bergstrom, P., & Riedl, J. (1994). In R. Furuta & C. Neuwirth (Eds.). *GroupLens: An open architecture for collaborative filtering of Netnews. Proceedings of the International Conference on Computer Supported Cooperative Work (CSCW '94)* (pp. 175–186). New York, NY: ACM Press.

Sarkar, C., Wohn, D. Y., Lampe, C., & DeMaagd, K. (2012). *A quantitative explanation of governance in an online peer-production community*. Paper presented at the 30th International Conference on Human Factors in Computing Systems, Austin, TX.

Sarwar, B. M., Karypis, G., Konstan, J. A., & Riedl, J. (2000). Analysis of recommender algorithms for E-commerce. In *Proceedings of the 2nd ACM Conference on Electronic Commerce* (pp. 158–167). New York, NY: ACM Press.

Smock, A. D., Ellison, N. B., Lampe, C., & Wohn, D. Y. (2011). Facebook as a toolkit: A uses and gratification approach to unbundling feature use. *Computers in Human Behavior, 27*(6), 2322–2329. doi:10.1016/j.chb.2011.07.011.

Snyder, M., & Omoto, A. M. (2008). Volunteerism: Social issues perspectives and social policy implications. *Social Issues and Policy Review, 2*(1), 1–36.

Steinfield, C., DiMicco, J. M., Ellison, N. B., & Lampe, C. (2009). Bowling online: Social networking and social capital within the organization. In *Proceedings of the Fourth International Conference on Communities and Technologies* (pp. 245–254). New York, NY: ACM Press.

Steinfield, C., Ellison, N., & Lampe, C. (2008). Social capital, self esteem, and use of online social network sites: A longitudinal analysis. *Journal of Applied Developmental Psychology, 29*(6), 434–445.

Suh, B., Chi. E. H., Kittur, A., & Pendleton, B. A. (2008). Lifting the veil: Improving accountability and social transparency in Wikipedia with wikidashboard. In *Proceedings of the Twenty-Sixth Annual SIGCHI Conference on Human Factors in Computing Systems (CHI '08)* (pp. 1037–1040). New York, NY: ACM Press.

Suls, J., Martin, R., & Wheeler, L. (2002). Social comparison: Why, with whom, and with what effect? *Current Directions in Psychological Science, 11*(5), 159–163.

Terveen, L., Hill, W., Amento, B., McDonald, D., & Creter, J. (1997). PHAOKS: A system for sharing recommendations. *Communications of the ACM, 40*(3), 59–62.

Valenzuela, S., Park, N., & Kee, K. F. (2009). Is there social capital in a social network site?: Facebook use and college students' life satisfaction, trust and participation. *Journal of Computer-Mediated Communication, 14*(4), 875–901.

von Ahn, L. (2006). Games with a purpose. *IEEE Computer Magazine, 39*(6), 92–94.

von Ahn, L., & Dabbish, L. (2008). General techniques for designing games with a purpose. *Communications of the ACM, 51*(8), 58–67.

Weisz, J. D. (2010). *Collaborative online video watching*. Doctoral dissertation, Carnegie Mellon University.

Wohn, D. Y., Velasquez, A., Bjornrud, T., & Lampe, C. (2012). Habit as an explanation of participation in an online peer-production community. In *Proceedings of the 30th International Conference on Human Factors in Computing Systems (CHI)* (pp. 2905–2914). New York, NY: ACM Press.

Field Deployments: Knowing from Using in Context

Katie A. Siek, Gillian R. Hayes, Mark W. Newman, and John C. Tang

Introduction

Field deployments enable researchers to study interactions with users in situ in both everyday life and extreme situations. Researchers deploying robust prototypes in the wild may use a variety of methods for collecting data, including both qualitative and quantitative empirical approaches. The complexity and scope of field deployments is shaped by choices about the target population (e.g., convenience samples vs. the general public), scale (e.g., a few local users vs. thousands of users over the Web), and duration (e.g., a few days to longitudinal studies over months). Although field deployments can be expensive, resource intensive, and time consuming, they enable collection of rich data from *real* usage that informs future designs, develops stakeholder buy-in, and provides empirical evidence of the emergence and co-construction of sociotechnical systems through the introduction of novel technologies into everyday experiences (Cherns, 1976). In this chapter, we describe field

K.A. Siek (✉)
Informatics, Indiana University, 919 E. 10th Street, Bloomington, IN 47408-3912, USA

University of Colorado, Boulder, CO, USA
e-mail: ksiek@indiana.edu

G.R. Hayes
Bren School of Information and Computer Sciences, University of California,
5072 Donald Bren Hall, Irvine, CA 92797-3440, USA
e-mail: hayesg@uci.edu

M.W. Newman
School of Information, University of Michigan, 105 S. State St., Ann Arbor, MI 48109, USA
e-mail: mwnewman@umich.edu

J.C. Tang
Microsoft Research, 1065 La Avenida, Mountain View, CA 94043, USA
e-mail: johntang@microsoft.com

J.S. Olson and W.A. Kellogg (eds.), *Ways of Knowing in HCI*,
DOI 10.1007/978-1-4939-0378-8_6, © Springer Science+Business Media New York 2014

deployments in relation to other methods commonly used in HCI research, provide background on how to accomplish a successful deployment study, and describe some of our own and others' work in this area so that individuals may become more expert in this approach.

Deployment as an HCI Research Method

Field studies enable researchers to collect empirical data in a relatively naturalistic way. For the purposes of this chapter, we define deployment studies as a type of field study, in which the focus is on the trial of a newly developed or created technology (often a prototype) in situ. We consider the two key defining elements of deployments when compared to other HCI techniques:

- They seek to evaluate the impacts novel technologies and particular populations, activities, and tasks have on each other.
- They seek to perform such evaluations within the intended context of use.

Lab-based studies, wherein representative users perform tasks that approximate real world actions with the technology in an artificial environment (the "lab"), can be easier and perhaps quicker to answer questions about the usability, efficiency, or perceived usefulness of a system. However, lab-based studies may tell us little about whether and how technologies will be adopted, used, adapted, or abandoned in real world use. Ecological gaps (Thomas & Kellogg, 1989) may occur in lab-based studies through the elimination or addition of influences in the lab that may not be present in the real world. In particular, lab-based studies are poorly suited to examining how a technology interacts with other aspects of the environment—including technology already in use, distractions and concurrent activities, social and organizational constraints on use, and so on. As an example, imagine a researcher who wishes to evaluate a new mobile tour guide application. By testing in the lab, she might be able to uncover problems with the application's usability, such as hard to understand instructions, unintuitive commands and navigation operations, and information that is poorly presented. She might also be able to learn speculative reactions from participants regarding how the system would or would not be useful when touring a new city. By conducting a field deployment study, however, she could learn much more about how the system actually supports moment-to-moment needs when touring the city, while also learning how the system integrates with other concurrent activities such as engaging with physical landmarks and attractions, interacting with travelling companions, avoiding traffic, shopping, dining, etc.

Understanding technology within the context of use is an important part of HCI research, and can be approached in a variety of ways including field observations, interviews, contextual inquiry, and participant-observation (to name but a few, many of which are covered elsewhere in this book). These methods study everyday practice to understand people's use of technology within particular contexts, often (in HCI, anyway) with an eye towards designing potential future interventions. Field deployments frequently make use of these methods to understand the relationship of the technology being introduced to the environment, but introducing new technology through the deployment means the very "everyday practice" that is the subject of study will be in flux. This dynamic, in turn, has implications for the application of such methods and for the analytical techniques used. For example, the timing of observations and interviews is of critical importance, and such methods may need to be repeated over time to understand the shifts in usage over time. They may also be applied in conjunction with quantitatively oriented methods, such as surveys and usage log analysis, to identify events worthy of further investigation or to quantify effects discovered through other means. In short, field deployments share goals and techniques with other field-based approaches, but their interventionist nature imposes constraints on the goals that can be pursued and the application of different techniques.

Field deployments typically begin with having developed a new prototype that responds to an identified user need. While the researchers define the technology and study design, the participants in the field define the context of use. Because much of the field deployment results and ecological validity are shaped by the participants, we highlight three types of participants in field deployments.

Convenience Deployment

Deploying a new technology within one's own lab, family, or social network is commonly referred to as "eating your own dog food" or a "friends and family" deployment. For the purposes of this chapter, we describe it as convenience deployment, borrowing from the notion of convenience sampling in other empirical work. When conducting a convenience deployment, researchers can presume relatively easy access to participants—hence the convenience. Greater familiarity with both the participants in the study and their environments can make building and maintaining the system simpler. However, this population is almost certainly not representative of the larger public and may be particularly inclined to support the research efforts with favorable feedback. For example, after a friend had used an application, he told the research team he was able to successfully log out of the application—even though later the research team discovered that he had an earlier version of the application in which there was no logout button. Likewise, in another project in which the research team deployed an intentionally invasive application in a shared workspace, the team had to recruit new researchers to interview participants who preferred not to be interviewed directly by the primary research team for fear of offending them (Hayes et al., 2007). Despite the challenges to generalizability, these kinds of studies are valuable for helping the research team assess their study design and expectations before a full deployment.

Semi-controlled Studies

Semi-controlled studies may involve people who know the research team and those who are unknown at the start of the study, but with whom the researchers have extensive communication and relationships throughout the study period. Creating a robust prototype that can withstand the abuses of participants who are unfamiliar with the research is a substantial challenge but necessary to get as naturalistic an experience as possible. Typically, in semi-controlled studies, participants are recruited to use the prototype for the purposes of the study and may only be allowed to use the technology for the duration of the study—regardless of their desire to continue after the study concludes. For example, when deploying a mobile health application, the research team might work with clinicians to support the analysis of health data collected through the application, but without this engagement from providers, long-term use of the application would not be possible. In this approach, some quantitative data may be collected—particularly as the numbers of research participants and measures grow. However, with a semi-controlled deployment, qualitative data may be just as, if not more, important. The intensity of this kind of intervention can be challenging and rewarding to both participants and researchers, and through it, relationships tend to grow, creating some of the same bias dynamics as in the convenience deployment model. While semi-controlled studies may suffer from some of the same challenges as convenience deployments—namely, acquiescence bias and lack of generalizability—they offer the research team a stronger position from which to argue that the deployment population and site possess carefully chosen characteristics that can be extrapolated to a wider community.

In the Wild

The phrasing "in the wild" implies getting as close to naturalistic usage as one can when introducing something novel. In this case, the technology is deployed almost exclusively to people unknown to the research team, who are not invested in the project nor in technology for technology's sake—and in fact may be inclined to be critical of it. The prototype in these types of studies must be robust and may even be at a level of "beta" testing for a commercial product. For example, IBM Research's deployment of the Beehive social networking prototype within the company attracted over 30,000 employee users after a year, leading to several research opportunities (DiMicco et al., 2008). Having a prototype robust enough to work at this level allows researchers to collect "real world" usage statistics for tens or tens of thousands of participants. Few research prototypes make it to this level of use, but many commercial products, particularly Web-based on-line services, are evaluated using this model.

Appropriate Research Questions for Deployment Studies

Field deployments help us better understand how people use systems in their everyday lives, thus the questions that we can answer revolve around real world use. Specifically, field deployments provide rich data about how closely a concept meets the target population's needs and how users accept, adopt, and appropriate a system in actual use over time.

Field studies can also be used to validate a concept or prototype—either of a system design based on established or discovered population needs, or an innovative technology that no one has yet seen or experienced. In these studies, having a fully functioning prototype is not necessary. A low fidelity prototype can often be informative. For example, a block of wood was used to validate the concept of the first Personal Digital Assistant in the lab (Morrow, 2002). Likewise, when validating a concept, the research team might engage with a pilot population before working with the final target population. For example, Amazon, Facebook, and Google regularly test new interfaces and services for a subset of their users before widely deploying changes. The pilot population does not always have to be from the target population, though the closer they are, the more likely the results are to be helpful in preparing for the full deployment. For example, in a study at UC Irvine, the researchers piloted an application for premature infants (Tang, Hirano, Cheng, & Hayes, 2012) with parents of full-term newborns before deploying the technology to parents of preterm infants.

A major challenge to understanding the potential use of novel technologies is in their long-term use. Behavior management specialists note that it can take up to 6 months for a new behavior, such as the use of a new technology, to really take hold (Prochaska & DiClemente, 1982). At the same time, the social pressures and initial interest that may have driven use early on may wear off over time, enabling other challenges and issues to emerge. Deployment studies allow study of long-term system acceptance, adoption, and appropriation.

Researchers who want to evaluate software they have created on a new device must carefully design study instruments to tease apart influences due to participants' acceptance of the software being evaluated, other software on the device, or the device itself. For example, Le Dantec and Edwards (2008) reported that the device used in the deployment was seen as a status symbol, making it more difficult to tease out participants' reactions to the software deployed on the device. Analysis of system and application log files, interviews, and field observations help to identify the interplay, if any, among these possible factors. If new technology is deployed within a target population, understanding in advance how readily different participants will adopt the technology (e.g., applying a Diffusion of Innovation model (Rogers, 2003)) could help with later interpretation of observed usage and preference patterns.

For HCI researchers, the role of system appropriation and "design in action" (Sengers & Gaver, 2006) may be more interesting than questions of acceptance and adoption. As users take up new technologies, they mold them to their own interests and designs. During field deployments that last weeks or months, researchers can observe these phenomena over time and explore interesting research questions around the factors that would ultimately impact user adoption and appropriation. They can also actively take part in supporting appropriation by redesigning the prototypes throughout the study based on feedback and observations, and iterating with the research participants to come to the final design.

How to Do It: What Constitutes Good Work?

Learning from field deployments involves putting a working prototype into actual use and collecting data around that use. The major steps involved in a field deployment study are:

- Finding a field setting in which to deploy the system
- Defining the goals of a field deployment study
- Recruiting participants and ethical considerations
- Designing data collection instruments
- Conducting the field study
- Ending the deployment
- Analyzing the data

Each of these steps is described in more detail in the following sections.

Finding a Field Site

Field deployment sites should exercises the prototype in actual use while allowing the researchers access to study and learn from the deployment. As noted in the three types of participants in field deployments described earlier, choosing the field participants affects what can be learned and the amount of work needed to connect with the field. We describe several examples that illustrate some of these tradeoffs.

In their work on developing tools to support distributed work teams, Tang and Isaacs (1993) first surveyed distributed teams in their own company. This survey identified problems that people experienced during remote collaborations and led to the development of new video-based prototypes to connect the team members. They then deployed those prototypes in a team within the same company that were distributed across coasts in the USA and not personally known to the researchers using a semi-controlled deployment study. As discussed earlier, using one's own work setting as a field setting usually affords a great degree of access and shared context but can raise questions of generalizability. However, their analysis focused on features of their participants' work (e.g., communication frequency, supplementing video with shared drawing), that they argued would be generally shared by other distributed teams.

More recently, in Venolia et al.'s (2010) work on using video proxies to support remote workers in a distributed team, they developed proxy prototypes that they used among their own distributed team for over a year. Then, they found four different distributed teams, who did not personally know the research team, to deploy the prototypes. The four teams exhibited different characteristics of distribution. Some were in the same time zone, whereas one had a 3-h time zone difference. Some remote participants had joined the team within the last 8 months, whereas one had been working with the team for 3 years. There was a range of seniority levels among the remote participants. This research extended beyond using the prototypes among themselves to study different distributed team configurations, while still leveraging the access and benefits of working within the same company.

When working with external partners to carry out a deployment, it is often necessary to invest time to develop and maintain relationships before, during, and after the deployment. Sometimes a key individual can provide the needed access. For example, in Connelly et al.'s (2012) work, a nurse realized that the low-literacy dialysis patients for whom she cared were interested in monitoring their nutrition, but that their literacy levels would make it nearly impossible to monitor themselves with current tools. The nurse worked with computing researchers to design an appropriate sociotechnical solution. In this case, computing researchers connected with community partners that already had established relationships with the researcher's institution (e.g., a university hospital or community center where employees regularly volunteer). In other cases, however, researchers must actively search for community partners and set up meetings to discuss possible collaborations. Researchers at the University of Colorado sponsored luncheons where multiple community partners in the same subject area were invited. The researchers got an opportunity to learn about issues the community partners regularly face and the community partners benefited by learning about possible technology solutions.

In many cases, researchers must be dedicated to creating a long-term meaningful community partnership to ensure that all of the stakeholders feel valued in the collaborative research project. Sometimes it is a good idea to complete smaller projects for the community partner to begin the collaboration effort. For example, a community partner may need a Web site redesigned—although this is not research, it is important to the community partner and could be a perfectly scoped project for an undergraduate research assistant. From the community partner's view, research projects typically ebb and flow with intense busy periods during studies and then down periods when results are analyzed and the system is iterated on. Researchers may want to consider how to maintain a regular presence with the community partner during these low interaction periods. This could include having regular meetings to update the entire research team on the project's progress. When the community partner members are not explicitly part of the research team, researchers may want to regularly volunteer at the community partner site so the community members do not see the research team as simply a group who appears when the researchers need something.

Defining the Goals of a Field Deployment Study

Field deployments can have a wide variety of goals and answer diverse research questions. A field deployment study offers the opportunity to learn about how people interact with the prototype and conduct activities to improve the prototype design. While there may be a blurry line between field deployment research and actual product usability testing, conducting field deployments is fundamentally about increasing our understanding about the prototype system in ways that generalize beyond the specific deployment. At the same time, field deployments can tell us a great deal about the populations and contexts studied that we may not have learned from merely observing their existing practices and technology use.

Separating Adoption from Use

In deployment studies, one is often asking two questions simultaneously: (1) Will people use this prototype? (2) If they do, will they enjoy it, will they see benefits from it, etc.?

In general, attracting voluntary adoption and use is an extraordinarily high bar for any technology to meet, and few research prototypes can attain the level of robustness, completeness, or aesthetic appeal that would be necessary for real-world adoption. As a result, many field deployments feature artificial inducements for adoption and use in order to focus on other factors such as the usefulness of specific system features, the appropriateness of the system in the given social context, the ability of the system to be appropriated for particular participant needs and practices, or the impacts of using the system on other factors, such as users' behavior changes, work productivity, etc.

For example, in the Estrellita project at UC Irvine (Tang et al., 2012), the researchers struggled with whether to compensate participants for submitting data. Often, in medical studies, researchers compensate participants for adhering to the protocol, because the primary goal of the research is to determine efficacy. Whether or not people would adhere to the protocol without additional compensation is left for a later time, after efficacy has been established. Thus, in this study, the Estrellita team decided to compensate participants for each week they had actively collected data through the system. The researchers did not, however, go as far as many medical studies and require a certain threshold of adherence to the data collection regimen beyond having done something. There is no right answer to these kinds of decisions, but they must be factored in to the study design. Compensation schemes should be reported in publications so that readers can carefully analyze the results in light of the compensation scheme.

The Guiding Light: The Users' Needs or Research Questions?

Field deployments inherently require close relationships between the researchers and those who are participating in the research. This closeness can be seen as a contaminant by those who favor controlled trials. We argue that these relationships can provide insights and additional research questions that may broaden the research contribution. Developing research questions in deployment studies is often not a one-time endeavor. Researchers must continuously reflect on the field deployment process, often altering the prototype and research questions along the way. However, these changes may not be reflected in their empirical scholarly articles, which focus on representing research questions and answers clearly and succinctly. Field deployments are useful both in iteratively improving the design and in understanding the social activity surrounding the use of the prototype.

The Context of the Researcher's Affiliation

If the goals of the study include understanding adoption, use, and perceptions of the prototype's qualities (e.g., usefulness, usability, desirability), researchers must be aware of the impact that the perceived relationship between the researcher and the prototype can have on the results of the deployment. As in lab studies, if the prototype is seen as the fruits of the researcher's labor, whether that be a product being developed in industry or a new technology developed in an academic setting, the users may respond with a bias toward validating the researcher's goals. On the other hand, prototypes that address a pressing user need, especially in sensitive contexts such as healthcare, may elicit user behavior that encourages further development of the prototype, regardless of its current utility. Deploying novel technologies in an industrial context may be particularly challenging in this regard. A managerial or corporate mandate may require employees to use the system, further complicating a relationship in which researchers already have substantial shared attributes with participants. For example, many companies have a relatively homogeneous computing environment, in which common applications (e.g., e-mail, calendaring, corporate workflow) and networking infrastructure can be assumed. Likewise, shared corporate culture can limit the generalizability of findings beyond that environment, even in multinational corporations.

Academic and industrial researchers focused on settings outside the corporation often collaborate with community partners or researchers in other departments during field deployments. These team members each have their own work cultures and associated expectations. Researchers work in heterogeneous environments in which they must sometimes provide additional resources to make the environments more homogeneous (e.g., providing participants with the same mobile phone and data plan). Because multiple stakeholders—academic researchers and community partners—are working on the same project, there may be bureaucratic issues related to intellectual property or human subjects research protections.

Participant Recruitment

When researchers work with community partners for a field deployment, they have to coordinate among multiple groups within their own and with partner institutions. In addition, they have to avoid disrupting the partners' functioning environments. Researchers and community partners should discuss their expectations for collaborations, meetings, recruitment processes, project summaries, and publications early on in the coordination process. Some field sites get more attention from researchers than others, such as a nearby school, hospital, or low socioeconomic status community. Thus, additional communication among researchers at a particular institution could help ensure sites are not overburdened with studies.

The commitment required from research participants tends to be particularly intense with deployment studies. Thus, recruitment and retention in studies can be challenging as participants consider the impact of long-term studies on their everyday lives. Likewise, this commitment can tax researchers. For example, if the target user group needs resources, such as mobile phones, and the researcher has a limited supply, then the researcher will have to consider a rolling deployment, in which the participants begin the study in different cohort groups and still have the same—or at least a similar—experience. Such study designs can introduce problems when comparing across cohorts, however. For example, in a nutrition monitoring study, the researcher should be careful that the rolling deployment time periods selected have comparable culturally influenced and seasonal eating habits. It would be difficult to compare usage patterns if one cohort used the system in September and October and another cohort used the system in November and December (when Western cultures have more eating related celebrations).

Permission to Proceed

Beyond the agreement of any individuals at a field site, researchers often must coordinate with administrators to get permission to conduct the research at the site.

This agreement process can include meetings, presentations, project documentation, and approval by any ethics boards at these sites in addition to institutional approval at the researcher's home institutions. Submitting to multiple ethics boards requires careful planning and coordination to ensure everything is consistent with all of the submissions and any required changes from one board are approved by them all. If at any time there is an issue in this process, we recommend calling a representative from each ethics board to discuss the challenges that the researcher is facing. For example, in one of Hayes's projects, an institutional review board (IRB) required a collaboration letter from the community partner, but the community partner would not provide a collaboration letter until they received a notification that the institution's IRB had approved the study, resulting in deadlock. To proceed, the research team organized a conference call with the ethics board representatives and agreed that the IRB protocols could be submitted at the same time. This was then followed by another call initiated by the institutional IRB to note that the study would be approved pending the community partner's ethic board approval.

Compensation

A final challenge in recruitment and study design is compensation. In all studies, one must be careful to compensate an amount that is appropriate for the socioeconomic status of the participants. What is appropriate for high earners might be coercive to someone earning much less. Beyond these issues, however, deployment studies bring additional considerations for compensation. They can be intensely involved and require more effort on the part of the participants than they—and even the researchers—might initially realize. Part of this challenge stems from our thoughts about compensating people based on the amount of time they spend doing study activities. Whereas in a lab study, use of the system for an hour might take place in a predictable way, an hour of system use in a field study might take place over days or weeks in small chunks, something that is difficult to compensate for appropriately. At the same time, however, use of the system can itself be a substantial benefit that encourages people to stay engaged with the study despite the intensity of the research activities.

Designing Data Collection Instruments

Fundamental to any research project are questions of what to measure and how. Field deployments are ecumenical in the types of methods employed, but they do afford some techniques not seen in other types of HCI knowledge generation. In particular, field deployments allow for long-term repeated measures studies and for analysis of usage logs. Qualitatively, researchers may conduct repeated interviews, use surveys and other tools to garner user responses, and observe prototype usage. Quantitatively, researchers may wish to measure time on task, efficiency and productivity, task load (e.g., through tools like the NASA-TLX (Hart, 2006)), and changes in user perceptions over time. Often, field deployments employ multiple methods to enable triangulation on particular issues, including several presented in this book such as surveys (Chap. "Surveys Research in HCI"), log analysis (Chap. "Understanding Behavior Through Log Data and Analysis"), sensor data collection (Chap. "Sensor Data Streams"), social network analysis (Chap. "Social Network Analysis in HCI"), and retrospective analysis (Chap. "Looking Back: Retrospective Study Methods for HCI").

Regardless of the methods chosen, data collection over extended periods of time can become a burden on participants. Repeated surveys may begin to irritate participants over time, and if those surveys are tied to system use, they can even stop using the prototype simply to avoid logging their use or answering questions about it. Thus, measures that minimize explicit user intervention may be preferable for sustained engagement. These methods can include observation, log analysis, task-based measures, and implicit measures. Finally, a combination of participant fatigue and the long-term participant–researcher relationships may make participants eager to please. Thus, researchers must pay special attention to when and how data are collected. For example, it is key that the use of a prototype tool be time-stamped to avoid the situation in which use is considered high, but participants are only using the tool directly before an interview or meeting with the researchers—sometimes referred to as Parking Lot Compliance (Stone, Shiffman, Schwartz, Broderick, & Hufford, 2003).

Implicit measures, such as the number of e-mails or phone calls initiated in a deployment study of a novel communication tool, can be analyzed to support a variety of claims. However, logging activities can become burdensome if data collection is not automated. For example, in one study in which participants were asked to document face-to-face visits in a log, there were some inaccuracies in the recorded data (Tang & Isaacs, 1993). One participant even shared an anecdote that although he would usually greet his colleague at the beginning of the workday, he sometimes skipped doing so during the study simply to avoid having to log it. If researchers choose to automate data collection, we encourage researchers to develop scripts prior to the study so they can easily analyze the data continuously during and after the deployment.

Field deployments can make use of both within and between subjects study designs. It can be difficult to recruit enough participants for statistically significant results with either design, but of course, this challenge is more acute for between-subjects comparative trials. Because participant numbers can be small in deployment studies, it can be hard to measure changes in major outcomes (e.g., improved educational performance or behavior change). Thus, it can be particularly important to examine intermediate and process-based outcomes. For example, if participants describe in interviews becoming more aware of an activity that the system monitors, this outcome is indicative of the potential for future behavior change even if such change has not yet occurred in the scope of the study. Likewise, research teams must consider incremental evaluation during the deployment, such as periodic interviews, to inform future questions about why things are happening.

Because field deployments are costly in terms of time and resources for all stakeholders, researchers should ensure the study has some value for the target population and that they are collecting enough data to yield some deeper understanding of the system and target population. This is one way to accomplish the Belmont report's ethical goal to "maximize the potential benefits and minimize the potential harms" as discussed in the chapter "Research and Ethics in HCI." An easy way to check on the data collection methods is to revisit the research questions and expectations with community partners. Researchers can ask themselves, "What data do I need to answer these research questions?" and "Given this data that I plan to collect, what could I say to my research community or community partner?" Study design reviews, where researchers—some of whom may not be intimately involved in the study—meet to discuss the system and study design, may be helpful in identifying evaluation holes or alternative methods for data collection.

As in any project, researchers learn as they conduct field studies. Thus, we recommend conducting small deployments or pilots before starting a larger field deployment. For example, the researcher and research team could pilot the study on themselves in the form of a convenience deployment, as discussed earlier. If the researcher and her colleagues are not willing to use the system or participate in the study as it was designed, participants should not have to go through it. Once the researchers have "eaten their own dog food," then the system and study design can be iterated on and the team can decide if further small deployments are necessary (e.g., a friends and family deployment) before the true community partner deployment is initiated.

The Field Deployment

Deployment studies are messy. Events and constraints in the real world will have an impact on the data being collected. Researchers need to be responsive in adjusting the data collection, study plan, or other factors as conditions evolve during the deployment. We encourage transparently describing ways that the field deployment may have evolved in response to real world events in the setting. This not only emphasizes the ecological validity of how real usage affects the use and the study of the technology, but also helps the reader interpret the data presented in the study. In a field deployment of a video conferencing system by Tang, Isaacs, and Rua (1994), they included a section that discusses the messiness of the data collection that got favorable comments from our readers.

During the entire field deployment, the research team must prepare for and be dedicated to incremental, continuous analysis to understand what is being done and to inform future questions. Incremental analysis can be done on any of the data collected by the researchers—from checkpoint interviews to automatically captured usage logs. These future questions and insights can be integrated into the current field deployment or used for reflection after the deployment to improve the system and study design. The incremental analysis, however, does not replace the final analysis where the research team looks at the complete data sets. During the final analysis, researchers can revisit the questions and insights that were made during the field deployment to verify or challenge them.

No matter how organized a research team is for a field deployment, the study may still feel chaotic and the data messy. This chaos results from the all of the possible variables and situations that cannot be controlled when participants use a system in situ. In these engagements, participants may be lost to follow up for a variety of reasons—moving, growing tired of the study, stolen equipment, and more. Likewise, research questions often change as a result of what is learned. For example, as a system designed for teaching students began to be used by the classroom staff as a communication tool, the research team added research questions around changes in communication patterns in the classroom (Cramer, Hirano, Tentori, Yeganyan, & Hayes, 2011). This shift in research questions resulted in changes to both what data were being collected, including video of teachers talking even when no children were present, and how already collected data were analyzed, such as by re-coding video of classroom activities when children were present with a new emphasis on teacher communication.

Developing and Supporting a Working Prototype

In addition to monitoring the data collection and continually reassessing research questions, the prototypes themselves often require continual monitoring and support. Members of the research team must be constantly on-call to field inquiries from participants, community partners, and other stakeholders to troubleshoot problems, provide support, and generally make the field deployment run smoothly. Moreover, prototypes often change during the course of a deployment as the researchers learn more about the system in use. Making substantive improvements to prototypes to address newly identified requirements, usability problems, and software bugs is standard fare when conducting field deployments, and resources need to be allocated accordingly.

Even when plans are made to offer support, issues can arise. In one situation at the University of Colorado, the study participants did not have access to computers and Internet, thus we provided participants with a mobile phone number that they could call 24-h a day in case they had any questions or problems throughout the deployment. Because no calls were received, the research team was surprised to learn of problems participants encountered during a checkpoint interview. The participants, who all lived in a city 45 min away, said they could not call because the study line was in a different area code, prompting the researchers to use a phone number in the same area code as the participants in future studies.

Regardless of plans to offer support, deployed prototypes must be robust enough to be independently used, usually without the luxury and handholding of technical support nearby. This robustness requires a considerable amount of investment up front to make a robust working prototype, and a commitment to providing responsive technical support throughout the deployment to address issues that invariably arise. Researchers should pilot test the prototype in realistic settings before embarking on a field deployment study. Mechanisms for responding to issues that arise in the deployment, such as what participants should do when something goes wrong, should be worked out in advance.

Changes to the Deployment

In formal lab-based experiments or clinical trials, changes cannot be made to the intervention midstream without invalidating data collected before the change. When using field deployments in HCI research, however, changes to the prototype and/or data collection regime are often necessary and the rationale for those changes and their impacts can add value to the research. Researchers should document any issues that they encounter, ensure there are checkpoints throughout the study for discussing ideas and results, and make changes in response to anything that comes up unexpectedly. Reporting this messiness historically has been challenging in HCI publications, but is becoming more commonplace and should be encouraged. If a deployment is not working—either the system or study—then the research team can stop the deployment, pull out the system, and reevaluate the research goals, study design, and strategy. Some deployments may last a long time—we have seen examples where research teams decided to stop a deployment after 3.5 years to reevaluate the study and update the system (Cramer et al., 2011; Hirano et al., 2010). It is better to err on the side of caution than push forward with a deployment that could frustrate and not benefit participants, alienate community partners, and waste researchers' time and resources.

Ending the Field Deployment

At some point, the field deployment will end—either because the deployment is not meeting the needs of the group or because the study has run its course. When a field deployment is drawing to a close, the research team has to consider what, if any, impact the end of the deployment will have on the community partner and target population. For example, has the deployment improved any of the processes or activities of the stakeholders? If so, what are the ethical implications of removing the system? Occasionally, the IRB or community ethics board may add a requirement to the study protocol noting that if a participant indicates that the system is beneficial, then the participant must be provided the system. In our experiences, this requirement has varied between providing the system to the participant for free or giving the participant the option to purchase the hardware and software. In the former case, the research team has to learn about any processes they have to follow by the funding agencies and employment site to legally give hardware and software away that has been purchased by the employment site. Some research teams may consider providing the software as part of the participant incentive. Selling the prototype at least makes the costs visible to the participants and may factor into a decision about continuing use.

An awkward moment can arise at the end of the study when the researchers plan to withdraw the prototype out of daily use, particularly when the prototype has proven to be valuable in meeting users' needs. For example, in the study of the video proxy prototypes (Venolia et al., 2010), the study plan was to observe each group 2 weeks before deploying the prototype, 6 weeks while using the prototype, and for 2 weeks after withdrawing the prototype. When it came time to remove the prototype, one group taped a note on it asking the researchers not to remove the prototype because they wanted to use it for an important meeting that afternoon. The researchers adjusted the study schedule to accommodate the group's request, which they noted as an extra bit of data documenting the value of the prototype to the group.

Yet, research prototypes are often not built to withstand the rigors of daily use, nor can the research team maintain the level of technical support needed to keep the prototype working indefinitely. Even with explicit agreements at the beginning of the study, if a prototype has become an integral part of the participants' lives, it can be difficult to remove it.

In cases in which the continued operation of the prototype is viable, the research team must also consider what, if any, support will be available. Post-study support for a research system is difficult because researchers move on to the next project and do not typically have spare resources to support technology left over from previous projects. Often prototypes will cease to work after a time due to platform software updates and technology evolution. Thus, the research team should let the target population know clearly what kind and for how long support will be available if the system is kept in use after the conclusion of the study.

After the field deployment is completed, the research team must also consider what type of relationship they want with the community partner. At the very least, the researchers must ensure that the community partner is debriefed to discuss study implications and insights. If the research team would like to continue the collaboration with the community partner, then they should revisit prior agreements to ensure all parties are comfortable with continued collaboration and future research endeavors.

Reporting the Data

Field deployments provide rich, diverse, and messy data sets for researchers to analyze. Although analysis methods are not unique to field deployments because they are native to the theories and methods employed for a particular study, researchers must provide rich detail on what data was used from the field deployment to help the research community assess the results. We encourage researchers to not only report on their study and analysis methods, but also report on any data cleaning that may have happened. For example, was Parking Lot Compliance data included in the analysis? If so, why was that decision made? Although Parking Lot Compliance data may be artificial in that participants generated the data in a short amount of time because of some study event, it can also provide the researchers with insights into the participants' lives. For example, in a photo elicitation study, participants took pictures of the contents in their cabinets and refrigerator before they met with researchers. During our initial study assessment, researchers at the University of Colorado marked the data resulting from Parking Lot Compliance, but continued to reflect on the data. In a short period of time, we found that these pictures were incredibly telling about how the local dietary culture overrode the participant's ethnic culture because the participant had recently moved into the area and went from having fresh fruits and milk in her refrigerator to having soda and prepackaged foods in 6 weeks (Khan, Ananthanarayan, & Siek, 2011). It would also benefit the research community to discuss any issues the research team encountered with recruitment, retention, and engagement of the target population to help the community gauge the difficulties in conducting a field deployment in these environments.

Because every field deployment is different, researchers' experiences may vary depending on the theories, methods, environments, and populations of their work. This "how to" section therefore provides insights from our collective experience into how to start, implement, and finish a field deployment study but is not entirely prescriptive. Researchers should plan for potential challenges that they may encounter and take advantage of the particular opportunities available in their sites and with the technologies they wish to deploy.

Becoming More Expert in Field Deployment Studies

Field deployments require a multiple methods approach to design the deployment, evaluate it, and analyze the data. For researchers who work with community partners, it may be beneficial to consider the power dynamic between partners and researchers in the design process to better frame their goals, methods, and communication expectations. Will researchers and community partners have an *equal partnership* in every step of the research process, such as in Participatory Design and Community-Based Participatory Design (Israel, Eng, Schulz, & Parker, 2005)? Or will researchers work collaboratively *with* the target population to improve a social construct with Action Research (see Chap. "Knowing by Doing: Action Research as an Approach to HCI" (Hayes, 2011), and (Lewin, 1946))? Perhaps researchers will design an intervention *for* the target population with User Centered Design. Although there are many other theoretical lenses that researchers could consider in their field deployments, we briefly discuss the major approaches the HCI community has employed.

Researchers can—and often should—combine qualitative (see Chaps. "Reading and Interpreting Ethnography" and "Grounded Theory Method") and quantitative methods (see Chaps. "Surveys Research in HCI" and "Sensor Data Streams") when designing field deployment studies. Researchers should reflect on their intellectual commitments and preferred approaches for data collection and interpretation. For example, do they believe that their assertions can and should be verified with collected data—a positivist approach that motivates more controlled study designs? Or do they resonate more with the belief that social life cannot be studied the same way as natural world phenomena, leading to non-positivistic approaches such as those used in naturalistic inquiry (Lincoln & Guba, 1985)? Deployments can be somewhat methods-agnostic, enabling the research team to choose those empirical methods that most resonate with their own traditions and beliefs and to connect the methods most closely to the questions being asked. Ultimately, the unifying characteristic of deployments is the study of system use in context, regardless of the empirical approach used for evaluation and understanding.

While field deployments are used throughout HCI, they have played an especially important role in CSCW and Ubicomp research because it is difficult to replicate the important factors that will affect system use in the lab. Beyond the papers already discussed on using video to support distributed teams (Tang & Isaacs, 1993; Tang et al., 1994; Venolia et al., 2010), there are a number of CSCW studies based on field deployments. Erickson et al. (1999) developed an awareness tool called Babble that gave a visual representation of colleagues' presence and activity in CMC conversations. They not only used Babble within their own group, but deployed it in other working groups at IBM. Their deployment helped them understand how teams used Babble and led to their research insight on social translucence for negotiating making contact through CMC. Similar to the Beehive work cited earlier (DiMicco et al., 2008), Bluemail was a prototype e-mail tool that was deployed within IBM for over a year and used by over 13,000 employees. Analysis of the usage data over that deployment enabled research around different e-mail foldering patterns by country (Tang et al., 2009) and strategies for refinding e-mail messages (Whittaker, Matthews, Cerruti, Badenes, & Tang, 2011). Brush, Inkpen, and Tee (2008) looked at how seven pairs of families used a prototype calendar and photo sharing system over a 5-week deployment. Given the importance of the social context on the use of CSCW tools, field deployments are an effective way of studying how they are used.

The Ubicomp literature has a rich history of case studies paired with lessons learned, to describe research focused on dynamic, in situ experiences afforded by mobile interactions (Scholtz & Consolvo, 2004). The case study approach provides researchers with examples of study designs that could be adapted, with the additional benefit of becoming aware of possible issues prior to deployment. Scott Carter and Jennifer Mankoff (2005) explored what prototype fidelity is needed to evaluate Ubicomp systems in certain contexts. Consolvo et al. (2007) have reflected on their experiences with in situ data collection with respect to evaluator-initiated, context-triggered, and user initiated data collection—an important consideration for application and evaluation instrument design. For in situ system evaluation, Hazlewood, Stolterman, and Connelly (2011) consider the difficult question of when are people participants in ambient display environments—when is someone interacting with a truly pervasive system or simply an innocent, collocated bystander? Finally, Favela, Tentori, and Gonzalez (2010) challenge the community to reflect on the ecological validity—that is, how realistic the evaluation environment is—of our evaluations and argue for a middle ground between a controlled experiment and truly in situ assessments to better understand the perceived pervasiveness of deployed prototypes.

We encourage the community to report on the messiness of their deployments and data analysis, but a researcher may still wonder what the limits are to reporting. We find the principles listed in the Statement on Reporting of Evaluation Studies in

Health Informatics (STARE-HI) beneficial when considering what to include in our publications (Talmon et al., 2009). STARE-HI describes everything that researchers should report in each section of their paper. Although STARE-HI was designed with the health informatics community in mind, the evaluation study aspect of the framework is applicable to any field deployment.

Example Field Deployment Papers

In addition to the above CSCW and Ubicomp case studies, we briefly recommend three exemplar papers that utilize field deployments to better understand user needs (Le Dantec & Edwards, 2008), build on user needs to design a prototype system (Tentori & Favela, 2008), and evaluate prototype systems (Consolvo et al., 2008). We chose these based on the study design, data analysis, well thought out implications informed by the results, and the respect the authors showed for the target population. Readers should note how the paper authors carefully frame their findings to show how each result is a valuable contribution to the field—even in the cases where reviewers may think a result is banal because it is obvious—a common criticism in field deployments. We counter that we cannot reliably *know* that a result exists until we rigorously study the phenomenon, thus the need for field deployments.

Le Dantec and Edwards (2008) show how researchers can employ a photo elicitation study to better understand the needs of an underserved population—in this case, urban homeless. The authors investigated 28 homeless people's technology perceptions and informational needs by providing participants with disposable cameras to document their lives for 2 weeks. Participants' images were used as reflective probes during an interview at the end of the study. The qualitative results provide the reader with a well-defined picture of the target population. The authors conclude with sociotechnical interventions for homeless people that may put into question some of the assumptions that the HCI community has about this population, such as homeless people's mobile phone usage. They also have an informative discussion detailing the challenges they encountered while working with the target population. In this case, the authors deployed a technology that is already well-understood and robust to help shed light on the context into which that technology was being deployed.

Tentori and Favela (2008) illustrate a complete design cycle for an activity-aware application in a hospital environment that facilitates collaboration and coordination. The authors sought to create design guidelines and tools to collect contextual data describing specific activities in a hospital setting. To this end, the authors report on findings from their 196 h shadowing study with 15 medical professionals. A major contribution of the paper is how the authors designed their guidelines and tools—detailing scenarios and application features that the community can use in the design of their own activity-aware applications. The authors conclude the paper with example applications that uses the proposed tools to address the target population's needs. This same research team followed on this work with additional studies, both controlled and naturalistic deployments, reflecting in a final paper on the degree of ecological validity, generalizability, and pervasiveness of these various deployments (Favela et al., 2010).

Consolvo et al. (2008) designed a physical activity self-monitoring application. Their paper provides readers with an excellent example of the iteration that research teams must go through before deploying an application for a longer period of time. The UbiFit System paper studied how effective an ever-present mobile phone display with abstract physical activity progress visualizations (e.g., a pink flower to represent cardio training) would motivate a target population to partake in physical activity. The research team evaluated UbiFit with a comparative study of 28 participants over a 3-month period during the winter holiday season. The results provide readers with insights on how to design studies to evaluate the effectiveness of specific application features. In addition, the authors highlight how quantitative and qualitative data complement each other to further bolster overall arguments about the interface effectiveness to motivate physical activity. The authors conclude with a refreshingly candid discussion about the target population's experiences in the study and their own study design limitations.

Personal Testimonials

Katie's Story: But Can They Do That?!

During my Ph.D. research I investigated how to empower low-literacy dialysis patients to better manage their nutrient and fluid intake. When we proposed a mobile application that could provide dialysis patients with the ability to scan or select food icons from an interface to receive real-time feedback on what they were consuming, we were inundated with "But can they do that?" type questions that ranged from using technology to scan food items to understanding what patients really consumed in their everyday lives. Soon after completing lab studies that showed they could complete tasks on the technology, we were asked, "Sure, but can they do that *all day long*?" Thus, began my introduction to field deployments—the only way I could really find out if patients could complete these tasks in their everyday lives.

I typically work with underserved populations, so it is important for me to choose methods and technologies that relate to existing practices to help participants better envision their part in the study. I have shadowed participants and conducted photo elicitation studies to understand their needs. I also evaluate new prototypes by utilizing system logs, application logs, checkpoint interviews, validated survey instruments, and best practices for the fields we are collaborating with (e.g., 24 h food recalls). It is typical for the study facilitators to share about themselves in reciprocation of how much the participants' share with them. I enjoy these informal conversations because they develop deeper relationships with participants.

Gillian's Story: "There Is Nothing So Practical as a Good Theory"—Kurt Lewin, 1951

As an undergraduate student, I interned at the NIAID in Tony Faucci's AIDS lab. For a wanna-be AIDS researcher, there was literally no better place to be in the USA. After 2 weeks of watching my petri dishes and carefully noting their progress, I was bored senseless. My mentor took me down to the lab where we drew blood from volunteers who are HIV+. I was invigorated talking to them about their stories and only regretted that the mechanisms of disease eradication we were studying in the lab were years if not decades from human use. I returned to Vanderbilt with a sense of renewed interest in research and a sense that I really ought to change my major from Molecular Biology to something that would let me get out there with people. As much as it might not seem so on the outside, I found computer science to be the place that let me do that.

As a graduate student, then, I was always interested in how what we were learning and theorizing and postulating could be used in the real world. Ever a pragmatist and a bleeding heart, I sought opportunities to do "applied" work, never seeing applied as a dirty word (see my other chapter in this book, "Knowing by Doing: Action Research as an Approach to HCI"). At the same time, I became deeply interested in how building, creating, making, and sharing technologies can be used to further our understanding of the world. It is the synergy of creating tools and knowledge at the same time that brought me to love deployment studies as an approach to HCI research.

Mark's Story: When Is It Worth the Pain?

I employ field deployments as one of many research methods, along with qualitative field studies, system building, and lab-based user studies. My interest in field deployments as a way of knowing stems directly from my use of multiple methods and the challenge of choosing which methods to use to answer specific questions. Like many HCI researchers who build systems, I dream of seeing my work in the hands of real users in real situations, but my experiences with deployments (e.g., (Newman, Ducheneaut, Edwards, Sedivy, & Smith, 2007; Zheng et al., 2010)) have led me to wonder about whether and when the pain of deploying is outweighed by the knowledge that might be gained. To help address this question for my own benefit and for the benefit of the HCI research community, I assembled a panel at HCIC 2011 that led to this book chapter.

John's Story: Field Deployments as the Intersection of Design and Ethnography

I see field deployments as an integral part of the design process. I was trained in the Design tradition to begin any design process with "need-finding" to identify what unmet user needs inspire designing something new. Yet, at the time, design schools did not offer any systematic methods for observing the world to identify these needs. Then I was introduced to Lucy Suchman's (1987) research on applying ethnographic methods to understand how users interact with technology. I saw these methods as a rich approach for need-finding by observing ways in which technology was not fully meeting the user's needs in current work practice. Studies of existing work practice are the starting point for designing and building some new technology to address user's unmet needs.

For me, field deployments are interesting because they sit at the intersection of designing something new and ethnographically observing it in use. Upon building a working prototype, placing that prototype into actual use in a field deployment exercises the design to help examine how it is meeting the user's needs. Field deployments provide an opportunity to validate the design, identify improvements for the next design iteration, and come to a better understanding of the user's activities that could lead to other design explorations.

Exercises

1. How are field deployments different from experiments? From quasi-experiments?
2. How does ending a field deployment compare with ending an engagement in Action Research? What can be done to mitigate the problems generated by the end?

References

Brush, A. J. B., Inkpen, K., & Tee, K. (2008). SPARCS: Exploring sharing suggestions to enhance family connectedness. *Proceedings of the Conference on Computer-Supported Cooperative Work (CSCW 2008)* (pp. 629–638).

Carter, S., & Mankoff, J. (2005). Prototypes in the wild: Lessons learned from evaluating three Ubicomp systems. *IEEE Pervasive Computing, 4*(4), 51–57.

Cherns, A. (1976). The principles of sociotechnical design. *Human Relations, 29*(8), 783–792.

Connelly, K., Siek, K. A., Chaudry, B., Jones, J., Astroth, K., & Welch, J. L. (2012). An offline mobile nutrition monitoring intervention for varying literacy patients receiving hemodialysis: A pilot study examining usage and usability. *Journal of the American Medical Informatics Association, 19*(5), 705–712. doi:10.1136/amiajnl-2011-000732.

Consolvo, S., Harrison, B., Smith, I., Chen, M. Y., Everitt, K., Froehlich, J., et al. (2007). Conducting in situ evaluations for and with ubiquitous computing technologies. *International Journal of Human Computer Interaction, 22*(1–2), 103–118.

Consolvo, S., Klasnja, P., McDonald, D. W., Avrahami, D., Froehlich, J., LeGrand, L., et al. (2008). Flowers or a robot army?: Encouraging awareness & activity with personal, mobile displays. *Proceedings of the 10th International Conference on Ubiquitous Computing* (pp. 54–63).

Cramer, M., Hirano, S. H., Tentori, M., Yeganyan, M. T., & Hayes, G. R. (2011). Classroom-based assistive technology: Collective use of interactive visual schedules by students with autism. *Proceedings of the SIGCHI Conference on Human Factors in Computing Systems (CHI '11)* (pp. 1–10).

DiMicco, J., Millen, D. R., Geyer, W., Dugan, C., Brownholtz, B., & Muller, M. (2008). Motivations for social networking at work. *Proceedings of the Conference on Computer-Supported Cooperative Work (CSCW) 2008*, San Diego, CA (pp. 711–720).

Erickson, T., Smith, D. N., Kellogg, W. A., Laff, M., Richards, J. T., & Bradner, E. (1999). Socially translucent systems: Social proxies, persistent conversation, and the design of "Babble". *Proceedings of the SIGCHI Conference on Human Factors in Computing Systems (CHI 1999)* (pp. 72–79).

Favela, J., Tentori, M., & Gonzalez, V. M. (2010). Ecological validity and pervasiveness in the evaluation of ubiquitous computing technologies for healthcare. *International Journal of Human Computer Interaction, 26*(5), 414–444.

Hart, S. G. (2006). NASA-Task Load Index (NASA-TLX); 20 years later. In *Proceedings of the Human Factors and Ergonomics Society 50th Annual Meeting* (pp. 904-908). Santa Monica: HFES.

Hayes, G. R. (2011, August). The relationship of action research to human-computer interaction. *ACM Transactions on Computer-Human Interaction, 18*(3), Article 15, 20 pages.

Hayes, G. R., Poole, E. S., Iachello, G., Patel, S. N., Grimes, A., Abowd, G. D., et al. (2007). Physical, social, and experiential knowledge in pervasive computing environments. *IEEE Pervasive Computing, 6*(4), 56–63.

Hazlewood, W. R., Stolterman, E., & Connelly, K. (2011). Issues in evaluating ambient displays in the wild: Two case studies. *Proceedings of the 29th International Conference on Human Factors in Computing Systems* (pp. 877–886).

Hirano, S., Yeganyan, M., Marcu, G., Nguyen, D., Boyd, L. A., & Hayes, G. R. (2010). vSked: Evaluation of a system to support classroom activities for children with autism. In *Proceedings of the SIGCHI Conference on Human Factors in Computing Systems (CHI '10)* (pp. 1633–1642).

Israel, B. A., Eng, E., Schulz, A. J., & Parker, E. A. (2005). *Methods in community-based participatory research for health.* San Francisco, CA: Jossey-Bass.

Khan, D. U., Ananthanarayan, S., & Siek, K. A. (2011). Exploring everyday health routines of a low socioeconomic population through multimedia elicitations. *Journal of Participatory Medicine, 3*, e39. 10 pages.

Le Dantec, C. A., & Edwards, W. K. (2008). Designs on dignity: perceptions of technology among the homeless. *Proceedings of the Twenty-Sixth Annual SIGCHI Conference on Human Factors in Computing Systems* (pp. 627–636). doi:10.1145/1357054.1357155

Lewin, K. (1946). Action research and minority problems. *Journal of Social Issues, 2*(4), 34–46.

Lincoln, Y. S., & Guba, E. G. (1985). *Naturalistic inquiry*. Beverly Hills, CA: Sage.

Morrow, D. S. (2002). *Transcript of a video history interview with Jeff Hawkins, Founder, Chairman and Chief Product Officer Handspring* [Interview transcript]. Retrieved from Computerworld Honors Program International Archives site: http://www.cwhonors.org/archives/histories/hawkins.pdf

Newman, M. W., Ducheneaut, N., Edwards, W. K., Sedivy, J. Z., & Smith, F. (2007). Supporting the unremarkable: Experiences with the obje display mirror. *Personal and Ubiquitous Computing, 11*(7), 523–536.

Prochaska, J. O., & DiClemente, C. C. (1982). Transtheoretical therapy: Toward a more integrative model of change. *Psychotherapy: Theory, Research, & Practice, 19*(3), 276–288.

Rogers, E. M. (2003). *Diffusion of innovations*. New York, NY: Free Press.
Scholtz, J., & Consolvo, S. (2004). Toward a framework for evaluating ubiquitous computing applications. *IEEE Pervasive Computing Magazine, 3*, 82–88.
Sengers, P., & Gaver, B. (2006). Staying open to interpretation: Engaging multiple meanings in design and evaluation. *Proceedings of the 6th Conference on Designing Interactive Systems (DIS '06)*, New York, NY, USA (pp. 99–108). doi:10.1145/1142405.1142422.
Stone, A. A., Shiffman, S., Schwartz, J. E., Broderick, J. E., & Hufford, M. R. (2003). Patient compliance with paper and electronic diaries. *Controlled Clinical Trials, 24*(2), 182–199.
Suchman, L. (1987). *Plans and situated actions: The problem of human-machine communication (learning in doing: Social cognitive and computational perspectives)*. New York, NY: Cambridge University Press.
Talmon, J., Ammenwerth, E., Brender, J., de Keizer, N., Nykänen, P., & Rigby, M. (2009). STARE-HI–Statement on reporting of evaluation studies in Health Informatics. *International Journal of Medical Informatics, 78*(1), 1–9.
Tang, K. P., Hirano, S. H., Cheng, K. C., & Hayes, G. R. (2012). Balancing caregiver and clinician needs in a mobile health informatics tool for preterm infants. In *6th International Conference on Pervasive Computing Technologies for Healthcare* (pp. 1–8).
Tang, J. C., & Isaacs, E. (1993). Why do users like video? Studies of multimedia supported collaboration. *Computer Supported Cooperative Work, 1*(3), 163–196.
Tang, J. C., Isaacs, E. A., & Rua, M. (1994). Supporting distributed groups with a montage of lightweight interactions. *Proceedings of the Conference on Computer-Supported Cooperative Work (CSCW) '94*, Chapel Hill, NC (pp. 23–34).
Tang, J. C., Matthews, T., Cerruti, J. A., Dill, S., Wilcox, E., Schoudt, J., et al. (2009, February). Global differences in attributes of email usage. *International Workshop on Intercultural Collaboration (IWIC 2009)*, Stanford, CA, (pp. 185–194).
Tentori, M., & Favela, J. (2008). Collaboration and coordination in hospital work through Activity-aware Computing. *International Journal of Cooperative Information Systems, 17*(4), 1–30.
Thomas, J. C., & Kellogg, W. A. (1989). Minimizing ecological gaps in interface design. *IEEE Software, 6*(1), 78–86.
Venolia, G., Tang, J., Cervantes, R., Bly, S., Robertson, G., Lee, B., et al. (2010). Embodied social proxy: Mediating interpersonal connection in hub-and-satellite teams. *Proceedings of the SIGCHI Conference on Human Factors in Computing Systems (CHI 2010)* (pp. 1049–1058).
Whittaker, S., Matthews, T., Cerruti, J. A., Badenes, H., & Tang, J. C. (2011). Am I wasting my time organizing email?: A study of email refinding. *Proceedings of the SIGCHI Conference on Human Factors in Computing Systems (CHI 2011)* (pp. 3449–3458).
Zheng, K., Newman, M. W., Veinot, T. C., Kim, H., Meadowbrooke, C. C., & Perry, E. E. (2010). Using online peer-mentoring to empower young adults with end-stage renal disease: A feasibility study. *Proceedings of the American Medical Informatics Association Annual Symposium*, Washington, DC, USA.

Science and Design: The Implications of Different Forms of Accountability

William Gaver

Fools rush in where Angels fear to tread.

Alexander Pope

When I was working towards my PhD in psychology and cognitive science, I ran a series of experiments investigating whether people could hear the length and material of struck wood and metal bars. My curiosity was motivated by J.J. Gibson's (1979) ecological theories of perception. If, as he argued, our visual perception has evolved to "pick up" information about the world conveyed by the structure of light, then, I surmised, our hearing might well be attuned to auditory information about sound-producing events. In my pursuit of an experimental demonstration of this, I spent months finding just the right kinds of metal and wooden bars, experimenting with recording conditions to capture just the right set of sounds, tinkering with experimental instructions and response scales, and running numerous "pilot studies". Finally, when I had everything working well, I collected my experimental data and spent more months trying out different analysis methods until I found several that seemed to give clarity to the data—and finally, the experiment was done.

Writing up the study, I used the canonical structure for reporting experiments. I set the scene both theoretically and in terms of related work, using that to motivate a set of hypotheses, describing my methods, stimuli and procedure, and then reported the data and discussed how they reflected on my initial hypotheses. What I did not do—of course!—was talk about all the work done to achieve the final data set: the shopping I did in specialist hardwood stores, the improvising of foam mounts that would let the bars sound when struck, the ways I tried to get participants to listen to the right things, and so on. Instead, I told the story the way I had been taught, as a linear narrative from theory to experiment to data and back to theory, in which each step was logically connected to the previous ones and to those that followed.

W. Gaver (✉)
Interaction Research Studio, Goldsmiths, University of London,
Lewisham Way, London SE14 6NW, UK
e-mail: w.gaver@gold.ac.uk

J.S. Olson and W.A. Kellogg (eds.), *Ways of Knowing in HCI*,
DOI 10.1007/978-1-4939-0378-8_7,

Flash forward 20 years, and I am a designer working on another project. As part of a larger consortium that included computer scientists and sociologists, a team from my studio—itself quite interdisciplinary—developed a system called the Local Barometer and deployed it to a volunteer household. This involved installing an anemometer in the back garden so we could measure wind speed and direction outside the house, and using this to control an algorithm that searched for online advertisements originating upwind from the home. Text and images from the advertisements we found were displayed on a series of six small devices designed to be positioned on various shelves, racks and tables around the home, after some processing to remove overtly commercial references, emphasise the resemblance to poetry, and adjust aspect ratios. The notion was that the system, which had been inspired by a wide range of influences, might raise awareness of the sociocultural landscape around the home—but we were not committed to this idea either as a hypothesis or goal; instead we treated the notion as a potentially disposable guide for our thinking about the design.

Once we had everything set up and running in the household, we gave our primary contact, R, a "user manual" and explained how the system worked. But we avoided telling him about how we thought he or his friends might use it or what our ideas were in developing it, since the point of the exercise was to see how they would interpret this situation on their own, without our help. Over the following month, we used a variety of means to see what R had made of the Barometers. Detailed reports were made by an ethnographer on our team, who visited the house, observed how R interacted with the system and had many long conversations with R about it. Another source of information was unexpected: the Barometers had a technical flaw (faulty garbage collection in the operating system of the mobile telephones we used for their implementation), which meant that they had to be rebooted every few days. R eventually learned to restart the devices himself, but until then our regular "service" visits provided opportunities for informal chats about the devices that seemed particularly revealing because their ostensible purpose had nothing to do with assessment. Finally, we captured yet another perspective by hiring a professional filmmaker to make a documentary video about R's experiences with the devices. To ensure what independence we could, we did not tell the filmmaker about the devices or our intentions for them, but let him learn about them from R himself. Moreover, we were never present during filming, and explicitly told the filmmaker that we did not want a promotional piece, but instead his own potentially critical account.

Characterising Science and Design

In many ways, the two projects I have just described are quite similar. In each case, I was involved in devising and implementing a physical situation (vibrating bars, and the Local Barometer), which involved a great many pragmatic and exploratory activities. In both cases too, what I made was influenced by, and meant to be informative to, a body of ideas about people and the world (ecological psychology in the first case, designing for ludic engagement[1] in the second) that served not only to describe existing things but also to suggest new avenues for exploration. Also in both cases, I created the physical situation in order to put it before people unconnected with my profession (the "participants") to see what they would make of it. Finally, in each case I pursued these activities as a form of research—in other words, I did what I did to learn something new, and presented an account of the process and results to an academic research community (see Gaver et al. 2008; Gaver 1988).

Yet the two projects were also different in ways that I will suggest are important. The impact sound experiments were motivated by the possibility of applying Gibson's thinking about light and vision to questions concerning sound and hearing, not in an analogical or metaphorical way but as a logical extension of his analysis to a new domain. In contrast, the Local Barometer was inspired by a wide range of influences, all helping to shape the final result but without the closely linked reasoning that led to my recordings of impact sounds. Similarly, I had fairly specific hypotheses about the impact sound experiments: I expected, on the basis of theory and my analysis of sound-producing events, that people would be able to hear both the material and length of struck bars with a good degree of accuracy. In contrast, our expectations for the Local Barometer were much more nebulous—we hoped that people would find the system engaging, and told our stories about sociocultural texture, but in reality we had little idea how people might use or think about the system in their day-to-day lives. Nor did the vagueness of our expectations worry us: on the contrary, the prospect of inciting surprising forms of engagement was what motivated the study. In addition, although I constructed the apparatus and thus the sounds used in the impact sound experiment, they were interesting precisely because they were representative of phenomena that are wide-spread and well-known, and in that sense there was nothing new about them at all. In contrast, the Local Barometer was interesting precisely because it *was* novel: to our knowledge, it represented a form of electronic threshold between the home and its local environment that had not existed before.

In this chapter, I want to explore the differences between doing these kinds of projects—which I take as typical of research through science on the one hand, and through design on the other—in more detail.[2] To be sure, I am mindful of the perils of trying to characterise science or design as if either were a unitary endeavour. After all, disciplines that identify themselves as branches of the sciences range from particle physics to library sciences, and involve vastly different mixtures of quantitative and qualitative theory, experimentation and empirical observation, taxonomic classification, procedural know-how, and long apprenticeships. Equally, activities self-identified as design vary from those that rely explicitly on individual and group

[1] Ludic engagement refers to forms of interaction that are not utilitarian or task-oriented, but exploratory, provisional and curiosity-driven: playful in the broadest sense (see Gaver, 2009).

[2] Many others have discussed whether and how design and science are distinct approaches, as well as whether they should be or not. I do not present a survey here, but see e.g. (Cross, Naughton, & Walker, 1981; Louridas, 1999; Schön, 1999; Cross, 2007; Stolterman, 2008; Gaver, 2012; and particularly Nelson & Stolterman, 2003).

creativity to so-called design science, and practices ranging from work done directly for commercial clients, to that done in the design departments of large organisations, to entrepreneurial work with no client other than eventual buyers, to practices verging on the artistic whose "clients" might include galleries and collectors. I do not want to argue about which of these constitute "real" science, or "real" design. Instead, I appeal here to design and science as categories identified not by a set of definitional criteria, but by features that each tend to have in common. From this point of view, a given activity is counted as a science, or as form of design, depending on its similarity to canonical examples of each. What I want to do here is characterise what I think are fundamental distinctions between science and design identified in this way. Given my appeal to a definition of science and design based on family resemblance, the test of these distinctions is not whether they hold for all examples of self-defined science and design, but rather whether they are recognisable for the kinds of activities we most readily identify as one or the other—a matter which readers will have to decide for themselves.[3]

With those provisos, it is time to rush in, and discuss the differences between science and design.

A Matter of Accountability

Reflecting on my experiences working as a scientist, and later as a designer, a core difference in pursing research from these traditions has to do with the issues that must be addressed in defending each kind of work from the criticisms and questions of colleagues.

Presenting scientific research such as the impact sound study, I would expect to be asked a series of questions, all of which amount to variations on a single one: "how do you know what you say is true?" These are questions about process, including conceptual and practical moves and the linkage between the two. How did my experiments operationalise the theory I was testing? Did I control for any potential confounds? How many participants were there? Were the stimuli presented in random order, or perhaps using a Latin Square design? Would an alternative explanation render my results inconclusive? And so on. How interesting my results were—whether they were counter-intuitive, or shed new light on a phenomenon or

[3] To make matters worse, I am purposely not distinguishing design in general from "research through design" in what follows. Such a distinction is neither simple nor productive, in my view. For instance, people have suggested that research through design is different from "real" design in not having a client, or clear problem to solve. But researchers do have their clients, including research funders, academic audiences, and the people who might encounter their work, and these are not so different from the managers, colleagues, other departments, purchasers and end users that "real" designers have to please. Equally, many "real" designers do not solve problems so much as they explore new configurations of materials and form in an endless conversation with each other and the surrounding culture, while practitioners of research through design commonly *do* address problems, such how to reflect new aspects of human experience.

theory, or simply displayed a pleasing sense of elegance and order—were secondary concerns. To be sure, the topicality, novelty or potential benefits of a given line of research might help it attract notice and support, but scientific research fundamentally stands or falls on the thoroughness with which activities and reasoning can be tied together. You just cannot get in the game without a solid methodology. The most astonishing finding is without scientific merit if its methodology is suspect. Conversely, the most pedestrian result is scientifically valid if it can be shown to be the result of a meticulous approach.

The situation is different for design. The basic question here is "*does it work?*" The issue of whether something "works" goes beyond questions of technical or practical efficacy to address a host of social, cultural, aesthetic and ethical concerns. Is it plausible to think that people will engage with a system that is not guided by predefined tasks? Can you really scrape information from the Web that way? Does the form and colour fit the context, with the appropriate functional, social, cultural and aesthetic connotations? Does the design tend to stereotype the people and places it addresses? To be sure, questions of process might enter the discussion—how did you come to think of your user group in such a way? Why did you choose to use that form of input?—but such questions are not grounds in and of themselves for judging a design successful or unsuccessful. Instead, they are asked to elicit answers providing resources for better appreciating a design's intentions and plausibility. They may help critics to "get it", perhaps by allowing interpretation from other perspectives, or by reassuring clients that an idea responds to needs of potential customers—or they may fail to help a design that is slow to convince. Still, it is perfectly possible, even common, for a compelling, eye-opening design to emerge from a process that is idiosyncratic and even a bit mad. We talk of "inspired" ideas with more enthusiasm than we talk of "informed" ones. And successful designs validate new methods and conceptual perspectives, rather than the other way around. In design, even the most meticulous methodology will not redeem a bad design, and even the most hare-brained processes will not ruin a good one.

The distinct sorts of questions asked of science and design manifest the different kinds of *accountability* that apply to each—that is, the expectations of what activities must be defended and how, and by extension the ways narratives (accounts) are legitimately formed about each endeavour. Science is defined by epistemological accountability, in which the essential requirement is to be able to explain and defend the basis of one's claimed knowledge. Design, in contrast, works with *aesthetic accountability*, where "aesthetic" refers to how satisfactory the composition of multiple design features are (as opposed to how 'beautiful' it might be). The requirement here is to be able to explain and defend—or, more typically, to demonstrate—that one's design *works*.

In suggesting that science is epistemologically accountable, and design aesthetically accountable, I do not mean to suggest that other concerns are completely irrelevant to these pursuits. As I have suggested, the topicality, intrigue and potential impact of a given scientific research project can have a huge influence on whether it is lauded at conferences and attracts multimillion dollar funding, or languishes in the back corridors of some university. But before questions of timeliness, interest and relevance even arise—the prerequisite for them making any sense at all—the scientific validity of the project must be established. The most eloquent narrative about potential impacts (increasingly demanded by funding agencies) will not redeem a proposal judged to be unscientific by reviewers; equally, the most feted, faddish, and even effective diet plan may be derided as unscientific if there is insufficient evidence to validate it. The epistemological accountability of scientific projects is *essential*, while interest and impact are not definitional of science. And the converse is true for design's accountability to "working": the ability to talk a convincing game about the mind-blowing conceptual flights and hundreds of person-hours behind a given design may help draw attention to it, but this will not make it a valid design if it is incoherent, unfinished or implausible. Its aesthetic accountability—its ability to integrate functional, formal, material, cultural and emotional concerns (for instance)—is *essential*, while arguments based on process are, at best, secondary.

Mechanisms of Progress

The different systems of accountability for science and design—the need to be able to defend one's knowledge, on the one hand, and that one's productions work, on the other—parallel the different strategies the two endeavours use to proceed.

For science, the logic of day-to-day research—what Kuhn (1970) called "normal science"—revolves around an iterative process of using theory to understand observations of the world, and observations to test and extend theory. Theory, usually taking the form of an ontology of entities and the causal connections amongst them, embodies an explanation of phenomena of interest and potentially allows their prediction. There are two basic pathways to theory expansion. The researcher may gather observations of a body of phenomena that appears theoretically salient, or which simply happens to seem interesting. Gathering repeated observations allows induction of new hypotheses that may modify relevant theory. The more stereotypically "scientific" route, however, goes the other way, relying on theory's nature not only to explain phenomena and their relations that have already been observed but also, through its mechanism of entities and connections, to have implications about things that have not yet been seen. Where those implications are not so close to established fact as to be axiomatic, or where the theory is unclear in its implications (and note that identifying either condition relies on the scientists' experience and skill) a set of hypotheses may emerge about a possible state of affairs suggested by the theory. So, for example, thinking about how ecological psychology might be applied to auditory perception led me to hypothesise that people might be able to hear the physical attributes of sound sources. In order to test hypotheses such as these, they need to be operationalised in the form of a set of experiments or observations that simultaneously reflect the hypothesis and can yield unequivocal data. Operationalised hypotheses allow salient phenomena to be assessed empirically to see whether they fit the theory. This typically involves the situating of general hypotheses in particular contexts (e.g. specifying that everyday listening might apply to hearing attributes of impact events), the contrivance of experiments or other data-collecting activities, and the analysis of data, all deployed not only to determine whether the observed phenomena agree with the theory but also to elaborate the theory or even modify it.

There is a set of core values that characterises the pursuit of scientific knowledge, whether through induction or hypothesis testing, which the methods developed for pursuing knowledge in these ways seek to realise. Perhaps most important is that scientific knowledge should be *replicable*, able to be reproduced by others, both to allow it to be built upon and as a fundamental guarantee of its epistemological accountability. This means it should be *objective*, with a truth-value independent of individual experimenters. It should be *generalisable*, in the sense that scientific phenomena are expressed and understood abstractly enough that instances of them can be found in a wide variety of circumstances. Scientific theory is ideally *causal*, explaining the connections amongst related phenomena as a matter of necessity rather than correlation or coincidence. Theory should not only *explain* phenomena that have already been observed but also *predict* new ones. And so on. Perhaps the most essential value is *definiteness*. Being able to say *what* you know—precisely, and ideally quantifiably—and *how* you know, and *when* or under what conditions what you know is known to be true—these are the hallmarks of science.

Of course, as those versed in the sociology of science, science and technology studies, and similar fields have shown, these values are not simply given or received; they have to be *achieved* in the doing of science. Latour (1987), for instance, points to the "Janus faces" of science: if one looks at science after the fact, then the account above may fit, but if one looks at science as it is happening, things look very different. As numerous empirical studies have shown, scientists do not proceed in any simple mechanical way from theory to hypotheses to tests to conclusions. As my own introduction illustrated, a huge amount of work behind the scenes is done to produce the simplest experimental demonstration. Moreover, a great deal of post hoc rationalisation goes into aligning empirical data, hypotheses and theory. Scientists rarely or ever explain their methods sufficiently to allow replication, and anyway few scientists ever bother to replicate work done elsewhere. On top of that, the success of any given scientific endeavour will depend on the way that (from the perspective of the received account of the scientific method I gave above) "extra-scientific" agencies can be marshalled: for example, whether or not a given line of research will be supported by employers and funding bodies, find sympathetic reviewers and take a significant place in webs of citation depends on the technical resources to demonstrate its merit, as well as its authors' reputations, social-professional networks and potential for reciprocal influence (c.f. Latour, 1987). In the end, the so-called "scientific method" outlined above is an *achievement*, an account hewn from processes that are far more complex and embedded in the pragmatic politics of science than it admits.

Nonetheless, the core values of replicability, objectivity, generalisability and so on remain central in this process, because they serve to *guide* the efforts, to provide a goal for what should be achieved. Even if the "scientific method" is a simplification of what science in action is actually like, it is a simplification that is upheld as an integral ritual in the doing of science. Whether or not science actually proceeds according to the logic from theory to hypothesis to data to analysis to theory, this is how it is *presented*, in academic articles, in conferences, in job talks and in funding applications. It was not by accident that when I wrote up the struck bar experiments, I omitted details about how they came to be—the shopping for wood, the cutting and sanding, the fabrication of a mechanism to strike them—and the way they came to be analysed—playing with the parameter space of a time-varying Fourier analysis, the different 2- and 3D visualisations I tried, and so forth. Nor did I leave these details out to save space or avoid boring readers. No, I did not report those details because they were irrelevant to the clear causal flow between logic, materiality, events and their interpretation I needed to establish, and thus needed to be omitted from the project's official history muddy lest that flow be muddied. For it is in terms of that stream of logic that scientific research is formally *assessed* by reviewers. Conference committee meetings may give rise to any number of discussions about how boring or wrong-headed a piece of scientific research is, but when it comes time make a formal decision then methodological weaknesses, not aesthetic (or cultural, or political) shortcomings, are the resources panel members use to justify rejection. Funding agencies and recruitment committees may turn down an applicant on the grounds that a given line of research is outside their scope, or that its impact will be minor, but the surest path to rejection is by failing to establish epistemological accountability, whether because of faulty reasoning, a misjudged method or a simple lack of clarity about the logic and activities used to pursue a topic. Researchers know this, of course. They know they must outline a research plan that follows scientific logic if they want to gain funding, and they know they must present a completed piece of research according to the logic of science if they want to be published. And because of this, no matter how much extraneous backstage activity may go unreported, and no matter how post hoc the account may be, then unless they are out-and-out frauds there will be, running through scientific researchers' day to day research activities, the skeleton of the scientific method presented above. As a post-hoc rationalisation, the logic of scientific method may seem to be a fiction, or even a lie, but if so it is a lie by omission not commission, and a fiction that guides and constrains real scientific activity.

For design, the logic of activity is different. The designer encounters a world, which crucially includes designed artefacts as well as people and physical phenomena, and has the job of fashioning something new that works for that world. A significant step on this journey is the development of a proposal, or proposals, about what might be built. Proposals may vary widely in their specificity, from evocative and unrealisable sketches, to abstract representations of intention, to relatively complete specifications or scenarios. In each case, the role of design proposals is both to create and constrain. On the one hand, they suggest things that might be made, things that have not hitherto existed. Simultaneously, their collection implicitly limits the myriad possibilities for design offered by a given situation by focusing attention on one or a few more-or-less concrete configurations. For instance, the Local Barometers came about when, after some time exploring ideas for a project in which we knew, simply, that we would develop new technological artefacts for the home, one of us started exploring notions of information being carried into the home by the wind. Once a proposal is agreed on, this serves as a brief for further elaboration and refinement of what the artefact *will* and *will not* be. Typically this involves a combination of progressively more focused design explorations and proposals, including what Schön (1983) calls a "conversation with materials", as a myriad of decisions are made (Stolterman, 2008) and the artefact that will actually be built is resolved. Finally, the finished artefact is assessed through some combination of critique, commercial success or failure, and empirical study of what people do with it and how it might affect their lives, until accounts about it settle down, and it is ready to take its place in the world and its artefacts to serve as a context for new designs.

It is tempting to see parallels between the basic mechanisms of science and design progress described here. Are not design proposals like hypotheses, suggesting possibilities that might be investigated—in the case of a scientific hypothesis, the possibility that a certain supposition may be true; in the case of a design proposal, the possibility that a certain artefact (or kind of artefact) might "work"? And are not products like experiments, contrived to allow empirical test of the conjectures embodied by design proposals? For that matter, are not scientific experiments themselves designed products, artefacts that must be invented and refined just like a new chair or an interactive website? Of course they are—and yet, like any analogy, the focus on similarities between science and design obscures as well as reveals. Worse: the analogy of science and design is positively dangerous, because it obscures the very features that give each endeavour its specificity and potency. For where scientific hypotheses sprout from a ground of theoretical or empirical confidence, design proposals are inventions that spring up under the influence of a potentially unlimited number of influences that include, but are by no means limited to, theoretical frameworks or empirical observation. Where scientific hypotheses are uncertain because they project tentatively from truths confidently held towards those that are conjectural, design proposals are vague because they are tools for imagining things that do not yet exist. Scientific studies are contrived to *control and hold apart* the factors that potentially cause phenomena of interest; designs are arranged as configurations in which elements *merge and blend* like ingredients in a recipe. Finally, scientific activities seek to discover, explain and predict things that are held to pre-exist in the world, whereas design is fundamentally bent on creating the new.

Design and the New

Science uncovers what exists, and design creates the new. This might seem the most profound difference between the two endeavours, and given how many other commentators suggest that this is the case, it may seem strange that I have not highlighted it before now. And indeed, this distinction does seem to underlie much of what is different about science and design. Science is based on realism, a deep assumption that things exist apart from our thinking of them, and further that they interact in non-arbitrary ways, with the complexity we normally experience resulting from a smaller number of underlying principles. The goal of science is figure out how the world works by dismantling its complications, teasing out its separate elements, and figuring out how they interlock to operate together. The fundamental assumption of design, in contrast, is that new things can—and should—be made. The goal of design is to make the world work in new ways by producing new complications, assembling elements in new ways, and crafting them to work together. Design can work with the world as found: it does not have to concern itself with realism in any deep way, nor does it have to get to the bottom of how things really work. It may do so, of course, and often designers are adept at finding radical new ways of understanding materials, people and processes in the course of their work. But this is not a requirement for good design, because design is not responsible for explaining the world as it is, but for producing new artefacts that work.

Design's concern for creating new things leads to a different set of values than those for science. Good science is characterised by replicability, objectivity, generality and causal explanations. Successful design artefacts, in contrast, are characterised by *working*—by functioning efficiently and effectively, by solving problems neatly or reconfiguring them insightfully, by using materials and production processes in elegant ways, and so forth. Beyond this, some designs—perhaps many of those that succeed in the ways just described—embody other values that make them as powerful in opening new understandings of the world as scientific discoveries are. Think of Durrell Bishop's answering machine (Crampton-Smith, 1995), in which messages are represented by RFID tagged marbles, allowing them to be manipulated, relocated and used with other devices. This opened the world of tangible computing by showing how the affordances of the physical world could be harnessed to communicate those of the digital one. Or remember the way the iPod superseded portable media players not only through its elegant product design but also by its ability to merge and detach from an online world of commercial and non-commercial media (Levy, 2006). Or consider the Brainball (Hjelm and Browall, 2000), in which winning a contest requires being more relaxed than one's opponent, simultaneously demonstrating neurological interaction and playfully subverting competition to create an entertaining and thought-provoking game. Such designs have *individuality*: they possess their own character, which is not only original but also integrated and with a clear personality or style. They *resonate*, reminding and energising and speaking to a wide and potentially incommensurable variety of influences, issues, artefacts, phenomena and perspectives, both natural and cultural. They are *evocative*, stimulating new possibilities for design, whether similar, compatible, extended or even counterbalancing. And perhaps most of all, they are *illuminating*, reaching out beyond their immediate functionality to suggest new ways to perceive and inhabit the world.

Designs' values are deeply bound to its fundamental undertaking of realising new possibilities, but it is only this in combination with its aesthetic accountability that distinguishes it most surely from science. Design's concern with the new, and science's with the existing, may distinguish them from each other, but it is not enough to distinguish them from other kindred disciplines. Literary fiction, poetry, the arts, documentary filmmaking, and at least some strands of the humanities can all offer insights into "what is" while eschewing scientific methods and embracing aesthetic accountability. Engineering and other forms of applied science, on the other hand, routinely use scientific theories, methods and findings to construct "what might be" using a form of epistemological accountability ("how do you know it won't fall down?"). It is the *combination* of accountability plus orientation to what exists that best captures the differences between science and design. And in my view, it is their contrasting accountability that allows their most characteristic—and productive—differences to be best appreciated.

Design Methods and Productive Indiscipline

The reason I suggest that design's aesthetic accountability is more useful as a focus than its orientation towards the new in understanding how it operates differently from science is because of the methodological implications of that accountability. Sciences' epistemological accountability, its commitment to being able to answer questions about how one knows one's assertions are true, constrains its methods towards those that tend to be empirical, specified in advance, standardised, replicable, independent of the observer and (ideally) quantifiable. Design's aesthetic accountability, in contrast, means that its methods do not necessarily have to have any such characteristics. They may, of course—aesthetic accountability does not imply that scientific methods are out of bounds—but equally, design may thrive on information that is fictional as well as factual, and on reasoning and activities that are improvised, unrepeatable and highly personal. Design methods often exhibit a *productive indiscipline* thanks to their freedom from epistemological accountability. That is, design processes are not bound to particular theoretical or methodological rationales, but can borrow from all disciplines or none. Even more subversive, from a scientific point of view, all this can be left unclear. Knowing whether something is true or not, whether things have changed or whether a view is idiosyncratic or widely shared may simply not matter when it comes to design. On the contrary, in some cases a lack of knowledge (and meta-knowledge) leads to just the sort of conceptual space in which imagination seems to thrive. We might say that, where science relies on epistemological *accountability*, design can often work from a kind of epistemological *ambiguity*.

Many of the design methods used in my studio illustrate the kind of fluid flow between certainty and speculation that design allows. In this section, I discuss some of these methods, organised according to the typical project trajectory we use for describing our projects. In this account, design projects typically progress through four stages linked by different sorts of activity. Most projects start with the identification of a *context* for design, which is elaborated, specified and investigated through further work. This informs the development of numerous *proposals* for what might be made, which act as landmarks to create and expand a space of possibilities for design. A turning point is reached when a specific direction is chosen, at which point activity turns towards the refinement of a realised *prototype*. Finally, this is assessed using various means to reach *conclusions* about its success and, more importantly, the lessons to be learned from the project. This trajectory is typical of many, many descriptions of design process, and like most it implies a kind of waterfall model in which stages are encountered sequentially (with the possibility of iteration). In reality, design projects rarely proceed in such an orderly manner. Important insights from contextual studies may seem to disappear from the following design proposals, only to become salient late in the development of a prototype. Proposals may be inspired by seemingly unrelated sources, or may spin off to different projects, or become the context for new proposals. Development work might transmute to contextual research. And so on. Nonetheless, this trajectory is not entirely fictional—our projects do tend to proceed through these phases in this order, and moreover it plays a role in how we attend to their progress—and thus for the sake of organisation I will describe some of our approaches in the sequence it suggests.

Exploring Context

Our design projects are almost always set in an explicit or implied context, and early design activities will usually be concerned with better understanding the people and situations for which we are designing. This involves elaborating and enriching information about the setting, but can also require particularising or specifying examples of a context that is initially only broadly or vaguely defined. Considered instrumentally, our goal in this phase is twofold: first, to build an understanding of the setting that is rich enough to allow design ideas to be checked for plausibility and likely problems, and second, to find inspiration for design directions. These two objectives can pull in different directions. On the one hand, trying to ensure that designs will be appropriate and fit for purpose suggests gathering as complete and veridical account as we can. Inspiration, on the other hand, often comes from particularly striking facts about the context, or idiosyncratic views on it, even if these are unrepresentative or unconfirmed. Balance is crucial: too little contextual information can lead to free-floating speculation, but too comprehensive an account can smother creative ideas and lead to predictable responses. Thus in our approaches we tend to gather a great deal of eclectic material about the contexts for which we design, but to appreciate both gaps and questionable perspectives as leading to the kind of interpretative speculation that leads naturally to invention.

For instance, the Local Barometer described at the beginning of this chapter was developed for a project initially defined as exploring how merging the digital and the physical could produce new technological products for the home. In order both to enrich and focus the topic, our initial external research included academic publications from disciplines spanning the sciences and engineering, psychoanalysis and social science, the humanities, cultural studies and philosophy. We looked to examples from design and from the contemporary arts, including artists such as Sophie Calle (Calle & Auster, 1999), Ilya Kabakov (1998) and Gillian Wearing (e.g. Ferguson, De Salvo, & Slyce, 1999) who make social and cultural interventions, or those such as Gregory Crewdson (Crewdson & Moody, 2002) and Gordon Mata-Clark (Diserens, 2003) who offer surprising new views of the domestic. We looked to a range of sources from the popular or tabloid press, dealing with topics such as journalists who search peoples' trash for intimate information, as well as niche publications such as pamphlets about how to hide money and weapons in the home (e.g. US Government, 1971). Taken together, these resources allowed us to amass a multifaceted appreciation of "the home" in which academic respectability was less important than developing a richness and narrative depth that we felt nurtured our design.

In addition, we ran a Domestic Probes study with 20 volunteer households from the greater London area to uncover orientations and activities that might undermine any stereotypes we might bring to the project. The Probes, an approach invented by Tony Dunne and myself for an earlier project (Gaver, Dunne, & Pacenti, 1999), are, from my point of view, all but defined by their unscientific nature. We recruited volunteers by advertising in a variety of popular publications including a local newspaper, a publication of classified advertisements, and a magazine for the "horse and hound set", taking people on a first come, first served basis and making no attempt to achieve demographic representativeness (though our volunteers ended up represented a wide range of ages, backgrounds and socioeconomic status). We gave each household a Probe package containing a dozen tasks designed to be intriguing, but seldom clear about what information we were asking for or how the results might be interpreted. These included, for instance, a disposable camera, repackaged to remove it from its commercial origins, with requests for pictures such as "a view from your kitchen window", "a social gathering", "the spiritual centre of your home", and "something red". A drinking glass was included with instructions indicating that it should be held to the ear to listen to interesting sounds, with observations to be written directly on the glass with a special pen enclosed in the package. Pages with graphics including, for instance, a cricket game, a wooded slope, and Dante's Heaven and Hell were provided for people to diagram their friends and family circles, a knowing perversion of a traditional social science approach (e.g. Scott, 2000). Finally, a small digital recorder was repackaged with instructions to pull a tab when waking from a vivid dream, at which point a red LED lit up and the participant had 10 s to tell us about the dream before the device shut off, offering no facility for replaying or editing the dream, only the choice of whether or not to return it to us.

Tasks such as these provide a puzzle to participants about how to react, and their responses—hundreds of photographs, notes and drawings—defy easy summary or analysis. Our probes are purposely designed this way, not least to disrupt assumptions for all of us about the roles of researchers and "subjects".[4] Moreover, we emphasised their atypical nature with reassurance that not all materials need to be

[4] Others design "probes" to avoid such disruption; see Boehner et al. (2007).

completed, and by suggesting that participants should feel free to tell us stories—or simply lie—if they wished. By blocking expectable lines of questioning, and even approaches to answering, the Probes force both participants and us to struggle for communication, and in so doing produce surprising angles and perspectives on our participants. At best, the returns achieve a balance between inspiration and information. They are fragmentary, elusive and unreliable, but they are also real, offering numerous small glimpses into the facts of peoples' lives. Taken together, their ambiguity—their very lack of scientific validity—evoke for us as designers the kind of grounded curiosity, empathy and conjecture that we find useful in our work.

Developing a Design Space

A natural and desirable (though sadly not inevitable) consequence of an evolving understanding of a design situation is the emergence of speculation about what might be made that will work in that situation. Like most designers, we externalise ideas through sketching, but as we move towards sharing them with one another we usually develop more finished design proposals using combinations of collage, diagrams, computer drawing and rendering, and written annotations. Once we have amassed enough of these—and 50 is not unusual for a given project—they are often collected into a workbook, arranged into a set of post-hoc categories to indicate the shared themes that are beginning to become clear.

One might imagine that the proposals would be based fairly directly on the returns of Probe studies. Having collected Probe returns, we might use them to draw up a contextual account featuring a set of key issues, recommendations or requirements, which could lead relatively directly to a set of designs. This is not what happens. Not only are Probe returns difficult to analyse or summarise, but we also prefer to avoid mediating representations of the returns, or summaries of contextual research in general. Instead, proposals emerge seemingly spontaneously, and may reflect any number of influences including ones that seem completely unrelated to anything that has gone before. This does not suggest that Probes are irrelevant or a waste of resources. They can help us better understand the context for which we are designing, help us in assessing whether it is plausible that given proposals will work, and even inspire ideas relatively directly. But they are not *responsible* for doing so. Freed from epistemological accountability ("how do you know this is the right design proposal?") we can pursue ideas without worrying about explicitly justifying them with previous research.

Design proposals are seldom detailed or elaborate. Instead, they are often comprised of an evocative image or two, annotated with captions ranging from a few words to a few paragraphs. Rarely, if ever, do they include technological details or sequential scenarios of use, or even much in the way of detailed functionality. Succinct as they are, however, sketch proposals can be remarkably rich in pointing towards configurations of motivations, functionality, technologies, emotional or cultural qualities and the anticipated effects or experiences that make up a direction for design. Moreover, when gathered together as a workbook, collections of proposals allow a *design space* to emerge, making clear a bounded range of possibilities characterised by a range of dimensions we are interested in exploring. Individual proposals play a dual role in this process: they both represent (more or less) specific configurations for further development, but also, and often more importantly, landmarks in a space from which other ideas may be developed.

Similarly to the Probes, workbooks balance concrete factuality with an openness to reinterpretation. As externalisations of design ideas, and moreover ones which are presented slightly more formally than sketches, they have a reality that is relatively free from an identifiable authorial hand, and thus available for critique and change. Achieving this may require specifying aspects previously left unconsidered, or they may inherit the connotations of the resources used to create them (e.g. images used for collage, or the style of renderings). At the same time, proposals are often indicative rather than detailed representations. Collages may use images that hint at dimensionality, appearance and materials, while retaining enough of their (unconnected) origins to indicate they are not to be taken as literal representations. Renderings and illustrations are often diagrammatic, clearly leaving elements unspecified or unresolved. The basic concepts themselves are often "placeholders", gesturing towards design directions rather than specifying them. Thus proposals are deeply provisional, allowing a great deal of room for elaboration, change or development (Gaver, 2011). Because they are aesthetically rather than epistemologically accountable, they do not need to be part of a longer chain of argument from an initial setting to a final design, but need only "work" in the sense of suggesting potentially topical and compelling possibilities.

Refinement and Making

Workbooks are a means of developing understanding of what actually to make. After some period—often months—of developing a design space through contextual research, one or several collections of proposals, and associated technical experimentation, it becomes time to focus efforts around one or a few directions to take forward. This might involve the progressive development of an existing proposal, but often a new proposal will emerge which integrates and consolidates the thinking embodied in a number of other ones. At best, when this happens there is a kind of "audible click" as consensus quickly forms around an agreed direction for development, and other proposals and possibilities are deferred or fade away. The new proposal serves as a design brief, and from this point efforts are turned towards detailing, refining, and making the new design.

The evolution of a design from a proposal to an actual artefact is, literally, a slow process of materialisation. At first the work tends to be symbolic in nature, involving tens or hundreds of sketches and later diagrams and CAD renderings. This is soon accompanied by physical explorations, as form models are made from cardboard or foam, and fabricated using rapid prototyping machines. Materials such as plastics or wood or metal are sourced and tested for their aesthetic and functional properties. Components such as displays or buttons are gathered and evaluated for their appearance and tactility, with many being abandoned and a select few retained. Computational and electronic experiments are performed, not in the scientific sense of testing hypotheses, but in the sense of trying out a set of arrangements to see whether they hold promise. New processes of making are explored and refined. Over time, larger and more complex configurations are constructed, as when a computer display is mounted in a cardboard form study, until the first working models appear and a final specification is finally drawn.

During this process, hundreds of decisions are made—in the case of the Local Barometers, for instance, to base the design on a partially deconstructed mobile phone, to use a variety of shapes to afford different placements in the home, to scroll text vertically, to use brightly coloured card over a plastic structure, to use separate devices for text and images, and so on—and the final design resolves as its features slowly become definite (c.f. Stolterman, 2008). Each decision embodies the designers' judgements about a potentially multitudinous range of concerns, from functionality and cost to emotional tone and cultural connotation. Moreover, each is made in context of the other decisions that have been or will need to be made, and is situated in the circumstances of development, including both those of the setting for which it is devised and those of the designers that make it. In the end, the final design, if well made, is the result of a tightly woven web of judgements that are contingent and situated, and shaped by an indefinite mix of practical, conceptual, cultural and personal considerations. Yet the result, a highly finished product, is an "ultimate particular" (Stolterman, 2008), as definite and precise as any scientific theory.

The practice of resolving a design from an agreed proposal to a finished product is an essential aspect of design, bringing into play the full range of expertise and skills of its designers. Nonetheless, this aspect of design is seldom reported in detail (though see Jarvis, Cameron, & Boucher, 2012), perhaps because the myriad of decisions involved are difficult to organise to provide a coherent account. It is through the process of making that a great deal of understanding—of a domain, of people, of conceptual issues—is both exercised and furthered. It is the product itself, however, that typically serves both as the report of that understanding and as the means by which it is assessed.

Assessment and Learning

As decisions combine with one another to form a complete and highly finished design, it is as if an elaborate theory is constructed, embodied by the emerging design, about the important factors and configurations in designing just such a device in just such circumstances, a theory as definite as the physical components used to construct it. Moreover, the design will imply, with a varying degree of specificity, how people are expected to engage with it and what their resulting experiences may be. This is not like a scientific theory, however. Instead it is dependent and localised.

It does not arise by necessity from any preceding contextual research or design space explorations, though of course it would be likely to reflect them. Moreover, the "theory" embodied in a design is not articulated by it. Not only is it impossible to "read" an artefact unequivocally for its conceptual import, but designers themselves also may be unable to explicate the full rationale for their decisions, many or most of which are a matter of "feel" rather than explicit reasoning (c.f. Carroll & Kellogg, 1989).

Nonetheless, the "theory" of a new design begs to be tested by exposure to the people who might use it. For specifically targeted designs—a potato masher, for instance, or a word processer—laboratory based "user testing" based on scientific experimentation may seem adequate. Even in these cases long-term, naturalistic field tests may be better at uncovering the subtle aesthetic, social and cultural aspects of the experiences they offer; the ways they are talked about, displayed or hidden away, used and misused. In the case of the designs we produce in our studio, certainly, the best way we can find out what a design is really "for" is to allow people to use them in their everyday environments over long periods of time, since our designs are purposely left open to multiple interpretations. Deploying our designs allows us both to discover the questions we should ask about their use, and some of the answers to those questions.

Because this form of assessment does not involve the testing of specific hypotheses so much as the discovery of multiple possible forms of engagement, it benefits from a variety of views rather than a single summary judgement. Given that designs can be appreciated from a number of different perspectives, and that different people may find different ways to engage and make meaning with them—or fail to do so—multiple, inconsistent and even incompatible accounts may all be equally true. For instance, the Local Barometers were variously regarded as aesthetically intriguing artefacts for the home, as representing a dangerous form of subliminal advertising, as offering gifts, as unique and cutting-edge designs, and as annoying bits of broken electronics. To focus on one of these accounts over the others, or to amalgamate them without regard to the way they accumulate, combine and change over time and from place to place, would not produce a more general or abstract account, but one that flattens the experiences afforded by the designs.

Thus we use a number of tactics to gather multiple accounts, and invite distinctive perspectives to reach the full range of possible orientations. Our methods range in the degree of technical specialism they require. A great deal of information comes simply from informal chats with the volunteers who live with our designs, and especially those occasioned by unconnected activities such as routine maintenance or visits to document the devices in situ. For academic credibility, ethnographic observations and interviews often form the backbone of our assessment, with their mixture of empirical observation, interpretation and storytelling providing a coherent account of how people orient to and engage with the products we make, and some indication of the range of those engagements. (See chapter "Reading and Interpreting Ethnography") Finally, we often draw on the specialised expertise of "cultural commentators" (Gaver, 2007) who are independent of our Studio, and whose disciplines and institutional ties are independent of our own. I have already described how we commissioned a documentary film of the Plane Tracker. We have also solicited the interest of independent journalists to write stories of deployments, ideally to be published as an indication that they have been written for their purposes, not ours. We have hired a poet to write about two of our prototypes in their settings, with results that were sometimes opaque, and sometimes extraordinarily moving. From the mundane to the artistic, each of these forms of description provides a new point of view, a new approach into the ways that designs are experienced and used.

What Designers Know

What can designers claim to know during and after this process? Given the indiscipline that I argue characterises design processes, productive or not, what can we claim to have learned?

From a scientific point of view, the answer is "not a lot". The complex, idiosyncratic and interpretative nature of design means that there is little epistemological accountability in the results. The processes used in the course of design may be replicable, but the ways designers respond are not. Equally, an individual design artefact is of course replicable, but only after the fact, with a different history from the original (not least because of the original's existence): identical designs are seldom if ever produced independently, nor would designers want them to be (Fallman & Stolterman, 2010). Granted, certain design themes or tropes may be exercised repeatedly, but this is a much weaker and more contingent form of replicability, however, than that found in science. Replicability, a key characteristic of scientific knowledge, is largely unavailable through design.

Similarly, the understandings achieved through design research are of limited generality, or at least become increasingly dilute the more they are generalised. This contrasts with regularities found in the sciences—such as the law of gravity, for instance, or the 7 ± 2 limit of short term memory, or Fitt's Law—which remain equally specific over a wide range of domains and scales. Because design is a matter of integrating myriad considerations, any given abstraction tends to be situated and contingent, and alters as it is applied to new domains or new scales (c.f. Louridas, 1999). Moreover, design is in constant conversation with itself, changing the ground on which it operates, so that many of the approaches that have succeeded in the past will find that their motivating circumstances have changed or simply fallen out of fashion. The result is that design theories tend to be indicative and aspirational, rather than explanatory of stable phenomena. This does not necessarily make them ineffectual, but the "knowledge" they embody is of a different order than scientific knowledge.

What design *does* have to offer, what we *can* know, are the artefacts that design produces—not only the finished designs themselves but also the probes and probe returns, the sketches and workbooks, the technical experiments and form models. These are real, tangible things that have the definitiveness and detail that eludes attempts to conceptualise design. As the result of the many judgements that designers make to produce them, they embody a host of ideas about the conceptual, material, social, technical and philosophical issues they address. Moreover, they realise those ideas in material form: they serve as existence proofs of particular configurations of perspectives and stances. Of course, for many of the sketches and proposals and form studies produced in the course of design, these existence proofs are themselves unproven; they may not be viable or desirable or even technically possible. Nonetheless, they exist: they establish a position and thus help define a space for a design. Moreover, when a design is complete and well finished, and found to "work" by appropriate criteria, then it can serve as a landmark for future design, an example of what can be done and a way to go about doing it. It concretises the kind of truth that design can produce, and that designers can use to inspire their own work.

The truths embodied by the artefacts of design do not speak for themselves, however. The features of interest, the commitments of the designer, the configurations that count, all may remain opaque or open to an indefinite number of contrasting interpretations. Thus designed artefacts are typically accompanied by explanatory comments, whether in the form of their designers' descriptions, advertisements pointing out their unique features, user manuals that explain how they are to be used, or critical reviews that compare them to other related designs. Much the same thing happens in presenting research through design: in the explanatory and conceptual accounts we give of our work, we point out new achievements, relate the designs we produce to theoretical work, and situate them in a context of related research. We *annotate* our designs, commenting on them to explain how they work and are related to matters of concern.[5]

Designed artefacts are too complex to be fully annotated, however. With the hundreds of detailed decisions that go into their making, ranging from their philosophical or political commitments to the speed of scrolling deemed optimal, it is practically impossible to comment completely on every detail of a design, much less on exactly how these are configured together. Moreover, a great deal of design knowledge is tacit and unspoken, or the product of hand-eye-mind coordination that is exceedingly difficult to articulate. Gathering a number of designs to form a *portfolio* can help to focus on a set of themes, features and configurations. Related by their concern with common issues, groups of designs define a space of possibilities, define a set of salient design dimensions within that domain, and take positions—some successful, some less so—within that space. Appropriate annotations can highlight and explain those dimensions and configurations, and moreover by maintaining a link with a portfolio of design artefacts, annotations can avoid the dilution that comes from unanchored generality.

Annotated portfolios may capture best what we can know through design. Attempts to abstract and generalise the knowledge produced through design runs afoul of its situated, multilayered, configured and contingent nature. A great deal of what designers learn is tacit, part of their lived experience, and shared in the culture of their fellow designers. Their knowledge is manifested, however, in the form of

[5] Much of this section is based on Gaver (2012), Bowers (2012) and Gaver and Bowers (2012).

the artefacts they produce. It may not be possible to read these artefacts unequivocally, but any ambiguity in their interpretation may be useful in inspiring new designs. Moreover, when articulated through annotation, the knowledge they encapsulate may be exposed, extended, and linked to the concerns of a domain of research. This gives the learning produced by research through design a different nature from that produced by science, which, though not completely separable from the experiments, observations and measurements that give rise to it, can nonetheless be crafted to travel much further without distortion (Latour, 1987). Design knowledge is most trustworthy when it stays close to designed artefacts.

Design as (In)Discipline

Neuroscience, sociology, fine arts, literature, computer science, experimental psychology and theology have all met, at one time or another, in design. Design provides a useful meeting point both because one of its core activities is the synthesis of diverse concerns, and because it is more concerned with creating things that work than battling over facts. But while design can benefit from, and contribute to, a wide variety of academic discourses, it need not inherit their forms of discipline. Unconstrained by epistemological accountability, design often exhibits a *productive indiscipline*, borrowing from all disciplines or none to claim extraordinary methodological freedom. This does not imply that design is undisciplined, however. The relaxation of truth claims in many of the processes of design may suggest a kind of free-for-all, in which anything goes and there is no basis for discrimination. But aesthetic accountability—the responsibility to make things that work—is a demanding discipline of its own. It may benefit from a kind of playfulness of thinking that thrives on the methodological indiscipline I have described, but it also requires the ability to fit together ideas, materials, technologies, timings, situations, people and cultures. Designers need to have enough self-indulgence to become passionate about their ideas, while maintaining the ability to take a critical perspective on the things they are producing. They have the liberty to eschew traditional methods, but in avoiding the responsibilities these imply they also relinquish the reassurance that comes with following well-understood paths. Most of all, designers have to wait until late in the process to discover if their designs *work*, if all the bets they have made along the way, the myriads of decisions they have made, have finally paid off.

It can be tempting to avoid, or at least mitigate, the uncertainties that come with aesthetic accountability by imposing methodological frameworks to design, as an a approximation of the step-by-step assurance that scientific methods seem to offer. Both in research through design and in design education, it may seem legitimate to structure the space of potential processes by introducing methods that have been used successfully before—brainstorming, personas, probes—in the hopes this will optimise the chances of producing successful work. Such an approach may indeed be useful in introducing students to the overall "feel" of doing design, and in reducing the overhead for more experienced designers of developing bespoke approaches to projects. The danger, however, is that in avoiding the terrors that come with indiscipline one also loses its advantages: the possibility of situating methods to the particularities of a project or people, and to find idiosyncratic and personal approaches to projects that can lead to innovative results.

Conclusion

Distinguishing science and design in terms of their different forms of accountability appears the clearest way of understanding the tenor of these two forms of endeavour. The need for science to defend the basis of its claims each step of the way provides it with a remarkable mechanism for achieving clarity, replicability and generalisable abstraction. To be sure, the actual doing of science can be far more messy and bound in worldly power-politics than such an account suggests, but the rituals surrounding the presentation of scientific work as empirically accountable has allowed it to transcend its pragmatic realities and to produce a body of methods, theories and analytic tools that is arguably our most effective means for producing generalisable knowledge of the world. In its adherence to aesthetic accountability, in contrast, design is arguably our best strategy for producing things that that *work*, not only in the sense of being functional but meaningful and inspiring as well. Freed from the shackles of certainty, designers are at liberty to speculate, experiment, dream and improvise—as long as they do so in ways that are accountable as design. The processes themselves are not effective at producing new *facts*, in the scientific sense, mired as they are in interpretation, ambiguity, imprecision and contingency. But they can be powerful in producing new *understandings*, based in experience, interpretation, and particular settings. Moreover, design produces new *artefacts*, each embodying its own truths, just as real as those discovered through science, which can be articulated and extended and used as the foundation for yet newer creations.

One of my purposes in describing science and design in these ways is to emphasise that these two endeavours should be seen as distinct from one another, each with its own logic, motivation and values.[6] It would be a mistake to compare the two approaches to the detriment of either. Design is not a poor cousin of science. Instead, it is an independent approach with its own expertise and knowledge (c.f. Stolterman, 2008; Nelson & Stolterman, 2003). Research through design, similarly, should not be seen as an attempt to bring the principles of science to design, but as an autonomous approach that uses projection and making as tools for learning about people, technologies and the world.

[6] Of course science and design may be intertwined in practice; what my argument here suggests is the importance of being clear about the form of accountability claimed for different aspects of the process and results.

Exercises

1. Which of the methods in this book would be characterised by "science?" Which "design?" Which methods, if any, are hard to classify?
2. Can "design" research answer questions about causal relationships? Justify your answer.

Acknowledgements This discussion is an updated version of keynote addresses delivered to DIS'00 and HCIC'10. The research was supported by European Research Council's Advanced Investigator Award no. 226528, "ThirdWave HCI". I am grateful to John Bowers, Eric Stolterman, Kirsten Boehner, Anne Schlottmann, Wendy Kellogg, Judy Olson and John Zimmerman for their comments on this chapter, though it must be admitted that few if any of them would fully agree with the result.

References

Boehner, K., Vertesi, J., Sengers, P., Dourish, P. (2007). How HCI interprets the Probes. *Proc CHI'07*.

Bowers, J. (2012). The logic of annotated portfolios: Communicating the value of research through design. In *Proc. DIS'12* (pp. 68–77).

Calle, S., & Auster, P. (1999). *Sophie Calle: Double game*. London: Violette Limited.

Carroll, J., & Kellogg, W. (1989) Artifact as theory-nexus: Hermeneutics meets theory-based design. In *Proc. CHI'89* (pp. 7–14).

Crampton-Smith, G. (May/June 1995). The hand that rocks the cradle. *I.D. Magazine*, 60–65.

Crewdson, G., & Moody, R. (2002). *Twilight: Photographs by Gregory Crewdson*. New York: Harry N Abrams Inc.

Cross, N. (2007). *Designerly ways of knowing*. Basel: Birkhäuser.

Cross, N., Naughton, J., & Walker, D. (1981). Design method and scientific method. *Design Studies, 2*(4), 195–201.

Diserens, C. ed. (2003). Gordon Matta-Clark. London & New York: Phaidon.

Fallman, D., & Stolterman, E. (2010). Establishing criteria of rigour and relevance in interaction design research. *Digital Creativity, 21*(4), 265–272.

Ferguson, R., De Salvo, D. M., & Slyce, J. (1999). *Gillian wearing*. London: Phaidon.

Gaver, W. (1988). Everyday listening and auditory icons. Doctoral dissertation (University Microfilms No. 8908009).

Gaver, W. (2007). Cultural commentators: Non-native interpretations as resources for polyphonic assessment. *International Journal of Human-Computer Studies, 65*(4), 292–305.

Gaver, W. (2009). Designing for homo ludens, still. In T. Binder, J. Löwgren, & L. Malmborg (Eds.), *(Re)searching the digital bauhaus* (pp. 163–178). London: Springer.

Gaver, W. (2011). Making spaces: How design workbooks work. In *Proc. CHI'11*.

Gaver, W. (2012). What should we expect from research through design?. In *Proc. CHI'12* (pp. 937–946).

Gaver W., Boucher A., Law A. Pennington S., Bowers J., Beaver J., et al. (2008). Threshold devices: Looking out from the home. In *Proc. CHI 2008*.

Gaver, B., Dunne, T., & Pacenti, E. (1999). Cultural probes. *Interactions, 6*(1), 21–29.

Gibson, J. J. (1979). *The ecological approach to visual perception*. Hillsdale, NJ: Lawrence Erlbaum.

Hjelm, S. I., & Browall, C. (2000). Brainball-using brain activity for cool competition. In *Proceedings of NordiCHI* (pp. 177–188).

Jarvis, N., Cameron, D., & Boucher, A. (2012). Attention to detail: Annotations of a design process. In *Proceedings of NordiChi*.
Kabokov, I. (1998). *A palace of projects*. London: Artangel.
Kuhn, T. (1970). *The structure of scientific revolutions*. London: University of Chicago Press.
Latour, B. (1987). *Science in action*. Cambridge, MA: Harvard University Press.
Levy, S. (2006). *The perfect thing: How the iPod shuffles commerce, culture, and coolness*. New York: Simon and Schuster.
Louridas, P. (1999). Design as bricolage: Anthropology meets design thinking. *Design Studies, 20*(6), 517–535.
Nelson, H. G., & Stolterman, E. (2003). *The design way: Intentional change in an unpredictable world: Foundations and fundamentals of design competence*. Englewood Cliffs, NJ: Educational Technology.
Schön, D. (1983). *The reflective practitioner: How professionals think*. New York: Basic Books.
Schön, D. A. (1999). *The reflective practitioner*. New York: Basic Books.
Scott, J. P. (2000). *Social network analysis: A handbook* (2nd ed.). Thousand Oaks, CA: Sage Publications.
Stolterman, E. (2008). The nature of design practice and implications for interaction design research. *International Journal of Design, 2*(1), 55–65.
US Government. (1971). *Hiding and storing stuff safely*.

Research Through Design in HCI

John Zimmerman and Jodi Forlizzi

Introduction

Many researchers have struggled to connect RESEARCH and DESIGN. Voices in the HCI research community coming from different disciplinary backgrounds have noted this challenge. Erik Stolterman stated that scientific research drives towards the *existing* and the *universal*, while design works in pursuit of the *non-existing* and in the creation of an *ultimate particular* (Stolterman, 2008). Design and scientific research, then, seem headed in opposite directions. Jane Fulton-Suri, reflecting on her training in social science followed by her experience working at a design consultancy, identified a gap between design with its focus on the future and social science research with its focus on the past and present. Alan Blackwell, discussing research from an engineering perspective, explained that research contributions must be *novel*, but not necessarily good. Design contributions, however, must be good but not necessarily novel (Blackwell, 2004). These observations reveal an uneasy tension between design and research within HCI. However, the repeated reflections and speculations on a connection hint at an underlying desire to discover a way to link these two together.

Research through Design (RtD) is an approach to conducting scholarly research that employs the methods, practices, and processes of design practice with the intention of generating new knowledge. People carrying out research using RtD generally reject the idea that research is synonymous with science. Instead, RtD frames design inquiry as a distinctly separate activity from engineering inquiry and scientific inquiry. RtD draws on design's strength as a reflective practice of continually reinterpreting and reframing a problematic situation through a process of making and critiquing artifacts that function as proposed solutions (Rittel & Webber, 1973;

J. Zimmerman (✉) • J. Forlizzi
HCI Institute and School of Design, Carnegie Mellon University,
5000 Forbes Avenue, Pittsburgh, PA 15213, USA
e-mail: johnz@cs.cmu.edu

J.S. Olson and W.A. Kellogg (eds.), *Ways of Knowing in HCI*,
DOI 10.1007/978-1-4939-0378-8_8,

Schön, 1983). RtD asks researchers to investigate the speculative future, probing on what the world could and should be.

On the surface, RtD can look suspiciously like design practice. However, it is generally more systematic and more explicitly reflective in its process of interpreting and reinterpreting a conventional understanding of the world, and it generally requires more detailed documentation of the actions and rationale for actions taken during the design process. An RtD project gets documented such that other researchers can reproduce the process; however, there is no expectation that others following the same process would produce the same or even a similar final artifact. The most important distinction between RtD and design practice is the intention that design researchers bring to bear on a problematic situation. In the practice of RtD, design researchers focus on how design actions produce new and valuable knowledge. This knowledge can take many different forms including: novel perspectives that advances understanding of a problematic situation; insights and implications with respect to how specific theory can best be operationalized in a thing; new design methods that advance the ability of designers to handle new types of challenges; and artifacts that both sensitize the community and broaden the space for design action The focus on producing these types of knowledge make RtD quite different than commercial practice with its focus on making a commercially successful product.

We see RtD as one way to respond to an interesting challenge noted by Jack Carroll and Wendy Kellogg in the early days of HCI. They noted, with great frustration, that in HCI, the *thing* proceeds *theory* instead of theory driving the creation of new things (1989). They noted that the mouse needed to be invented before studies could be done that showed this was a good design. As another example, people developed many, many different direct manipulation interfaces, such as Sketchpad (Sutherland, 1963), long before Ben Shneiderman wrote about the value of direct manipulation (1983). The practice of RtD in HCI implores researchers to become more active and intentional constructors of the world they desire, constructors of the world that they believe to be better than the one that currently exists. In response to this challenge, we see RtD as a way for many new things to enter into HCI that can spawn new theory. At the same time, these new things can be informed by current theory, creating an ongoing dialog between what is and what might be. Building on Nigel Cross' observation that knowledge resides in designers, in their practices, and in the artifacts they produce (Cross, 1999), RtD provides a research approach for these types of knowledge to be generated and disseminated within HCI.

A recent book by Ilpo Koskinen, Zimmerman, Binder, Redstrom, and Wensveen (2011) provides a detailed investigation of current RtD practices in the interaction design research community. The authors provide a brief history of RtD's evolution in different design research communities, detailing how three distinct practices emerged that they refer to as Lab, Field, and Showroom. The *Lab* practice comes mainly from the Netherlands. It combines design action with experimental evaluation processes traditionally used in psychology. It focuses on creating novel and much more aesthetically appealing ways for people to interact with things. The *Field* practice comes out of the Scandinavian tradition of participatory design and out of user centered-design practices in the USA. It merges research practices from sociology and anthropology with design action. In this practice, design researchers map out a problematic situation and offer design ideas to improve the state of the world. The *Showroom* practice borrows methods from art, fashion, and design. Here, researchers design provocative things that challenge the status quo. Critical designs force people to reconsider the world they inhabit and to notice aspects too often overlooked. The knowledge produced includes the characterization of the issue being critiqued, the approach used to draw the viewer's attention to the underlying issue, and the process used to arrive at the problem framing and the final artifact form.

History of RtD and Its Connection to HCI

The histories of RtD and of RtD in HCI are strongly connected to events in the design research community and to the emergence of interaction design as a design discipline distinct from architecture, industrial design, and communication design. The term "research through design" comes from Christopher Frayling (1993). He provided a descriptive framework for research in the arts as being:

1. *Research into design*—research into the human activity of design. Well-known examples include Herbert Simon's work on design as an artificial science (Simon, 1996), Harold Nelson and Erik Stolterman's work on the "way" of design (Nelson & Stolterman, 2012), and Donald Schön's work on design as a reflective practice (Schön & Bennett, 1996).
2. *Research for design*—research intended to advance the practice of design. This includes almost all design research including any work that proposes new methods, tools, or approaches; or any work that uses exemplars, design implications, or problem framings to discuss improving the practice of design.
3. *Research through design*—a type of research practice focused on improving the world by making new things that disrupt, complicate or transform the current state of the world. This research approach speculates on what the future could and should be based on an empathic understanding of the stakeholders, a synthesis of behavioral theory, and the application of current and near current technology. The knowledge produced functions as a proposal, not a prediction (Zimmerman, Stolterman, & Forlizzi, 2010).

Research *into design* and research *for design* both refer to the outcome of a research project; the type of knowledge that is produced. Research *through design* differs in that it is an approach to doing research. It can result in knowledge *for design* and *into design*.

Many design researchers have produced valuable frameworks for discussing what research is with respect to design (i.e., Buchanan, 2001; Cross, 1999); however, Frayling's framework, particularly his description of RtD, has been increasingly important in interaction design research (Basballe & Halskov, 2012). While Frayling coined the term "research through design," he is not the practice's point of origin. In fact, looking over the history of this research approach, it is possible to see it emerge from several different places. Here we tell three different origin stories: Rich Interaction, Participatory Design, and Critical Design. These are based on Koskinen et al.'s (2011) framework of Lab, Field, and Showroom.

Rich Interaction Design (Lab)

In the 1990s, researchers at the Technical Universities in the Netherlands drew a clear distinction between design, which was taught to students in the industrial design programs, and the scientific research performed by the faculty to investigate human perception, consumer preferences, emotional reactions, and the design process. A small group of researchers and designers at the Technical University in Delft observed that with the transition from mechanical interaction to electronic interaction, interaction possibilities should increase with the freedom from mechanical constraints. Yet almost all new products reduce interaction to pushing a labeled button. They viewed this trend as a failure to account for people as sensual beings; a failure to focus on perceptual motor skills as a source of interaction inspiration, due to an overreliance on cognition as a singular theoretical stance in interaction design, and a failure to consider aesthetics as a critical component of interaction. Drawing on theories from perceptual psychology, ecological psychology, and phenomenology, they worked from the perspective that interaction design should engage all of the senses, not just the visual. Following this theoretical stance, they created a new approach to interaction research with the goal of designing systems that would more fully engage people's bodies to richly express themselves and people's full range of senses as channels for input and feedback from interactive systems, a new research space they dubbed Rich Interaction (Frens, 2006a, 2006b).

These Dutch design researchers wanted to invent entirely new methods for people to interact with things. To do so, they combined aspects of experimental psychology and aspects of design practice. In general, they would start from a psychological theory with the goal of making it actionable through design. They would then conduct a series of design workshops to understand how to do it. These workshops brought together many designers who, working with rough materials, would rapidly invent new ways for people to interact with systems. Workshops investigated things like how consumer products like alarm clocks might function differently if they took into account the user's emotional state (Djajadiningrat et al., 2002), or how a vending machine might behave if it were polite or unfriendly (Ross et al., 2008). Outcomes of the workshops functioned as semi-articulated hypotheses of potentially better forms for interaction. Following the workshops, the researchers would select and refine an idea into a more detailed hypothesis. In order to validate these hypotheses, researchers make several slightly different versions of a single product and conduct controlled lab studies around the interaction.

Philip Ross' dissertation work at the Technical University in Eindhoven provides a good example. He carried out design workshops where design teams created

Fig. 1 Adaptive, interactive lamp by Philip Ross

candy vending machines based on different ethical stances including Confucianism, Kantian rationalism, vitalism, romanticism, and Nietzschean ethics (Koskinen et al., 2011). The workshop revealed that an ethical stance could drive design inspiration, and it revealed many aspects of Kantian rationalism embedded in the machines people interacted with everyday. Based on this finding, Ross hypothesized that ethics could be imbued into the design of interactive products by using a specific perspective to drive a design process. Over the course of his PhD studies, he developed many different lamp forms and interactive behaviors that explored subtle aspects of ethical stances.

This RtD practice grew out of academia through the collocation of psychologists and designers. Unlike more user-centered design processes popular at this time, this approach allowed the designer to brainstorm freely to create new innovations. This approach blends design methods to envision the unimagined and both analytic and experimental methods to evaluate the novel design offerings and to generate frameworks that described how rich interaction works. This approach is still practiced today by researchers in the Designing Quality in Interaction group at the Technical University in Eindhoven.

Figure 1 shows an image of Ross' final lamp (Ross & Wensveen, 2010). The lamp reacts by changing the intensity and direction of the light based on the way a person touches and strokes it. The lamp also uses an understanding of the situation at hand, such as the fact the user wants to read, to augment its behavior. This can range from helpful, where it provides light for the reading material, to playful, where it moves the light to draw the user's attention away from reading towards the lamp. The work investigates how designers can use ethics as a lens for investigating aesthetics in the behavior of adaptive products, broadening the space for innovation and creativity in the design process.

Fig. 2 Still from Maypole project showing a child using a working prototype of a digital camera designed for children

Participatory Design and User-Centered Design (Field)

The participatory design movement began in Scandinavia as a reaction to the disruptive force of information technology as it entered the workplace and caused breakdowns in traditional roles and responsibilities. Workers believed that IT systems often reduced their voice and diminished their craft skills. In addition, companies and software developers noted that many of these systems, while designed to increase productivity, often resulted in productivity losses because the people designing the systems lacked a detailed understanding of how the work was really performed (i.e., Kuhn, 1997; Orr, 1986).

Participatory design embraced a Marxist philosophy. It focused on developing a new approach to the design of work automation that increased democracy and protected workers. Design work was performed by interdisciplinary teams consisting of behavioral scientists, technologists, and designers, who brought a theoretical understanding of people and work, knowledge about the capabilities of technology, and skills at conceiving an improved future state. In addition, the design teams included workers selected by their peers, who brought domain expertise on work practices within an organization. Working together, these teams followed a rapid prototyping approach, iteratively conceiving of new work and workplace designs by starting with low-fidelity prototypes and working towards higher fidelity until a final concept emerged. These design teams socially prototyped the new work practices before committing to the technology that could bring about this future. Figure 2 shows an image from the Maypole project (Giller et al., 1999). Maypole was a

2-year European research project funded by i3net, the European Network for Intelligent Information Interfaces. One goal of the project was to develop communication concepts for children aged 8–12 and others in their social networks. The methods in this project extended the legacy of participatory design, and focused on interaction with users in real world settings. The image shows a child using a digital camera. Mobile technology to do what designers envisioned did not exist at that time, so researchers cobbled together a camera and output screen, tethered together and carried in a backpack. Making working prototypes like this camera, and giving them to users in the field allowed researchers to begin to investigate how new technology might create new practices around image making and messaging. Insights on what these new practices might be were then used to inform the design of new mobile technologies.

This new approach to the design of technical systems was intended to result in commercial products. What made this research and not design practice was the focus on developing a new design methodology that borrowed research practices from anthropology and sociology and combined them with design. Over time, researchers began to also see the artifacts they produced as a research contribution. For example, in the Maypole project, researchers made advances on how to conduct participatory design with children, which advanced this method over previous work that focused on workers. In addition, the user intentions for using a digital camera as well as the interactive behaviors created during the design of the camera could be generalize and applied to future cameras, once the technology allowed a digital camera of an appropriate size. Thi observation, that insights from making and the resulting artifact were also research contributions, helped refocused this research practice from being method specific to being more an investigation of the speculative and desirable future.

Critical Design (Showroom)

In RtD that follows a critical design approach (Dunne & Raby, 2001), design researchers make provocative artifacts that force people to think, to notice, and to reconsider some aspect of the world. The term "critical design" was first used by Tony Dunne to describe a philosophy about design that refutes the status quo (Dunne, 1999). However, the idea of designs that critique the current state of the world is much older and can be found in many design and art movements such as in the work of the Pre-Raphaelites or the Memphis Design Group. The approach emphasizes that design has other objectives than to help people and to improve the world.

This research approach draws from historical design practice. In the 1990s, in many design schools, a movement towards conceptual design rather than finished artifacts began to take place. It is possible to characterize much of the work done in fashion, conceptual architecture, and conceptual design as a type of critical design. Critical design offers a research approach that allows designers to draw on the strengths and traditions of design. The research involves a process of problem

Fig. 3 The prayer companion by the Interaction Research Studio at Goldsmiths, University of London

selection, exploration through the generation of many possible forms, and iterative refinement of a final form that approaches showroom quality. Knowledge is captured as the designer or design team engages in reflective writing that describes the process, the artifact, and the intended influence. While popularized at the Royal College of Art in the UK, this RtD approach has been taken up at a number of design schools.

Figure 3 shows the Prayer Companion. This design from the Interaction Research Studio at Goldsmiths University provided a well-known example of critical design in HCI. The device displays electronic news feeds in a convent of cloister nuns, providing a connection from the outside world as a source of prayer topics. The design raises interesting questions on many topics including the role of computation as a material embedded in sacred artifacts. The project mixed in elements of participatory and user centered design by installing the device into a convent and reflecting on how the nuns came to view and understand it.

Authors' Connections to RtD in HCI

The authors both participated in the development and growth of RtD as a research practice in HCI. Their interest in this topic was pragmatic. Both worked as practicing interaction designers in industry before becoming professors at Carnegie Mellon. Jodi worked at the design consultancy e-Lab in Chicago, performing fieldwork to inform the design of a variety of products and services. John worked at Philips Research. He collaborated with technical researchers to give more commercial forms to their research; forms intended to help product managers better understand what the technology could and should be. In accepting joint appointments at Carnegie Mellon's HCI Institute and at the School of Design (Jodi in 2000 and John in 2002), they were two of the first interaction design researchers working directly in the space of HCI research. However, what it meant to be an academic interaction design researcher at this time was still largely undefined. Carnegie Mellon had hired them both to help define the role of design research within HCI and to help develop new ways for interdisciplinary researchers in HCI to integrate design thinking into HCI research and education.

In the early 2000s, the term design research within the HCI research community generally meant the upfront research done in the practice of design. Daniel Fallman described this as *research-oriented design*; design work in HCI that is informed by upfront research as opposed to designers working in isolation in a studio (Fallman, 2003). At this time the term "design" within the HCI research community was synonymous with the term "practice." Design was not viewed as a discipline that could produce knowledge, but was instead used as a term to help distinguish research from practice.

To better understand how design might best fit into HCI research, the authors, along with other collaborators, held a workshop at CHI 2004 with the intention of bringing researchers from a variety of disciplines together to discuss and advance the role of design (Zimmerman, Evenson, Baumann, & Purgathofer, 2004). Out of this workshop came a desire to move design thinking into HCI research by making a place for RtD. To do this, design researchers in HCI needed to convince the HCI research community to accept RtD research contributions as both valid and valuable. They needed the community to see the speculative artifacts designed in this process as more than an integration of known technologies. They needed this research community to see these things as rigorous speculations on the possible future that reveal new and important insights on how people understand and engage with new technology and the appropriate roles technology might play as it continues to move into more and more aspects of people's lives.

Interestingly, Bill Gaver presented a paper describing the Drift Table (Gaver et al., 2004) at the same CHI conference. Both the artifact and the paper describing its design and evaluation challenged the HCI community's institutionalized belief that interaction designs must require an explicit user intention; that everything should have a "right" way to be used. The design opened a new research space, which Gaver and his team called "ludic interaction." In addition, it provided a great example of how design and design thinking can expand the scope and role of HCI. The paper was presented as an "Experience Report," a track within CHI designed for practitioners to share their design cases. While clearly intended as a research contribution, the paper was not a part of the peer-reviewed technical papers section of the conference. Nevertheless, the Drift Table has been held up as one of the earliest and best examples of RtD published within HCI.

Following the workshop, the authors, working with their colleague Shelley Evenson, began a project focused on bringing RtD to HCI. We began by first detailing how knowledge is produced in the design of commercial products and services (Zimmerman, Evenson, & Forlizzi, 2004). The work explicitly noted the different kinds of knowledge produced at different points in typical design process (Fig. 4). Based on this map from commercial practice, we embarked on a project to create a model of RtD that could work for the HCI research community. The intention of this model was not to define a singular type of design research in HCI. Instead, it was

Project process by phase / Research knowledge production by phase

	define	discover	synthesize	construct	refine	reflect
Project process by phase	• team building • technical assessment • hypothesize	• contexts • benchmarking • user needs	• process maps • opportunity map • frameworks • personas • scenarios	• features and functions • behavior • design language • interactions and flow models • collaborative design	• evaluation • scoping • interation • specification	• post mortem • opportunity map • benchmarking • market acceptance
Research knowledge production by phase	• prototypical user model • prototypical user needs • client's needs	• user mental models • user process models • user's relation to context • summary of current products meeting needs (lite review)	• relationships needs of users, client, and context • identify gaps (opportunities for new product or service)	• examples of process and flow models that users will and will not accept • insights into high-level guidelines for interaction • evaluation of widget performance and its relationship to software reuse • improved interaction flow models		• opportunities for improving design process • acceptance of design in the market place • new assessment of gaps (opportunities for new products and services)

Fig. 4 Knowledge opportunities in the design process

intended to open a door to create an initial foothold for RtD research contributions to gain entry into HCI research venues.

We began by interviewing leading HCI researchers and leading HCI designers, probing to understand how they define design, on how they see design producing knowledge, and on how design might or might not fit into HCI research. One of our favorite comments came from a psychologist who shared that: "designers make things, not knowledge." Another, also from a psychologist, asked: "Why do you want to make the right thing? Why not conduct a two-by-two experiment?" This comment in particular helped us see a critical disconnect between design thinking with its focus actively constructing a subjectively preferred future and scientific thinking with its focus on universal truth that remains true through time. Following the interviews, the team created a model of RtD in HCI, evaluating and iteratively refining it through many presentations and individual meetings with HCI researchers, design researchers working in HCI, and design researchers working in the design research community. The results of this work came to the HCI research community as a paper and presentation at CHI 2007 (Zimmerman, Forlizzi, & Evenson, 2007) and to the design research community through an article in a special issue of the journal *Design Issues* on design research, edited by this team (Forlizzi, Zimmerman, & Evenson, 2008).

In this model (Fig. 5), interaction design researchers following an RtD approach work to integrate three types of knowledge in the design of new things: how, true, and real (Zimmerman et al., 2007). The types of knowledge build on the definitions of "real, true and ideal" knowledge introduced by Nelson and Stolterman in the *Design Way* (2012). From engineers, design researchers take "how" knowledge; the latest technical possibilities. From behavioral scientists, they take "true" knowledge, models and theories of human behavior. From anthropologists they take "real" knowledge; thick descriptions of how the world currently works. Based on these three types of inputs—how, true and real knowledge—design researchers ideate many possible visions of a preferred future state by imagining new products, services, systems, and environments that address challenges and opportunities and that advance the current state of the world to a preferred state. In a sense, the design

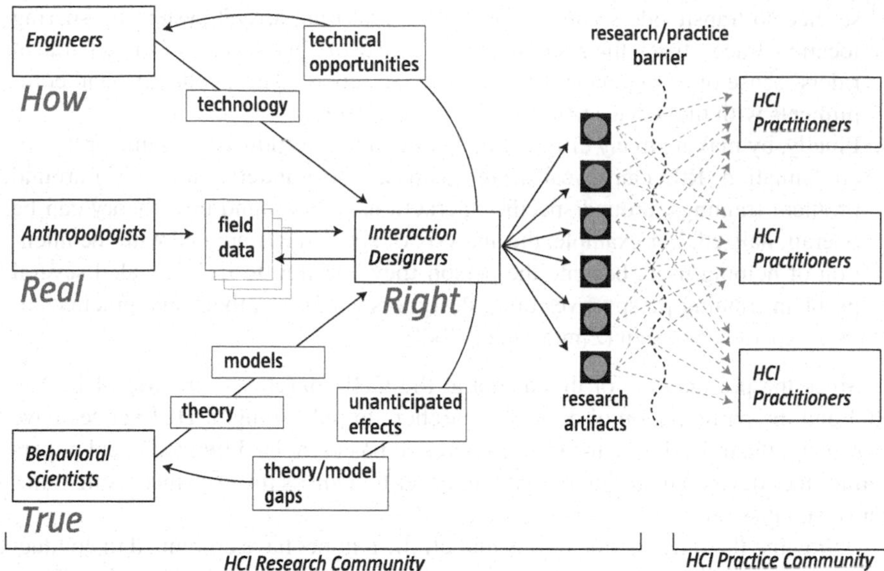

Fig. 5 Our model of research through design within HCI

researchers attempt to make the "right thing," an artifact that can improve the state of the world. This model builds on the idea of design as a process of repeatedly reframing a problem through a process of proposing possible solutions (Buchanan, 1995; Rittel & Webber, 1973; Schön, 1983; Simon, 1996).

The model illustrates four types of research outputs to different disciplines within HCI.

1. RtD can produce *technical opportunities* that feed back to engineers. These are places where if a nearly possible technical advance existed, it would benefit the world. For example, the fieldwork, concept generation, and speed dating studies by Davidoff et al. revealed that parents would benefit from computational systems that learned routine pick-up and drop-offs in order to help parents to not forget their children (Davidoff, Lee, Dey, & Zimmerman, 2007; Davidoff, Lee, Yiu, Zimmerman, & Dey, 2006; Davidoff, Zeibart, Zimmerman, & Dey, 2011; Davidoff, Zimmerman, & Dey, 2010).
2. RtD can expose gaps in current behavioral theory. For example, theory on product attachment explains why parents grow to cherish some of the books they read to their children. This theory, however, fails to explain why this attachment does not seem to develop when parents read these same stories to their children using an eReader. Something in the digital quality of this material possession makes it different (Odom, Zimmerman, & Forlizzi, 2011).
3. By making things and placing them into the world, RtD can change the current state, creating new situations and new practices for anthropologists and design researchers to investigate. For example, the Tiramisu project produced a mobile

service so transit riders could coproduce a real-time arrival system by sharing location traces from their smart phones. Use of this service changed transit riders' sense of engagement, causing them to report issues, concerns, and compliments with the service (Zimmerman et al., 2011).
4. Finally, by making many different things intended to address the same problematic situation, RtD can reveal design patterns (Alexander et al., 1977) around problem framings, around specific interactions, and around how theory can be operationalized. For example, the analysis of six artifacts created with the intention of helping users become the person they "desired to be," revealed several problem framing perspectives other designers can use in their own practices of user experience design (Zimmerman, 2009).

Since the presentation of this model at the CHI conference, the use of RtD in HCI and the participation of design researchers in publishing at HCI venues have grown significantly. Today the CHI conference has organized two technical papers committees devoted to design research in order to address the ever-increasing number of submissions.

Following the introduction of this model, the authors have continued to conduct research on and with RtD. They have both conducted a number of research projects following this approach and published those at both HCI and interaction design research venues. In addition, they have worked to formalize this research approach. Specifically, they have investigated how RtD can lead to theory and how researchers can better evaluate what makes a better or worse RtD contribution (Forlizzi, Zimmerman, & Stolterman, 2009; Zimmerman et al., 2010; Zimmerman & Forlizzi, 2008).

RtD's Contributions to HCI

RtD offers many contributions to HCI. Two that we focus on here are

1. the reflective practice of *reframing* the underlying situation and goal of the project during the design process and
2. a shift to investigating the future as a way of *understanding the world that should be brought into being*.

Below we provide an overview of two design cases. The first talks about the design of the Reverse Alarm Clock, and describes how the process of reframing helped to advance the understanding of what domestic technology should do and how it might be best situated in the home. The second addresses the design and investigations of Snackbot, a snack delivery robot. This RtD investigation of a speculative future where robots bring workers snacks addresses both the need for guidelines on how robots should socially engage with people and the complex issues surrounding the sedentary work practices of today's office workers and the growing obesity problem.

Reverse Alarm Clock

The reverse alarm clock project provides an RtD research example of how problems and project goals are continually questioned and reframed during the research process. This project began with a fieldwork observation that busy parents in dual-income families sometimes yell at their children in the morning. The goal of getting everyone out of the home on time is important to parents, but it is not often important to young children. In the stress of the morning rush, a lost shoe or a slow eater can push parents beyond the edge of their patience. When they yell at their children, they feel they are failing as a parent by starting their child's day off with such a negative tone. Parents who have their sleep disrupted, by small children who wake at night, have even less emotional reserve to maintain their patience during the morning rush.

Based on this observation, the project team was motivated to design a system that could help small children to stay in bed at night. The result of this effort was the reverse alarm clock (Fig. 6). In its final form, the clock consisted of four parts:

Display: The display expresses time as three states, each associated with a rule: When the moon is up, the child must be in bed; when the moon sets, the child can get out of bed if they wish; and when the sun rises, the child must get up.

Treasure box: This box goes next to the child's bed. At night the child places a token into the box, selecting the music to play at wakeup, when the sun rises. After placing the token in the box, the child presses a button on the top of the box, causing the moon to rise.

Controller: A circular dial, called the controller, hangs on the wall near the light switch at the entry to the child's room. It has two levers: one to set the moonset time and one to set the sunrise time.

Bed sensor: A sensor mat goes under the mattress to detect if the child climbs out of bed. If a child leaves the bed while the moon is up, nothing happened. If the child leaves the bed after the moon has set, the sun instantly rises and the wake up music starts to play.

The initial problem framing viewed children getting up at night as a contributing factor for why parents might yell in the morning. Guided by this framing, the team investigated why children get up at night. Through literature and interviews with sleep specialists they learned that young children have no sense of duration. In viewing the situation from the child's perspective the team began to see parents as inconsistent. Sometimes a child would wake and visit their parents, and their parents would be glad to see them. Other times the child would wake and visit their parents, and the parents would be upset and insist they go back to bed. Taking the child's view, the problem became one of usability. The child needed better feedback that could help them predict how their parents might react to their action. This new framing motivated many of the initial designs for the display.

The design team then expanded their scope from the wakeup and began looking at bedtime. While often viewed by busy parents as a stressful time, the team also

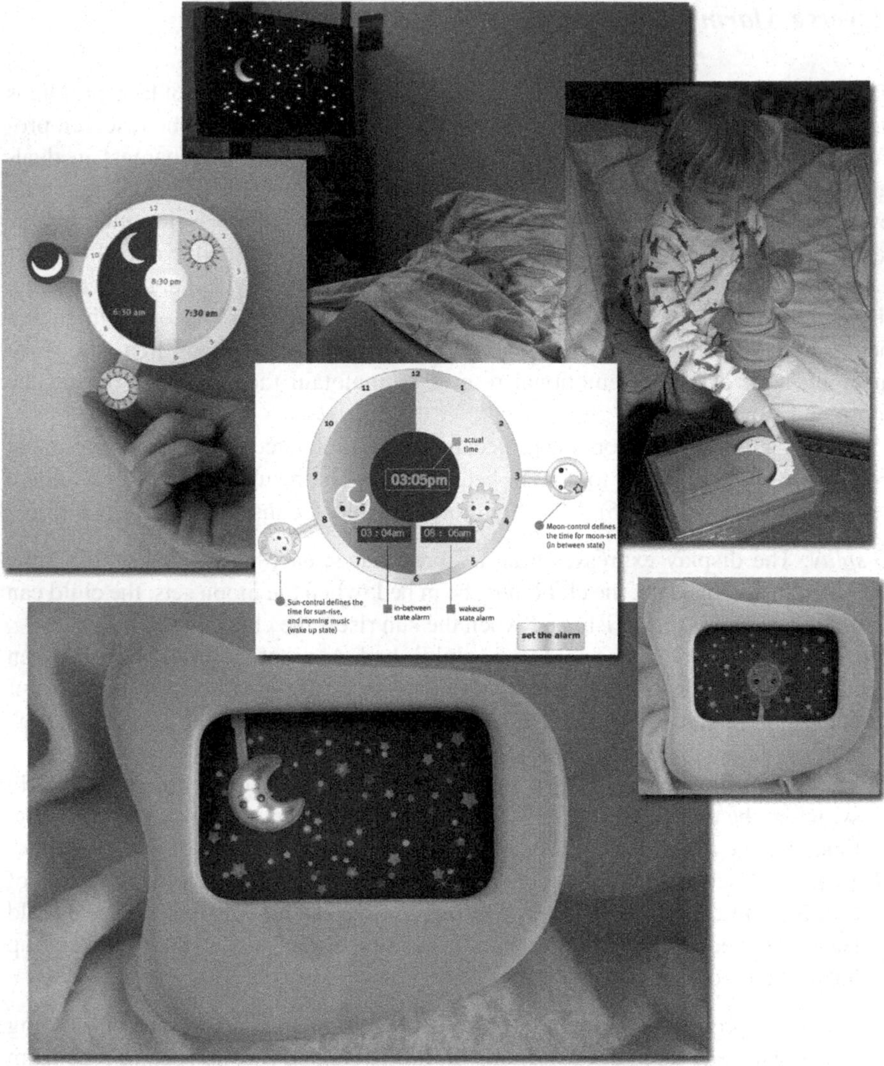

Fig. 6 (*Clockwise from top*) Experience prototyping of an early version of the display with a 3-year-old child. Experience prototyping of the treasure box. Final design showing the sun after it has risen. Final design showing the moon before it sets. Early prototype of the controller. (*Center*) Screen based version of controller used for the field trial, which ran on a laptop in the child's room

observed a sweet and intimate interaction between parent and child when parents would read bedtime stories. The team members noted how their own parents kept and cherished specific books they had read together at bedtime. In watching these bedtime routines unfold, the team noted that when reading to children, parents stopped multicontexting—a term introduced by Darrah, English-Lueck, and

Freeman (2001) to describe how busy parents inhabit both work and home roles simultaneously—and fully engaged in their role as parent. These observations helped to shift the focus away from this simply being a problem of parenting consistency. The team expanded their scope, looking for opportunities to connect the intimate ritual of bedtime with the design. This perspective eventually led the team to add the treasure box as a way of connecting the product with this intimate, nightly routine.

The team advanced the design through a process of scenario building, body storming (Buchenau & Suri, 2000), and rough prototyping. Through critique and reflection on the many sketches and prototypes created, the team began to play with the complex issue of control. Smart home research shows that busy parents are much more interested in gaining control over their life than gaining increased control of their stuff (Davidoff, Lee, Yiu, Zimmerman, & Dey, 2006). Other smart home research showed that parents did not want new technology that automated parenting; that took over parenting responsibilities (Davidoff, Lee, Dey, & Zimmerman, 2007). These insights helped to reframe the project goal. The team began to look for ideas where parents gained control of a situation and for opportunities to support parents in parenting as opposed to automating parenting tasks.

This new framing led to the controller and to the design of the display as a set of rules. The controller provides a type of relative control for parents. They can set a very different moonset time for a Tuesday and for a Saturday while the child's experience with the display remains consistent. In terms of supporting parenting, the linking of the display to a set of rules framed the clock as a tool parents use to help teach their children to make good decisions, a long-term goal of parenting. When the child has a bad dream or feels sick and seeks comfort in their parent's bed, the parents' reaction to the situation helps shape the child's understanding of when rules must be followed and when they might be suspended. From this new framing, the clock's role is to provide information a child can understand, and the parent provides the guidance to the child, helping them interpret the meaning of this information in different situations.

Once the team arrived at a final form, they continued the design process by making three versions of the clock and conducting a field trial with three families. They recruited families who had problem sleepers; children that repeatedly got up at night and woke their parents. In early discussions with parents, one mother described moving her three-year-old from the crib to a toddler bed as the "worst three weeks" of her life. The clock functioned well, helping to reduce the frequency with which children woke their parents at night and reducing the length of time it took to get children back into bed. More importantly, however, discussions with parents both before and after the installation help change the team's idea of how the product should situate in a family's life. Up to this point the design had been focused on solving a problem for families that were experiencing disrupted sleep. It was seen as a solution once the problem had occurred. In talking to these parents about the transition from crib sleeping and co-sleeping to having a child sleep in a toddler bed, the team was struck by the sense of celebration and achievement bound up in this transition. These new beds were presented to children as a sign of achievement; a sort of

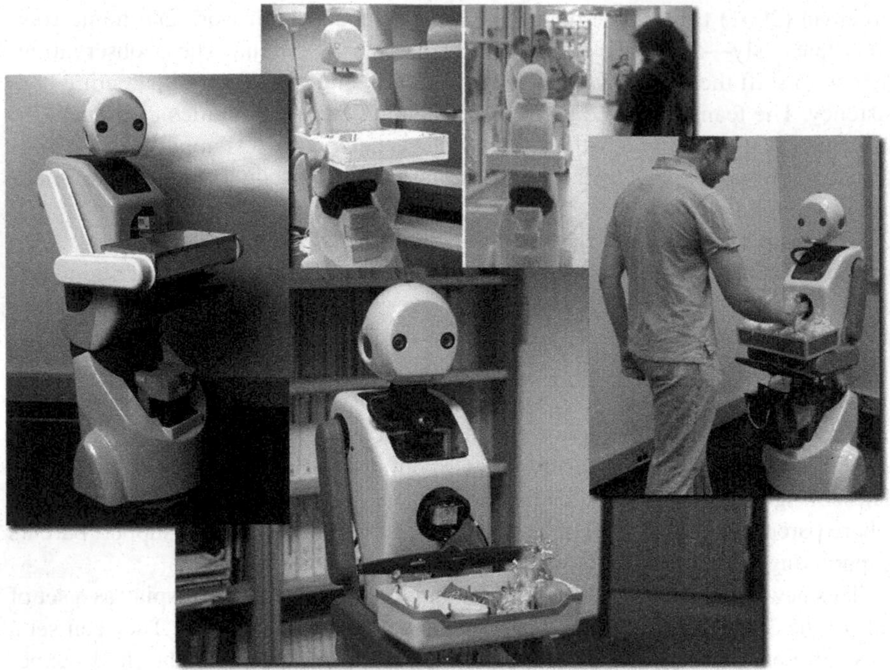

Fig. 7 (*Clockwise from top*) Exploring different physical forms in the context of use. person collecting the snack they ordered from Snackbot. Final design showing robot holding a tray of healthy and less healthy snacks. Nearly final version of Snackbot investigating color, arms, and tray

graduation. Seeing families use the clock and listening to the transition stories again reframed the goal of this project. Instead of a focus on solving a problem, the design team noted that the transition point, where the child moves to a toddler bed was the ideal time to introduce the clock into a child's life. It should be part of the transition in the same way as the toddler bed, as a celebrated artifact of this transition. In stepping back, the team saw that many technologies designed for the home could also follow a similar path, of being introduced during life stage transitions as opposed to being introduced in reaction to a breakdown in the family.

Snackbot

Snackbot (Fig. 7) is an example of how prototyping a future state, which is core to RtD research, can open up new ideas for how future technology products can benefit people's lives.

This project began with the desire to develop a robot so that the research team could study human–robot interaction (HRI) over time. Many questions about long-term HRI are unanswered. How do people's perception and attitudes towards a robot evolve over time? What interaction design strategies will reinforce a positive long-term relationship between people and a robot? Will employees engage with a robot as the design team intends, or will they appropriate the robot in new ways, as have happened with other technologies? Could robots deliver services that are beneficial to people over the long term? How should robotic products and services be designed?

As designers, we had three design goals for development of the Snackbot robot. The first was to *develop the robot holistically*. Rather than advancing technology *per se* or focusing on one aspect of design or interaction, such as a dialog system, we took a design approach that considered the robot at a human–robot–context systems level. The second goal was to *simultaneously develop a robotic product and service*. By this we mean that the robot as a product would have to be more than sociable and attractive; it would need to deliver something useful to people. The third goal was to *develop interaction designs that would help to evoke social behavior*. Because the robot was meant to serve as a research platform that would be used by people over time, decisions about functions and features were made supporting the interest of promoting sociability. In its final form, the Snackbot delivery service (Fig. 7) consisted of the following components:

Robot. Snackbot (Lee et al., 2009) is a 4.5-foot tall, anthropomorphic wheeled robot. The robot can make head movements to each side, and up and down, and can animate its LED mouth display to smile, frown, or show a neutral expression, and employ speech output.

Snack ordering Web site. Participants could order snacks using a custom designed snack ordering Web site and database. Customers specified the snack type, delivery day, and their office number. Only those registered in the study could order snacks through the Web site.

Snacks. Snackbot delivered six different snacks—apples, bananas, oranges, Reese's peanut butter cups, Snickers candy bars, and chocolate chip cookies. We chose a mixture of snacks that were not always available in the workplace.

Robot control interface. A GUI allowed an operator to control the robot's navigation, nonverbal movements, and dialog system remotely. The interface showed the video feed from the robot, the robot's location on the building map, its head position, and a number of dialog scripts. The operator could see a participant's actions through the video/audio feed on the interface.

Operator. An operator transformed the orders on the Web site to a delivery schedule, loaded the snacks on the robot's tray, initialized the robot at the start of each delivery run, and localized it.

The development of the robot, Web site, and GUI was an iterative, user-centered design process, and was guided by frequent meetings of our interdisciplinary research and design team. Once the system was complete, we could begin to systematically explore what aspects of the design could be modified to increase trust and rapport with the robot, increase the likelihood that customers would use the service again, and inspire customers to make healthy snack choices or to converse about personal topics with Snackbot.

For example, in one study, we focused on operationalizing behavioral economics theories to help people make healthy snack choices by varying the design and location of how the snacks were presented on the Web site and on the robot (Lee, Kiesler, & Forlizzi, 2011). We learned that some strategies are more persuasive than others, especially for those with less healthy lifestyles. We can design to support convenience or minimize it, and to support choice influence by making healthy choices look good and unhealthy choices look stale or unappetizing. We learned that we can even leverage social influence in helping people to make better decisions.

In another study, we varied the robot's behavior to offer a set of customers personalized snack delivery service (Lee, Kiesler, Forlizzi, & Rybski, 2012). Customers receiving personalized service were offered the option of a special mystery snack in lieu of their order, or an opportunity to do a neck stretch with the robot. The robot also chatted with customers about things it remembered about them, such as what snacks they had ordered in the past, and what kind of work they did. Customers in the control condition received typical social chat and did not have the option to receive a mystery snack as their order. Despite the knowledge of Snackbot as a machine, participants receiving snacks delivered by the robot developed a variety of social relationships with the robot. Beyond one-on-one interaction, the robot created a ripple effect in the workplace, triggering new behaviors among employees, including politeness, protection of the robot, mimicry, social comparison, and even jealousy. The design implications of this work will assist in development of many kinds of future technology for the workplace.

Snackbot as an RtD effort continues to explore many questions about the social use of robotic technology within an organization. However, additionally, the robot and the service it provides create a rich context for understanding and advancing research in the design of technology services. In this work, we are creating new understanding of how people will adopt technology over time, and will develop rapport, trust, and liking of assistive robots and technology services. Without this research platform, we would make far less educated judgments about an improved future state.

How to Do It

To carry out an RtD research project, we suggest a team follow five simple steps:

1. Select
2. Design
3. Evaluate
4. Reflect and disseminate
5. Repeat

Select involves choosing a research problem worthy of investigation. Teams should first decide if they want to focus on a problem or a design opportunity. They need to select a new material to play with, a context and target population to understand and empathize with, a societal issue or insight, and/or a theoretical framing they wish to apply to interaction. Selecting is an iterative process of trying many different things until the team agrees. Other important factors in terms of selecting include if the research problem lends itself to investigation via RtD. Is this a wicked, messy problem space that can best be addressed through the application of design thinking? For example, does it have multiple agendas driven by different stakeholders and entrenched interests that prevent an optimal solution? Finally, the teams will want to consider the skills the research team possesses, as well as the desires and concerns of the people or institutions that are funding the research project.

Next, the team should consider which RtD practice to follow (Lab, Field, or Showroom) or if they wish to mix two of the practices together. Once they make a selection, we suggest a literature review to discover examplars of this kind of RtD research. One or two exemplars can provide scaffolding to guide the process.

After choosing the exemplar(s), the team can begin *design* activities. They should conduct a literature review to understand the state of the art and the questions and concerns of other researchers working in this space. They can then continue by conducting fieldwork, by holding a design workshop, by playing with a new material, or by exploring ideas in the studio. In these early stages of the project, the team is searching to understand what the state of the world is and how they might offer a new perspective, a new problem framing, which provides a path to a preferred future.

Once they have an initial framing, the team can explore by creating new product/service ideas and then select and iteratively evolve and refine an idea into a completed form. Through their process of making and critiquing, the team should *evaluate* and continually challenge their initial framing. In a sense, each new concept they generate will offer a different framing through its embodiment of a solution, and part of the critique is to explicate the proposal that is embedded in the designer and in their solution. Throughout this process, the team should document their design moves, the rationale (Moran & Carroll, 1996) for these moves, and how different hunches did and did not work out. In addition, the team should reflect on how their framing of the situation evolves and work to capture the reasons their framing changes.

When the team has an artifact they like, they should evaluate it based on the concerns of the specific RtD practice they selected (Lab, Field, or Showroom) and on concerns specific to their research question. Work following a Lab practice will most likely result in several similar artifacts that can be assessed in a lab study. Work following the Field practice will most likely place a working prototype into the field and assess if it produces the intended behaviors and outcomes (See "Field Deployments: Knowing from Using in Context"). In addition, the researchers will look for the creation of new practices as people attempt to incorporate this new thing into their lives. Finally, work following the Showroom practice will likely involve the installation of a working system in a gallery or in some other place where people outside of the research team can experience the design and can begin to question the world around them.

Following the evaluations, the team should reflect on what they have learned and then work to disseminate the research. Dissemination can happen in terms of publication in peer-reviewed venues such as conferences or journals. It might also take the form of a video or demonstration. Finally, for some RtD projects, the work may result in a working system that remains in use by people long after the research project has ended, provoking designers to think about the next problematic situation and solution.

The final step in the process is to repeat. Koskinen et al. (2011) note that RtD researchers who produce the best research results do so by repeatedly investigating the same situation. It is through the development of research programs much more than through individual projects that the best results emerge (See Koskinen et al., 2011: Chap. 10: Building Research Programs).

Exemplars we recommend looking at include:

Rich Interaction Design (Lab)

1. Joep Frens' investigation of rich interaction through the design of camera that rejects the convention of buttons with labels. (Frens, 2006a, 2006b).
2. Philip Ross' work on ethics and aesthetics through the design of a lamp. (Ross, 2008; Ross & Wensveen, 2010).
3. Camille Mousette's investigation of sketching with haptics (Although this lacks the formal lab studies of these other examples, it does have workshops run with designers to see if the methods of sketching could be transferred to these designers) (Moussette, 2012).

Participatory Design and User-centered Design (Field)

1. Hutchinson et al.'s work on technology probes produced a fascinating design method where researchers design and implement a working system which they then place into the hands of users for long periods of time in order to observe appropriation (Hutchinson et al., 2003).
2. Scott Davidoff et al.'s work on a smart home system that helps busy parents with the logistics around picking up and dropping off children (Davidoff, Lee, Dey, & Zimmerman, 2007; Davidoff, Lee, Yiu, Zimmerman, & Dey, 2006; Davidoff, Zeibart, Zimmerman, & Dey, 2011; Davidoff, Zimmerman, & Dey, 2010)
3. John Zimmerman et al.'s work on Tiramisu, a service design project that allows transit riders to coproduce real-time arrival information with their transit service (Yoo, Zimmerman, & Hirsch, 2013; Yoo, Zimmerman, Steinfeld, & Tomasic, 2010; Zimmerman et al., 2011).
4. Sara Ljungblad's work on connecting people's marginal practices to design innovation and how she integrated lomography into camera phones (2007). The work does seem to foreshadow commercial products like Instagram.

Critical Design (Showroom)

1. Bill Gaver et al.'s Drift Table, which challenges the HCI assumption that interaction design must have a clear intention for how it should be used (Gaver et al., 2004).

2. Bill Gaver et al.'s Prayer Companion, which pushed digital information into sacred places (Gaver et al., 2010).
3. Eric Paulos et al.'s work on Jetsam, a system that publically displayed items people threw into public trashcans (Paulos & Jenkins, 2005).
4. James Pierce et al.'s work to get people to rethink their ability to produce, store, and share energy (Pierce & Paulos, 2010).

Exercises

1. Generate three questions that can be answered by RtD. How would you characterize them?
2. How does research through design compare with design?

References

Alexander, C., Ishikawa, S., Silverstein, M., Jacobson, M., Fiksdahl-King, I., & Angel, S. (1977). *A Pattern Language: Towns, Buildings, Construction*. New York: Oxford University Press.

Basballe, D., & Halskov, K. (2012). Dynamics of Research through Design. In *Proceedings of the Conference on Designing Interactive Systems* (pp. 58–67). ACM Press.

Blackwell, A. (2004). *Accessible abstractions for homes, schools and offices*. Seminar Talk given on September 8, 2004 at Carnegie Mellon's HCI Institute.

Buchanan, R. (1995). Wicked Problems in Design Thinking. In V. Margolin & R. Buchanan (Eds.), *The Idea of Design* (pp. 3–20). Cambridge, MA: MIT Press.

Buchanan, R. (2001). Design research and the new learning. *Design Issues, 17*(4), 3–23.

Buchenau, M., & Suri, J. F. (2000). Experience Prototyping. In *Proceedings of the Conference on Designing Interactive Systems* (pp. 424–433). ACM Press.

Carroll, J. M., & Kellogg, W. A. (1989). Artifact as Theory-Nexus: Hermeneutics Meets Theory-based Design. In *Proceedings of the Conference on Human Factors in Computing Systems* (pp. 7–14). ACM Press.

Cross, N. (1999). Design Research: A Disciplined Conversation. *Design Issues, 15*(2), 5–10.

Darrah, C. N., English-Lueck, J., & Freeman, J. (2001). Families at work: An ethnography of dual career families. *Report for the Sloane Foundation* (Grant Number 98-6-21).

Davidoff, S., Lee, M. K., Dey, A. K., & Zimmerman, J. (2007). Rapidly Exploring Application Design through Speed Dating. In *Proceedings of the Conference on Ubiquitous Computing* (pp. 429–446). Springer.

Davidoff, S., Lee, M. K., Yiu, C. M., Zimmerman, J., & Dey, A. K. (2006). Principles of Smart Home Control. In *Proceedings of the Conference on Ubiquitous Computing* (pp. 19–34). Springer.

Davidoff, S., Zeibart, B., Zimmerman, J., & Dey, A. (2011). Learning Patterns of Pick-ups and Drop-offs to Support Busy Family Coordination. In *Proceedings of the Conference on Human Factors in Computing Systems* (1175–1184). ACM Press.

Davidoff, S., Zimmerman, J., & Dey, A. (2010). How Routine Learners can Support Family Coordination. In *Proceedings of the Conference on Human Factors in Computing Systems* (pp. 2461–2470). ACM Press.

Djajadiningrat, T., Overbeeke, K., & Wensveen, S. (2002). But how, Donald, tell us how?: On the creation of meaning in interaction design through feedforward and inherent feedback. In *Proceedings of the Conference on Designing Interactive Systems* (pp. 285–291). ACM Press.

Dunne, A. (1999). *Hertzian Tales: Electronic Products, Aesthetic Experience, and Critical Design*. Cambridge, MA: MIT Press.

Dunne, A., & Raby, F. (2001). *Design Noir: The Secret Life of Electronic Objects*. Basel, Switzerland: Birkhäuser.

Fallman, D. (2003). Design-oriented Human-computer Interaction. In *Proceedings of the Conference on Human Factors in Computing Systems* (pp. 225–232). ACM Press.

Forlizzi, J., Zimmerman, J., & Evenson, S. (2008). Crafting a Place for Interaction Design Research in HCI. *Design Issues, 24*(3), 19–29.

Forlizzi, J., Zimmerman, J., & Stolterman, E. (2009). From Design Research to Theory: Evidence of a Maturing Field. In *Proceedings of the Conference on the International Association for Societies of Design Research*. IASDR Press.

Frayling, C. (1993). Research in Art and Design. *Royal College of Art Research Papers, 1*(1), 1–5.

Frens, J. W. (2006a). *Designing for Rich Interaction: Integrating Form, Interaction, and Function*. Doctoral Dissertation, Eindhoven University of Technology, Eindhoven, The Netherlands.

Frens, J. (2006b). A Rich User Interface for a Digital Camera. *Personal and Ubiquitous Computing, 10*(2), 177–180.

Gaver, W., Blythe, M., Boucher, A., Jarvis, N., Bowers, J., & Wright, P. (2010). The Prayer Companion: Openness and Specificity, Materiality and Spirituality. In *Proceedings of the Conference on Human Factors in Computing Systems* (pp. 2055–2064). ACM Press.

Gaver, W. W., Bowers, J., Boucher, A., Gellerson, H., Pennington, S., Schmidt, A., et al. (2004). The Drift Table: Designing for Ludic Engagement. In *Extended Abstracts of the Conference on Human Factors in Computing Systems* (pp. 885–900). ACM Press.

Giller, V., Tscheligi, M., Sefelin, R., Mäkelä, A., Puskala, A., & Karvonen, K. (1999). Maypole Highlights: Image makers. *Interactions, 6*(6), 12–15.

Hutchinson, H., Mackay, W., Westerlund, B., Bederson, B. B., Druin, A., Plaisant, C., et al. (2003). Technology probes: Inspiring design for and with families. In *Proceedings of CHIo3* (pp. 17–23). New York, NY: ACM Press.

Koskinen, I., Zimmerman, J., Binder, T., Redstrom, J., & Wensveen, S. (2011). *Design Research through Practice: From the Lab, Field, and Showroom*. Boston: Morgan Kaufmann.

Kuhn, S. (1997). Design for People at Work. In T. Winograd (Ed.), *Bringing Design to Software* (pp. 273–305). Menlo Park, CA: Addison Wesley Publishing Company.

Lee, M. K., Forlizzi, J., Rybski, P. E., Crabbe, F., Chung, W., Finkle, J., et al. (2009). The Snackbot: Documenting the Design of a Robot for Long-term Human-robot Interaction. In *Proceedings of the Conference on Human-Robot Interaction* (pp. 7–14). ACM Press.

Lee, M. K., Kiesler, S., & Forlizzi, J. (2011). Mining behavioral economics to design persuasive technology for healthy choices. In *Proceedings of thethe Conference on Human Factors in Computing Systems* (pp. 325–334). ACM Press.

Lee, M. K., Kiesler, S., Forlizzi, J., & Rybski, P. (2012). Ripple effects of an embedded social agent: A field study of a social robot in the workplace. In *Proceedings of the Conference on Human Factors in Computing Systems* (pp. 695–704). ACM Press.

Ljungblad, S. (2007). Designing for New Photographic Experiences: How the Lomographic Practice Informed Context Photography. In *Proceedings of the Conference on Designing Pleasurable Products and Services* (pp. 357–374). ACM Press.

Moran, T., & Carroll, J. (1996). *Design Rationale: Concepts, Techniques, and Use*. Mahwah, NJ: Lawrence Erlbaum Associates.

Moussette, C. (2012). Simple Haptics: Sketching Perspectives for the Design of Haptic Interactions. Doctoral Dissertation, Umea University, Umea, Sweden.

Nelson, H. G., & Stolterman, E. (2012). *The Design Way: Intentional Change in a Unpredictable World*. Cambridge, MA: MIT Press.

Odom, W., Zimmerman, J., & Forlizzi, J. (2011). Teenagers and their Virtual Possessions: Design Opportunities and Issues. In *Proceedings of the Conference on Human Factors in Computing Systems* (pp. 1491–1500). ACM Press.

Orr, J. (1986). Narratives at Work: Story telling as Cooperative Diagnostic Activity. In *Proceedings of the Conference on Computer Supported Cooperative Work* (pp. 62–72). ACM Press.

Paulos, E., & Jenkins, T. (2005). Urban Probes: Encountering our Emerging Urban Atmospheres. In *Proceedings of Conference on Human Factors in Computing Systems* (pp. 341–350). ACM Press.
Pierce, J., & Paulos, E. (2010). Materializing Energy. In *Proceedings of Conference on Designing Interactive Systems* (pp. 113–122). ACM Press.
Rittel, H. W. J., & Webber, M. M. (1973). Dilemmas in a General Theory of Planning. *Policy Sciences, 4*(2), 155–169.
Ross, P. R. (2008). *Ethics and Aesthetics in Intelligent Product and System Design.* Doctoral Dissertation, Eindhoven University of Technology, Eindhoven, The Netherlands.
Ross, P. R., Overbeeke, C. J., Wensveen, S. A., & Hummels, C. M. (2008). A Designerly Critique on Enchantment. *Journal of Personal and Ubiquitous Computing, 12*(5), 359–371.
Ross, P. R., & Wensveen, S. A. G. (2010). Designing Aesthetics of Behavior in Interaction: Using Aesthetic Experience as a Mechanism for Design. *International Journal of Design, 4*(2), 3–13.
Schön, D. (1983). *The Reflective Practitioner*. London: Temple Smith.
Schön, D., & Bennett, J. (1996). Reflective Conversation with Materials. In T. Winograd (Ed.), *Bringing Design to Software* (pp. 171–189). Menlo Park, CA: Addison Wesley Publishing Company.
Shneiderman, B. (1983). Direct Manipulation: A Step Beyond Programming Languages. *IEEE Transactions on Computers, 16*(8), 57–69.
Simon, H. (1996). *The Sciences of the Artificial* (3rd ed.). Cambridge, MA, USA: MIT Press.
Stolterman, S. (2008). The Nature of Design Practice and Implications for Interaction Design Research. *International Journal of Design, 2*(1), 55–65.
Sutherland, I. E. (1963). Sketchpad: A Man-machine Graphical Communication System. In *Proceedings of the Spring Joint Computer Conference* (pp. 329–346). Spartan Press.
Yoo, D., Zimmerman, J., & Hirsch, T. (2013). Probing Bus Stop for Insights on Transit Co-design. In *Proceedings of the Conference on Human Factors in Computing Systems* (pp. 1797–1806). ACM Press.
Yoo, D., Zimmerman, J., Steinfeld, A., & Tomasic, A. (2010). Understanding the Space for Co-design in Riders' Interactions with a Transit Service. In *Proceedings of the Conference on Human Factors in Computing Systems* (pp. 1797–1806). ACM Press.
Zimmerman, J. (2009). Designing for the Self: Making Products that Help People Become the Person they Desire to be. In *Proceedings of the Conference on Human Factors in Computing Systems* (pp. 395–404). ACM Press.
Zimmerman, J., Evenson, S., Baumann, K., & Purgathofer, P. (2004). Workshop on the Relationship Between Design and HCI. In *Extended Abstracts of the Conference on Human Factors in Computing Systems* (pp. 1741–1742). ACM Press.
Zimmerman, J., Evenson, S., & Forlizzi, J. (2004). Discovering and Extracting Knowledge in the Design Project. In *Proceedings of Future Ground.* Design Research Society.
Zimmerman, J., & Forlizzi, J. (2008). The Role of Design Artifacts in Design Theory Construction. *Artifact, 2*(1), 41–45.
Zimmerman, J., Forlizzi, J., & Evenson, S. (2007). Research through Design as a Method for Interaction Design Research in HCI. In *Proceedings of the Conference on Human Factors in Computing Systems* (pp. 493–502). ACM Press.
Zimmerman, J., Stolterman, E., & Forlizzi, J. (2010). An Analysis and Critique of Research through Design: Towards a Formalization of a Research Approach. In *Proceedings of the Conference on Designing Interactive Systems* (pp. 310–319). ACM Press.
Zimmerman, J., Tomasic, A., Garrod, C., Yoo, D., Hiruncharoenvate, C., Aziz, R., et al. (2011). Field Trial of Tiramisu: Crowd-sourcing Bus Arrival Times to Spur Co-design. In *Proceedings of Conference on Human Factors in Computing Systems* (pp. 1677–1686). ACM Press.

Experimental Research in HCI

Darren Gergle and Desney S. Tan

Experimental Research in HCI

The experimental method is a technique used to collect data and build scientific knowledge, and it is one of the primary methodologies for a wide range of disciplines from biology to chemistry to physics to zoology, and of course human–computer interaction (HCI).

In this chapter, we learn about the basics of experimental research. We gain an understanding of critical concepts and learn to appreciate the ways in which experiments are uniquely suited to answer questions of causality. We also learn about best practices and what it takes to design, execute, and assess good experimental research for HCI.

A Short Description of Experimental Research

At its heart, experimental research aims to show how the manipulation of one variable of interest has a direct causal influence on another variable of interest (Cook & Campbell, 1979). Consider the research question, "How does the frame rate of a video affect human perception of fluid movement?"

Breaking this down, we can examine several of the elements necessary for good experimental research. The first has to do with the notion of *causality*. Our example question implicitly posits that a change in one variable, in this case frame rate, causes variation in another variable, the perception of fluid movement. More

D. Gergle (✉)
Northwestern University, 2240 Campus Drive, Evanston, IL 60208, USA
e-mail: dgergle@northwestern.edu

D.S. Tan
Microsoft Research, One Microsoft Way, Redmond, WA 98052, USA
e-mail: desney@microsoft.com

J.S. Olson and W.A. Kellogg (eds.), *Ways of Knowing in HCI*,
DOI 10.1007/978-1-4939-0378-8_9,

generally, we often think of two variables, *X* and *Y*; and establishing the notion of causality, which implies that changes in *X* lead to changes in *Y*.

The second thing to note is the idea of *variables*. The researcher needs to manipulate the levels or degree of one or more variables, known as the *independent variables*, while keeping constant other extraneous factors. In this example, our independent variable is frame rate, and we could show the same video at different frame rates, while controlling for other factors such as brightness, screen size, etc. It is also important that we are able to measure the effect that these manipulations have on one or more *dependent variables*. In this case, our dependent variable may be a rating score that captures human perception of fluid movement.

The third thing to note is that our initial question could be formally stated as a *hypothesis* regarding the predicted relationship between frame rate and perception of fluid movement. For example, "An increase in frame rate will increase human perception of fluid movement." The formulation of a hypothesis is important in that it clearly states the parameters of the experiment and communicates the expected relationship. The observed data are then subjected to statistical analysis to provide evidence for or against the hypothesized relationship.

Finally, true experiments require *random assignment* of participants to experimental conditions. Random assignment is critical in establishing equivalent participant groups (with some probability) on both measured and unmeasured characteristics at the outset of the study. This safeguards against systematic biases in assignment of the participants to the experimental conditions, and increases the likelihood that differences across the groups result solely from the treatment to which they are assigned. Without random assignment there exists a risk that attributes of the participants drive the changes in the dependent variable.

Returning to our frame rate example, imagine running a study in which one group of participants watches a video at a low frame rate and a second group watches the same video at a much higher frame rate. You cleverly devise a way to measure perception of fluid movement, recruit participants to come to the lab, and assign the first ten arrivals to the high frame rate condition and the next ten arrivals to the low frame rate condition. After collecting and analyzing your data you find—counter to your hypothesis—that the individuals in the high frame rate condition rated the video as less fluid. Upon further reflection you realize that the participants that showed up first did so because they have a personality type that makes them the kind of person to arrive early. It just so happens that this personality trait is also associated with greater attention to detail and as a result they rate things more critically than the late arrivals. When you do not make use of random assignment, you increase the risk of such confounds occurring.

History, Intellectual Tradition, Evolution

To gain a deeper sensitivity to the role experimental research plays in HCI today, it is helpful to trace its roots, which go back to the development and formalization of the scientific method. Aristotle is often credited in developing initial ideas toward the search for "universal truths," and the scientific method was popularized and experienced a major emergence with the work of Galileo and others in what is known as the Scientific Revolution of the sixteenth through eighteenth centuries. In a nutshell, scientific inquiry aims to understand basic relations that exist between circumstances and behaviors, with the ultimate goal of aggregating this understanding into a formal body of knowledge.

While experimental research was originally developed as a paradigm for the physical sciences to establish scientific principles and laws, starting in the late nineteenth and early twentieth centuries, psychologists such as Wilhelm Wundt and G. Stanley Hall developed experimental laboratories to investigate human thought and behavior. It quickly became apparent that humans posed a particular challenge for measurement. If humans behaved in a systematic and consistent fashion like the physical world, the application of the scientific method to questions of human behavior would be straightforward. But they do not; individuals vary in their behavior from one moment to the next, and across individuals there can be enormous variability.

As a result of this, researchers in psychology, sociology, cognitive science and information science, as well as the social sciences more broadly, developed new research techniques that were more appropriate for dealing with the vagaries of human behavior in a wide variety of contexts. Most of this early research stayed close to the ideals of the traditional sciences by applying the techniques to support systematic knowledge production and theoretical development regarding human behavior.

As the field of HCI evolved, it became clear that experimental research was useful not only for generating hypothesis-driven knowledge and theoretical advancement but also for informing practical and applied goals. In a recent piece entitled, "Some Whys and Hows of Experiments in Human–Computer Interaction," Hornbæk (2011, pp. 303–305) further argues that experimental research is suitable for investigating process details in interaction as well as infrequent but important events by virtue of the ability to recreate them in a controlled setting. He also highlights the benefits of sidestepping problems with self-reports that stem from faulty human judgments and reflections regarding what lies behind our behaviors and feelings during interaction.

Using an approach known as A/B testing, controlled online experiments are used at large Internet companies such as Google, Microsoft, or Facebook to generate design insights and stimulate innovation (Kohavi, Henne, & Sommerfield, 2007; Kohavi & Longbotham, 2007; Kohavi, Longbotham, & Walker, 2010). Accordingly, some HCI research is more theoretically driven (e.g., Accot & Zhai, 1997; Gergle, Kraut, & Fussell, 2013; Hancock, Landrigan, & Silver, 2007; Wobbrock, Cutrell, Harada, & MacKenzie, 2008), while other research is more engineering-driven with the goal to demonstrate the utility of a technology from a more applied perspective (e.g., Gutwin & Penner, 2002; Harrison, Tan, & Morris, 2010; MacKenzie & Zhang, 1999; Nguyen & Canny, 2005).

Experimental techniques are also widely used in usability testing to help reveal flaws in existing designs or user interfaces. Whether evaluating if one user interface design is better than another; showing how a new recommender system algorithm influences social interaction; or assessing the quality, utility, or excitement engendered by a new device when we put it to use in the world, good experimental research practices can be applied to make HCI more rigorous, informative and innovative. In fact, many of the benefits of experimental research and its techniques can be seen in HCI studies ranging from tightly controlled laboratory experiments (e.g., MacKenzie & Zhang, 1999; Veinott, Olson, Olson, & Fu, 1999) to "in the wild" field experiments (e.g., Carter, Mankoff, Klemmer, & Matthews, 2008; Cosley, Lam, Albert, Konstan, & Riedl, 2003; Evans & Wobbrock, 2012; Koedinger, Anderson, Hadley, & Mark, 1997; Oulasvirta, 2009).

Advantages of Experimental Research

As a methodology, experimentation has a number of advantages over other HCI research methods. One of the most commonly recognized advantages hinges on its *internal validity*,[1] or the extent to which the experimental approach allows the researcher to minimize biases or systematic error and demonstrate a strong causal connection. When done properly it is one of the few methodologies by which cause and effect can be convincingly established.

In Rosenthal and Rosnow's terms, experimental research focuses on the identification of causal relationships of the form "*X* is responsible for *Y*." This can be contrasted with two other broad classes of methodologies: descriptive studies that aim to capture an accurate representation of what is happening and relational studies that intend to capture the relationship between two variables but not necessarily a causal direction (see Rosenthal & Rosnow, 2008, pp. 21–32).

The experimental method uses precise control of the levels of the independent variable along with random assignment to isolate the effect of the independent variable upon a dependent variable. It also permits the experimenter to build up models of interactions among variables to better understand the differential influence of a variable across a range of others.

It also makes use of quantitative data that can be analyzed using inferential statistics. This allows for statistical and probabilistic statements about the likelihood of seeing the results, and discussion about the size of the effect in a way that is meaningful when comparing to other hypothesized sources of influence.

Experimental research also provides a systematic process to test theoretical propositions and advance theory. A related advantage is that experiments can be replicated and extended by other researchers. Over time, this increases our confidence in the findings and permits the generalization of results across studies, domains, and to wider populations than initially studied. This supports the development of more universal principles and theories that have been examined by a number of independent researchers in a variety of settings.

[1] Much of what makes for good experimental design centers on minimizing what are known as threats to internal validity. Throughout this chapter we address many of these including construct validity, confounds, experimenter biases, selection and dropout biases, and statistical threats.

Limitations of Experimental Research

In general, experimental research requires well-defined, testable hypotheses, and a small set of well-controlled variables. However, this may be difficult to achieve if the outcomes depend on a large number of influential factors or if carefully controlling those factors is impractical. If an important variable is not controlled for, there is a chance that any relationship found could be misattributed.

While an advantage of experimental research is internal validity, the flipside is that these benefits may come at the risk of low *external validity*. External validity is the degree to which the claims of a study hold true for other contexts or settings such as other cultures, different technological configurations, or varying times of the day. A side effect of controlling for external factors is that it can sometimes lead to overly artificial laboratory settings. This increases the risk of observing behavior that is not representative of more ecologically valid settings.

That said, when designing a study there are ways to bolster external validity. Olson and colleagues' paper on group design processes (Olson, Olson, Storrøsten, & Carter, 1993) exemplifies three ways to increase external validity when designing an experiment. First, they chose a task that was a good match for the kinds of activities they had observed in the field—designing an automatic post office—and they tested the task with real software developers to ensure it was an accurate portrayal of everyday work activities. Second, they chose participants for the study that were as close as possible to those they studied in the field. In this case they chose MBA students with at least 5 years of industry experience and who had already worked together on group projects. Third, they assessed the similarity of the behaviors between the laboratory study and their fieldwork on several key measures such as time spent on specific aspects of design and characteristics of the discussions (see Olson et al., 1993, pp. 333–335 and Fig. 4).

Another common challenge for HCI researchers is that they often want to show that their system is "just as good" as another system on some measures while having advantages in other areas. A common mistake is to treat a lack of significance as proof that no difference exists. To effectively establish that things are "just as good" a form of equivalence testing is needed; effect sizes, confidence intervals, and power analysis[2] techniques can be used to show that the effect either does not exist or is so small that it is negligible in any practical sense (for details see Rogers, Howard, & Vessey, 1993).

Furthermore, it should be recognized that hypotheses are never really "proven" in an absolute sense. Instead, we accrue evidence in support of or against a given hypothesis, and over time and repeated investigation support for a position is strengthened. This is critical and points to the importance of replication in experimental work. However, replication is often less valued (and thus harder to publish) in HCI than the novelty of invention. We argue, along with several colleagues, that

[2] G*Power 3 is a specialized software tool for power analysis that has a wide number of features and is free for noncommercial use. It is available at http://www.gpower.hhu.de

as the field matures, replication and extension should become more valued outcomes of HCI research (Wilson, Mackay, Chi, Bernstein, & Nichols, 2012).

Finally, because experimental research is often taught early in educational programs and hence is a familiar tool, it is sometimes force-fit into situations where research questions might have been more appropriately addressed using less formal instantiations of the experimental method or by using other methodologies (for a critique and response, see Lieberman, 2003; Zhai, 2003). A poorly executed experiment may have the veneer of "scientific validity" because of the methodological rigor, but ultimately provides little more than well-measured noise.

How to Do It

In HCI, we often want to compare one design or process to another, decide on the importance of a possible problem or solution, or evaluate a particular technology or social intervention. Each of these challenges can be answered using experimental research. But how do you design an experiment that provides robust findings?

Hypothesis Formulation

Experimental research begins with the development of a statement regarding the predicted relationship between two variables. This is known as a research hypothesis. In general, hypotheses clarify and clearly articulate what it is the researcher is aiming to understand. A hypothesis both *defines the variables involved* and *the relationship between them*, and can take many forms: A causes B; A is larger, faster, or more enjoyable than B; etc.

A good hypothesis has several characteristics. First, the hypothesis should be *precise*. It should clearly state the conditions in the experiment or state the comparison with a control condition. It should also describe the predicted relationship in terms of the measurements used.

Second, the hypothesis should be *meaningful*. One way it can be meaningful is by leading to the development of new knowledge, and in doing so it should relate to existing theories or point toward new theories. Hypotheses in the service of applied contributions can also be meaningful as they reveal something about the design under investigation and can convince us that a new system is more efficient, effective, or entertaining than the current state-of-the-art.

Third, the described relationship needs to be *testable*. You must be able to manipulate the levels of one variable (i.e., the independent variable) and accurately measure the outcome (i.e., the dependent variable). For example, you could be highly influenced by "The Truman Show" (Weir, 1998) and hypothesize that "we are living in a large fish tank being watched by other humans with which we can have no contact." While the statement may or may not be true, it is not testable and therefore it is speculation and not a scientific hypothesis.

Finally, the predicted relationship must be *falsifiable*. A common example used to demonstrate falsifiability examines the statement, "Other inhabited planets exist in the universe." This is testable as we could send out space probes and show that there are other inhabited planets. However, the lack of detection of inhabited planets cannot falsify the statement. You might argue, "what if every single planet is observed?", but it could be that the detection mechanisms we use are simply not sensitive enough. Therefore, while this statement could be true, and even shown to be true, it is not falsifiable and thus is not an effective scientific hypothesis. You must be able to disprove the statement with empirical data.

Evaluating Your Hypothesis

Once you have established a good hypothesis, you need to demonstrate the degree to which it holds up under experimental scrutiny. Two common approaches for doing this are hypothesis testing and estimation techniques.

Hypothesis Testing

Hypothesis testing, specifically null hypothesis significance testing, is widely used. In the context of HCI, this approach often aims to answer the question "Does it work?" or "Are the groups different?"

The first step in null hypothesis significance testing is to formulate the original research hypothesis as a *null hypothesis* and an *alternative hypothesis*.[3] The null hypothesis (often written as H_0) is set up as a falsifiable statement that predicts no difference between experimental conditions. Returning to our example from the beginning of the chapter, the null hypothesis would read, "Different frame rates *do not* affect human perception of fluid movement." The alternative hypothesis (often written as H_A or H_1) captures departures from the null hypothesis. Continuing with the example, "different frame rates *do* affect human perception of fluid movement."

The second step is to decide on a significance level. This is a prespecified value that defines a tolerance for rejecting the null hypothesis when it is actually true (also known as a Type I error). More formally, this is stated as alpha (α) and it captures the conditional probability, Pr(reject $H_0|H_0$ true). While a somewhat arbitrary choice, the convention of $\alpha = 0.05$ is often used as the threshold for a decision.

The third step is to collect the data (this is a big step that is addressed later in the chapter) and then apply the appropriate statistical test to obtain a p value.

[3] Here we present the Neyman–Pearson approach to hypothesis testing as opposed to Fisher's significance testing approach. Lehmann (1993) details the history and distinctions between these two common approaches.

The p value tells you the probability of obtaining the observed data, or more extreme data, if the null hypothesis were true. More formally, Pr(observed data|H_0 true). Therefore, a low p value indicates that the observed results are unlikely if the null hypothesis were true.

The final step compares the observed p value with the previously stated significance level. If $p < \alpha$, then you reject the null hypothesis. Thus, by rejecting the null hypothesis that "Different frame rates do not affect human perception of fluid movement," we bolster the evidence that different frame rates may affect human perception of fluid movement (i.e., we gather additional support for the alternative hypothesis).

While methodologically straightforward to apply, you should recognize concerns with this methodology, so as not to accidentally misinterpret results. These concerns center on its dichotomous "accept" or "reject" outcome, widespread misinterpretation and faulty reporting of results, and inattention to the magnitude of effects and their practical significance (Cohen, 1994; Cumming, 2012, pp. 8–9; Johnson, 1999; Kline, 2004). Several common misunderstandings stem from a misinterpretation of statistical results such as the mistaken belief that a p value indicates the probability of the result occurring because of sampling error or that $p < .05$ means the chances of a Type I error occurring are less than 5 %. Other common mistakes stem from faulty conclusions drawn after accepting or rejecting the null hypothesis such as suggesting that the failure to reject the null hypothesis is proof of its validity, or the common misperception that a smaller p value means a larger effect exists. Finally, researchers should not lose sight of the fact that statistical significance does not imply substantive significance or practical importance. For a detailed description of these and other common mistakes see (Kline, 2013, pp. 95–103).

Estimation Techniques

While the notion of a null hypothesis can be useful to understand the basic logic of the experimental methodology, null hypothesis testing is rarely adequate for what we really want to know about the data. To address some of the challenges of traditional hypothesis testing approaches, contemporary methods rely on *estimation techniques* that focus on establishing the magnitude of an effect through the application of confidence intervals and effect sizes (for recent coverage see Cumming, 2012; Kline, 2013, pp. 29–65).[4] Accessible and thorough descriptions on various estimation techniques can be found in (Cumming, 2012; Cumming & Finch, 2001; Ellis, 2010; Kelley & Preacher, 2012). Bayesian statistics are another alternative that provide greater capability to estimate and compare likelihoods for various hypotheses. For introductions to the Bayesian approach see (Kline, 2013, pp. 289–312; Kruschke, 2010).

[4] We return to effect sizes and confidence intervals in the section "What constitutes good work," where we describe how they can be used to better express the magnitude of an effect and its real world implications.

Estimation techniques retain the notion of a research hypothesis and accruing evidence for or against it, but the emphasis is on quantifying the magnitude of an effect or showing how large or small differences are between groups, technologies, etc. In the context of HCI, estimation approaches aim to answer more sophisticated questions such as, "How well does it work across a range of settings and contexts?" or "What is the size and relative importance of the difference between the groups?" In other words, it aims to quantify the effectiveness of a given intervention or treatment and focuses the analysis on the size of the effect as well as the certainty underlying the claim. This approach may be more appropriate for applied disciplines such as HCI (Carver, 1993) as it shifts the emphasis from statistical significance to the size and likelihood of an effect, which are often the quantities we are more interested in knowing.

Variables

The choice of the right variables can make or break an experiment and it is one of the things that must be carefully tested before running an experiment. This section covers four types of variables: independent, dependent, control variables, and covariates.

Independent Variable

The *independent variable* (*IV*) is manipulated by the researcher, and its conditions are the key factor being examined. It is often referred to as *X*, and it is the presumed cause for changes that occur in the dependent variable, or *Y*.

When choosing an IV, a number of factors should be taken into account. The first is that the researcher can establish *well-controlled variation* in its conditions or levels. This can be accomplished by manipulating the stimuli (e.g., the same movie recorded at different frame rates), instructions (e.g., posing a task as cooperative vs. competitive), or using measured attributes such as individual differences (e.g., selecting participants based on gender or education levels[5]). A group in the condition that receives the manipulation is known as the treatment group, and this group is often compared to a control group that receives no manipulation.

The second is the ability to provide a clear *operational definition* and confirm that your IV has the intended effect on a participant. You need to clearly state how the IV was established so that other researchers could construct the same variable and replicate the work. In some cases, this is straightforward as when testing different input devices (e.g., trackpad vs. mouse). In other cases it is not. For example, if

[5] When using measures such as education level or test performance, you have to be cautious of regression to the mean and be sure that you are not assigning participants to levels of your independent variable based on their scores on the dependent variable or something strongly correlated with the DV (also known as sampling on the dependent variable) (Galton, 1886).

you vary exposure to a warning tone, the operational definition should describe the frequency and intensity of the tone, the duration of the tone, and so on. This can become especially tricky when considering more subjective variables capturing constructs such as emotional state, trustworthiness, etc. A challenge that must be addressed in the operational definition is to avoid an *operational confound*, which occurs when the chosen variable does not match the targeted construct or unintentionally measures or captures something else.

A *manipulation check* should be used to ensure that the manipulation had the desired influence on participants. It is often built into the study or collected at the conclusion. For example, if you were trying to experimentally motivate participants to contribute to a peer-production site such as OpenStreetMap,[6] a manipulation check might assess self-reported motivation at the end of the study in order to validate that your manipulation positively influenced motivation levels. Otherwise the measured behavior could be due to some other variable.

A third important factor to consider is the *range* of the IV (i.e., the difference between the highest and lowest values of the variable). Returning to the example of motivating OpenStreetMap contributions, the range of values you choose is important in determining whether or not motivation levels actually change for your participants. If you gave the "unmotivated" group one dollar, and the "motivated" group two dollars, the difference may not be enough to elicit a difference in cooperative behavior. Perhaps one dollar versus ten dollars may make a difference. It is important that the ranges are realistic and practically meaningful.

Another critical aspect to variable selection is choosing meaningful or interesting variables for your study. In practice this can be even more difficult than addressing the aspects described above. Good variables should be theoretically or practically interesting; they should help to change our way of thinking; they should aim to provide deeper understanding, novel insight, or resolve conflicting views in the literature. Knowing what others have studied and recognizing the gaps in the prior literature can help to achieve this goal.

Dependent Variable

The *dependent variable* (*DV*), often referred to as *Y*, is the outcome measure whose value is predicted to vary based upon the levels of the IV. Common types of dependent variables used in HCI research are self-report measures (e.g., satisfaction with an interface), behavioral measures (e.g., click-through rates or task completion times), and physiological measures (e.g., skin conductance, muscle activity, or eye movements). Picking a good DV is crucial to a successful experiment, and a key element of a good DV is the extent to which it can accurately and consistently capture the effect you are interested in measuring.

Reliability is important when choosing a DV. A measure is perfectly reliable if you get the same result every time you repeat the measurement under identical

[6] http://www.openstreetmap.org

conditions. There are many steps that help to increase the reliability of a DV[7] and decrease the variability that occurs due to measurement error. For each of your DVs, try to:

- *Clearly specify the rules for quantifying your measurement*: Similar to the construction of the IV, you need to be able to detail exactly how your DV was constructed and recorded. This includes formulating coding and scoring rules for the quantification of your measure, or detailing the calculations used when recording the value of your DV. If you cannot clearly articulate your rules you will likely introduce noise into your measure.
- *Clearly define the scope and boundaries of what you are going to measure*. You need to articulate the situations, contexts, and constraints under which you collect your data. For example, suppose you want to measure online content sharing by counting how many times in a session people perform link sharing to external web content. What counts as "a session?" What counts for "link sharing?" Does it have to be original content or can it be a copy of someone else's post? Does it have to be the actual link to a URL or could it be a snippet of content?

Validity is another important consideration when choosing your DV. It is not enough to know that a measure is reliable. It is also important to know that a measure captures the construct it is supposed to measure—if it does so it is considered a valid measure. The following lists ways to assess the validity of your measures, in order from weakest to strongest[8]:

- *Face validity* is the weakest form of validity. It simply means that your measure appears to measure what it is supposed to measure. For example, imagine you propose to measure online satisfaction with a web purchasing process by counting the number of positive emoticons that are present in the purchase comments. You feel that the more a person uses positive emoticons, the more satisfied they were, so "on its face" it is a valid measure.
- *Concurrent validity* uses more than one measure for the same construct and then demonstrates a correlation between the two measures at the same point in time. The most common way to examine concurrent validity is to compare your DV with a gold-standard measure or benchmark. However, concurrent validity can suffer from the fact that the secondary variable or benchmark for comparison may have the same inaccuracies as the DV under investigation.
- *Predictive validity* is a validation approach where the DV is shown to accurately predict some other conceptually related variable later in time. The prototypical example is the use of high-school GPA to predict first year's GPA in undergraduate classes.

---

[7] When developing new measures it is important to assess and report their reliability. This can be done using a variety of test–retest assessments.

[8] Sara Kiesler and Jonathon Cummings provided this structured way to think about dependent variables and assessing forms of reliability and validity.

- Best practice is to make use of *standardized* or *published* measures when available.[9] The major benefit is that a previously validated and published measure has been through a rigorous evaluation. However, the challenge in using preexisting measures is to make sure that they accurately capture the construct you want to measure.

The *range* of the DV is another important aspect to consider. A task that is so easy that everyone gets everything correct exhibits a "ceiling effect"; while a task so difficult that nobody gets anything correct exhibits a "floor effect." These effects limit the variability of measured outcomes, and as a result the researcher may falsely conclude there is no influence of the IV on the DV.

Related to range is the *sensitivity* of the dependent variable. The measure must be sensitive enough to detect differences at an appropriate level of granularity. For example, an eye tracker with an accuracy of 2° will not be able to capture a potentially meaningful and consistent difference of ½°.

The final thing to consider when selecting a DV is *practicality*. Some data are more accessible than others and therefore are more viable for a given study. Some practical aspects to consider: How often do the events occur? Will the cost of collecting the data be prohibitive? Can you access all of the data? Will your presence influence the behavior under observation?

Control Variable

In addition to independent and dependent variables, there are a number of potential variables that must remain constant; otherwise you run the risk of fluctuations in an unmeasured variable masking the effect of the independent variable on the dependent variable. A *control variable* is a potential IV that is held constant. For example, when running reaction time studies you need to control lighting, temperature, and noise levels and ensure that they are constant across participants. Holding these variables constant is the best way to minimize their effects on the dependent variable. Unlike an independent variable, a control variable is not meant to vary but rather stay constant in order to "control" for its influence on the DV. For any given experiment there are an infinite number of external variables, so researchers make use of theory, prior literature and good discretion to choose which variables to control.

Covariate

While a good experiment does its best to control for other factors that might influence the dependent variable, it is not always possible to do so for all extraneous

[9] It should be noted that numerous surveys and questionnaires published in the HCI literature were not validated or did not make use of validated measures. While there is still some benefit to consistency in measurement, it is less clear in these cases that the measures validly capture the stated construct.

variables. *Covariates* (or, somewhat confusingly, "control variables" in the regression sense) are additional variables that may influence the value of the dependent variable but that are not controlled by the researcher and therefore are allowed to naturally vary. These are often participant baseline measures or demographic variables for which there is theoretical rationale or prior evidence suggesting a correlation to the dependent variable. The idea is that they need to be controlled because random assignment is not perfect, particularly in small samples, and therefore experimental groups may not have been completely equivalent before the treatment. When this is the case, covariates can be used to control for potential confounds and can be included in the analysis as statistical controls.

Research Designs

Up to this point we have discussed the basic components of experimentation. In this section we examine various research designs that bring together these components in ways to best accrue evidence for a research hypothesis. While there are several texts that provide extensive coverage of experimental designs, we focus on designs most commonly used in HCI research. We examine randomized experiments (also known as "true experiments") and quasi-experiments and discuss the differences between the two designs.

Randomized Experiments

We begin by examining a class of experiments known as randomized experiments (Fisher, 1925). Their distinguishing feature is that participants are *randomly assigned* to conditions, as this results in groups that, on average, are similar to one another (Shadish, Cook, & Campbell, 2002, p. 13). In order to keep from conflating attributes of the participants with the variables under investigation, randomized, unbiased assignment of participants to the various experimental conditions is required for all of these study designs. This can often be done through a coin toss, use of a table of random numbers, or a random number generator.[10]

We begin by describing single-factor designs that allow us to answer questions about the relationship between a single IV and a single DV. We then move on to examine more advanced designs for multiple IVs and a single DV (known as *factorial designs*) as well briefly discuss those designs involving multiple IVs and multiple DVs.

---

[10] Lazar and colleagues (Lazar, Feng, & Hochheiser, 2010, pp. 28–30) provide a step-by-step discussion of how to use a random number table to assign participants to conditions in various experimental designs. In addition, numerous online resources exist to generate tables for random assignment to experimental conditions (e.g., http://www.graphpad.com/quickcalcs/randomize1.cfm).

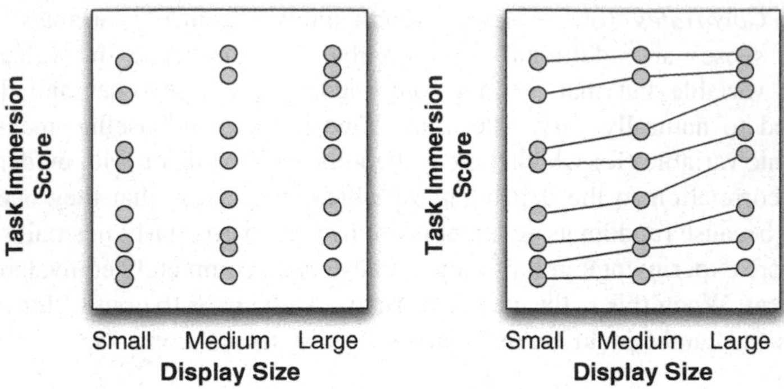

Fig. 1 Example demonstrating the ability to more easily detect differences with within-subjects design (*right*) as compared to a between-subjects design (*left*) when there are large individual differences in participants' scores

Between-Subjects Design

The *between-subjects design* is one of the most commonly used experimental designs and is considered by many to be the "gold standard" of randomized experimental research. Participants are randomly assigned to a single condition (also known as a level of the IV).

Consider, as an example, a rather simple research question that aims to assess the effect that *display size has on task immersion*. Your independent variable is display size, and it has three conditions: small, medium, and large. You also have a single dependent variable: a behavioral measure of task immersion. Let us also assume that you have 24 participants enrolled in the study. In a between-subjects design, you would assign eight participants to the small display size condition, eight to the medium display size condition, and the remaining eight to the large display size condition.

Most of the benefits of a between-subjects design derive from the fact that each participant is only exposed to a single condition. As a result, there is no concern that the participant will learn something from their exposure to one condition that will influence measurement of another condition. This is particularly useful for scenarios where the participant may learn or develop competencies that could affect their performance in another condition.

If fatigue is likely to be an issue, between-subjects designs have the advantage of shorter duration because the subjects are only exposed to a single experimental condition. Between-subjects designs also afford lengthier experimental tasks for the same reason.

However, there are also a number of drawbacks to the between-subjects design. The biggest disadvantage occurs when there are large individual differences in performance as measured by the DV. This can translate into a failure to detect a difference when there is one (i.e., a Type II error) because the higher individual variance makes it difficult (relatively speaking) to achieve a statistically significant result. Figure 1 demonstrates this difference. Looking at the data in the left-hand panel from 24 different individuals (eight in each of the conditions) you would be hard pressed to suggest there is a difference in performance across the groups. However, consider the same spread of data drawn from eight individuals who participated in each condition as shown in the right-hand panel (this is a within-subjects design and is discussed in the next section). In this figure each individual's data points are connected by a line and it is easy to see that in all cases the score increases, even though there is a great deal of variability across participants in their baseline levels of task immersion.

Another disadvantage is that the groups of participants assigned to the various conditions may not be equivalent and may systematically vary along an unseen dimension—and this is why random assignment is a crucial requirement of all true experiments. In addition, there are a number of practical challenges with between-subjects designs such as the need for a larger number of participants to examine an equivalent number of experimental conditions.

Within-Subjects Design

A within-subjects design is one in which participants are assigned to all conditions (i.e., all levels of the IV) or have repeated exposure to a single condition (known as a repeated measures design). Returning to our research question regarding display size and task immersion, each of the 24 participants would be exposed to the small, medium, and large display sizes.

The main advantage of within-subjects designs stems from the fact that the same participant is examined under numerous conditions, which effectively allows them to serve as their own control. When there is a large amount of individual variation, a within-subjects design is a more sensitive design for capturing differences across conditions because you can look at differences within a person experiencing the conditions. If everyone, independent of level of performance, is better on one than the other, then you can still find significant differences. The general rule of thumb is that when there are large individual differences with respect to the dependent variable, a within-subjects design will be more effective.

Within-subjects designs can also be highly efficient. The number of participants required to show a significant difference among experimental conditions is reduced compared to a between-subjects design. For example, if you have three conditions, you would need three times the number of participants in a between-subjects design as you would in a within-subjects design. In factorial designs, which we discuss later, the multiplier can be even greater. This efficiency can be particularly helpful when studying populations that are high-risk, rare (e.g., participants with rare disabilities or in an isolated locale) or difficult to recruit in large numbers or for long periods of time (e.g., celebrities, high-level executives, and medical surgeons).

The major disadvantage to within-subjects design is that once participants are exposed to a condition they may be altered in a way that will impact their behavior in other conditions. For example, if a participant learns something in the first exposure that influences their performance, there is no way to have them "unlearn" what

Table 1 Summary table for choosing a between-subjects design or a within-subjects design

Choose…	
Between-subjects design	Within-subjects design
• When there are small individual differences, but large expected differences across conditions	• When there are large individual differences (i.e., high variance across participants with respect to the dependent variable(s) of interest)
• When learning and carryover effects are likely to influence performance	• When tasks are unlikely to be affected by learning and carryover effects are unlikely to occur
• When fatigue may be an issue	• When working with rare or hard to reach populations

was just gained. This is particularly problematic with studies that involve learning or insight solutions where you suddenly understand something that was previously perplexing. More generally, these problems are known as *order effects*, since the results may be influenced by the order in which participants go through the conditions.

Another challenge for within-subjects designs has to do with fatigue. For tasks that are physically or cognitively challenging, having the subject perform several repeated tasks is not an ideal solution. If participants become tired, the data can be influenced by the fatigue. Spreading the testing out over time (e.g., hours or days) can resolve the fatigue issue but can introduce unwanted extraneous influences, not to mention the practical issues of researcher time and scheduling.

Learning and fatigue are issues that often come up in HCI research. For example, consider a study examining information retrieval in two different websites. If the participants learn about the basic structure of the website in the first trial, they will carry over this knowledge to the same task on the second site. These types of problems are more generally known as *carryover effects*, and there are several ways to minimize their impact that are described in the following sections. For a summary of factors to consider when choosing between a between-subjects design and a within-subjects design, see Table 1.

Counterbalancing. Counterbalancing helps minimize carryover and order effects by controlling the presentation order of conditions across participants so that each condition appears in each time period an equal number of times. In our display size study this means we would want the small, medium, and large display size conditions to appear in each presentation position an equal number of times.

Complete counterbalancing requires that the participants are balanced across all possible treatment orders. In a simple experiment with few conditions, this is relatively easy. Table 2 shows our three-level experiment with its six possible orderings. However, as the number of conditions increases, the potential orderings grow at a rate of $n!$, where n is the number of conditions.

Since complete counterbalancing is only feasible for small numbers of conditions—with only five conditions there are 120 different orderings needed—researchers have developed a compromise approach where each treatment occurs equally often in each position. *Latin square designs*[11] (Cochran & Cox, 1957;

[11] There are numerous online resources for obtaining Latin square tables (e.g., http://statpages.org/latinsq.html).

Table 2 Complete counterbalancing for a 3-level IV (A,B,C), within-subjects experiment

Participant	First treatment	Second treatment	Third treatment
1	A (small display)	B (medium display)	C (large display)
2	A	C	B
3	B	A	C
4	B	C	A
5	C	A	B
6	C	B	A

Table 3 A Latin square design for a 4-level IV (A,B,C,D), within-subjects experiment

Participant	First treatment	Second treatment	Third treatment	Fourth treatment
1	A	B	C	D
2	B	C	D	A
3	C	D	A	B
4	D	A	B	C

Fisher & Yates, 1953; Kirk, 1982; Rosenthal & Rosnow, 2008, pp. 192–193) are a form of *partial counterbalancing* that ensure that each condition appears in each position an equal number of times. Table 3 presents a simple Latin square for four conditions.

A common question that arises regarding Latin square designs is what to do with the next cluster of participants. One option would be to continue to use the same Latin square over and over again for each new cluster of participants (e.g., 1–4, 5–8, 9–12, and so on). If using this approach, be sure to test whether the partial counterbalancing is systematically related to the effects of the conditions. An alternative is to generate new Latin squares for each additional cluster of participants. This has the advantage of reducing the likelihood that the partial counterbalancing correlates with the results, but the disadvantage is that this correlation cannot be tested in a straightforward way (for details on these approaches see Kirk, 2013, Chaps. 14–16).

Even better than standard Latin square designs are *balanced Latin square designs* where each condition precedes and follows each other condition equally often. This can help to minimize sequential effects.[12] For example, in Table 3 notice that A precedes B in three of the four rows. A better design can be seen in Table 4 where A precedes B an equal number of times as B precedes A. A balanced Latin square (Bradley, 1958; Williams, 1949) can be constructed for an even number of conditions using the following algorithm for the first row of the square: 1, 2, n, 3,

[12] This approach only balances for what are known as first-order sequential effects. There are still a number of ways in which repeated measurement can be systematically affected such as nonlinear or asymmetric transfer effects. See (Kirk, 2013, Chap. 14) or other literature on Latin square or combinatorial designs for more details.

Table 4 A balanced Latin square design for a 4-level IV (A,B,C,D), within-subjects experiment

Participant	First treatment	Second treatment	Third treatment	Fourth treatment
1	A	B	D	C
2	B	C	A	D
3	C	D	B	A
4	D	A	C	B

$n-1$, 4, $n-2$, …, where n = the number of conditions. Each subsequent row is constructed by adding 1 to the value of the preceding row (or subtracting 1 if the value is equal to n).[13]

Factorial Designs

Up to this point we have focused on experiments that examine a single independent variable at a time. However, in many studies you will want to observe multiple independent variables at the same time, such as gender, display size and task complexity. In such a design each variable is called a *factor*, and the designs that make use of many factors are *factorial designs*.[14] Factorial designs can be either between-subjects, within-subjects, or both in what is known as *mixed factorial designs*[15] (or split-plot designs).

The number of factors and their conditions can be multiplied to yield the total number of conditions you will have for a given experiment. A study with two factors, each with two conditions would yield four total conditions. The name for such a design would be a 2×2 factorial. There is no theoretical limit to the number of factors that can be included in a study; however, there are practical limitations since each additional factor can drastically increase the number of participants needed and the analysis and interpretation become correspondingly complex. For example, a 3×3×4×2 design would yield 72 different configurations that would each require enough participants to have a well-powered experiment. If you were using a between-subjects design and including 10 participants in each condition, you would need 720 participants! If you used a mixed factorial or within-subjects design you could reduce the overall number of participants needed, but you would have to be careful about fatigue, ordering and carryover effects.

[13] If your experiment has an odd number of conditions, then two balanced Latin squares are needed. The first square is generated using the same method described in the text, and the second square is a reversal of the first square.

[14] As a side note, Latin square designs are a within-subject version of a general class of designs known as fractional factorial designs. Fractional factorial designs are useful when you want to explore numerous factors at once but do not have the capacity to run hundreds or thousands of participants to cover the complete factorial (see Collins, Dziak, & Li, 2009).

[15] In practice, mixed factorial designs are often used when examining different groups of participants (e.g., demographics, skills). For example, if you are interested in differences in user experience across three different age groups, a between-subjects factor may be age group (teen, adult, elderly), while a within-subjects factor may be three different interaction styles.

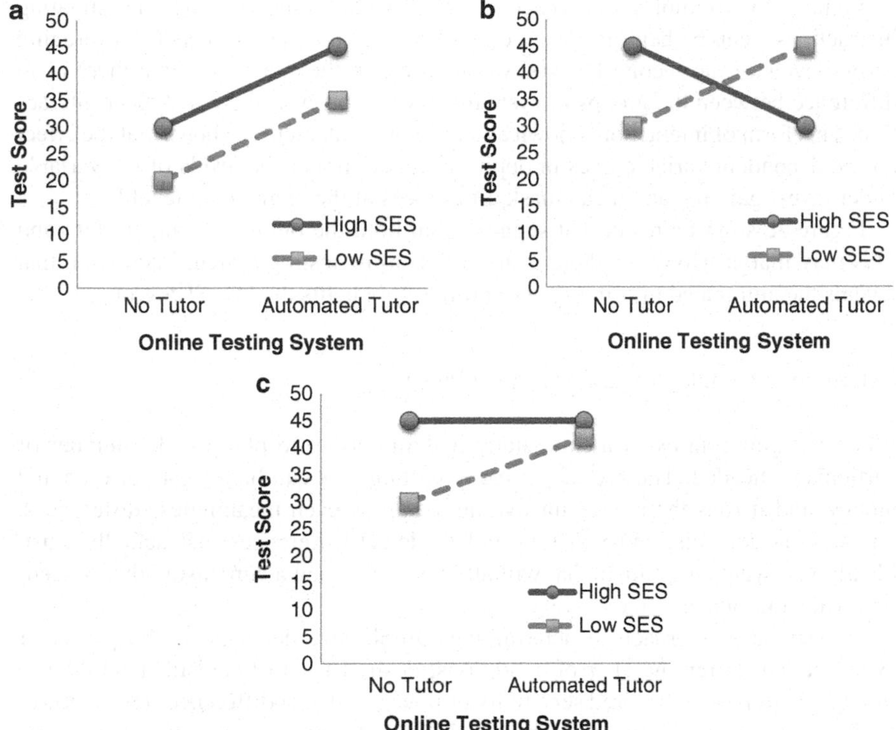

Fig. 2 Three sample outcomes from a 2×2 factorial design showing (**a**) two main effects, no interaction, (**b**) no main effects but a crossover interaction, (**c**) two main effects and an interaction (Color figure online)

Main effects and interactions. A major strength of factorial designs is that they allow you to build up a more complex understanding of the simultaneous relationship between several independent variables and the dependent variable. In other words, you can examine both *main effects* and *interactions*. A main effect is the influence of a single independent variable upon the dependent variable. An interaction occurs when the effect of one independent variable on the dependent variable varies according to the levels of another independent variable.

Figure 2 illustrates a subset of the possible outcomes for a 2×2 factorial design that examines test performance for two different groups—low socioeconomic status (SES) and high SES—using one of two different online testing systems (one with an automated tutor and one without). For this design there are two potential main effects: SES and online testing system. There is also an SES×online testing system interaction.

Figure 2a shows what a graph might look like with a main effect of SES where high SES scores higher than low SES (i.e., the red line is higher than the blue line) and a main effect of online testing system where the automated tutor scores higher than the no tutor system (i.e., the average of the two points on the left is lower than the average of the two points on the right).

Figure 2b exemplifies another possibility and shows why investigating interactions[16] can be helpful. If you only examined the main effects (by averaging across levels of the second IV) you would come to the conclusion that there is no difference between the groups or systems tested. However, there is a clear interaction. This form of interaction, known as a crossover interaction, shows that the effect on the dependent variable goes in opposite directions for the levels of the variable under investigation—and it can mask differences at the main effect level.[17]

Figure 2c shows a result that suggests that both the online tutoring system and SES may matter. However, there is an SES × online testing system interaction that reveals the automated tutoring system primarily benefits the low SES group.

Determining Sample Size and Statistical Power

When designing an experimental study it is important to plan for the number of participants needed. The use of too many participants can be a waste of time and money, and it runs the risk of uncovering small or even meaningless differences. Too few participants, and you may fail to detect differences that actually exist. Ideally you want an estimate that will allow you to reach a conclusion that is accurate with sufficient confidence.

A systematic approach to determining sample size depends on the particular experimental design, number of conditions, desired level of statistical confidence ($p < .05$ is often used), desired sensitivity or power to detect differences (80 % power is often used), a good estimate of the variability in the measurements, and an understanding of what a meaningful difference is in the context of your experiment.

Bausell and Li (2002) and Cohen (1988) provide excellent coverage of the topic, and Kenny (1987, Chap. 13) provides a nice example for studies with a small number of experimental conditions. There are also numerous web resources for determining appropriate sample sizes such as http://www.statsoft.com/textbook/power-analysis/. Most statistical software packages also provide tools to generate visual representations called power curves that can be particularly useful when you are less confident of your measurement estimates.

Quasi-Experimental Designs

In HCI research true random assignment may be impractical, infeasible, or unethical. For example, consider a study that compares performance in a classroom with a new technological innovation versus a traditional classroom without it. In this case, the students are not randomly assigned but instead are preselected based on the classroom to which they were previously assigned. When this is the case, there is a

[16] Note that common transformations of the data (e.g., logarithmic or reciprocal transformations) can affect the detection and interpretation of interactions. Such transformations are performed when the data deviate from the distributional requirements of statistical tests, and researchers need to be cautious when interpreting the results of transformed data.

[17] For factorial designs with more factors, higher-order interactions can mask lower-order effects.

risk that other factors may come into play when measuring the dependent variable. For instance, the teacher of the technological innovation class may also be a better teacher and that may be the primary reason for a performance enhancement.

Quasi-experimental designs[18] aim to address the internal validity threats that come about from a lack of randomization. The designs tend to vary along two primary dimensions: those with or without control or comparison groups; and those with or without pre- and post-intervention measures.

Non-equivalent Groups Design

The non-equivalent groups design is one of the most commonly applied quasi-experimental designs in HCI. The goal is to measure changes in performance that result from some intervention. However, this design lacks the random assignment of participants to experimental groups. This is why it is called "non-equivalent" groups—because the two groups are not equivalent in a way that they would be if random assignment had been used. In many ways it is structured like a typical pre-test/post-test design with a control or comparison group:

Group A: Obs_1–[Intervention]– Obs_2
Control Group: Obs_1 Obs_2

The ideal outcome from such a design is that there is little difference in the pre-intervention measure (pre-test) but large differences in the post-test measure. In other words, the more likely that the groups are equivalent at pre-test time (Obs_1), the more confidence we can have in the differences that appear post intervention (Obs_2). However, there are still a number of threats to internal validity. One is that there are latent attributes of Group A that are not revealed in the pre-testing but that interact with the intervention in some way. Another is that the groups are receiving uneven exposure over time between the pre-test and post-test. Returning to the classroom example, if the teacher in the classroom with the technological innovation also exposes students to something else related to the dependent variable, then we run the risk of misattributing the changes in the dependent variable.

Interrupted Time-Series Design

The interrupted time-series is another popular quasi-experimental design.[19] It infers the effects of an independent variable by comparing multiple measures obtained

[18] For more detailed coverage of quasi-experimental designs see (Cook & Campbell, 1979; Shadish et al., 2002).

[19] Time-series approaches have particular statistical concerns that must be addressed when analyzing the data. In particular, they often produce data points that exhibit various forms of autocorrelation, whereas many statistical analyses require that the data points are independent. There are numerous books and manuscripts on the proper treatment of time-series data, many of which reside in the domain of econometrics (Gujarati, 1995, pp. 707–754; Kennedy, 1998, pp. 263–287).

before and after an intervention takes place. It is often used when there is a naturally occurring event that takes place or in field studies where it is infeasible to have a control group.

The basic form of an interrupted time-series design relies on a series of measurements with knowledge of when an intervention, treatment or event occurred, followed by another series of measurements:

Group A: Obs_1–Obs_2–Obs_3–[Intervention]–Obs_4–Obs_5–Obs_6

If the intervening event or treatment had an effect, then the subsequent series of observed values should experience a quantifiable discontinuity from the preceding measurements. While the easiest change to see is an immediate shift from a flat line, there are numerous ways in which the changes can manifest including intercept or slope changes.[20]

However, there are some major threats to internal validity that must be assessed with time-series designs in HCI. The primary concern hinges on whether another influential event took place at the same time as the intervention (e.g., a major press release about your online news system broke at the same time you implemented a new algorithm aiming to improve online contributions), or whether there was significant mortality or drop out that occurred between the first set of measures and the second (e.g., the participants that were not contributing much dropped out completely for the later stages of the study).

Strengthening Causal Inferences from Quasi-Experimental Designs

For both non-equivalent groups and interrupted time-series designs, there are a number of concerns that arise regarding internal validity, most of which result from the lack of random assignment or use of a control group. To address these concerns, a number of variations have been developed.

The first integrates *treatment removal* into the design.[21] If the intervention is reversible, then the research design can include this to bolster the causal evidence. The first part of the study is the same as the interrupted time-series design, but the second half includes a removal of treatment followed by additional measures:

Group A: Obs_1–Obs_2 [+Intervention] Obs_3–Obs_4 [–Intervention] Obs_5–Obs_6

Naturally, you can extend this design to have *multiple additions and deletions*. If the dependent variable is sensitive to the intervention you should see it respond to each addition and deletion of the treatment, increasing the likelihood that you have identified a causal effect.

[20] For a detailed discussion of interrupted time-series designs see (Shadish et al., 2002, pp. 171–206).

[21] These are also known as A-B-A or withdrawal designs, and are similar to many approaches used for small-N or single-subject studies with multiple baselines. For further details see (Shadish et al., 2002, pp. 188–190).

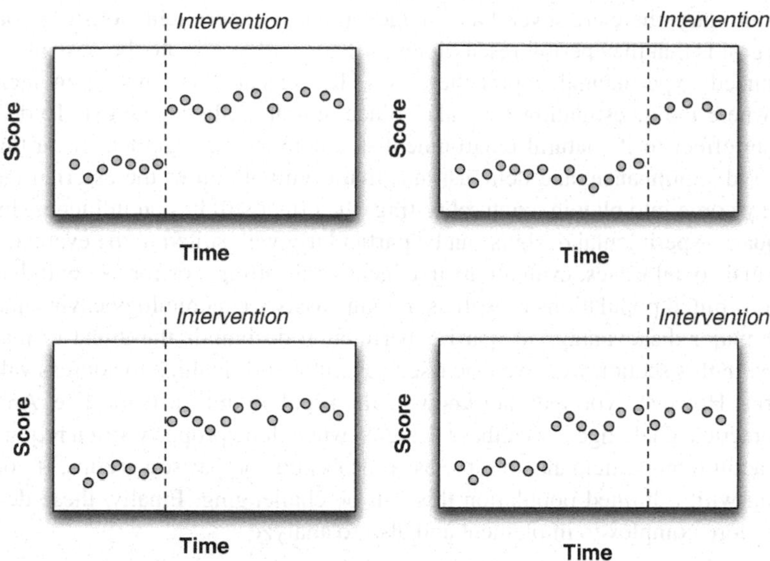

Fig. 3 An illustration of the benefit of time-series with switching replications for detecting or minimizing the potential influence of exogenous factors. The *top* two figures illustrate a discontinuity in the time-series that occurs inline with the intervention, while in the *bottom* two figures the discontinuity in the data occurs at the same time point regardless of the intervention (i.e., it is more likely due to an exogenous influence)

A second variation uses *switching replications* (Shadish et al., 2002, pp. 146–147). Switching replications make use of more than one group in order to introduce interventions at different times:

Group A: Obs_1–[Intervention]–Obs_2–Obs_3–Obs_4–Obs_5–Obs_6
Group B: Obs_1–Obs_2–Obs_3–[Intervention]–Obs_4–Obs_5–Obs_6
Group C: Obs_1–Obs_2–Obs_3–Obs_4–Obs_5–[Intervention]–Obs_6

If the treatment truly causes a shift in the dependent variable, then you should see the shift whenever the intervention takes place (see top panel of Fig. 3), whereas if the change in the dependent variable was caused by another external factor (e.g., the aforementioned press release), then the shift would occur at the same time regardless of the timing of the intervention (see bottom panel of Fig. 3). Introducing the intervention at different times helps to counter internal validity arguments regarding the influence of simultaneous events, history, or even mortality issues.

Finally, you can couple the approaches of interrupted time-series and non-equivalent control group designs. This design can offer some of the strongest support for causal inferences:

Group A: Obs_1–[Intervention]–Obs_2–Obs_3–Obs_4–Obs_5–Obs_6
Group B: Obs_1–Obs_2–Obs_3–[Intervention]–Obs_4–Obs_5–Obs_6
Group C: Obs_1–Obs_2–Obs_3–Obs_4–Obs_5–[Intervention]–Obs_6
Control Group: Obs_1–Obs_2–Obs_3–Obs_4–Obs_5–Obs_6

In summary, there are several advantages to quasi-experimental designs. One of the biggest is that they permit research investigations that may not be possible using randomized experimental approaches. For HCI researchers, this often includes cases where the investigation best takes place in a naturalistic context. To demonstrate an effect in its natural environment is a convincing argument regarding its real-world significance, and demonstrates that even with all of the external factors that may come into play in a natural setting, the effect still has an influence. In this way, quasi-experimental designs can be particularly well suited to the evaluation of contextual social issues, evaluations in educational settings, or for use with hard to reach or limited populations as well as in many assistive technology environments.

The major disadvantage of quasi-experimental designs is the threat to internal validity. In this section we have discussed several ways in which to address validity concerns. However, you may not know of the problem until it is too late. Another more practical challenge is that these designs, when done properly, often require the use of additional participants to serve as controls and comparison groups. If you are working with a limited population this can be challenging. Finally, these designs can be more complex to implement and also to analyze.

Statistical Analysis

Just as important as the research design is planning the statistical analysis ahead of time in a way that ensures you can draw the appropriate conclusions from your experiments. Once the data have been collected, descriptive and inferential statistical analysis methods are used to assess confidence in the findings. A detailed treatment of statistics is beyond the scope of this chapter and the reader is instead directed to the references at the end of the chapter.

Over the years, however, we have found that having a pointer of where to look for the right statistical tests is just as important both when designing an experiment and when evaluating the results of a study. There are numerous flow charts available online for choosing the right statistical test for a given experimental design (e.g., http://abacus.bates.edu/~ganderso/biology/resources/stats_flow_chart_v2003.pdf).

What Constitutes Good Work?

So what ultimately constitutes good experimental research? As Robert Abelson describes in his seminal book, "Statistics as Principled Argument," it's M.A.G.I.C. Abelson (1995) suggests that a persuasive argument using experimental results relies upon the Magnitude, Articulation, Generality, Interestingness, and Credibility of your research. While the MAGIC acronym was originally developed to describe data analysis and its presentation, it can also be useful when thinking about what constitutes good experimental research.

The MAGIC Criteria

Magnitude. The magnitude of your research has to do with understanding the size of the effect being reported and whether it is big enough to have "real world" implications. Assessing magnitude requires more than just obtaining a *statistically significant* difference between experimental conditions. In fact, as previously discussed, a common mistake is to report the *p* value as indicative of an effect's magnitude. The *p* value, critically, depends on two things: the size of the difference between the two groups[22] and the size of the sample. Thus, you can achieve a significant result with a small sample when there is a really big difference between your groups; alternatively, you can also achieve a significant result with very small differences between the groups, if you have a large enough sample. As a result, a better (i.e., smaller) *p* value does not mean it is a "more significant" or "bigger" effect. Reporting *p* values will tell you if there is a significant difference between the groups under investigation; it will not in and of itself tell you whether the difference is meaningful.

The concept of *effect size* can help to determine whether the difference is meaningful. Effect sizes are used to quantify the size of the mean difference between groups (Abelson, 1995, pp. 45–52; Cohen, 1988; Grissom & Kim, 2005; Rosenthal & Rosnow, 2008, pp. 55–58). They can be reported either in original units (i.e., the raw score) or in standardized forms, the latter of which can also be used when the variable's units do not have an inherent scale or meaning. Effect size is a cleaner measure of magnitude and should not be confused with statistical significance. Unfortunately, most HCI researchers have not yet embraced the use of effect sizes even though it is now mandated in many other scientific venues (e.g., American Psychological Association, 2010, p. 34). However, exemplary papers do exist, especially those performing meta-analyses on topics such as self-disclosure in digital environments (Weisband & Kiesler, 1996) or examining the influence of humanlike faces in embodied agents on interaction experience (Yee, Bailenson, & Rickertsen, 2007), as well as individual experimental studies that compare effect sizes across conditions (Gergle et al., 2013).

Another way HCI researchers can better express magnitude is to report *confidence intervals* (Cumming & Finch, 2001; Smithson, 2003). Confidence intervals provide a more intuitive and meaningful description of the mean difference between the groups. Instead of providing a single number, they identify the range in which the true difference is likely to fall. Confidence intervals, and their corresponding confidence limits, are an intuitive way of specifying not just an estimate of the difference but also the likely minimum and maximum values of the difference. A good example drawn from a field experiment can be seen in Oulasvirta and colleagues' research (Oulasvirta, Tamminen, Roto, & Kuorelahti, 2005).

Finally, there is a more practical side to magnitude that is determined by the choice of experimental design and manipulations. Consider a study that shows a

[22] We use a two-condition example for ease of exposition.

large effect with a rather subtle manipulation vs. one that shows a large effect with an extreme manipulation. For example, demonstrating an increase in contributions to an online peer-production system by providing a graphical badge on a person's profile page (subtle) vs. paying them $100 to contribute more content (not-so-subtle). To the extent that you can produce the same size effects with the former, your results have greater magnitude, and oftentimes, practical importance.

Articulation. Articulation refers to the degree of detail that is reported about the research findings. Consider the following three descriptions which range from least to most detailed in discussing the results of a 3 (Input Style) × 2 (Gender) factorial experiment: (a) "there was a significant performance difference between the three UI input styles"; (b) "There was a significant performance difference between input styles and also a significant performance difference by gender"; or (c) "There were significant differences between all types of input styles with style 1 being 75 % faster than style 2, which in turn was 18 % faster than style 3. Moreover, these performance differences across input styles were even stronger for females than for males, and females overall were 7.2 % faster than males." While the various statements are reporting the same general trend in findings, the last statement does so with much greater articulation. For a discussion of ways to enhance reporting of results with respect to articulation see Abelson (1995, pp. 104–131).

Generality. Generality represents the extent to which the research results apply outside the context of the specific study. One aspect of this is external validity, or the degree to which the results can be generalized to other situations, people, or times.

The sample and the population from which it is drawn often limits generality. For example, if you are only studying Facebook users, you cannot make claims that generalize to the entire world's population—especially given that a significant majority of the world does not actually use Facebook in a significant way. You can, however, make claims about the smaller population of Facebook users. Similarly, US college students, often easily recruited because they are required to serve in experiments as part of a course requirement, are not indicative of people in the rest of the world in many, many, ways.

Another limitation often comes from the choice of experimental and statistical controls employed in a study. In HCI, it is often the case that a highly controlled laboratory study with participants who have no history together may not be generalizable to the real-world field environment where the environment can be noisy and chaotic, people have prior relational histories, motivation can widely vary, etc. Using a wider range of contextual variations within studies, and a systematic program of replication and extension along with the application of meta-analysis (for an introduction to the technique, see Borenstein, Hedges, & Higgins, 2009; for HCI examples, see McLeod, 1992; Weisband & Kiesler, 1996; Yee et al., 2007) across numerous studies, are ways to broaden the scope of your findings and improve the generality of your research.

Interestingness. While the first three criteria can be treated in a more objective fashion, the last two have more subjective elements. Interestingness has to do with the importance of the research findings, and this can be achieved in various ways. Here we focus on three dimensions of interestingness: theoretical, practical, and novelty.[23]

The *theoretical* dimension centers on experimental HCI research that seeks to inform. Theoretical contributions often consist of new or refined concepts, principles, models, or laws. For experimental work to be interesting on a theoretical dimension, the findings have to change what theorists think. If we consider theory as our best encapsulation of why things work as they do, then challenging that assumption or refining it in order to make our theories more complete or correct is a hallmark of good theoretical research. The extent to which the theory must change, or the number of theories that are influenced by your findings, are two key ways in which importance is assessed.

There are numerous experimental and quasi-experimental studies that make contributions on the theoretical dimension. For example, work by Zhu and colleagues challenges the traditional notion of online leadership, and suggests that it may be a more egalitarian construct than previously assumed (Zhu, Kraut, & Kittur, 2012; see also Keegan & Gergle, 2010). Dabbish and colleagues (Dabbish, Kraut, & Patton, 2012) used an innovative online experiment to reveal the communication behaviors and theoretical mechanisms by which commitment to online groups occurs. Finally, several classic studies in the domain of Fitts' Law have advanced the theory by demonstrating trajectory-based steering laws (Accot & Zhai, 1997; Wobbrock et al., 2008).

The *practical* dimension centers on experimental HCI research that seeks to solve everyday problems and issues. Practical contributions can take the form of the development of useful new metaphors, design guidelines or design patterns, new products or services, and design checklists or best practices. This type of work may take a more pragmatic and sometimes atheoretical approach to design and development. In these cases, experimental research techniques often focus on evaluating or verifying the utility of a new design or practice. Some excellent examples of this approach are provided in Kohavi and colleagues' work on using web experiments to inform design choices (Kohavi, Henne, & Sommerfield, 2007; Kohavi & Longbotham, 2007; Kohavi, Longbotham, & Walker, 2010).

The *novelty* dimension centers on experimental HCI research that seeks to invent. This often includes the design, development, and deployment of new systems; new infrastructures and architectures; and new tools or interaction techniques. While not all novel contributions of this type in the HCI literature require experimental support, many are accompanied by an experimental demonstration of their utility and how well they perform in new settings or relative to existing best practices or state-of-the-art algorithms or systems.

---

[23] While we separate these three areas in order to discuss the relative contributions that are made in each, it is not to suggest that these are mutually exclusive categories. In fact, some of the most influential work has all three dimensions. For a more nuanced discussion of the integration of theoretical (basic) and practical (applied) research in an innovation context see Stokes (1997) *Pasteur's Quadrant*.

Gutwin and Penner's work on telepointer traces (Gutwin & Penner, 2002), Wigdor and colleagues' LucidTouch system (Wigdor, Forlines, Baudisch, Barnwell, & Shen, 2007), or Zhai and Kristensson's work on the SHARK shorthand gesturing system (Kristensson & Zhai, 2004; Zhai & Kristensson, 2003) all make use of elements of experimental design[24] to rigorously demonstrate the utility of their novel designs and systems.

Credibility. Credibility is established by convincing the readers and reviewers that your work has been performed competently and with regard to common pitfalls and traps—it serves to bolster the plausibility of your claims. Much of what we have discussed throughout this chapter is aimed at establishing and supporting the credibility of your work. Doing things correctly, according to preestablished best practices and guidelines, is the easiest way to convince others of the credibility of experimental research. Dealing with internal and external validity, choosing a sample and understanding its limits, recognizing potential confounds, reporting on large and meaningful effects, performing appropriate analyses and correctly reporting and representing your findings are all keys to establishing credible experimental research.

Writing Up Experimental Research

In order for experimental HCI research to have an impact, it needs to be communicated to other researchers. While a detailed discussion of writing and dissemination is beyond the scope of this chapter—and several excellent guides already exist (e.g., Bem, 2003)—the following provides a brief description of the central elements required when reporting experimental research.

The general form of an experimental research article follows the hour-glass writing form. It is broad at the beginning and end, and narrow in the middle. Keep in mind that the main goal of your research paper is to motivate and detail your argument, demonstrate what you did, and convince the reader of your contribution. It is not a chronology of everything you did from day one, nor is it a detailed description of every single fact you uncovered. It is a pointed argument. The following presents a standard structure for an experimental research piece, and we focus on elements that we feel are often misreported or problematic in HCI related venues:

Introduction. The introduction should answer the question, "What is the problem?" and "Why should anyone care?"[25] It should provide an overview of the work and

[24] Not all of these studies are strict randomized experiments. For example, the SHARK evaluation does not make use of a control or comparison group. However, many use experimental research techniques to effectively demonstrate the feasibility of their approach.

[25] The framing questions in this section are drawn from Judy Olson's "10 questions that every graduate student should be able to answer." The list of questions and related commentary can be found here: http://beki70.wordpress.com/2010/09/30/judy-olsons-10-questions-and-some-commentary/

develop the central argument for the paper. It should identify the problem, provide rationale for why it matters and requires further research, describe and situate the research in the context of related literature, and end with the specific goals of the study often stated in the form of hypotheses or research questions. Be sure to state the research questions early, and walk the reader through your argument. Use plain English. Provide examples. Be concrete.

Method. The method section should aim to answer the question, "What did I do?" It should begin with a detailed description of who the *participants* were (e.g., age, gender, SES, education level, and other relevant demographic variables). It is also important to know about the motivations used to achieve participant involvement. Was it done for course credit? Were the participants paid? If so, did it depend on their performance? etc.

The *sampling procedure* should then be discussed. For example, were the participants drawn from a randomized national sample or perhaps snowball sampling was used? Next, the approach used to *assign participants* to experimental conditions should be described. Were the participants randomly assigned, was some form of paired assignment used, or were preexisting groups used (e.g., classrooms)?

The next area to include in the method is a description of the *experimental design* and the *experimental conditions*. The type of design should be clearly articulated (e.g., between- or within-subjects, mixed factorial design, or interrupted time series). The dependent and independent variables should also be described. This should be followed by a description of the *stimuli* and *materials* used to collect the data.

Finally, the written *procedure* should provide a detailed description of the processes used to collect the data. Describe any particular machinery, software, or measurement instruments. Discuss how the participants were handled before, during, and after the study, and detail the presentation order of materials. This should be followed by a description of the analysis where you detail what statistical comparisons were planned, discuss how missing data were treated, state how your dependent variable was captured, scored, annotated, etc.

The rule of thumb for the amount of detail that should go into the method section is that it should be enough for another researcher to be able to replicate the study if they chose to do so.

Results. The results section should aim to answer the question, "What did I find?" It should present the analyses performed and the major findings. You should present the results in a way that best supports the central argument being proposed in the paper, and be explicit when addressing central research questions and hypotheses.

The results section should focus on the most important findings or DVs. Remember, you are presenting results that are relevant to the central argument of the paper (both those that support and contradict your argument). Be sure to state each finding in a clear form without the use of jargon, and then support it with statistics. Remember that the statistics are not the focal point of the results section. The statement of the finding is the important part, and the statistics should be used to bolster the reader's confidence in that statement. Show the most relevant findings in tables and figures, and be sure to point out the figures and tables in the accompanying prose. It can also be useful to interpret as you present, although you need to be clear about what are actual results and what are interpretations of the results. Finally, end the results section with a reminder of the purpose of the experiment and provide a light summary of the results with respect to the central argument.

Discussion. The discussion section should aim to answer the question, "What does all of this mean?" and "Why does it matter?" Remember, this section (and the paper as a whole) should be a pointed argument. The discussion section is where you can contextualize your results both with respect to the central research questions and hypotheses and in relation to prior work in the area.

In this section you should start by reviewing the evidence you have garnered toward your position and discuss the evidence against it. Be sure not to oversell your findings. You should also be sure to discuss the limitations of the current study or approach and address possible alternative explanations for your findings.

Once you have discussed your results in detail, you can begin to talk about the broader implications of the work whether they are for design, policy, or future work. You can describe the ways in which new experiments can be performed to address open questions or describe new directions that need to be addressed given the findings you have revealed.

Conclusion. Finally, you should conclude the paper with a restatement of your work. The conclusion is, in many ways, like the introduction of the paper. This is often a single paragraph that reminds the reader of the initial goals of the work, what you found, what it means, and why it matters—both for the particular questions under investigation as well as more broadly.

Personal Story about How the Authors Got into this Method

In this section we describe our personal experiences with research we conducted together with colleagues Randy Pausch and Peter Scupelli exploring the cognitive effects of physically large displays.

At the time we began our large display work (around 1999), LCD manufacturing was becoming significantly more efficient, creating a supply of cheaper and larger displays. Furthermore, projectors and digital whiteboards were becoming commonplace in conference rooms, and researchers were exploring the extension of these large displays into more traditional office spaces. Although researchers had articulated the qualitative benefits that large displays had on group work, little research had been done to quantify the benefits for individual users, which we had anecdotally noticed in our various new display setups. We thus set out to compare and understand the effects that physical display size (i.e., traditional desktop displays vs. large wall displays) had on task performance (Tan, Gergle, Scupelli, & Pausch, 2006).

We began this work armed with theoretical foundations articulated in prior research. Quite a bit of work suggested wider fields of view offered by larger displays were beneficial across a variety of tasks (e.g., Czerwinski, Tan, & Robertson, 2002). This work pointed not only to pragmatic benefits of large displays such as ease of viewing, which facilitated better social interaction, but also to an increased sense of presence, for example, in virtual environments. However field of view was a function of two variables: display size and distance from the user. We set out to isolate and understand the effects of physical display size specifically.

To do this, we had to keep constant the visual angle subtended from the user to each of the small and large displays by adjusting the viewing distances appropriately, hence varying only the physical display size. In fact, we considered fastening users' heads in place to prevent movement that may have caused field of view differences, but this was uncomfortable and we ran various pilot studies showing that small movements of the head did not account for any of the effects seen, so our main experiments were run without this constraint. We were careful to hold other displays factors such as screen resolution, refresh rate, color, brightness, and contrast constant across the displays so that we could isolate any effects to the display size with minimal confounds.

In the beginning, we conducted exploratory experiments with a wide variety of tasks to uncover areas of interest. We found something interesting—display size did not seem to affect reading comprehension tasks (remember, we could not "prove" equivalence between the conditions, so this is not a definitive statement, but helped us focus our efforts elsewhere), but that users performed significantly better on a spatial orientation task in which they had to perform imagined rotations of a boat. We hypothesized that this was due to the way the images were perceived in each display condition and thus the strategy with which users performed the task. In pilot studies, we tried using questionnaires as well as structured interviews to determine the strategy users employed, but found that users were not able, either implicitly or explicitly, to articulate their cognitive strategy. Hence, we designed a series of experiments to probe this more deeply.

Returning to theoretical foundations suggested two cognitive strategies that could have been employed for spatial rotations: an egocentric rotation, in which users take a first-person view and imagine rotating their bodies within an environment, or an exocentric rotation, in which users take a third-person view and imagine objects rotating around other objects in space. Evidence in the psychology literature suggests that egocentric rotations, where appropriate, are much more efficient. We thus hypothesized that (a) we could bias users into adopting one or the other of the strategies by manipulating the instructions; (b) that the egocentric strategy was indeed more efficient than the exocentric one for our task; (c) when no explicit strategy was provided, the display size would serve (just as explicit instructions may) to implicitly bias the user towards a particular cognitive strategy. We refer the reader to (Tan, Gergle, Scupelli, & Pausch, 2003) for experimental design and results. The experiment supported these hypotheses.

Building on this, we then set out to understand if there were specific tasks that would benefit more or less from this effect. Namely, we hypothesized that large displays bias users into using egocentric strategies and do not increase performance on "intrinsically exocentric" tasks for which egocentric strategies are not useful. Hence we selected a set of stimuli and tasks (i.e., the well-validated Card, Cube, and Shepard Metzler tasks) that we believed from prior work to be explicitly exocentric tasks, and results did not show effects as seen with the first round of tasks. Note that this was exactly the test of equivalence that we instruct readers to be cautious with. In fact, we did not demonstrate equivalence with this experiment, merely the lack of an observed effect, which we carefully treated as converging evidence to an already sizable set of evidence collected through the other experiments.

Finally, we extended the results from controlled (and contrived) tasks to a set of experiments that demonstrated more ecologically valid findings to demonstrate the robustness of the effects (Tan, Gergle, Scupelli, & Pausch, 2004). We increased the complexity of task and spatial abilities used as well as adding user interaction in rich dynamic three-dimensional virtual environments.

We showed in the first of these experiments that (a) users perform better in mental map formation and memory tasks when using physically large displays due to the increased likelihood that they adopt egocentric strategies; (b) users perform better in the path integration task when they are interactively moving themselves through the virtual environment; (c) the effects induced by physical display size are independent of those induced by interactivity. We also then demonstrated in a separate experiment that even in an environment crafted with cues such as distinct landmarks and rich textures to be realistic and memorable (i.e., navigating an out-of-the-box Unreal Tournament 2003 world), users perform better in mental map formation and memory tasks when using physically large displays due to the increased likelihood that they adopt egocentric strategies. More recently, we demonstrated how a theoretical understanding provides strong predictive power when we demonstrated how large display influences on egocentric and exocentric perspectives ultimately manifest in language differences (e.g., in the use of local and remote references) in a collaborative task (Bao & Gergle, 2009).

In this section, and throughout the chapter, we have described and discussed many critical concepts that need to be considered when using experimental research to answer research questions and reveal causal relations. Through thoughtful development of research questions and careful research design, experimental research can be a powerful way of knowing and we feel it is an important methodology for any HCI researcher's toolbox.

References for Becoming More Expert in Experimental Research

Throughout the chapter we have provided numerous citations to works that address the various issues in more depth. In addition to these citations (which can be used as model papers or authoritative sources), there are several exceptional texts that merit attention.

David W. Martin has written an excellent book for newcomers (or even old-timers who need a refresher) to experimental research, "Doing Psychology Experiments" (Martin, 2004). While the work is focused on psychology, Martin writes in an extremely accessible fashion and focuses on applied questions closer to that found in HCI research than many more esoteric theoretical treatments of the topic.

A seminal and extremely thorough, although more technical, treatment of experimental design can be found in Rosenthal & Rosnow's "Essentials of Behavioral Research: Methods and Data Analysis" (Rosenthal & Rosnow, 2008).

A central challenge in HCI is how to assess the quality of a new design in a real world context outside of the laboratory. When doing so we often lose the ability to have the refined control that most experimentalists strive to obtain. Education researchers have struggled with a similar problem for decades, and there are several excellent books and papers on quasi-experiments and ways to achieve the best possible control in such environments. These approaches are often more balanced between internal and external validity, and find ways to accrue evidence toward a causal argument without the strict control of the laboratory experiment. An older but particularly good treatment on the subject of quasi-experimental design is provided by Campbell and Stanley (Campbell, Stanley, & Gage, 1963), and a more recent and comprehensive authoritative guide is provided by Shadish, Cook, and Campbell (Shadish et al., 2002).

Statistical Analysis

For an introduction to statistical analysis, Weiss' introductory statistics textbook is extremely well-written, comprehensive, and detailed with accessible examples throughout (Weiss, 2008).

In addition, an excellent and comprehensive resource for reviewing what should be included in research papers when using various statistical techniques from both a writer's and reviewer's perspective is available in Hancock and Mueller's (2010) edited volume, "The Reviewer's Guide to Quantitative Methods in the Social Sciences." It covers everything from ANOVA to factor analysis to hierarchical linear modeling to inter-rater reliability to structural equation modeling, and beyond, in a concise and accessible fashion.

There have also been a number of HCI researchers who have focused on experimental methods and statistics directly for HCI researchers. An excellent self-guided tutorial can be found in Wobbrock's statistical analysis techniques for HCI researchers (Wobbrock, 2011). In this work Wobbrock focuses exclusively on methods and approaches that are common for HCI researchers. He also draws out examples of common problems and challenges found in HCI research.

Exercise

1. Generate an example of an interaction in a multi-variable experiment.
2. In your own research, what experiment could you run? How many subjects? What would they do? What are the materials you'd need? What data would you collect?

Acknowledgements We would like to thank Wendy Kellogg, Robert Kraut, Anne Oeldorf-Hirsch, Gary Olson, Judy Olson, and Lauren Scissors for their thoughtful reviews and comments on the chapter.

References

Abelson, R. P. (1995). *Statistics as principled argument*. Hillsdale, NJ: L. Erlbaum Associates.

Accot, J., & Zhai, S. (1997). Beyond Fitts' law: Models for trajectory-based HCI tasks. In *Proceedings of the SIGCHI conference on human factors in computing systems* (pp. 295–302). New York, NY: ACM.

American Psychological Association. (2010). *APA manual (publication manual of the American Psychological Association)*. Washington, DC: American Psychological Association.

Bao, P., & Gergle, D. (2009). What's "this" you say?: The use of local references on distant displays. In *Proceedings of the SIGCHI conference on human factors in computing systems* (pp. 1029–1032). New York, NY: ACM.

Bausell, R. B., & Li, Y.-F. (2002). *Power analysis for experimental research: A practical guide for the biological, medical, and social sciences*. Cambridge, NY: Cambridge University Press.

Bem, D. J. (2003). Writing the empirical journal article. In J. M. Darley, M. P. Zanna, & H. L. Roediger III (Eds.), *The compleat academic: A practical guide for the beginning social scientist* (2nd ed.). Washington, DC: American Psychological Association.

Borenstein, D. M., Hedges, L. V., & Higgins, J. (2009). *Introduction to meta-analysis*. Chichester: Wiley.

Bradley, J. V. (1958). Complete counterbalancing of immediate sequential effects in a Latin square design. *Journal of the American Statistical Association, 53*(282), 525–528.

Campbell, D. T., Stanley, J. C., & Gage, N. L. (1963). *Experimental and quasi-experimental designs for research*. Boston, MA: Houghton Mifflin.

Carter, S., Mankoff, J., Klemmer, S., & Matthews, T. (2008). Exiting the cleanroom: On ecological validity and ubiquitous computing. *Human–Computer Interaction, 23*(1), 47–99.

Carver, R. P. (1993). The case against statistical significance testing, revisited. *The Journal of Experimental Education, 61*(4), 287–292.

Cochran, W. G., & Cox, G. M. (1957). *Experimental designs*. New York, NY: Wiley.

Cohen, J. (1988). *Statistical power analysis for the behavioral sciences*. Hillsdale, NJ: L. Erlbaum Associates.

Cohen, J. (1994). The earth is round (p < .05). *American Psychologist, 49*(12), 997–1003.

Collins, L. M., Dziak, J. J., & Li, R. (2009). Design of experiments with multiple independent variables: A resource management perspective on complete and reduced factorial designs. *Psychological Methods, 14*(3), 202–224.

Cook, T. D., & Campbell, D. T. (1979). *Quasi-experimentation: Design & analysis issues for field settings*. Chicago: Rand McNally.

Cosley, D., Lam, S. K., Albert, I., Konstan, J. A., & Riedl, J. (2003). Is seeing believing?: How recommender system interfaces affect users' opinions. In *Proceedings of the SIGCHI conference on human factors in computing systems* (pp. 585–592). New York, NY: ACM.

Cumming, G. (2012). *Understanding the new statistics: Effect sizes, confidence intervals, and meta-analysis*. New York, NY: Routledge.

Cumming, G., & Finch, S. (2001). A primer on the understanding, use, and calculation of confidence intervals that are based on central and noncentral distributions. *Educational and Psychological Measurement, 61*(4), 532–574.

Czerwinski, M., Tan, D. S., & Robertson, G. G. (2002). Women take a wider view. In *Proceedings of the SIGCHI conference on human factors in computing systems* (pp. 195–202). New York, NY: ACM.

Dabbish, L., Kraut, R., & Patton, J. (2012). Communication and commitment in an online game team. In *Proceedings of the SIGCHI conference on human factors in computing systems* (pp. 879–888). New York, NY: ACM.

Ellis, P. D. (2010). *The essential guide to effect sizes: Statistical power, meta-analysis, and the interpretation of research results*. Cambridge, NY: Cambridge University Press.

Evans, A., & Wobbrock, J. O. (2012). Taming wild behavior: The input observer for text entry and mouse pointing measures from everyday computer use. In *Proceedings of the SIGCHI conference on human factors in computing systems* (pp. 1947–1956). New York, NY: ACM.

Fisher, R. A. (1925). *Statistical methods for research workers*. Edinburgh: Oliver and Boyd.

Fisher, R. A., & Yates, F. (1953). *Statistical tables for biological, agricultural and medical research*. Edinburgh: Oliver & Boyd.

Galton, F. (1886). Regression towards mediocrity in hereditary stature. *The Journal of the Anthropological Institute of Great Britain and Ireland, 15*, 246–263.

Gergle, D., Kraut, R. E., & Fussell, S. R. (2013). Using visual information for grounding and awareness in collaborative tasks. *Human–Computer Interaction, 28*(1), 1–39.

Grissom, R. J., & Kim, J. J. (2005). *Effect sizes for research: A broad practical approach*. Mahwah, NJ: Lawrence Erlbaum Associates.

Gujarati, D. N. (1995). *Basic econometrics*. New York, NY: McGraw-Hill.

Gutwin, C., & Penner, R. (2002). Improving interpretation of remote gestures with telepointer traces. In *Proceedings of the ACM SIGCHI conference on computer supported cooperative work* (pp. 49–57). New York, NY: ACM.

Hancock, J. T., Landrigan, C., & Silver, C. (2007). Expressing emotion in text-based communication. In *Proceedings of the SIGCHI conference on human factors in computing systems* (pp. 929–932). New York, NY: ACM.

Hancock, G. R., & Mueller, R. O. (2010). *The reviewer's guide to quantitative methods in the social sciences*. New York, NY: Routledge.

Harrison, C., Tan, D., & Morris, D. (2010). Skinput: Appropriating the body as an input surface. In *Proceedings of the SIGCHI conference on human factors in computing systems* (pp. 453–462). New York, NY: ACM.

Hornbæk, K. (2011). Some whys and hows of experiments in Human–Computer Interaction. *Foundations and Trends in Human–Computer Interaction, 5*(4), 299–373.

Johnson, D. H. (1999). The insignificance of statistical significance testing. *The Journal of Wildlife Management, 63*, 763–772.

Keegan, B., & Gergle, D. (2010). Egalitarians at the gate: One-sided gatekeeping practices in social media. In *Proceedings of the ACM SIGCHI conference on computer supported cooperative work* (pp. 131–134). New York, NY: ACM.

Kelley, K., & Preacher, K. J. (2012). On effect size. *Psychological Methods, 17*(2), 137–152.

Kenny, D.A. (1987). Statistics for the social and behavioral sciences. Canada: Little, Brown and Company.

Kennedy, P. (1998). *A guide to econometrics*. Cambridge, MA: The MIT Press.

Kirk, R. E. (1982). *Experimental design: Procedures for the behavioral sciences* (2nd ed.). Monterey, CA: Brooks/Cole.

Kirk, R. E. (2013). *Experimental design: Procedures for the behavioral sciences*. Thousand Oaks, CA: Sage.

Kline, R. B. (2004). *Beyond significance testing: Reforming data analysis methods in behavioral research*. Washington, DC: American Psychological Association.

Kline, R. B. (2013). *Beyond significance testing: Statistics reform in the behavioral sciences*. Washington, DC: American Psychological Association.

Koedinger, K. R., Anderson, J. R., Hadley, W. H., & Mark, M. A. (1997). Intelligent tutoring goes to school in the big city. *International Journal of Artificial Intelligence in Education, 8*, 30–43.

Kohavi, R., Henne, R. M., & Sommerfield, D. (2007). Practical guide to controlled experiments on the web: Listen to your customers not to the hippo. In *Proceedings of the ACM SIGKDD international conference on knowledge discovery and data mining* (pp. 959–967). New York, NY: ACM.

Kohavi, R., & Longbotham, R. (2007). Online experiments: Lessons learned. *Computer, 40*(9), 103–105.

Kohavi, R., Longbotham, R., & Walker, T. (2010). Online experiments: Practical lessons. *Computer, 43*(9), 82–85.

Kristensson, P.-O., & Zhai, S. (2004). SHARK^2: A large vocabulary shorthand writing system for pen-based computers. In *Proceedings of the ACM symposium on user interface software and technology* (pp. 43–52). New York, NY: ACM.

Kruschke, J. K. (2010). What to believe: Bayesian methods for data analysis. *Trends in Cognitive Sciences, 14*(7), 293–300.

Lazar, J., Feng, J. H., & Hochheiser, H. (2010). *Research methods in human-computer interaction*. Chichester: Wiley.

Lehmann, E. L. (1993). The fisher, Neyman-Pearson theories of testing hypotheses: One theory or two? *Journal of the American Statistical Association, 88*(424), 1242–1249.

Lieberman, H. (2003). The tyranny of evaluation. Retrieved August 15, 2012, from http://web.media.mit.edu/~lieber/Misc/Tyranny-Evaluation.html

MacKenzie, I. S., & Zhang, S. X. (1999). The design and evaluation of a high-performance soft keyboard. In *Proceedings of the SIGCHI conference on human factors in computing systems* (pp. 25–31). New York, NY: ACM.

Martin, D. W. (2004). *Doing psychology experiments*. Belmont, CA: Thomson/Wadsworth.

McLeod, P. L. (1992). An assessment of the experimental literature on electronic support of group work: Results of a meta-analysis. *Human–Computer Interaction, 7*(3), 257–280.

Nguyen, D., & Canny, J. (2005). MultiView: Spatially faithful group video conferencing. In *Proceedings of the SIGCHI conference on human factors in computing systems* (pp. 799–808). New York, NY: ACM.

Olson, J. S., Olson, G. M., Storrøsten, M., & Carter, M. (1993). Groupwork close up: A comparison of the group design process with and without a simple group editor. *ACM Transactions on Information Systems, 11*(4), 321–348.

Oulasvirta, A. (2009). Field experiments in HCI: Promises and challenges. In P. Saariluoma & H. Isomaki (Eds.), *Future interaction design II*. New York, NY: Springer.

Oulasvirta, A., Tamminen, S., Roto, V., & Kuorelahti, J. (2005). Interaction in 4-second bursts: The fragmented nature of attentional resources in mobile HCI. In *Proceedings of the SIGCHI conference on Human factors in computing systems* (pp. 919–928). New York, NY: ACM.

Rogers, J. L., Howard, K. I., & Vessey, J. T. (1993). Using significance tests to evaluate equivalence between two experimental groups. *Psychological Bulletin, 113*(3), 553.

Rosenthal, R., & Rosnow, R. L. (2008). *Essentials of behavioral research: Methods and data analysis* (3rd ed.). New York, NY: McGraw-Hill.

Shadish, W. R., Cook, T. D., & Campbell, D. T. (2002). *Experimental and quasi-experimental designs for generalized causal inference*. Boston, MA: Houghton Mifflin.

Smithson, M. (2003). *Confidence intervals*. Thousand Oaks, CA: Sage.

Stokes, D. E. (1997). *Pasteur's quadrant: Basic science and technological innovation*. Washington, DC: Brookings Institution Press.

Tan, D. S., Gergle, D., Scupelli, P., & Pausch, R. (2003). With similar visual angles, larger displays improve spatial performance. In *Proceedings of the SIGCHI conference on human factors in computing systems* (pp. 217–224). New York, NY: ACM.

Tan, D. S., Gergle, D., Scupelli, P. G., & Pausch, R. (2004). Physically large displays improve path integration in 3D virtual navigation tasks. In *Proceedings of the SIGCHI conference on human factors in computing systems* (pp. 439–446). New York, NY: ACM.

Tan, D. S., Gergle, D., Scupelli, P., & Pausch, R. (2006). Physically large displays improve performance on spatial tasks. *ACM Transactions on Computer Human Interaction, 13*(1), 71–99.

Veinott, E. S., Olson, J., Olson, G. M., & Fu, X. (1999). Video helps remote work: Speakers who need to negotiate common ground benefit from seeing each other. In *Proceedings of the SIGCHI conference on human factors in computing systems* (pp. 302–309). New York, NY: ACM.

Weir, P. (1998). *The Truman show*. Drama, Sci-Fi.

Weisband, S., & Kiesler, S. (1996). Self disclosure on computer forms: Meta-analysis and implications. In *Proceedings of the SIGCHI conference on human factors in computing systems* (pp. 3–10). New York, NY: ACM.

Weiss, N. A. (2008). *Introductory statistics*. San Francisco, CA: Pearson Addison-Wesley.

Wigdor, D., Forlines, C., Baudisch, P., Barnwell, J., & Shen, C. (2007). Lucid touch: A see-through mobile device. In *Proceedings of the ACM symposium on user interface software and technology* (pp. 269–278). New York, NY: ACM.

Williams, E. J. (1949). Experimental designs balanced for the estimation of residual effects of treatments. *Australian Journal of Chemistry, 2*(2), 149–168.

Wilson, M. L., Mackay, W., Chi, E., Bernstein, M., & Nichols, J. (2012). RepliCHI SIG: From a panel to a new submission venue for replication. In *Proceedings of the ACM conference extended abstracts on human factors in computing systems* (pp. 1185–1188). New York, NY: ACM.

Wobbrock, J. O. (2011). Practical statistics for human-computer interaction: An independent study combining statistics theory and tool know-how. *Presented at the Annual workshop of the Human-Computer Interaction Consortium (HCIC '11)*. Pacific Grove, CA.

Wobbrock, J. O., Cutrell, E., Harada, S., & MacKenzie, I. S. (2008). An error model for pointing based on Fitts' law. In *Proceedings of the SIGCHI conference on human factors in computing systems* (pp. 1613–1622). New York, NY: ACM.

Yee, N., Bailenson, J. N., & Rickertsen, K. (2007). A meta-analysis of the impact of the inclusion and realism of human-like faces on user experiences in interfaces. In *Proceedings of the SIGCHI conference on human factors in computing systems* (pp. 1–10). New York, NY: ACM.

Zhai, S. (2003). Evaluation is the worst form of HCI research except all those other forms that have been tried. Retrieved February 18, 2014, from http://shuminzhai.com/papers/EvaluationDemocracy.htm

Zhai, S., & Kristensson, P.-O. (2003). Shorthand writing on stylus keyboard. In *Proceedings of the SIGCHI conference on human factors in computing systems* (pp. 97–104). New York, NY: ACM.

Zhu, H., Kraut, R., & Kittur, A. (2012). Effectiveness of shared leadership in online communities. In *Proceedings of the ACM SIGCHI conference on computer supported cooperative work* (pp. 407–416). New York, NY: ACM.

Survey Research in HCI

Hendrik Müller, Aaron Sedley, and Elizabeth Ferrall-Nunge

Short Description of the Method

A survey is a method of gathering information by asking questions to a subset of people, the results of which can be generalized to the wider target population. There are many different types of surveys, many ways to sample a population, and many ways to collect data from that population. Traditionally, surveys have been administered via mail, telephone, or in person. The Internet has become a popular mode for surveys due to the low cost of gathering data, ease and speed of survey administration, and its broadening reach across a variety of populations worldwide. Surveys in human–computer interaction (HCI) research can be useful to:

- Gather information about people's habits, interaction with technology, or behavior
- Get demographic or psychographic information to characterize a population
- Get feedback on people's experiences with a product, service, or application
- Collect people's attitudes and perceptions toward an application in the context of usage
- Understand people's intents and motivations for using an application
- Quantitatively measure task success with specific parts of an application
- Capture people's awareness of certain systems, services, theories, or features
- Compare people's attitudes, experiences, etc. over time and across dimensions

H. Müller (✉)
Google Australia Pty Ltd., Level 5, 48 Pirrama Road, Pyrmont, NSW 2009, Australia
e-mail: hendrik82@gmail.com

A. Sedley
Google, Inc., 1600 Amphitheatre Parkway, Mountain View, CA 94043, USA
e-mail: asedley@gmail.com

E. Ferrall-Nunge
Twitter, Inc., 1355 Market Street, Suite 900, San Francisco, CA 94103, USA
e-mail: enunge@gmail.com

J.S. Olson and W.A. Kellogg (eds.), *Ways of Knowing in HCI*,
DOI 10.1007/978-1-4939-0378-8_10,

While powerful for specific needs, surveys do not allow for observation of the respondents' context or follow-up questions. When conducting research into precise behaviors, underlying motivations, and the usability of systems, then other research methods may be more appropriate or needed as a complement.

This chapter reviews the history of surveys and appropriate uses of surveys and focuses on the best practices in survey design and execution.

History, Intellectual Tradition, Evolution

Since ancient times, societies have measured their populations via censuses for food planning, land distribution, taxation, and military conscription. Beginning in the nineteenth century, political polling was introduced in the USA to project election results and to measure citizens' sentiment on a range of public policy issues. At the emergence of contemporary psychology, Francis Galton pioneered the use of questionnaires to investigate the nature vs. nurture debate and differences between humans, the latter of which evolved into the field of psychometrics (Clauser, 2007). More recently, surveys have been used in HCI research to help answer a variety of questions related to people's attitudes, behaviors, and experiences with technology.

Though nineteenth-century political polls amplified public interest in surveys, it was not until the twentieth century that meaningful progress was made on survey-sampling methods and data representativeness. Following two incorrect predictions of the US presidential victors by major polls (Literary Digest for Landon in 1936 and Gallup for Dewey in 1948), sampling methods were assailed for misrepresenting the US electorate. Scrutiny of these polling failures; persuasive academic work by statisticians such as Kiaer, Bowley, and Neyman; and extensive experimentation by the US Census Bureau led to the acceptance of random sampling as the gold standard for surveys (Converse, 1987).

Roughly in parallel, social psychologists aimed to minimize questionnaire biases and optimize data collection. For example, in the 1920s and 1930s, Louis Thurstone and Rensis Likert demonstrated reliable methods for measuring attitudes (Edwards & Kenney, 1946); Likert's scaling approach is still widely used by survey practitioners. Stanley Payne's, 1951 classic "The Art of Asking Questions" was an early study of question wording. Subsequent academics scrutinized every aspect of survey design. Tourangeau (1984) articulated the four cognitive steps to survey responses, noting that people have to comprehend what is asked, retrieve the appropriate information, judge that information according to the question, and map the judgement onto the provided responses. Krosnick & Fabrigar (1997) studied many components of questionnaire design, such as scale length, text labels, and "no opinion" responses. Groves (1989) identified four types of survey-related error: coverage, sampling, measurement, and non-response. As online surveys grew in popularity, Couper (2008) and others studied bias from the visual design of Internet questionnaires.

The use of surveys for HCI research certainly predates the Internet, with efforts to understand users' experiences with computer hardware and software. In 1983, researchers at Carnegie Mellon University conducted an experiment comparing

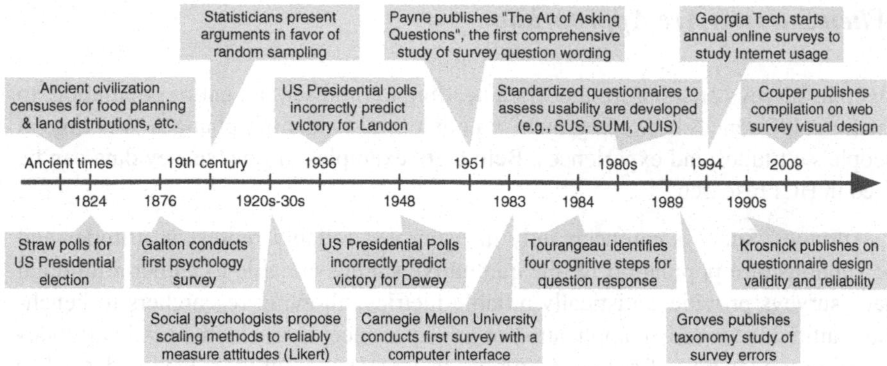

Fig. 1 Summary of the key stages in survey history

computer-collected survey responses with those from a printed questionnaire, finding less socially desirable responses in the digital survey and longer open-ended responses than in the printed questionnaire (Kiesler & Sproull, 1986). With the popularization of graphical user interfaces in the 1980s, surveys joined other methods for usability research. Several standardized questionnaires were developed to assess usability (e.g., SUS, QUIS, SUMI, summarized later in this chapter). Surveys are a direct means of measuring satisfaction; along with efficiency and effectiveness, satisfaction is a pillar of the ISO 9241, part 11, definition of usability (Abran et al., 2003). User happiness is fundamental to Google's HEART framework for user-centric measurement of Web applications (Rodden, Hutchinson, & Fu, 2010). In 1994, the Georgia Institute of Technology started annual online surveys to understand Internet usage and users and to explore Web-based survey research (Pitkow & Recker, 1994). As the Internet era progressed, online applications widely adopted surveys to measure users' satisfaction, unaddressed needs, and problems experienced, in addition to user profiling. See a summary of key stages in survey history in Fig. 1.

What Questions the Method Can Answer

When used appropriately, surveys can help inform application and user research strategies and provide insights into users' attitudes, experiences, intents, demographics, and psychographic characteristics. However, surveys are not the most appropriate method for many other HCI research goals. Ethnographic interviews, log data analysis, card sorts, usability studies, and other methods may be more appropriate. In some cases, surveys can be used with other research methods to holistically inform HCI development. This section explains survey appropriateness, when to avoid using surveys, as well as how survey research can complement other research methods.

When Surveys Are Appropriate

Overall, surveys are appropriate when needing to represent an entire population, to measure differences between groups of people, and to identify changes over time in people's attitudes and experiences. Below are examples of how survey data can be used in HCI research.

Attitudes. Surveys can accurately measure and reliably represent attitudes and perceptions of a population. While qualitative studies are able to gather attitudinal data, surveys provide statistically reliable metrics, allowing researchers to benchmark attitudes toward an application or an experience, to track changes in attitudes over time, and to tie self-reported attitudes to actual behavior (e.g., via log data). For example, surveys can be used to measure customer satisfaction with online banking immediately following their experiences.

Intent. Surveys can collect peoples' reasons for using an application at a specific time, allowing researchers to gauge the frequency across different objectives. Unlike other methods, surveys can be deployed while a person is actually using an application (i.e., an online intercept survey), minimizing the risk of imperfect recall on the respondent's part. Note that specific details and the context of one's intent may not be fully captured in a survey alone. For example, "Why did you visit this website?" could be answered in a survey, but qualitative research may be more appropriate in determining how well one understood specific application elements and what users' underlying motivations are in the context of their daily lives.

Task success. Similar to measuring intent, while HCI researchers can qualitatively observe task success through a lab or a field study, a survey can be used to reliably quantify levels of success. For example, respondents can be instructed to perform a certain task, enter results of the task, and report on their experiences while performing the task.

User experience feedback. Collecting open-ended feedback about a user's experience can be used to understand the user's interaction with technology or to inform system requirements and improvements. For example, by understanding the relative frequency of key product frustrations and benefits, project stakeholders can make informed decisions and trade-offs when allocating resources.

User characteristics. Surveys can be used to understand a system's users and to better serve their needs. Researchers can collect users' demographic information, technographic details such as system savviness or overall tech savviness, and psychographic variables such as openness to change and privacy orientation. Such data enables researchers to discover natural segments of users who may have different needs, motivations, attitudes, perceptions, and overall user experiences.

Interactions with technology. Surveys can be used to understand more broadly how people interact with technology and how technology influences social interactions with others by asking people to self-report on social, psychological, and demographic variables while capturing their behaviors. Through the use of surveys, HCI researchers can glean insights into the effects technology has on the general population.

Awareness. Surveys can also help in understanding people's awareness of existing technologies or specific application features. Such data can, for example, help researchers determine whether low usage with an application is a result of poor awareness or other factors, such as usability issues. By quantifying how aware or unaware people are, researchers can decide whether efforts (e.g., marketing campaigns) are needed to increase overall awareness and thus use.

Comparisons. Surveys can be used to compare users' attitudes, perceptions, and experiences across user segments, time, geographies, and competing applications and between experimental and control versions. Such data enable researchers to explore whether user needs and experiences vary across geographies, assess an application's strengths and weaknesses among competing technologies and how each compares with their competitors' applications, and evaluate potential application improvements while aiding decision making between a variety of proposed designs.

When to Avoid Using a Survey

Because surveys are inexpensive and easy to deploy compared to other methods, many people choose survey research even when it is inappropriate for their needs. Such surveys can produce invalid or unreliable data, leading to an inaccurate understanding of a population and poor user experiences. Below are some HCI research needs that are better addressed with other methods.

Precise behaviors. While respondents can be asked to self-report their behaviors, gathering this information from log data, if available, will always be more accurate. This is particularly true when trying to understand precise user behaviors and flows, as users will struggle to recall their exact sequence of clicks or specific pages visited. For behaviors not captured in log data, a diary study, observational study, or experience sampling may gather more accurate results than a survey.

Underlying motivations. People often do not understand or are unable to explain why they take certain actions or prefer one thing over another. Someone may be able to report their intent in a survey but may not be aware of their subconscious motivations for specific actions. Exploratory research methods such as ethnography or contextual inquiry may be more appropriate than directly asking about underlying motivations in a survey.

Usability evaluations. Surveys are inappropriate for testing specific usability tasks and understanding of tools and application elements. As mentioned above, surveys can measure task success but may not explain why people cannot use a particular application, why they do not understand some aspect of a product, or why they do not identify missteps that caused the task failure. Furthermore, a user may still be able to complete a given task even though he or she encountered several confusions, which could not be uncovered through a survey. Task-based observational research and interview methods, such as usability studies, are better suited for such research goals.

Survey research

What are the reasons for trends or distributions?

Small-sample qualitative method

Is the data just anecdotal or representative?

Fig. 2 Employing survey research either before or after research using other methods

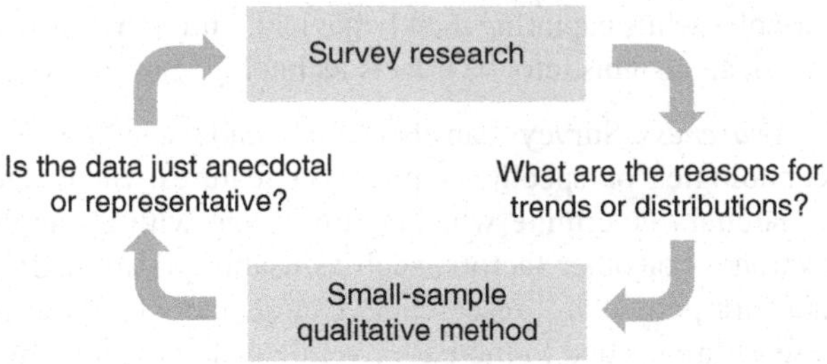

Using Surveys with Other Methods

Survey research may be especially beneficial when used in conjunction with other research methods (see Fig. 2). Surveys can follow previous qualitative studies to help quantify specific observations. For many surveys, up-front qualitative research may even be required to inform its content if no previous research exists. On the other hand, surveys can also be used to initially identify high-level insights that can be followed by in-depth research through more qualitative (meaning smaller sample) methods.

For example, if a usability study uncovers a specific problem, a survey can quantify the frequency of that problem across the population. Or a survey can be used first to identify the range of frustrations or goals, followed by qualitative interviews and observational research to gain deeper insights into self-reported behaviors and sources of frustration. Researchers may interview survey respondents to clarify responses (e.g., Yew, Shamma, & Churchill, 2011), interview another pool of participants in the same population for comparison (e.g., Froelich et al., 2012), or interview both survey respondents and new participants (e.g., Archambault & Grudin, 2012).

Surveys can also be used in conjunction with A/B experiments to aid comparative evaluations. For example, when researching two different versions of an application, the same survey can be used to assess both. By doing this, differences in variables such as satisfaction and self-reported task success can be measured and analyzed in parallel with behavioral differences observed in log data. Log data may show that one experimental version drives more traffic or engagement, but the survey may show that users were less satisfied or unable to complete a task. Moreover, log data can further validate insights from a previously conducted survey. For example, a social recommendation study by Chen, Geyer, Dugan, Muller, and Guy (2009) tested the quality of recommendations first in a survey and then through logging in a large field deployment. Psychophysiological data may be another objective accompaniment to survey data. For example, game researchers have combined surveys with data such as facial muscle and electrodermal activity (Nacke, Grimshaw, & Lindley, 2010) or attention and meditation as measured with EEG sensors (Schild, LaViola, & Masuch, 2012).

How to Do It: What Constitutes Good Work

This section breaks down survey research into the following six stages:

1. Research goals and constructs
2. Population and sampling
3. Questionnaire design and biases
4. Review and survey pretesting
5. Implementation and launch
6. Data analysis and reporting

Research Goals and Constructs

Before writing survey questions, researchers should first think about what they intend to measure, what kind of data needs to be collected, and how the data will be used to meet the research goals. When the survey-appropriate *research goals* have been identified, they should be matched to *constructs*, i.e., unidimensional attributes that cannot be directly observed. The identified constructs should then be converted into one or multiple survey questions. Constructs can be identified from prior primary research or literature reviews. Asking multiple questions about the same construct and analyzing the responses, e.g., through factor analysis, may help the researcher ensure the construct's validity.

An example will illustrate the process of converting constructs into questions. An overarching research goal may be to understand users' happiness with an online application, such as Google Search, a widely used Web search engine. Since happiness with an application is often multidimensional, it is important to separate it into measurable pieces—its constructs. Prior research might indicate that constructs such as "overall satisfaction," "perceived speed," and "perceived utility" contribute to users' happiness with that application. When all the constructs have been identified, survey questions can be designed to measure each. To validate each construct, it is important to evaluate its unique relationship with the higher level goal, using correlation, regression, factor analysis, or other methods. Furthermore, a technique called *cognitive pretesting* can be used to determine whether respondents are interpreting the constructs as intended by the researcher (see more details in the pretesting section).

Once research goals and constructs are defined, there are several other considerations to help determine whether a survey is the most appropriate method and how to proceed:

- Do the survey constructs focus on results which will directly address research goals and inform stakeholders' decision making rather than providing merely informative data? An excess of "nice-to-know" questions increases survey length and the likelihood that respondents will not complete the questionnaire, diminishing the effectiveness of the survey results.

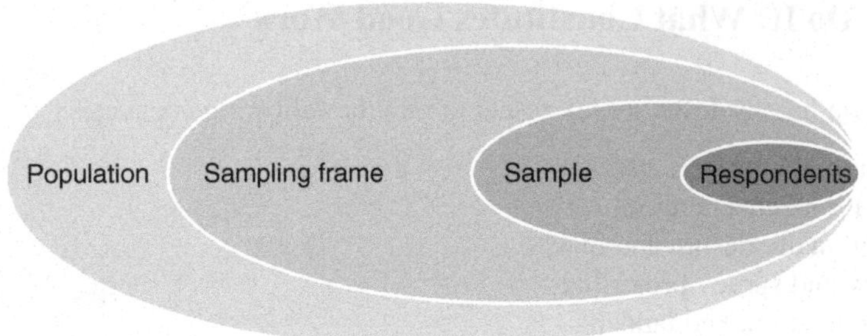

Fig. 3 The relationship between population, sampling frame, sample, and respondents

- Will the results be used for longitudinal comparisons or for one-time decisions? For longitudinal comparisons, researchers must plan on multiple survey deployments without exhausting available respondents.
- What is the number of responses needed to provide the appropriate level of precision for the insights needed? By calculating the number of responses needed (as described in detail in the following section), the researcher will ensure that key metrics and comparisons are statistically reliable. Once the target number is determined, researchers can then determine how many people to invite.

Population and Sampling

Key to effective survey research is determining who and how many people to survey. In order to do this, the survey's *population*, or set of individuals that meet certain criteria, and to whom researchers wish to generalize their results must first be defined. Reaching everyone in the population (i.e., a census) is typically impossible and unnecessary. Instead, researchers approximate the true population by creating a *sampling frame*, i.e., the set of people who the researcher is able to contact for the survey. The perfect sampling frame is identical to the population, but often a survey's sampling frame is only a portion of the population. The people from the sampling frame who are invited to take the survey are the *sample*, but only those who answer are *respondents*. See Fig. 3 illustrating these different groups.

For example, a survey can be deployed to understand the satisfaction of a product's or an application's users. In this case, the population includes everyone that uses the application, and the sampling frame consists of users that are actually reachable. The sampling frame may exclude those who have abandoned the application, anonymous users, and users who have not opted in to being contacted for research. Though the sampling frame may exclude many users, it could still include far more people than are needed to collect a statistically valid number of responses. However, if the sampling frame systematically excludes certain types of people (e.g., very dissatisfied or disengaged users), the survey will suffer from *coverage error* and its responses will misrepresent the population.

Probability Versus Non-probability Sampling

Sampling a population can be accomplished through probability- and non-probability-based methods. *Probability or random sampling* is considered the gold standard because every person in the sampling frame has an equal, nonzero chance of being chosen for the sample; essentially, the sample is selected completely randomly. This minimizes *sampling bias*, also known as *selection bias*, by randomly drawing the sample from individuals in the sampling frame and by inviting everyone in the sample in the same way. Examples of probability sampling methods include random digit telephone dialing, address-based mail surveys utilizing the US Postal Service Delivery Sequence File (DSF), and the use of a panel recruited through random sampling, those who have agreed in advance to receive surveys. For Internet surveys in particular, methods allowing for random sampling include intercept surveys for those who use a particular product (e.g., pop-up surveys or in-product links), list-based samples (e.g., for e-mail invitations), and pre-recruited probability-based panels (see Couper, 2000, for a thorough review). Another way to ensure probability sampling is to use a preexisting sampling frame, i.e., a list of candidates previously assembled using probability sampling methods. For example, Shklovski, Kraut, and Cummings' (2008) study of the effect of residential moves on communication with friends was drawn from a publicly available, highly relevant sampling frame, the National Change of Address (NCOA) database. Another approach is to analyze selected subsets of data from an existing representative survey like the General Social Survey (e.g., Wright & Randall, 2012).

While probability sampling is ideal, it is often impossible to reach and randomly select from the entire target population, especially when targeting small populations (e.g., users of a specialized enterprise product or experts in a particular field) or investigating sensitive or rare behavior. In these situations, researchers may use *non-probability sampling* methods such as volunteer opt-in panels, unrestricted self-selected surveys (e.g., links on blogs and social networks), snowball recruiting (i.e., asking for friends of friends), and *convenience samples* (i.e., targeting people readily available, such as mall shoppers) (Couper, 2000). However, non-probability methods are prone to high sampling bias and hence reduce representativeness compared to random sampling. One way representativeness can be assessed is by comparing key characteristics of the target population with those from the actual sample (for more details, refer to the analysis section).

Many academic surveys use convenience samples from an existing pool of the university's psychology students. Although not representative of most Americans, this type of sample is appropriate for investigating technology behavior among young people such as sexting (Drouin & Landgraff, 2012; Weisskirch & Delevi, 2011), instant messaging (Anandarajan, Zaman, Dai, & Arinze, 2010; Junco & Cotten, 2011; Zaman et al., 2010), and mobile phone use (Auter, 2007; Harrison, 2011; Turner, Love, & Howell, 2008). Convenience samples have also been used to identify special populations. For example, because identifying HIV and tuberculosis patients through official lists of names is difficult because of patient confidentiality, one study about the viability of using cell phones and text messages in HIV and tuberculosis education handed out surveys to potential respondents in health clinic waiting rooms (Person, Blain, Jiang, Rasmussen, & Stout, 2011). Similarly, a study of Down's syndrome patients' use of computers invited participation through special interest listservs (Feng, Lazar, Kumin, & Ozok, 2010).

Determining the Appropriate Sample Size

No matter which sampling method is used, it is important to carefully determine the target sample size for the survey, i.e., the number of survey responses needed. If the sample size is too small, findings from the survey cannot be accurately generalized to the population and may fail to detect generalizable differences between groups. If the sample is larger than necessary, too many individuals are burdened with taking the survey, analysis time for the researcher may increase, or the sampling frame is used up too quickly. Hence, calculating the optimal sample size becomes crucial for every survey.

First, the researcher needs to determine approximately how many people make up the population being studied. Second, as the survey does not measure the entire population, the required level of precision must be chosen, which consists of the margin of error and the confidence level. The *margin of error* expresses the amount of sampling error in the survey, i.e., the range of uncertainty around an estimate of a population measure, assuming normally distributed data. For example, if 60 % of the sample claims to use a tablet computer, a 5 % margin of error would mean that actually 55–65 % of the population use tablet computers. Commonly used margin of errors are 5 and 3 %, but depending on the goals of the survey anywhere between 1 and 10 % may be appropriate. Using a margin of error higher than 10 % is not recommended, unless a low level of precision can meet the survey's goals. The *confidence level* indicates how likely the reported metric falls within the margin of error if the study were repeated. A 95 % confidence level, for example, would mean that 95 % of the time, observations from repeated sampling will fall within the interval defined by the margin of error. Commonly used confidence levels are 99, 95, and 90 %; using less than 90 % is not recommended.

There are various formulas for calculating the target sample size. Figure 4, based on Krejcie and Morgan's formula (1970), shows the appropriate sample size, given the population size, as well as the chosen margin of error and confidence level for your survey. Note that the table is based on a population proportion of 50 % for the response of interest, the most cautious estimation (i.e., when higher or lower than 50 %, the required sample size declines to achieve the same margin of error). For example, for a population larger than 100,000, a sample size of 384 is required to achieve a confidence level of 95 % and a margin of error of 5 %. Note that for population sizes over about 20,000, the required sample size does not significantly increase. Researchers may set the sample size to 500 to estimate a single population parameter, which yields a margin of error of about ±4.4 % at a 95 % confidence level for large populations.

After having determined the target sample size for the survey, the researcher now needs to work backwards to estimate the number of people to actually invite to the

Confidence level	90%				95%				99%			
Size of population / Margin of error	10%	5%	3%	1%	10%	5%	3%	1%	10%	5%	3%	1%
10	9	10	10	10	9	10	10	10	9	10	10	10
100	41	73	88	99	49	80	92	99	63	87	95	99
1000	63	213	429	871	88	278	516	906	142	399	648	943
10,000	67	263	699	4035	95	370	964	4899	163	622	1556	6239
100,000	68	270	746	6335	96	383	1056	8762	166	659	1810	14227
1,000,000	68	270	751	6718	96	384	1066	9512	166	663	1840	16317
100,000,000	68	271	752	6763	96	384	1067	9594	166	663	1843	16560

Fig. 4 Sample size as a function of population size and accuracy (confidence level and margin of error)

survey, taking into account the estimated size for each subgroup and the expected response rate. If a subgroup's incidence is very small, the total number of invitations must be increased to ensure the desired sample size for this subgroup. The *response rate* of a survey describes the percentage of those who completed the survey out of all those that were invited (for more details, see the later sections on monitoring survey paradata and maximizing response rates). If a similar survey has been conducted before, then its response rate is a good reference point for calculating the required sample size. If there is no prior response rate information, the survey can be sent out to a small number of people first to measure the response rate, which is then used to determine the total number of required invitations.

For example, assuming a 30 % response rate, a 50 % incidence rate for the group of interest, and the need for 384 complete responses from that group, 2,560 people should be invited to the survey. At this point, the calculation may determine that the researcher may require a sample that is actually larger than the sampling frame; hence, the researcher may need to consider more qualitative methods as an alternative.

Mode and Methods of Survey Invitation

To reach respondents, there are four basic survey modes: mail or written surveys, phone surveys, face-to-face or in-person surveys, and Internet surveys. Survey modes may also be used in combination. The survey mode needs to be chosen carefully as each mode has its own advantages and disadvantages, such as differences in typical response rates, introduced biases (Groves, 1989), required resources and costs, audience that can be reached, and respondents' level of anonymity.

Today, many HCI-related surveys are Internet based, as benefits often outweigh their disadvantages. Internet surveys have the following major advantages:

- Easy access to large geographic regions (including international reach)
- Simplicity of creating a survey by leveraging easily accessible commercial tools

- Cost savings during survey invitation (e.g., no paper and postage, simple implementation, insignificant cost increase for large sample sizes) and analysis (e.g., returned data is already in electronic format)
- Short fielding periods, as the data is collected immediately
- Lower bias due to respondent anonymity, as surveys are self-administered with no interviewer present
- Ability to customize the questionnaire to specific respondent groups using skip logic (i.e., asking respondents a different set of questions based on the answer to a previous question)

Internet surveys also have several disadvantages. The most discussed downside is the introduction of *coverage error*, i.e., a potential mismatch between the target population and the sampling frame (Couper, 2000; Groves, 1989). For example, online surveys fail to reach people without Internet or e-mail access. Furthermore, those invited to Internet surveys may be less motivated to respond or to provide accurate data because such surveys are less personal and can be ignored more easily. This survey mode also relies on the respondents' ability to use a computer and may only provide the researcher with minimal information about the survey respondents. (See chapter on "Crowdsourcing in HCI Research.")

Questionnaire Design and Biases

Upon establishing the constructs to be measured and the appropriate sampling method, the first iteration of the survey questionnaire can be designed. It is important to carefully think through the design of each survey question (first acknowledged by Payne, 1951), as it is fairly easy to introduce biases that can have a substantial impact on the reliability and validity of the data collected. Poor questionnaire design may introduce *measurement error*, defined as the deviation of the respondents' answers from their true values on the measure. According to Couper (2000), measurement error in self-administered surveys can arise from the respondent (e.g., lack of motivation, comprehension problems, deliberate distortion) or from the instrument (e.g., poor wording or design, technical flaws). In most surveys, there is only one opportunity to deploy, and unlike qualitative research, no clarification or probing is possible. For these reasons, it is crucial that the questions accurately measure the constructs of interest.

Going forward, this section covers different types of survey questions, common questionnaire biases, questions to avoid, visual design considerations, reuse of established questionnaires, as well as visual survey design considerations.

Types of Survey Questions

There are two categories of survey questions—open- and closed-ended questions. Open-ended questions (Fig. 5) ask survey respondents to write in their own answers, whereas closed-ended questions (Fig. 6) provide a set of predefined answers to choose from.

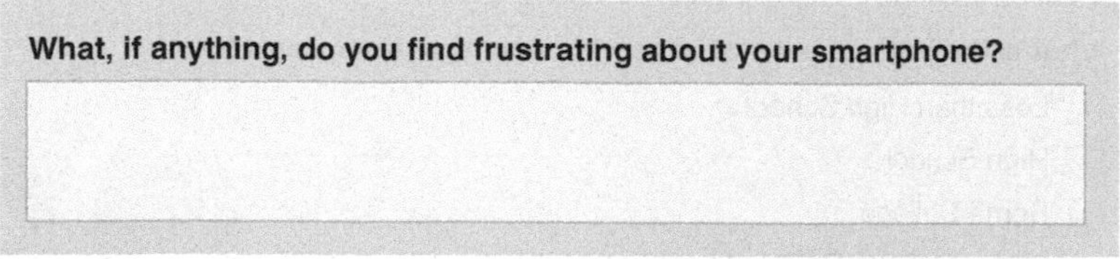

Fig. 5 Example of a typical open-ended question

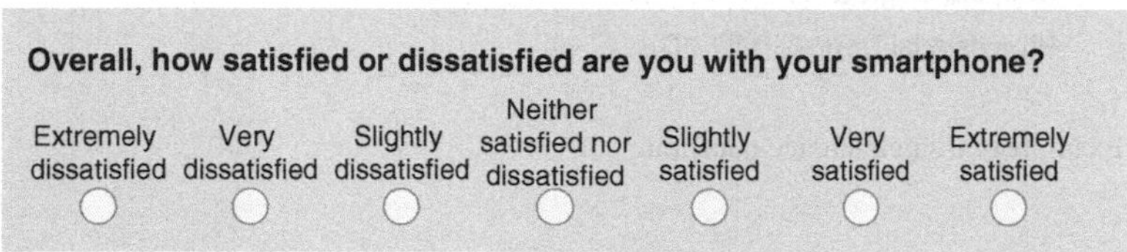

Fig. 6 Example of a typical closed-ended question, a bipolar rating question in particular

Open-ended questions are appropriate when:

- The universe of possible answers is unknown, e.g., "What is your favorite smartphone application?". However, once the universe of possible answers is identified, it may be appropriate to create a closed-ended version of the same question.
- There are so many options in the full list of possible answers that they cannot be easily displayed, e.g., "Which applications have you used on your smartphone in the last week?".
- Measuring quantities with natural metrics (i.e., a construct with an inherent unit of measurement, such as age, length, or frequency), when being unable to access information from log data, such as time, frequency, and length, e.g., "How many times do you use your tablet in a typical week?" (using a text field that is restricted to numeric input, the answers to which can later be bucketed flexibly).
- Measuring qualitative aspects of a user's experience, e.g., "What do you find most frustrating about using your smartphone?".

Closed-ended questions are appropriate when:

- The universe of possible answers is known and small enough to be easily provided, e.g., "Which operating system do you use on your smartphone?" (with answer options including "Android" and "iOS").
- Rating a single object on a dimension, e.g., "Overall, how satisfied or dissatisfied are you with your smartphone?" (on a 7-point scale from "Extremely dissatisfied" to "Extremely satisfied").
- Measuring quantities without natural metrics, such as importance, certainty, or degree, e.g., "How important is it to have your smartphone within reach 24 h a day?" (on a 5-point scale from "Not at all important" to "Extremely important").

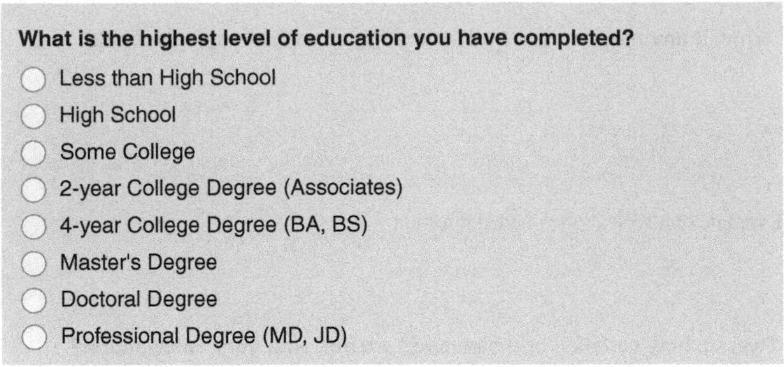

What is the highest level of education you have completed?

- ◯ Less than High School
- ◯ High School
- ◯ Some College
- ◯ 2-year College Degree (Associates)
- ◯ 4-year College Degree (BA, BS)
- ◯ Master's Degree
- ◯ Doctoral Degree
- ◯ Professional Degree (MD, JD)

Fig. 7 Example of a single-choice question

Which of the following apps do you use daily on your smartphone?
Select all that apply.

- ☐ Gmail
- ☐ Maps
- ☐ Calendar
- ☐ Facebook
- ☐ Hangouts
- ☐ Drive

Fig. 8 Example of a multiple-choice question

Types of Closed-Ended Survey Questions

There are four basic types of closed-ended questions: single-choice, multiple-choice, rating, and ranking questions.

1. *Single-choice questions* work best when only one answer is possible for each respondent in the real world (Fig. 7).
2. *Multiple-choice questions* are appropriate when more than one answer may apply to the respondent. Frequently, multiple-choice questions are accompanied by "select all that apply" help text. The maximum number of selections may also be specified to force users to prioritize or express preferences among the answer options (Fig. 8).
3. *Ranking questions* are best when respondents must prioritize their choices given a real-world situation (Fig. 9).
4. *Rating questions* are appropriate when the respondent must judge an object on a continuum. To optimize reliability and minimize bias, scale points need to be

Rank the following smartphone manufacturers in order of your preference:
Add a number to each row, 1 being the least preferred, 5 being the most preferred.

☐ Apple
☐ HTC
☐ Samsung
☐ Motorola
☐ Nokia

Fig. 9 Example of a ranking question

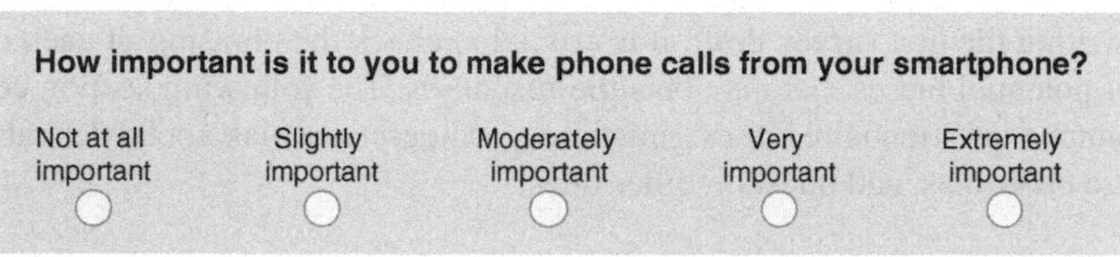

How important is it to you to make phone calls from your smartphone?
Not at all important
Slightly important
Moderately important
Very important
Extremely important

Fig. 10 Example of a rating question, for a unipolar construct in particular

fully labeled instead of using numbers (Groves et al., 2004), and each scale point should be of equal width to avoid bias toward visually bigger response options (Tourangeau, Couper, & Conrad, 2004). Rating questions should use either a unipolar or a bipolar scale, depending on the construct being measured (Krosnick & Fabrigar, 1997; Schaeffer & Presser, 2003).

Unipolar constructs range from zero to an extreme amount and do not have a natural midpoint. They are best measured with a 5-point rating scale (Krosnick & Fabrigar, 1997), which optimizes reliability while minimizing respondent burden, and with the following scale labels, which have been shown to be semantically equidistant from each other (Rohrmann, 2003): "Not at all …," "Slightly …," "Moderately …," "Very …," and "Extremely …." Such constructs include importance (see Fig. 10), interest, usefulness, and relative frequency. *Bipolar constructs* range from an extreme negative to an extreme positive with a natural midpoint. Unlike unipolar constructs, they are best measured with a 7-point rating scale to maximize reliability and data differentiation (Krosnick & Fabrigar, 1997). Bipolar constructs may use the following scale labels: "Extremely …," "Moderately …," "Slightly …," "Neither … nor …," "Slightly …," "Moderately …," and "Extremely …." Such constructs include satisfaction (see Fig. 6, from dissatisfied to satisfied), perceived speed (from slow to fast), ease of use (from difficult to easy), and visual appeal (from unappealing to appealing).

When using a rating scale, the inclusion of a midpoint should be considered. While some may argue that including a midpoint provides an easy target for respondents who shortcut answering questions, others argue that the exclusion of a midpoint forces people who truly are in the middle to choose an option that does not reflect their actual opinion. O'Muircheartaigh, Krosnick, and Helic (2001) found that having a midpoint on a rating scale increases reliability, has no effect on validity, and does not result in lower data quality. Additionally, people who look for shortcuts ("shortcutters") are not more likely to select the midpoint when present. Omitting the midpoint, on the other hand, increases the amount of random measurement error, resulting in those who actually feel neutral to end up making a random choice on either side of the scale. These findings suggest that a midpoint should be included when using a rating scale.

Questionnaire Biases

After writing the first survey draft, it is crucial to check the phrasing of each question for potential biases that may bias the responses. The following section covers five common questionnaire biases: satisficing, acquiescence bias, social desirability, response order bias, and question order bias.

Satisficing

Satisficing occurs when respondents use a suboptimal amount of cognitive effort to answer questions. Instead, satisficers will typically pick what they consider to be the first acceptable response alternative (Krosnick, 1991; Simon, 1956). Satisficers compromise one or more of the following four cognitive steps for survey response as identified by Tourangeau (1984):

1. *Comprehension* of the question, instructions, and answer options
2. *Retrieval* of specific memories to aid with answering the question
3. *Judgement* of the retrieved information and its applicability to the question
4. *Mapping* of judgement onto the answer options

Satisficers shortcut this process by exerting less cognitive effort or by skipping one or more steps entirely; satisficers use less effort to understand the question, to thoroughly search their memories, to carefully integrate all retrieved information, or to accurately pick the proper response choice (i.e., they pick the next best choice).

Satisficing can take weak and strong forms (Krosnick, 1999). Weak satisficers make an attempt to answer correctly yet are less than thorough, while strong satisficers may not at all search their memory for relevant information and simply select answers at random in order to complete the survey quickly. In other words, weak satisficers carelessly process all four cognitive steps, while strong satisficers typically skip the retrieval and judgement steps.

Respondents are more likely to satisfice when (Krosnick, 1991):

- Cognitive ability to answer is low.
- Motivation to answer is low.
- Question difficulty is high at one of the four stages, resulting in cognitive exertion.

To minimize satisficing, the following may be considered:

- Complex questions that require an inordinate amount of cognitive exertion should be avoided.
- Answer options such as “no opinion,” “don’t know,” “not applicable,” or “unsure” should be avoided, since respondents with actual opinions will be tempted and select this option (Krosnick, 2002; Schaeffer & Presser, 2003). Instead, respondents should first be asked whether they have thought about the proposed question or issue enough to have an opinion; those that haven’t should be screened out.
- Using the same rating scale in a series of back-to-back questions should be avoided. Potential satisfiers may pick the same scale point for all answer options. This is known as straight-lining or item non-differentiation (Herzog & Bachman, 1981; Krosnick & Alwin, 1987, 1988).
- Long questionnaires should be avoided, since respondents will be less likely to optimally answer questions when they become increasingly fatigued and unmotivated (Cannell & Kahn, 1968; Herzog & Bachman, 1981).
- Respondent motivation can be increased by explaining the importance of the survey topic and that their responses are critical to the researcher (Krosnick, 1991).
- Respondents may be asked to justify their answer to the question that may exhibit satisficing.
- Trap questions (e.g., “Enter the number 5 in the following text box:”) can identify satisficers and fraudulent survey respondents.

Acquiescence Bias

When presented with agree/disagree, yes/no, or true/false statements, some respondents are more likely to concur with the statement independent of its substance. This tendency is known as acquiescence bias (Smith, 1967).

Respondents are more likely to acquiescence when:

- Cognitive ability is low (Krosnick, Narayan, & Smith, 1996) or motivation is low.
- Question difficulty is high (Stone, Gage, & Leavitt, 1957).
- Personality tendencies skew toward agreeableness (Costa & McCrae, 1988; Goldberg, 1990; Saris, Revilla, Krosnick, & Shaeffer, 2010).
- Social conventions suggest that a “yes” response is most polite (Saris et al., 2010).
- The respondent satisfices and only thinks of reasons why the statement is true, rather than expending cognitive effort to consider reasons for disagreement (Krosnick, 1991).
- Respondents with lower self-perceived status assume that the survey administrator agrees with the posed statement, resulting in deferential agreement bias (Saris et al., 2010).

To minimize acquiescence bias, the following may be considered:

- Avoid questions with agree/disagree, yes/no, true/false, or similar answer options (Krosnick & Presser, 2010).
- Where possible, ask construct-specific questions (i.e., questions that ask about the underlying construct in a neutral, non-leading way) instead of agreement statements (Saris et al., 2010).
- Use reverse-keyed constructs; i.e., the same construct is asked positively and negatively in the same survey. The raw scores of both responses are then combined to correct for acquiescence bias.

Social Desirability

Social desirability occurs when respondents answer questions in a manner they feel will be positively perceived by others (Goffman, 1959; Schlenker & Weigold, 1989). Favorable actions may be overreported, and unfavorable actions or views may be underreported. Topics that are especially prone to social desirability bias include voting behavior, religious beliefs, sexual activity, patriotism, bigotry, intellectual capabilities, illegal acts, acts of violence, and charitable acts.

Respondents are inclined to provide socially desirable answers when:

- Their behavior or views go against the social norm (Holbrook & Krosnick, 2010).
- Asked to provide information on sensitive topics, making the respondent feel uncomfortable or embarrassed about expressing their actual views (Holbrook & Krosnick, 2010).
- They perceive a threat of disclosure or consequences to answering truthfully (Tourangeau, Rips, & Rasinski, 2000).
- Their true identity (e.g., name, address, phone number) is captured in the survey (Paulhus, 1984).
- The data is directly collected by another person (e.g., in-person or phone surveys).

To minimize social desirability bias, respondents should be allowed to answer anonymously or the survey should be self-administered (Holbrook & Krosnick, 2010; Tourangeau & Smith, 1996; Tourangeau & Yan, 2007).

Response Order Bias

Response order bias is the tendency to select the items toward the beginning (i.e., primacy effect) or the end (i.e., recency effect) of an answer list or scale (Chan, 1991; Krosnick & Alwin, 1987; Payne, 1971). Respondents unconsciously interpret the ordering of listed answer options and assume that items near each other are related, top or left items are interpreted to be "first," and middle answers in a scale without a natural order represent the typical value (Tourangeau et al., 2004). Primacy and recency effects are the strongest when the list of answer options is long (Schuman & Presser, 1981) or when they cannot be viewed as a whole (Couper et al., 2004).

To minimize response order effects, the following may be considered:

- Unrelated answer options should be randomly ordered across respondents (Krosnick & Presser, 2010).
- Rating scales should be ordered from negative to positive, with the most negative item first.
- The order of ordinal scales should be reversed randomly between respondents, and the raw scores of both scale versions should be averaged using the same value for each scale label. That way, the response order effects cancel each other out across respondents (e.g., Villar & Krosnick, 2011), unfortunately, at the cost of increasing variability.

Question Order Bias

Order effects also apply to the order of the questions in surveys. Each question in a survey has the potential to bias each subsequent question by priming respondents (Kinder & Iyengar, 1987; Landon, 1971).

The following guidelines may be considered:

- Questions should be ordered from broad to more specific (i.e., a funnel approach) to ensure that the survey follows conversational conventions.
- Early questions should be easy to answer and directly related to the survey topic (to help build rapport and engage respondents) (Dillman, 1978).
- Non-critical, complex, and sensitive questions should be included toward the end of the survey to avoid early drop-off and to ensure collection of critical data.
- Related questions need to be grouped to reduce context switching so that respondents can more easily and quickly access related information from memory, as opposed to disparate items.
- The questionnaire should be divided into multiple pages with distinct sections labeled for easier cognitive processing.

Other Types of Questions to Avoid

Beyond the five common questionnaire biases mentioned above, there are additional question types that can result in unreliable and invalid survey data. These include broad, leading, double-barreled, recall, prediction, hypothetical, and prioritization questions.

Broad questions lack focus and include items that are not clearly defined or those that can be interpreted in multiple ways. For example, "Describe the way you use your tablet computer" is too broad, as there are many aspects to using a tablet such as the purpose, applications being used, and its locations of use. Instead of relying on the respondent to decide on which aspects to report, the research goal as well as core construct(s) should be determined beforehand and asked about in a focused manner. A more focused set of questions for the example above could be "Which apps did you use on your tablet computer over the last week?" and "Describe the locations in which you used your tablet computer last week?".

Leading questions manipulate respondents into giving a certain answer by providing biasing content or suggesting information the researcher is looking to have confirmed. For example, "This application was recently ranked as number one in customer satisfaction. How satisfied are you with your experience today?". Another way that questions can lead the respondent toward a certain answer includes those that ask the respondent to agree or disagree with a given statement, as for example in "Do you agree or disagree with the following statement: I use my smartphone more often than my tablet computer." Note that such questions can additionally result in acquiescence bias (as discussed above). To minimize the effects of leading questions, questions should be asked in a fully neutral way without any examples or additional information that may bias respondents toward a particular response.

Double-barreled questions ask about multiple items while only allowing for a single response, resulting in less reliable and valid data. Such questions can usually be detected by the existence of the word "and." For example, when asked "How satisfied or dissatisfied are you with your smartphone and tablet computer?", a respondent with differing attitudes toward the two devices will be forced to pick an attitude that either reflects just one device or the average across both devices. Questions with multiple items should be broken down into one question per construct or item.

Recall questions require the respondent to remember past attitudes and behaviors, leading to recall bias (Krosnick & Presser, 2010) and inaccurate recollections. When a respondent is asked "How many times did you use an Internet search engine over the past 6 months?", they will try to rationalize a plausible number, because recalling a precise count is difficult or impossible. Similarly, asking questions that compare past attitudes to current attitudes, as in "Do you prefer the previous or current version of the interface?", may result in skewed data due to difficulty remembering past attitudes. Instead, questions should focus on the present, as in "How satisfied or dissatisfied are you with your smartphone today?", or use a recent time frame, for example, "In the past hour, how many times did you use an Internet search engine?". If the research goal is to compare attitudes or behaviors across different product versions or over time, the researcher should field separate surveys for each product version or time period and make the comparison themselves.

Prediction questions ask survey respondents to anticipate future behavior or attitudes, resulting in biased and inaccurate responses. Such questions include "Over the next month, how frequently will you use an Internet search engine?". Even more cognitively burdensome are *hypothetical* questions, i.e., asking the respondent to imagine a certain situation in the future and then predicting their attitude or behavior in that situation. For example, "Would you purchase more groceries if the store played your favorite music?" and "How much would you like this Website if it used blue instead of red for their color scheme?" are hypothetical questions. Other frequently used hypothetical questions are those that ask the respondent to prioritize a future feature set, as in "Which of the following features would make you more satisfied with this product?". Even though the respondent may have a clear answer to this question, their response does not predict actual future usage of or satisfaction with the product if that feature was added. Such questions should be entirely excluded from surveys.

Leveraging Established Questionnaires

An alternative to constructing a brand new questionnaire is utilizing questionnaires developed by others. These usually benefit from prior validation and allow researchers to compare results with other studies that used the same questionnaire. When selecting an existing questionnaire, one should consider their particular research goals and study needs and adapt the questionnaire as appropriate. Below are commonly used HCI-related questionnaire instruments. Note that as survey research methodology has significantly advanced over time, each questionnaire should be assessed for potential sources of measurement error, such as the biases and the to-be-avoided question types mentioned previously.

- *NASA Task Load Index (NASA TLX)*. Originally developed for aircraft cockpits, this questionnaire allows researchers to subjectively assess the workload of operators working with human–machine systems. It measures mental demand, physical demand, temporal demand, performance, effort, and frustration (Hart & Staveland, 1988).
- *Questionnaire for User Interface Satisfaction (QUIS)*. This questionnaire assesses one's overall reaction to a system, including its software, screen, terminology, system information, and learnability (Chin, Diehl, & Norman, 1988).
- *Software Usability Measurement Inventory (SUMI)*. This questionnaire measures perceived software quality covering dimensions such as efficiency, affect, helpfulness, control, and learnability, which are then summarized into a single satisfaction score (Kirakowski & Corbett, 1993).
- *Computer System Usability Questionnaires (CSUQ)*. This questionnaire developed by IBM measures user satisfaction with system usability (Lewis, 1995).
- *System Usability Scale (SUS)*. As one of the most frequently used scales in user experience, SUS measures attitudes regarding the effectiveness, efficiency, and satisfaction with a system with ten questions, yielding a single score (Brooke, 1996).
- *Visual Aesthetics of Website Inventory (VisAwi)*. This survey measures perceived visual aesthetics of a Website on the four subscales of simplicity, diversity, colorfulness, and craftsmanship (Moshagen & Thielsch, 2010).

Visual Survey Design Considerations

Researchers should also take into account their survey's visual design, since specific choices, including the use of images, spacing, and progress bars, may unintentionally bias respondents. This section summarizes such visual design aspects; for more details, refer to Couper (2008).

While objective images (e.g., product screenshots) can help clarify questions, context-shaping images can influence a respondent's mindset. For example, when asking respondents to rate their level of health, presenting an image of someone in a hospital bed has a framing effect that results in higher health ratings compared to that of someone jogging (Couper, Conrad, & Tourangeau, 2007).

The visual treatment of response options also matters. When asking closed-ended questions, uneven spacing between horizontal scale options results in a higher selection rate for scale points with greater spacing; evenly spaced scale options are recommended (Tourangeau, Couper, & Conrad, 2004). Drop-down lists, compared to radio buttons, have been shown to be harder and slower to use and to result in more accidental selections (Couper, 2011). Lastly, larger text fields increase the amount of text entered (Couper, 2011) but may intimidate respondents, potentially causing higher break-offs (i.e., drop-out rates).

Survey questions can be presented one per page, multiple per page, or all on one page. Research into pagination effects on completion rates is inconclusive (Couper, 2011). However, questions appearing on the same page may have higher correlations with each other, a sign of measurement bias (Peytchev, Couper, McCabe, & Crawford, 2006). In practice, most Internet surveys with skip logic use multiple pages, whereas very short questionnaires are often presented on a single page.

While progress bars are generally preferred by respondents and are helpful for short surveys, their use in long surveys or surveys with skip logic can be misleading and intimidating. Progress between pages in long surveys may be small, resulting in increased break-off rates (Callegaro, Villar, & Yang, 2011). On the other hand, progress bars are likely to increase completion rates for short surveys, where substantial progress is shown between pages.

Review and Survey Pretesting

At this point in the survey life cycle, it is appropriate to have potential respondents take and evaluate the survey in order to identify any remaining points of confusion. For example, the phrase "mobile device" may be assumed to include mobile phones, tablets, and in-car devices by the researcher, while survey respondents may interpret it to be mobile phones only. Or, when asking for communication tools used by the respondent, the provided list of answer choices may not actually include all possible options needed to properly answer the question. Two established evaluation methods used to improve survey quality are cognitive pretesting and field testing the survey by launching it to a subset of the actual sample, as described more fully in the remainder of this section. By evaluating surveys early on, the researcher can identify disconnects between their own assumptions and how respondents will read, interpret, and answer questions.

Cognitive Pretesting

To conduct a cognitive pretest, a small set of potential respondents is invited to participate in an in-person interview where they are asked to take the survey while using the think-aloud protocol (similar to a usability study). A cognitive pretest assesses question interpretation, construct validity, and comprehension of survey terminology and calls attention to missing answer options or entire questions (Bolton & Bronkhorst, 1995; Collins, 2003; Drennan, 2003; Presser et al., 2004). However, note that due to the testing environment, a cognitive pretest does not allow the researcher to understand contextual influences that may result in break-off or not filling out the survey in the first place.

As part of a pretest, participants are asked the following for each question:

1. "Read the entire question and describe it in your own words."
2. "Select or write an answer while explaining your thought process."
3. "Describe any confusing terminology or missing answer choices."

During the interview, the researcher should observe participant reactions; identify misinterpretations of terms, questions, answer choices, or scale items; and gain insight into how respondents process questions and come up with their answers. The researcher then needs to analyze the collected information to improve problematic areas before fielding the final questionnaire. A questionnaire could go through several rounds of iteration before reaching the desired quality.

Field Testing

Piloting the survey with a small subset of the sample will help provide insights that cognitive pretests alone cannot (Collins, 2003; Presser et al., 2004). Through field testing, the researcher can assess the success of the sampling approach, look for common break-off points and long completion times, and examine answers to open-ended questions. High break-off rates and completion times may point to flaws in the survey design (see the following section), while unusual answers may suggest a disconnect between a question's intention and respondents' interpretation. To yield additional insights from the field test, a question can be added at the end of each page or at the end of the entire survey where respondents can provide explicit feedback on any points of confusion. Similar to cognitive pretests, field testing may lead to several rounds of questionnaire improvement as well as changes to the sampling method. Finally, once all concerns are addressed, the survey is ready to be fielded to the entire sample.

Implementation and Launch

When all questions are finalized, the survey is ready to be fielded based on the chosen sampling method. Respondents may be invited through e-mails to specifically named persons (e.g., respondents chosen from a panel), intercept pop-up dialogs while using a product or a site, or links placed directly in an application (see the sampling section for more details; Couper, 2000).

There are many platforms and tools that can be used to implement Internet surveys, such as ConfirmIt, Google Forms, Kinesis, LimeSurvey, SurveyGizmo, SurveyMonkey, UserZoom, Wufoo, and Zoomerang, to name just a few. When deciding on the appropriate platform, functionality, cost, and ease of use should be taken into consideration. The questionnaire may require a survey tool that supports functionality such as branching and conditionals, the ability to pass URL parameters, multiple languages, and a range of question types. Additionally, the researcher may want to customize the visual style of the survey or set up an automatic reporting dashboard, both of which may only be available on more sophisticated platforms.

Piping Behavioral Data into Surveys

Some platforms support the ability to combine survey responses with other log data, which is referred to as piping. Self-reported behaviors, such as frequency of use, feature usage, tenure, and platform usage, are less valid and reliable compared to generating the same metrics through log data. By merging survey responses with behavioral data, the researcher can more accurately understand the relationship between respondent characteristics and their behaviors or attitudes. For example, the researcher may find that certain types of users or the level of usage may correlate with higher reported satisfaction. Behavioral data can either be passed to the results database as a parameter in the survey invitation link or combined later via a unique identifier for each respondent.

Monitoring Survey Paradata

With the survey's launch, researchers should monitor the initial responses as well as survey paradata to identify potential mistakes in the survey design. Survey paradata is data collected about the survey response process, such as the devices from which the survey was accessed, time to survey completion, and various response-related rates. By monitoring such metrics, the survey researcher can quickly apply improvements before the entire sample has responded to the survey. The American Association for Public Opinion Research specified a set of definitions for commonly used paradata metrics (AAPOR, 2011):

- Click-through rate: Of those invited, how many opened the survey.
- Completion rate: Of those who opened the survey, how many finished the survey.
- Response rate: Of those invited, how many finished the survey.
- Break-off rate: Of those who started, how many dropped off on each page.
- Completion time: The time it took respondents to finish the entire survey.

Response rates are dependent on a variety of factors, the combination of which makes it difficult to specify an acceptable response rate in HCI survey research. A meta-analysis of 31 e-mail surveys from 1986 to 2000 showed that average response rates for e-mail surveys typically fall between 30 and 40 %, with follow-up reminders significantly increasing response rates (Sheehan, 2001). Another review of 69 e-mail surveys showed that response rates averaged around 40 % (Cook, Heath, & Thompson, 2000). When inviting respondents through Internet intercept surveys (e.g., pop-up surveys or in-product links), response rates may be 15 % or lower (Couper, 2000). Meta-analyses of mailed surveys showed that their response rates are 40–50 % (Kerlinger, 1986) or 55 % (Baruch, 1999). In experimental comparisons to mailed surveys, response rates to Internet e-mail surveys were about 10 % lower (Kaplowitz, Hadlock, & Levine, 2004; Manfreda et al., 2008). Such meta reviews also showed that overall response rates have been declining over several decades (Baruch, 1999; Baruch & Holtom, 2008; Sheehan, 2001); however, this decline seems to have stagnated around 1995 (Baruch & Holtom, 2008).

Maximizing Response Rates

In order to gather enough responses to represent the target population with the desired level of precision, response rates should be maximized. Several factors affect response rates, including the respondents' interest in the subject matter, the perceived impact of responding to the survey, questionnaire length and difficulty, the presence and nature of incentives, and researchers' efforts to encourage response (Fan & Yan, 2010).

Based on experimentation with invitation processes for mail surveys, Dillman (1978) developed the "Total Design Method" to optimize response rates. This method, consistently achieving response rates averaging 70 % or better, consists of a timed sequence of four mailings: the initial request with the survey on week one, a reminder postcard on week two, a replacement survey to non-respondents on week four, and a second replacement survey to non-respondents by certified mail on week seven. Dillman incorporates social exchange theory into the Total Design Method by personalizing the invitation letters, using official stationery to increase trust in the survey's sponsorship, explaining the usefulness of the survey research and the importance of responding, assuring the confidentiality of respondents' data, and beginning the questionnaire with items directly related to the topic of the survey (1991). Recognizing the need to cover Internet and mixed-mode surveys, Dillman extended his prior work with the "Tailored Design Method." With this update, he emphasized customizing processes and designs to fit each survey's topic, population, and sponsorship (2007).

Another component of optimizing response rates is getting as many complete responses as possible from those who start the survey. According to Peytchev (2009), causes of break-off may fall into the following three categories:

- Respondent factors (survey topic salience and cognitive ability)
- Survey design factors (length, progress indicators, and incentives)
- Question design factors (fatigue and intimidation from open-ended questions and lengthy grid questions)

The questionnaire design principles mentioned previously may help minimize break-off, such as making surveys as short as possible, having a minimum of required questions, using skip logic, and including progress bars for short surveys.

Providing an incentive to encourage survey responses may be advantageous in certain cases. Monetary incentives tend to increase response rates more than non-monetary incentives (Singer, 2002). In particular, non-contingent incentives, which are offered to all people in the sample, generally outperform contingent incentives, given only upon completion of the survey (Church, 1993). This is true even when a non-contingent incentive is considerably smaller than a contingent incentive. One strategy to maximize the benefit of incentives is to offer a small non-contingent award to all invitees, followed by a larger contingent award to initial non-respondents (Lavrakas, 2011). An alternate form of contingent incentive is a lottery, where a drawing is held among respondents for a small number of monetary awards or other prizes. However, the efficacy of such lotteries is unclear (Stevenson, Dykema, Cyffka, Klein, & Goldrick-Rab, 2012). Although incentives will typically increase response rates, it is much less certain whether they increase the representativeness of the results. Incentives are likely most valuable when facing a small population or sampling frame, and high response rates are required for sufficiently precise measurements. Another case where incentives may help is when some groups in the sample have low interest in the survey topic (Singer, 2002). Furthermore, when there is a cost to contact each potential respondent, as with door-to-door interviewing, incentives will decrease costs by lowering the number of people that need to be contacted.

Data Analysis and Reporting

Once all the necessary survey responses have been collected, it is time to start making sense of the data by:

1. Preparing and exploring the data
2. Thoroughly analyzing the data
3. Synthesizing insights for the target audience of this research

Data Preparation and Cleaning

Cleaning and preparing survey data before conducting a thorough analysis are essential to identify low-quality responses that may otherwise skew the results. When taking a pass through the data, survey researchers should look for signs of poor-quality responses. Such survey data can either be left as is, removed, or presented separately from trusted data. If the researcher decides to remove poor data, they must cautiously decide whether to remove data on the respondent level (i.e., listwise deletion), an individual question level (i.e., pairwise deletion), or only beyond a certain point in the survey where respondents' data quality is declined. The following are signals that survey researchers should look out for at the survey response level:

- *Duplicate responses*. In a self-administered survey, a respondent might be able to fill out the survey more than once. If possible, respondent information such as name, e-mail address, or any other unique identifier should be used to remove duplicate responses.
- *Speeders*. Respondents that complete the survey faster than possible, speeders, may have carelessly read and answered the questions, resulting in arbitrary responses. The researcher should examine the distribution of response times and remove any respondents that are suspiciously fast.
- *Straight-liners and other questionable patterns*. Respondents that always, or almost always, pick the same answer option across survey questions are referred to as straight-liners. Grid-style questions are particularly prone to respondent straight-lining (e.g., by always picking the first answer option when asked to rate a series of objects). Respondents may also try to hide the fact that they are randomly choosing responses by answering in a fixed pattern (e.g., by alternating between the first and second answer options across questions). If a respondent straight-lines through the entire survey, the researcher may decide to remove the respondent's data entirely. If a respondent starts straight-lining at a certain point, the researcher may keep data up until that point.
- *Missing data and break-offs*. Some respondents may finish a survey but skip several questions. Others may start the survey but break off at some point. Both result in missing data. It should first be determined whether those who did not respond to certain questions are different from those who did. A non-response study should be conducted to assess the amount of non-response bias for each survey question. If those who did not answer certain questions are not meaningfully different from those who did, the researcher can consider leaving the data as is; however, if there is a difference, the researcher may choose to impute plausible values based on similar respondents' answers (De Leeuw, Hox, & Huisman, 2003).

Furthermore, the following signals may need to be assessed at a question-by-question level:

- *Low inter-item reliability*. When multiple questions are used to measure a single construct, respondents' answers to these questions should be associated with each other. Respondents that give inconsistent or unreliable responses (e.g., selecting "very fast" and "very slow" for separate questions assessing the construct of speed) may not have carefully read the set of questions and should be considered for removal.
- *Outliers*. Answers that significantly deviate from the majority of responses are considered outliers and should be examined. For questions with numeric values, some consider outliers as the top and bottom 2 % of responses, while others calculate outliers as anything outside of two or three standard deviations from the mean. Survey researchers should determine how much of a difference keeping or removing the outliers has on variables' averages. If the impact is significant, the researcher may either remove such responses entirely or replace them with a value that equals two or three standard deviations from the mean. Another way to describe the central tendency while minimizing the effect of outliers is to use the median, rather than the mean.

- *Inadequate open-ended responses.* Due to the amount of effort required, open-ended questions may lead to low-quality responses. Obvious garbage and irrelevant answers, such as “asdf,” should be removed, and other answers from the same respondent should be examined to determine whether all their survey responses warrant removal.

Analysis of Closed-Ended Responses

To get an overview of what the survey data shows, *descriptive statistics* are fundamental. By looking at measures such as the frequency distribution, central tendency (e.g., mean or median), and data dispersion (e.g., standard deviation), emerging patterns can be uncovered. The frequency distribution shows the proportion of responses for each answer option. The central tendency measures the “central” position of a frequency distribution and is calculated using the mean, median, and mode. Dispersion examines the data spread around the central position through calculations such as standard deviation, variance, range, and interquartile range.

While descriptive statistics only describe the existing data set, *inferential statistics* can be used to draw inferences from the sample to the overall population in question. Inferential statistics consists of two areas: estimation statistics and hypothesis testing. Estimation statistics involves using the survey’s sample in order to approximate the population’s value. Either the margin of error or the confidence interval of the sample’s data needs to be determined for such estimation. To calculate the margin of error for an answer option’s proportion, only the sample size, the proportion, and a selected confidence level are needed. However, to determine the confidence interval for a mean, the standard error of the mean is required additionally. A confidence interval thus represents the estimated range of a population’s mean at a certain confidence level.

Hypothesis testing determines the probability of a hypothesis being true when comparing groups (e.g., means or proportions being the same or different) through the use of methods such as *t*-test, ANOVA, or Chi-square. The appropriate test is determined by the research question, type of prediction by the researcher, and type of variable (i.e., nominal, ordinal, interval, or ratio). An experienced quantitative researcher or statistician should be involved.

Inferential statistics can also be applied to identify connections among variables:

- *Bivariate correlations* are widely used to assess linear relationships between variables. For example, correlations can indicate which product dimensions (e.g., ease of use, speed, features) are most strongly associated with users’ overall satisfaction.
- *Linear regression* analysis indicates the proportion of variance in a continuous dependent variable that is explained by one or more independent variables and the amount of change explained by each unit of an independent variable.

- *Logistic regression* predicts the change in probability of getting a particular value in a binary variable, given a unit change in one or more independent variables.
- *Decision trees* assess the probabilities of reaching specific outcomes, considering relationships between variables.
- *Factor analysis* identifies groups of covariates and can be useful to reduce a large number of variables into a smaller set.
- *Cluster analysis* looks for related groups of respondents and is often used by market researchers to identify and categorize segments within a population.

There are many packages available to assist with survey analysis. Software such as Microsoft Excel, and even certain survey platforms such as SurveyMonkey or Google Forms, can be used for basic descriptive statistics and charts. More advanced packages such as SPSS, R, SAS, or Matlab can be used for complex modeling, calculations, and charting. Note that data cleaning often needs to be a precursor to conducting analysis using such tools.

Analysis of Open-Ended Comments

In addition to analyzing closed-ended responses, the review of open-ended comments contributes a more holistic understanding of the phenomena being studied. Analyzing a large set of open-ended comments may seem like a daunting task at first; however, if done correctly, it reveals important insights that cannot otherwise be extracted from closed-ended responses. The analysis of open-ended survey responses can be derived from the method of *grounded theory* (Böhm, 2004; Glaser & Strauss, 1967) (see chapter on "Grounded Theory Methods").

An interpretive method, referred to as *coding* (Saldaña, 2009), is used to organize and transform qualitative data from open-ended questions to enable further quantitative analysis (e.g., preparing a frequency distribution of the codes or comparing the responses across groups). The core of such qualitative analysis is to assign one or several codes to each comment; each code consists of a word or a short phrase summarizing the essence of the response with regard to the objective of that survey question (e.g., described frustrations, behavior, sentiment, or user type). Available codes are chosen from a coding scheme, which may already be established by the community or from previous research or may need to be created by the researchers themselves. In most cases, as questions are customized to each individual survey, the researcher needs to establish the coding system using a deductive or an inductive approach.

When employing a *deductive* approach, the researcher defines the full list of possible codes in a top-down fashion; i.e., all codes are defined before reviewing the qualitative data and assigning those codes to comments. On the other hand, when using an *inductive* approach to coding, the codes are generated and constantly revised in a bottom-up approach; i.e., the data is coded according to categories identified by reading and re-reading responses to the open-ended question. Bottom-up, inductive coding is recommended, as it has the benefit of capturing categories the researcher may not have thought of before reading the actual comments; however, it requires more coordination if multiple coders are involved. (See "Grounded Theory Method" chapter for an analogous discussion.)

To measure the reliability of both the developed coding system and the coding of the comments, either the same coder should partially repeat the coding or a second coder should be involved. *Intra-rater reliability* describes the degree of agreement when the data set is reanalyzed by the same researcher. *Inter-rater reliability* (Armstrong, Gosling, Weinman, & Marteau, 1997; Gwet, 2001) determines the agreement level of the coding results from at least two independent researchers (using correlations or Cohen's kappa). If there is low agreement, the coding needs to be reviewed to identify the pattern behind the disagreement, coder training needs to be adjusted, or changes to codes need to be agreed upon to achieve consistent categorization. If the data set to be coded is too large and coding needs to be split up between researchers, inter-rater consistency can be measured by comparing results from coding an overlapping set of comments, by comparing the coding to a preestablished standard, or by including another researcher to review overlapping codes from the main coders.

After having analyzed all comments, the researcher may prepare descriptive statistics such as a frequency distribution of codes, conduct inferential statistical tests, summarize key themes, prepare necessary charts, and highlight specifics through the use of representative quotes. To compare results across groups, inferential analysis methods can be used as described above for closed-ended data (e.g., *t*-tests, ANOVA, or Chi-square).

Assessing Representativeness

A key criterion in any survey's quality is the degree to which the results accurately represent the target population. If a survey's sampling frame fully covers the population and the sample is randomly drawn from the sampling frame, a response rate of 100 % would ensure that the results are representative at a level of precision based on the sample size.

If, however, a survey has less than a 100 % response rate, those not responding might have provided a different answer distribution than those who did respond.

An example is a survey intended to measure attitudes and behaviors regarding a technology that became available recently. Since people who are early adopters of new technologies are usually very passionate about providing their thoughts and feedback, surveying users of this technology product would overestimate responses from early adopters (as compared to more occasional users) and the incidence of favorable attitudes toward that product. Thus, even a modest level of non-response can greatly affect the degree of non-response bias.

With response rates to major longitudinal surveys having decreased over time, much effort has been devoted to understanding non-response and its impact on data quality as well as methods of adjusting results to mitigate non-response error. Traditional survey assumptions held that maximizing response rates minimized non-response bias (Groves, 2006). Therefore, the results of Groves' 2006 meta-analysis were both surprising and seminal, finding no meaningful correlation between response rates and non-response error across mail, telephone, and face-to-face surveys.

Reporting Survey Findings

Once the question-by-question analysis is completed, the researcher needs to synthesize findings across all questions to address the goals of the survey. Larger themes may be identified, and the initially defined research questions are answered, which are in turn translated into recommendations and broader HCI implications as appropriate. All calculations used for the data analysis should be reported with the necessary statistical rigor (e.g., sample sizes, *p*-values, margins of error, and confidence levels). Furthermore, it is important to list the survey's paradata and include response and break-off rates (see section on monitoring survey paradata).

Similar to other empirical research, it is important to not only report the results of the survey but also describe the original research goals and the used survey methodology. A detailed description of the survey methodology will explain the population being studied, sampling method, survey mode, survey invitation, fielding process, and response paradata. It should also include screenshots of the actual survey questions and explain techniques used to evaluate data quality. Furthermore, it is often necessary to include a discussion on how the respondents compare to the overall population. Lastly, any potential sources of survey bias, such as sampling biases or non-response bias, should be outlined.

Exercises

1. What are the differences between a survey and a questionnaire, both in concept and design?
2. In your own research area, create a survey and test it with five classmates. How long do you think it will take a classmate to fill it out? How long did it take them?

Acknowledgements We would like to thank our employers Google, Inc. and Twitter, Inc. for making it possible for us to work on this chapter. There are many that contributed to this effort, and we would like to call out the most significant ones: Carolyn Wei for identifying published papers that used survey methodology for their work, Sandra Lozano for her insights on analysis, Mario Callegaro for inspiration, Ed Chi and Robin Jeffries for reviewing several drafts of this document, and Professors Jon Krosnick from Stanford University and Mick Couper from the University of Michigan for laying the foundation of our survey knowledge and connecting us to the broader survey research community.

References

Overview Books

Couper, M. (2008). *Designing effective Web surveys*. Cambridge, UK: Cambridge University Press.

Fowler, F. J., Jr. (1995). *Improving survey questions: Design and evaluation* (Vol. 38). Thousand Oaks, CA: Sage. Incorporated.

Groves, R. M. (1989). *Survey errors and survey costs*. Hoboken, NJ: Wiley.

Groves, R. M. (2004). *Survey errors and survey costs* (Vol. 536). Hoboken, NJ: Wiley-Interscience.

Groves, R. M., Fowler, F. J., Couper, M. P., Lepkowski, J. M., Singer, E., & Tourangeau, R. (2004). *Survey methodology*. Hoboken, NJ: Wiley.

Marsden, P. V., & Wright, J. (Eds.). (2010). *Handbook of survey research* (2nd ed.). Bingley, UK: Emerald Publishing Group Limited.

Sampling Methods

Aquilino, W. S. (1994). Interview mode effects in surveys of drug and alcohol use: A field experiment. *Public Opinion Quarterly., 58*(2), 210–240.

Cochran, W. G. (1977). *Sampling techniques* (3rd ed.). New York, NY: Wiley.

Couper, M. P. (2000). Web surveys: A review of issues and approaches. *Public Opinion Quarterly, 64*, 464–494.

Kish, L. (1965). *Survey sampling*. New York, NY: Wiley.

Krejcie, R. V., & Morgan, D. W. (1970). Determining sample size for research activities. *Educational and Psychological Measurement, 30*, 607–610.

Lohr, S. L. (1999). *Sampling: Design and analysis*. Pacific Grove, CA: Duxbury Press.

Questionnaire Design

Bradburn, N. M., Sudman, S., & Wansink, B. (2004). *Asking questions: The definitive guide to questionnaire design – for market research, political polls, and social and health questionnaires*. San Francisco, CA: Jossey-Bass. Revised.

Cannell, C. F., & Kahn, R. L. (1968). Interviewing. *The Handbook of Social Psychology, 2*, 526–595.

Chan, J. C. (1991). Response-order effects in Likert-type scales. *Educational and Psychological Measurement, 51*(3), 531–540.

Costa, P. T., & McCrae, R. R. (1988). From catalog to classification: Murray's needs and the five-factor model. *Journal of Personality and Social Psychology, 55*(2), 258.

Couper, M. P., Tourangeau, R., Conrad, F. G., & Crawford, S. D. (2004). What they see is what we get response options for web surveys. *Social Science Computer Review, 22*(1), 111–127.

Edwards, A. L., & Kenney, K. C. (1946). A comparison of the Thurstone and Likert techniques of attitudes scale construction. *Journal of Applied Psychology, 30*, 72–83.

Goffman, E. (1959). The presentation of self in everyday life, 1–17. Garden City, NY

Goldberg, L. R. (1990). An alternative description of personality: The big-five factor structure. *Journal of Personality and Social Psychology, 59*(6), 1216.

Herzog, A. R., & Bachman, J. G. (1981). Effects of questionnaire length on response quality. *Public Opinion Quarterly, 45*(4), 549–559.

Holbrook, A. L., & Krosnick, J. A. (2010). Social desirability bias in voter turnout reports tests using the item count technique. *Public Opinion Quarterly, 74*(1), 37–67.

Kinder, D. R., & Iyengar, S. (1987). *News That Matters: Television and American Opinion*. Chicago: University of Chicago Press.

Krosnick, J. A. (1991). Response strategies for coping with the cognitive demands of attitude measures in surveys. *Applied Cognitive Psychology, 5*, 213–236.

Krosnick, J. A. (1999). Survey research. *Annual review of psychology, 50*(1), 537–567.

Krosnick, J. A. (2002). The causes of no-opinion responses to attitude measures in surveys: They are rarely what they appear to be. In R. Groves, D. Dillman, J. Eltinge, & R. Little (Eds.), *Survey non-response* (pp. 87–100). New York: Wiley.

Krosnick, J. A., & Alwin, D. F. (1987). *Satisficing: A strategy for dealing with the demands of survey questions*. Columbus, OH: Ohio State University.

Krosnick, J. A., & Alwin, D. F. (1988). A test of the form-resistant correlation hypothesis ratings, rankings, and the measurement of values. *Public Opinion Quarterly, 52*(4), 526–538.

Krosnick, J. A., & Fabrigar, L. A. (1997). Designing rating scales for effective measurement in surveys. In L. Lyberg et al. (Eds.), *Survey measurement and process quality* (pp. 141–164). New York: Wiley.

Krosnick, J. A., Narayan, S., & Smith, W. R. (1996). Satisficing in surveys: Initial evidence. *New Directions for Evaluation, 1996*(70), 29–44.

Krosnick, J. A., & Presser, S. (2010). Question and questionnaire design. In P. V. Marsden & J. D. Wright (Eds.), *Handbook of survey research* (pp. 263–314). Bingley, UK: Emerald Group Publishing Limited.

Landon, E. L. (1971). Order bias, the ideal rating, and the semantic differential. *Journal of Marketing Research, 8*(3), 375–378.

O'Muircheartaigh, C. A., Krosnick, J. A., & Helic, A. (2001). Middle alternatives, acquiescence, and the quality of questionnaire data. In B. Irving (Ed.), *Harris Graduate School of Public Policy Studies*. Chicago, IL: University of Chicago.

Paulhus, D. L. (1984). Two-component models of socially desirable responding. *Journal of Personality and Social Psychology, 46*(3), 598.

Payne, S. L. (1951). *The art of asking questions*. Princeton, NJ: Princeton University Press.

Payne, J. D. (1971). The effects of reversing the order of verbal rating scales in a postal survey. *Journal of the Marketing Research Society, 14*, 30–44.

Rohrmann, B. (2003). Verbal qualifiers for rating scales: Sociolinguistic considerations and psychometric data. Project Report. Australia: University of Melbourne

Saris, W. E., Revilla, M., Krosnick, J. A., & Shaeffer, E. M. (2010). Comparing questions with agree/disagree response options to questions with construct-specific response options. *Survey Research Methods, 4*(1), 61–79.

Schaeffer, N. C., & Presser, S. (2003). The science of asking questions. *Annual Review of Sociology, 29*, 65–88.

Schlenker, B. R., & Weigold, M. F. (1989). Goals and the self-identification process: Constructing desired identities. In L. Pervin (Ed.), *Goal concepts in personality and social psychology* (pp. 243–290). Hillsdale, NJ: Erlbaum.

Schuman, H., & Presser, S. (1981). *Questions and answers in attitude surveys*. New York: Academic Press.

Simon, H. A. (1956). Rational choice and the structure of the environment. *Psychological Review, 63*(2), 129–138.

Smith, D. H. (1967). Correcting for social desirability response sets in opinion-attitude survey research. *Public Opinion Quarterly, 31*, 87–94.

Stone, G. C., Gage, N. L., & Leavitt, G. S. (1957). Two kinds of accuracy in predicting another's responses. *The Journal of Social Psychology, 45*(2), 245–254.

Tourangeau, R. (1984). *Cognitive science and survey methods. Cognitive aspects of survey methodology: Building a bridge between disciplines* (pp. 73–100). Washington, DC: National Academy Press.

Tourangeau, R., Couper, M. P., & Conrad, F. (2004). Spacing, position, and order: Interpretive heuristics for visual features of survey questions. *Public Opinion Quarterly, 68*(3), 368–393.

Tourangeau, R., Rips, L. J., & Rasinski, K. (2000). *The psychology of survey response*. Cambridge, UK: Cambridge University Press.

Tourangeau, R., & Smith, T. W. (1996). Asking sensitive questions the impact of data collection mode, question format, and question context. *Public Opinion Quarterly, 60*(2), 275–304.

Tourangeau, R., & Yan, T. (2007). Sensitive questions in surveys. *Psychological Bulletin, 133*(5), 859.

Villar, A., & Krosnick, J. A. (2011). Global warming vs. climate change, taxes vs. prices: Does word choice matter? *Climatic change, 105*(1), 1–12.

Visual Survey Design

Callegaro, M., Villar, A., & Yang, Y. (2011). A meta-analysis of experiments manipulating progress indicators in Web surveys. *Annual Meeting of the American Association for Public Opinion Research*, Phoenix

Couper, M. (2011). Web survey methodology: Interface design, sampling and statistical inference. Presentation at *EUSTAT-The Basque Statistics Institute*, Vitoria-Gasteiz

Couper, M. P., Conrad, F. G., & Tourangeau, R. (2007). Visual context effects in Web surveys. *Public Opinion Quarterly, 71*(4), 623–634.

Peytchev, A., Couper, M. P., McCabe, S. E., & Crawford, S. D. (2006). Web survey design paging versus scrolling. *Public Opinion Quarterly, 70*(4), 596–607.

Yan, T., Conrad, F. G., Tourangeau, R., & Couper, M. P. (2011). Should I stay or should I go: The effects of progress feedback, promised task duration, and length of questionnaire on completing Web surveys. *International Journal of Public Opinion Research, 23*(2), 131–147.

Established Questionnaire Instruments

Brooke, J. (1996). SUS-A quick and dirty usability scale. *Usability Evaluation in Industry, 189*, 194.

Chin, J. P., Diehl, V. A., & Norman, K. L. (1988, May). Development of an instrument measuring user satisfaction of the human-computer interface. In *Proceedings of the SIGCHI Conference on Human factors in computing systems* (pp. 213–218). New York, NY: ACM

Hart, S. G., & Staveland, L. E. (1988). Development of NASA-TLX (Task Load Index): Results of empirical and theoretical research. *Human Mental Workload, 1*, 139–183.

Kirakowski, J., & Corbett, M. (1993). SUMI: The software usability measurement inventory. *British Journal of Educational Technology, 24*(3), 210–212.

Lewis, J. R. (1995). IBM computer usability satisfaction questionnaires: Psychometric evaluation and instructions for use. *International Journal of Human-Computer Interaction, 7*(1), 57–78.

Moshagen, M., & Thielsch, M. T. (2010). Facets of visual aesthetics. *International Journal of Human-Computer Studies, 68*(10), 689–709.

Questionnaire Evaluation

Bolton, R. N., & Bronkhorst, T. M. (1995). Questionnaire pretesting: Computer assisted coding of concurrent protocols. In N. Schwarz & S. Sudman (Eds.), *Answering questions* (pp. 37–64). San Francisco: Jossey-Bass.

Collins, D. (2003). Pretesting survey instruments: An overview of cognitive methods. *Quality of Life Research an International Journal of Quality of Life Aspects of Treatment Care and Rehabilitation, 12*(3), 229–238.

Drennan, J. (2003). Cognitive interviewing: Verbal data in the design and pretesting of questionnaires. *Journal of Advanced Nursing, 42*(1), 57–63.

Presser, S., Couper, M. P., Lessler, J. T., Martin, E., Martin, J., Rothgeb, J. M., et al. (2004). Methods for testing and evaluating survey questions. *Public Opinion Quarterly, 68*(1), 109–130.

Survey Response Rates and Non-response

American Association for Public Opinion Research, AAPOR. (2011). *Standard definitions: Final dispositions of case codes and outcome rates for surveys.* (7th ed). http://aapor.org/Content/NavigationMenu/AboutAAPOR/StandardsampEthics/StandardDefinitions/StandardDefinitions2011.pdf

Baruch, Y. (1999). Response rates in academic studies: A comparative analysis. *Human Relations, 52*, 421–434.

Baruch, Y., & Holtom, B. C. (2008). Survey response rate levels and trends in organizational research. *Human Relations, 61*(8), 1139–1160.

Church, A. H. (1993). Estimating the effect of incentives on mail survey response rates: A meta-analysis. *Public Opinion Quarterly, 57*, 62–79.

Cook, C., Heath, F., & Thompson, R. L. (2000). A meta-analysis of response rates in Web- or Internet-based surveys. *Educational and Psychological Measurement, 60*(6), 821–836.

Dillman, D. A. (1978). *Mail and telephone surveys: The total design method.* New York: Wiley.

Dillman, D. A. (1991). The design and administration of mail surveys. *Annual Review of Sociology, 17*, 225–249.

Dillman, D. A. (2007). *Mail and Internet surveys: The tailored design method* (2nd ed.). Hoboken, NJ: Wiley.

Fan, W., & Yan, Z. (2010). Factors affecting response rates of the web survey: A systematic review. *Computers in Human Behavior, 26*(2), 132–139.

Groves, R. M. (2006). Non-response rates and non-response bias in household surveys. *Public Opinion Quarterly, 70*, 646–75.

Groves, R. M., Presser, S., & Dipko, S. (2004). The role of topic interest in survey participation decisions. *Public Opinion Quarterly, 68*(1), 2–31.

Kaplowitz, M. D., Hadlock, T. D., & Levine, R. (2004). A comparison of web and mail survey response rates. *Public Opinion Quarterly, 68*(1), 94–101.

Kerlinger, F. N. (1986). *Foundations of behavioral research* (3rd ed.). New York: Holt, Rinehart & Winston.

Kiesler, S., & Sproull, L. S. (1986). Response effects in the electronic survey. *Public Opinion Quarterly, 50*, 402–413.

Lavrakas, P. J. (2011). The use of incentives in survey research. *66th Annual Conference of the American Association for Public Opinion Research*

Lin, I., & Schaeffer, N. C. (1995). Using survey participants to estimate the impact of nonparticipation. *Public Opinion Quarterly, 59*(2), 236–258.

Lu, H., & Gelman, A. (2003). A method for estimating design-based sampling variances for surveys with weighting, poststratification, and raking. *Journal of Official Statistics, 19*(2), 133–152.

Manfreda, K. L., Bosnjak, M., Berzelak, J., Haas, I., Vehovar, V., & Berzelak, N. (2008). Web surveys versus other survey modes: A meta-analysis comparing response rates. *Journal of the Market Research Society, 50*(1), 79.

Olson, K. (2006). Survey participation, non-response bias, measurement error bias, and total bias. *Public Opinion Quarterly, 70*(5), 737–758.

Peytchev, A. (2009). Survey breakoff. *Public Opinion Quarterly, 73*(1), 74–97.

Schonlau, M., Van Soest, A., Kapteyn, A., & Couper, M. (2009). Selection bias in web surveys and the use of propensity scores. *Sociological Methods & Research, 37*(3), 291–318.

Sheehan, K. B. (2001). E-mail survey response rates: A review. *Journal of Computer Mediated Communication, 6*(2), 1–16.

Singer, E. (2002). The use of incentives to reduce non-response in household surveys. In R. Groves, D. Dillman, J. Eltinge, & R. Little (Eds.), *Survey non-response* (pp. 87–100). New York: Wiley. 163–177.

Stevenson, J., Dykema, J., Cyffka, C., Klein, L., & Goldrick-Rab, S. (2012). What are the odds? Lotteries versus cash incentives. Response rates, cost and data quality for a Web survey of low-income former and current college students. *67th Annual Conference of the American Association for Public Opinion Research*

Survey Analysis

Armstrong, D., Gosling, A., Weinman, J., & Marteau, T. (1997). The place of inter-rater reliability in qualitative research: An empirical study. *Sociology, 31*(3), 597–606.

Böhm, A. (2004). Theoretical coding: Text analysis in grounded theory. In *A companion to qualitative research*, London: SAGE. pp. 270–275.

De Leeuw, E. D., Hox, J. J., & Huisman, M. (2003). Prevention and treatment of item nonresponse. *Journal of Official Statistics, 19*(2), 153–176.

Glaser, B. G., & Strauss, A. L. (1967). *The discovery of grounded theory: Strategies for qualitative research*. Hawthorne, NY: Aldine de Gruyter.

Gwet, K. L. (2001). *Handbook of inter-rater reliability*. Gaithersburg, MD: Advanced Analytics, LLC.

Heeringa, S. G., West, B. T., & Berglund, P. A. (2010). *Applied survey data analysis*. Boca Raton, FL: Chapman & Hall/CRC.

Lee, E. S., Forthofer, R. N., & Lorimor, R. J. (1989). *Analyzing complex survey data*. Newbury Park, CA: Sage.

Saldaña, J. (2009). *The coding manual for qualitative researchers*. Thousand Oaks, CA: Sage Publications Limited.

Other References

Abran, A., Khelifi, A., Suryn, W., & Seffah, A. (2003). Usability meanings and interpretations in ISO standards. *Software Quality Journal, 11*(4), 325–338.

Anandarajan, M., Zaman, M., Dai, Q., & Arinze, B. (2010). Generation Y adoption of instant messaging: An examination of the impact of social usefulness and media richness on use richness. *IEEE Transactions on Professional Communication, 53*(2), 132–143.

Archambault, A., & Grudin, J. (2012). A longitudinal study of facebook, linkedin, & twitter use. In *Proceedings of the 2012 ACM Annual Conference on Human Factors in Computing Systems (CHI '12)* (pp. 2741–2750). New York: ACM

Auter, P. J. (2007). Portable social groups: Willingness to communicate, interpersonal communication gratifications, and cell phone use among young adults. *International Journal of Mobile Communications, 5*(2), 139–156.

Calfee, J. E., & Ringold, D. J. (1994). The 70 % majority: Enduring consumer beliefs about advertising. *Journal of Public Policy & Marketing, 13*(2).

Chen, J., Geyer, W., Dugan, C., Muller, M., & Guy, I. (2009). Make new friends, but keep the old: Recommending people on social networking sites. In *Proceedings of the 27th International Conference on Human Factors in Computing Systems (CHI '09)*, (pp. 201–210). New York: ACM

Clauser, B. E. (2007). The life and labors of Francis Galton: A review of four recent books about the father of behavioral statistics. *Journal of Educational and Behavioral Statistics, 32*(4), 440–444.

Converse, J. (1987). *Survey research in the United States: Roots and emergence 1890–1960*. Berkeley, CA: University of California Press.

Drouin, M., & Landgraff, C. (2012). Texting, sexting, and attachment in college students' romantic relationships. *Computers in Human Behavior, 28*, 444–449.

Feng, J., Lazar, J., Kumin, L., & Ozok, A. (2010). Computer usage by children with down syndrome: Challenges and future research. *ACM Transactions on Accessible Computing, 2*(3), 35–41.

Froelich, J., Findlater, L., Ostergren, M., Ramanathan, S., Peterson, J., Wragg, I., et al. (2012). The design and evaluation of prototype eco-feedback displays for fixture-level water usage data. In *Proceedings of the 2012 ACM Annual Conference on Human Factors in Computing Systems (CHI '12)* (pp. 2367–2376). New York: ACM

Harrison, M. A. (2011). College students' prevalence and perceptions of text messaging while driving. *Accident Analysis and Prevention, 43*, 1516–1520.

Junco, R., & Cotten, S. R. (2011). Perceived academic effects of instant messaging use. *Computers & Education, 56*, 370–378.

Katosh, J. P., & Traugott, M. W. (1981). The consequences of validated and self-reported voting measures. *Public Opinion Quarterly, 45*(4), 519–535.

Nacke, L. E., Grimshaw, M. N., & Lindley, C. A. (2010). More than a feeling: Measurement of sonic user experience and psychophysiology in a first-person shooter game. *Interacting with Computers, 22*(5), 336–343.

Obermiller, C., & Spangenberg, E. R. (1998). Development of a scale to measure consumer skepticism toward advertising. *Journal of Consumer Psychology, 7*(2), 159–186.

Obermiller, C., & Spangenberg, E. R. (2000). On the origin and distinctiveness of skepticism toward advertising. *Marketing Letters, 11*, 311–322.

Person, A. K., Blain, M. L. M., Jiang, H., Rasmussen, P. W., & Stout, J. E. (2011). Text messaging for enhancement of testing and treatment for tuberculosis, human immunodeficiency virus, and syphilis: A survey of attitudes toward cellular phones and healthcare. *Telemedicine Journal and e-Health, 17*(3), 189–195.

Pitkow, J. E., & Recker, M. (1994). Results from the first World-Wide web user survey. *Computer Networks and ISDN Systems, 27*(2), 243–254.

Rodden, R., Hutchinson, H., & Fu, X. (2010). Measuring the user experience on a large scale: User-centered metrics for web applications. In *Proceedings of the 28th International Conference on Human Factors in Computing Systems (CHI '10)* (pp. 2395–2398) ACM, New York, NY, USA

Schild, J., LaViola, J., & Masuch, M. (2012). Understanding user experience in stereoscopic 3D games. In *Proceedings of the 2012 ACM Annual Conference on Human Factors in Computing Systems (CHI '12)* (pp. 89–98). New York: ACM

Shklovski, I., Kraut, R., & Cummings, J. (2008). Keeping in touch by technology: Maintaining friendships after a residential move. In *Proceedings of the 26th Annual SIGCHI Conference on Human Factors in Computing Systems (CHI '08)* (pp. 807–816). New York: ACM

Turner, M., Love, S., & Howell, M. (2008). Understanding emotions experienced when using a mobile phone in public: The social usability of mobile (cellular) telephones. *Telematics and Informatics, 25*, 201–215.

Weisskirch, R. S., & Delevi, R. (2011). "Sexting" and adult romantic attachment. *Computers in Human Behavior, 27*, 1697–1701.

Wright, P. J., & Randall, A. K. (2012). Internet pornography exposure and risky sexual behavior among adult males in the United States. *Computers in Human Behavior, 28*, 1410–1416.

Yew, J., Shamma, D. A., & Churchill, E. F. (2011). Knowing funny: Genre perception and categorization in social video sharing. In *Proceedings of the 2011 Annual Conference on Human Factors in Computing Systems (CHI '11)* (pp. 297–306). New York: ACM

Zaman, M., Rajan, M. A., & Dai, Q. (2010). Experiencing flow with instant messaging and its facilitating role on creative behaviors. *Computers in Human Behavior, 26*, 1009–1018.

Crowdsourcing in HCI Research

Serge Egelman, Ed H. Chi, and Steven Dow

Introduction

Crowdsourcing involves recruiting large groups of people online to contribute small amounts of effort towards a larger goal. Increasingly, HCI researchers leverage online crowds to perform tasks, such as evaluating the quality of user generated content (Kittur, Suh, & Chi, 2008), identifying the best photograph in a set (Bernstein, Brandt, Miller, & Karger, 2011), transcribing text when optical character recognition (OCR) technologies fail (Bigham et al., 2010), and performing tasks for user studies (Heer & Bostock, 2010; Kittur, Chi, & Suh, 2008).

This chapter provides guidelines for how to use crowdsourcing in HCI research. We explore how HCI researchers are using crowdsourcing, provide a tutorial for people new to the field, discuss challenges and hints for doing crowdsourcing more effectively, and share three concrete case studies.

S. Egelman (✉)
Electrical Engineering & Computer Sciences, University of California, 731 Soda Hall, Berkeley, CA 94720, USA
e-mail: egelman@cs.berkeley.edu

E.H. Chi
Google, Inc., 1600 Amphitheatre Parkway, Mountain View, CA 94043, USA
e-mail: edchi@google.com

S. Dow
Human-Computer Interaction Institute, Carnegie Mellon University, 5000 Forbes Avenue, Newell Simon Hall, Pittsburgh, PA 15213, USA
e-mail: spdow@cs.cmu.edu

J.S. Olson and W.A. Kellogg (eds.), *Ways of Knowing in HCI*,
DOI 10.1007/978-1-4939-0378-8_11,

What Is Crowdsourcing?

Numerous online crowdsourcing platforms offer people micro-payments for completing tasks (Quinn & Bederson, 2011), but non-paid crowdsourcing platforms also exist. Non-paid crowd platforms typically offer some other value to users, such as embedding the task in a fun game (von Ahn & Dabbish, 2004) or engaging people in a cause, such as citizen science projects like Fold It, a protein folding effort (Hand, 2010). The increasing availability of crowdsourcing platforms has enabled HCI researchers to recruit large numbers of participants for user studies, to generate third-party content and quality assessments, and to build novel user experiences.

One canonical example of paid crowdsourcing from the crowdsourcing industry is business card data entry. Even very sophisticated algorithms utilizing OCR technology cannot deal with the great variety of different types of card designs in the real world. Instead, a company called CardMunch uploads business cards to Amazon's Mechanical Turk (MTurk)[1] to have them transcribed.[2] This way, a user who collects hundreds of business cards from a convention can have them transcribed very quickly and cheaply. Figure 1 shows the interface

Fig. 1 Example Business Card task in Amazon Mechanical Turk. Screenshot used by permission from LinkedIn.com

[1] http://www.mturk.com/.

[2] http://www.readwriteweb.com/archives/linkedin_updates_cardmunch_iphone_app.php.

used for transcription. This example illustrates the role crowdsourcing has played in merging computational algorithms with human intelligence. That is, in places where algorithms fall short, online crowds can supplement them with human computation.

There are many other examples of crowdsourcing that do not involve financial payments. One example is the online encyclopedia, Wikipedia,[3] where hundreds of thousands of contributors author and edit articles. Similarly, the Tiramisu project relies on GPS traces and problem reports from commuters to generate real-time arrival time predictions for a transit system (Zimmerman et al., 2011).

Another example of unpaid crowdsourcing is the reCAPTCHA project (von Ahn, Maurer, McMillen, Abraham, & Blum, 2008) where millions of Internet users translate small strings of scrambled text, typically to gain access or open a new user account. The purpose is twofold. The system verifies that the user is human, not some kind of automated algorithm. Moreover, the project aims to digitize old out-of-print books by giving users one known word and one unknown word. Over time and across many users, the system learns the probability distribution of the unknown words and eventually translates entire book collections. These non-paid crowdsourcing platforms demonstrate the variety of incentive mechanisms available.

A Brief History

Before the invention of electronic computers, organizations employed teams of "human computers" to perform various mathematical calculations (Grier, 2005). Within the past decade, this notion of human computation has once again gained popularity due to not just an increase in online crowdsourcing platforms but also because researchers have become better able to understand the limitations of *machine* computation.

In HCI research literature, the pioneering work of von Ahn and Dabbish first explored using game mechanisms in the "ESP Game" to gather labels for images (von Ahn & Dabbish, 2004). Kittur, Chi, & Suh, 2008) suggested the use of MTurk for user studies.

Since these two early works, a growing community of HCI researchers has emerged to examine and utilize crowdsourcing in its many forms. This is evident by both the presence of large workshops at top HCI conferences, such as the ACM CHI conference, as well as new workshops and conferences dedicated entirely to crowdsourcing, such as HCOMP and Computational Intelligence.

HCI researchers have explored crowdsourcing by:

1. Studying crowd platforms for intellectual tasks, e.g., Wikipedia and social search.
2. Creating "crowdsensing" applications, e.g., CMU's Tiramisu (Zimmerman et al., 2011) or Minnesota's Cyclopath (Priedhorsky & Terveen, 2008).

[3] http://www.wikipedia.org/.

3. Designing "games with a purpose," e.g., CMU's ESP Game (von Ahn & Dabbish, 2004) or U Washington's PhotoCity (Tuite, Snavely, Hsiao, Smith, & Popović, 2010).
4. Utilizing micro-task platforms (e.g., MTurk) for a variety of activities, ranging from user study recruitment to judgment gathering.

While HCI research has much to gain from studying existing large-scale online communities (such as Twitter, Google+, Reddit, or Wikipedia) or building new crowd-based platforms (e.g., Zimmerman et al., 2011), this chapter aims to provide a useful resource for people new to the domain. Given the wide variety of research in this space, this chapter focuses primarily on how HCI researchers can leverage general-purpose crowdsourcing platforms, which are often used for completing micro-tasks. They provide easy access to scalable, on-demand, inexpensive labor and can be used for many kinds of HCI research.

In the rest of this chapter, we first look at how crowdsourcing can be applied to typical HCI activities, such as conducting participant studies and recruiting independent judges. Second, we provide a number of considerations and tips for using crowds, including a short tutorial on Amazon's Mechanical Turk. Third, we share three case studies that explore how each of the authors has personally used crowdsourcing in his research. Finally, we explore new HCI applications for crowdsourcing research and provide links to additional crowdsourcing resources.

How HCI Researchers Can Leverage Crowds

For many common HCI research activities, the scale, diversity, availability, and affordability of online crowds provide value. This section covers several of the traditional HCI research activities that benefit from utilizing general-purpose crowdsourcing platforms. We describe more advanced uses of crowdsourcing later.

Conducting online surveys: Crowdsourcing provides a wonderful recruiting tool for surveys and questionnaires, because the ability to reach large populations allows researchers to select for specific demographics, as well as recruit diverse samples, as discussed in detail later.

To better select samples of workers, a number of researchers have been using MTurk to learn more about crowd workers themselves (Quinn & Bederson, 2011). For example, Ross, Irani, Silberman, Zaldivar, and Tomlinson (2010) learned that over the last few years the demographics of MTurk workers have been shifting from primarily US workers, to a split between US and Indian workers. This shift is partly due to the fact that MTurk started allowing people to receive payments in Indian Rupees.

Conducting experiments: Crowdsourcing provides a cheap and quick way to recruit participants for user studies or experiments. An early example of this was Kittur, Suh, and Chi's use of MTurk to conduct a user study about Wikipedia article quality

(2008). Heer and Bostock (2010) were able to replicate and extend previous studies of graphical perception focused on spatial encoding and contrast. Heer and Bostock estimated that their crowdsourced studies resulted in cost savings at a factor of six (ibid). Similarly, Egelman and colleagues performed several experiments to examine Internet users' security behaviors (Christin, Egelman, Vidas, & Grossklags, 2011; Egelman et al., 2010; Komanduri et al., 2011).

These researchers leveraged the time and cost savings of online crowds to examine many more experimental conditions than would have been possible in a laboratory setting. For instance, a week of recruiting might result in 100 laboratory participants, who would need to be paid at least $10 each to participate in a 10-min experiment. The same experiment posted on a crowdsourcing platform might yield over 1,000 online participants when paid $1 each. Toomim, Kriplean, Pörtner, and Landay (2011) have used MTurk to compare different user interfaces. They proposed that task completion rates by MTurk workers provide a new measure of the utility of user interfaces. Specifically, they hypothesized that a more usable UI leads to more workers finishing a task and for less money. With thousands of workers conducting tasks with a range of different UIs, the researchers were able to measure the relative dropout rates based on the quality of the UI and the payment amount.

Training of machine-learning algorithms: Other researchers have been using online crowds to gather training data for novel uses of machine learning. For example, Kumar, Kim, and Klemmer (2009) sought to develop software that will transform web content to new designs. The researchers recruited MTurk workers to help them tune an algorithm that converts one website's Document Object Model (DOM) into another. In the task, online workers were given two websites and then for any particular design element on one page, they were asked to find the corresponding element in the second page. With enough of these judgments, the machine-learning algorithm can "learn" the structural patterns that map content across different designs.

Analyzing text or images: The ESP Game was one of the first and best examples of crowdsourcing, where online participants "labeled" images as a secondary effect of playing a game (von Ahn & Dabbish, 2004). The game shows two online players the same image. To earn points, the players have to simultaneously guess the same word or phrase without communicating. The side product of this game interaction provides descriptive language for the image (i.e., "tags"). Since then, HCI researchers have adopted crowdsourcing to analyze text and images for various research goals. A number of researchers have used crowds to analyze/categorize texts, such as blog threads, Wikipedia entries, and tweets (André, Bernstein, & Luther, 2012).

For analyzing images, one early well-known example is the NASA Click workers,[4] who were unpaid volunteers from all corners of the Web that used a website to help identify and classify craters on Mars. This was also one of the earliest citizen science projects.

[4] http://beamartian.jpl.nasa.gov/.

Another creative user study using crowds was an experiment on the effect of emotional priming on brainstorming processes. Lewis, Dontcheva, and Gerber (2011) first used MTurk workers to judge the emotional affect of a set of images. These ratings allowed the researchers to select one positive, one negative, and one neutral image as the independent variable for a brainstorming experiment. The researchers found that priming with both positive and negative images can lead to more original idea generation than neutral imagery.

Gathering subjective judgments: A number of researchers have leveraged crowdsourcing to gather subjective quality judgments on content. For example, Kittur, Chi, and Suh's evaluation of Wikipedia article quality showed that MTurk workers generated ratings that correlated highly with expert Wikipedians' evaluations (2008).

Utilizing the subjective judgments of crowds, Dow et al. paid online crowds to judge banner ads created by participants in a design experiment (2010). They then conducted an experiment on the design process to examine whether creating and receiving feedback on multiple designs in parallel—rather than simply iterating serially—affects design results and exploration. Participants came to the lab and created web banner ads, and the resulting designs were launched online at Amazon MTurk to collect relative performance metrics, such as the quality and diversity of the ad designs. The judgments of online workers showed that the parallel process resulted in more diverse explorations and produced higher quality outcomes than the serial process.

Considerations and Tips for Crowdsourcing

In this section, we discuss some of the questions that researchers should be prepared to answer when deciding whether to use crowdsourcing. Many decisions are involved regarding what types of tasks and how workers should go about completing them. For instance,

- Are the tasks well suited for crowdsourcing?
- If it is a user study, what are the tradeoffs between having participants perform the task online versus in a laboratory?
- How much should crowd workers earn for the task?
- How can researchers ensure good results from crowdsourcing?

Here we breakdown these key questions, discuss the challenges of using online crowds, and offer tips to help overcome those challenges.

Finally, we illustrate how to use one particular crowdsourcing platform, Amazon's Mechanical Turk, and give an overview of other crowdsourcing platforms.

When Is Crowdsourcing Appropriate?

Crowdsourcing typically enables researchers to acquire a large amount of user data for a low-cost with fast turn-around times. However, while crowdsourcing can be used for many different things, and there are a wide variety of different crowdsourcing platforms, not every research project is well suited for crowdsourcing. Researchers must consider task complexity, task subjectivity, and what information they can (or need to) infer about their users when deciding whether they can collect sufficient data through crowdsourcing.

As with any research project, the researcher should start by writing down the questions that she hopes to answer. Next, she must determine what data she needs in order to answer those questions. Finally, she must decide whether a crowdsourcing platform is able to yield that data and whether it can do so reliably with the desired demographic.

For instance, on the one hand, when conducting a very short opinion survey that collects responses from as many people as possible in a very short amount of time, MTurk or Google's Consumer Survey[5] might be the most appropriate platform, because these platforms focus on reaching large samples of the general public. On the other hand, if a project requires advanced skills, a platform that focuses on domain experts, like oDesk[6] or 99designs,[4] might be more appropriate.

Crowdsourcing should generally be used for tasks that can be performed online with minimal supervision. Tasks that require real-time individual feedback from the researcher may not be appropriate for crowdsourcing. However, these guidelines are nuanced. For instance, while MTurk itself does not support many advanced ways of communicating with users, there is nothing preventing a researcher from using MTurk to redirect users to a website she controls wherein she can support more interaction with the workers.

There really are no hard rules as to what sorts of projects might benefit from a crowdsourcing approach. New crowdsourcing platforms and methodologies continue to enable researchers to conduct online tasks that were previously thought to be unsuited to crowdsourcing.

What Are the Tradeoffs of Crowdsourcing?

Just because a researcher believes she *can* use crowdsourcing to complete a particular research project does not mean that she *should*. While crowdsourcing presents many advantages over traditional laboratory or field experiments in which the researcher is directly interacting with participants, it also has drawbacks that researchers need to take into account.

[5] http://www.google.com/insights/consumersurveys/.

[6] http://www.odesk.com/.

In a laboratory or field experiment where subjects meet with researchers face-to-face, they may feel additional motivation to provide quality results due to the supervision (i.e., the "Hawthorne effect;" Landsberger, 1958). This is one tradeoff when performing unsupervised tasks online. For instance, unless there are clear quality controls, users may feel free to "cheat." Users who cheat rarely do so out of malice, but instead do so out of laziness. This is basic economics: if the same reward can be achieved for doing less work, many users will do so. In many crowdsourcing platforms, the researcher ultimately gets to decide which users receive remuneration. Therefore the issue is not so much preventing or minimizing cheating, but instead including quality controls so that the researcher may detect it and then reject those responses. We discuss this in more detail later in this section.

Another detriment to using crowdsourcing in experiments is the unavailability of qualitative observations. Unless the researcher has invested time in creating an environment that allows for detailed observations as the user completes the task, there is little way of gathering observational data on the steps the user took while submitting a response. On the other hand, supervised laboratory and field experiments provide researchers with opportunities to ask users follow-up questions, such as why a particular action was performed. (See Looking Back: Retrospective Study Methods for HCI in this volume.)

Finally, a benefit of crowdsourcing is that the low cost allows researchers to iteratively improve their experimental designs. When performing a laboratory or field experiment, pilot experiments are usually run on only a handful of participants due to the time and cost involved, which means that the opportunity to identify and correct potential pitfalls is drastically reduced. With crowdsourcing, because the cost is usually orders of magnitude lower per user, there is no reason why a researcher cannot run iterative pilot experiments on relatively large samples. Likewise, researchers can use the low cost as part of a quality control strategy: if multiple workers complete the same task, outliers can be detected and removed.

Who Are the Crowd Workers?

Prior to the availability of crowdsourcing platforms, HCI research involving diverse samples of human subjects was often prohibitively expensive. Researchers commonly recruited locally, using only coworkers or students recruited nearby. These convenience samples, while heavily biased, have been accepted in the research community because alternatives were not readily available. Of course, all research subject samples suffer from a bias: they include only those who are willing to participate in research studies. However, the advent of crowdsourcing has shown that much more diverse participant pools can be readily accessible (Kittur, Chi, & Suh, 2008). The ability to recruit participants from around the world raises other concerns; chief among them is being able to describe the participant demographics. Or put more succinctly, *who are these workers*?

For some types of HCI research, in which the goal is not to generalize findings to larger populations, participant demographics may not matter. For example, for purely creative endeavors, such as collecting user-generated artwork or designs, the locations or education levels of participants may not be of concern. Likewise, when ground truth is readily verifiable, such as using crowdsourcing for translation or transcription, demographics also may not matter. However, when the goal is to yield knowledge that is generalizable to a large population, such as rating photographs for emotional content, knowing the demographic might be crucial to the study's ecological validity.

Crowdsourcing suffers from the same shortcomings as other survey methods that involve collecting self-reported demographic data; survey respondents often omit responses to certain demographic questions or outright lie. Likewise, all research methods suffer from potential biases because the people who participate were only those who both saw the recruitment notice and decided to participate. When users are recruited using traditional methods, such as from a specific geographical area or due to a common interest (e.g., online forums), some amount of information is immediately known about the sample. However, crowdsourcing changes all of this because users are likely to come from more diverse backgrounds. As a first step in identifying workers, a researcher may want to think about limiting her sample to specific geographic areas. For instance, some studies have shown that the demographics of US-based MTurk users are similar to the demographics of US-based Internet users as a whole, though the former are slightly younger and more educated (Ipeirotis, 2010a; Ross et al., 2010). If the ability to restrict users by location is unavailable on the platform the researcher wishes to use, then the geolocations of the users' IP addresses may be a reasonable proxy for user locations.

Other demographic information, such as education level, age, or gender, may be harder to reliably collect. If demographic information is necessary, users should be asked to self-report it. As with traditional methods that collect self-reported demographics, this information suffers from the same shortcomings (i.e., users might omit it or provide incorrect information). The trustworthiness of self-reported demographics varies by platform. Third party services, such as CrowdFlower,[7] compile user statistics so that requesters can rely on having more demographic information, as well as a user's history of completing previous tasks. The bottom line is that researchers should be aware of the potential to reach a diverse sample, and think about the type of worker they wish to reach.

How Much Should Crowdworkers Be Paid?

Some crowdsourcing platforms reward users with intangible benefits, such as access to special content, the enjoyment of playing a game, or simply the knowledge that they are contributing to a community. For instance, users contribute to Wikipedia in

[7] http://www.crowdflower.com/.

order to extend the quality of publicly available knowledge; reCAPTCHA users transcribe words in order to prove that they are not computer programs trying to gain access to a system. However, on some platforms, many users expect monetary incentives to participate. This raises the question, *how much should workers earn for work?*

Payment amounts can have profound effects on experimental results. Pay too little, and one risks not attracting enough workers or only attracting a very specific demographic (i.e., those willing to work for very little). Pay too much, and one may quickly exhaust the budget or turn away potential workers who incorrectly estimate too much work is involved. Of course, the proper payment amount is governed by many factors, and the most important are: community standards for the platform being used, the anticipated amount of time to complete the task, and the type of work involved.

Knowing the target demographic is crucial for determining payment amounts. For instance, soliciting logos from users of the crowdsourcing platform for designs, 99designs, is likely to cost two orders of magnitude more than soliciting logo ideas from MTurk users. However, users on 99designs are often professional designers, and therefore payments and rewards are commensurate with experience and expertise. Of course, when using MTurk, one will likely have to filter through many more low-quality answers, potentially negating any cost differential (i.e., a researcher may pay one designer $100 on 99designs, whereas it may take paying 100 workers each $1 or more on MTurk to yield an acceptable design).

For tasks that do not leverage skilled workers, the rule of thumb is to offer payment relatively close to the prevailing minimum wage. This of course is a loaded term, especially when talking about workers who are based all over the world. Without explicitly restricting one's workers to a particular geographic location or socioeconomic class, the payment amount will add a selection bias to the sample. For instance, Christin et al. (2011) found that for the same task, when they increased the payment from $0.01 to $1.00, participants from the developed world—as a proportion of total participants—increased significantly. The obvious explanation for this is that when the payment was too low, participants from the developed world did not believe it was worth their time.

Prior to deploying a new task to be crowdsourced, researchers should always run pilot experiments to get a good idea of how long it will take to complete the task. Some crowdsourcing platforms even provide "sandbox" features that allow tasks to be tested in the experimental environment for free while the researcher prepares to deploy them. In these environments, one can modify a task while viewing it from the worker's perspective. When the researcher has a good estimate for the task's time commitment, the researcher can spend a few minutes surveying tasks of similar complexity that others are offering to get a better understanding of the current market rates. If the budget allows, researchers may want to consider pricing their tasks slightly higher than other similar tasks (e.g., 30 cents if other similar tasks are paying 25 cents). This may help them to reach a larger audience by making their tasks stand out. Paying too much, on the other hand, usually attracts noisy answers from participants trying to earn quick money. Gaming the system for an economic advantage is always irresistible for some workers.

For a given price, the complexity of the task also has a profound impact on workers' willingness to perform it. Researchers have found that cognitive tasks involving creative or personal contributions tend to require higher payment amounts. For example, researchers will likely need to pay users more to spend 10 min writing unique product reviews than to spend 10 min answering multiple-choice surveys.

How to Ensure Quality Work?

Because crowdsourcing deals with potentially broad and diverse audiences, it is important to be able to minimize poor-quality responses. Barring that, tasks should be designed to make poor-quality responses immediately identifiable, so that they can be easily removed post hoc. Some of the techniques for doing this come from survey design best practices that have existed for decades, described in more detail below. (See chapter on "Survey Research in HCI," this volume.)

The easiest way to increase work quality is by preventing workers with bad reputations from participating. Some crowdsourcing platforms allow requesters—those posting tasks—to leave feedback about each of their workers. Other platforms provide worker statistics, such as the percentage of accepted tasks that were completed to the satisfaction of the requester. It is then possible for requesters to set a threshold so that only workers who have exceeded a certain approval rating are eligible to participate in their tasks. However, the quality of reputation systems varies greatly across different crowdsourcing platforms.

The most important, yet hardest way of increasing the quality of workers' work is by carefully crafting the language on task instructions. As a general rule, instructions need to be as specific as possible, while also being succinct. Because workers come from very diverse backgrounds and are performing tasks unsupervised, the tasks need to be worded to avoid misunderstandings and minimize follow-up clarifications. Researchers might want to tailor their instructions based on participants' estimated reading levels (e.g., Flesch-Kincaid readability score); however, it generally takes several rounds of piloting and iterative changes in order to finalize task descriptions. This type of hardening of experimental procedure is also common in laboratory experiments, but can be somewhat heightened in crowdsourcing experiments since the researcher cannot be in a room with the subjects to clarify any confusion.

A researcher may design the most straightforward task but still get a significant number of fraudulent responses. If it is easy for workers to submit irrelevant responses in order to receive a payment, many invariably will. Kittur, Chi, and Suh showed that the key is in designing the task so that fraudulent or low-quality responses can be easily detected using well-established survey design techniques (2008). The easiest way of doing this is by adding additional questions to the task in which the ground truth is known (also referred to as "gold standard" questions). For instance, to help determine whether users read the questions, one might ask "how many letters are there in the word 'dog'?" or even something as simple as "please select 'false'."

Another way of detecting fraudulent responses is by including questions that require open-ended responses, which demonstrate that the worker read and understood the question, rather than selected a correct answer by chance (Fowler, 1995). For example, on a survey that consists of multiple-choice questions, a researcher should think about replacing one with a text-box response so that she can assess workers' diligence to the task. The text box does two things. First, it discourages would-be cheaters by increasing the effort required to provide a fake response so that it is closer to the effort required to provide a legitimate response. Second, free-text questions make it much easier to detect blatantly fraudulent responses, because the responses are usually either gibberish or off-topic, whereas fraudulent responses to multiple-choice questions are hard to separate out from the legitimate responses.

Finally, one of the greatest advantages of crowdsourcing is that it is relatively tolerant of mistakes because tasks can be altered, modified, and reposted very easily. If a researcher finds that she is having a hard time achieving the sample sizes that she requires, she can simply increase the payment amount and try again. If she is not yielding the type of data that she requires, she may want to reword the task or add additional instructions or questions and try again. Making modifications and redeploying research studies has previously been viewed as highly time-consuming and costly. But with crowdsourcing, researchers can iterate more easily with their experimental designs.

A Tutorial for Using Amazon's Mechanical Turk

To give an example of a general crowdsourcing platform, we provide a short tutorial of Amazon's Mechanical Turk (MTurk), the largest and most well-known platform for leveraging crowds of people. While many platforms exist for specific types of tasks, MTurk is the most popular one for general-purpose crowdsourcing, because it essentially supports any task that can be completed from within a web browser by an Internet user. For this reason, it has become widely used for research tasks ranging from surveys and behavioral experiments to creative design explorations.

The Basics

Like most crowdsourcing platforms, MTurk relies on two types of users: *workers* and *requesters*. A worker is someone who uses the platform for the purpose of completing tasks, whereas the requester is the person who posts and pays for those tasks, known as "HITs" (Human Intelligence Tasks) in MTurk. For the purpose of this example, assume a researcher wishes to recruit users to complete an online survey. To do this, she will need to post her survey to MTurk.

When creating a new HIT on MTurk, a researcher needs to consider and specify the following variables:

- Payment amount for each valid response.
- Total number of responses to be collected.
- The number of times a worker may complete the HIT.
- Time allotted for each worker to complete the HIT.
- Time before the HIT expires (regardless of the number of completed assignments).
- Time before results are automatically approved (i.e., if the requester does not approve/reject individual results in time).
- Qualification requirements (e.g., approval rate and geographic location).

Requesters have the choice between using Amazon's web interface, which allows for the creation of very basic web forms with minimal logic using their graphical interface or using MTurk's API to implement more complex features (such as embedding externally hosted content). Using the API means that one can use a common programming language (C/C++, Java, Python, Perl, etc.) to automate the process of posting HITs, approving workers' responses, and then ultimately compensating the workers. This way, developers can access crowds through their own software, without having to manually post tasks using the MTurk website. For this tutorial, we will assume the researcher uses Amazon's web interface and each question of her survey is a basic HTML web form element.

Qualification Tasks

If a researcher wants to target a particular type of worker, the naïve approach would be to add screening questions to the survey and then remove all respondents who do not meet the requirements post hoc. Of course, this is very costly because it involves compensating everyone who completes the survey earnestly, even those who the researcher does not want to ultimately include in her dataset. As another way of targeting specific types of users, MTurk offers "qualification HITs."

A qualification HIT can be used to screen potential workers before they are allowed to participate in future and more complex HITs. For instance, if a requester is trying to survey workers who are in the market for a new car, she might create a very quick qualification HIT wherein workers are surveyed about planned upcoming purchases. This survey is likely to be very short and pay relatively little; she might ask ten questions about future purchases and compensate workers $0.05 for their time. Based on workers' responses to this screening survey, the requester can then give selected workers a "qualification," which is a custom variable associated with their profile indicating that they completed the screening survey satisfactorily and are then eligible for follow-up HITs.

Finally, the requester adds a requirement to their main task that workers need to pass the qualification to be eligible to participate. This "real" survey is likely to be much longer and compensate workers much more, but since some irrelevant respondents are ineligible, the money is more efficiently spent. Using this method, the researcher may create a standing pool of eligible participants that she can approach again and again in the future for subsequent research tasks, by creating a list of all the workers to whom she has granted the qualification.

Beyond Basic Surveys

MTurk includes all that one needs to deploy basic surveys that use standard HTML elements (e.g., forms, radio buttons), but what happens when one wants to add more advanced logic or dynamic embedded content? Luckily, requesters are not limited to working with the interface elements that MTurk supports; they can also redirect workers to their own websites to complete HITs. For instance, for the aforementioned survey about car buyers, imagine the researcher wants workers to use a Flash applet to design their dream cars. To do this, the researcher would create the Flash applet on her own website and then direct the workers to this website in one of two ways. The first way of doing this is to make the HIT an "external question," where the HIT is hosted outside of MTurk and will therefore appear in an embedded frame. She may design what appears in the HIT's *iframe* as she sees fit, so long as she ensures that all data she wishes to collect gets sent as HTTP POST variables to a particular MTurk submission URL.

Of course, the easier way of directing workers to a different website to complete a task is by including a link in the HIT (e.g., "click here to open the survey in a new window"). The problem with this method is that the researcher will need to map users who completed the survey to workers on MTurk. To address this problem, a shared secret is needed. To give an example:

1. A worker visits MTurk and accepts the HIT.
2. In the HIT, the worker opens a new window for the survey, hosted on a separate website.
3. Once the worker completes the survey, the last page displays a secret word that the worker must submit to MTurk to receive compensation.
4. When the researcher downloads the MTurk results, there is no way of determining whether workers actually took the survey because it was on a different website. However, because the MTurk HIT asked them to submit the secret word shown on the last page of the survey, all responses not containing this secret word can be rejected (because there is no evidence they completed the survey).

This method has one obvious flaw: workers may talk to one another. There are several very popular online forums for MTurk workers to discuss recently completed HITs,[8,9] so it would be trivial for one of them to reveal the secret word to other workers. One way around this is to create a unique—or reasonably unique—shared secret for each of the workers. For instance, some survey websites allow

[8] http://turkopticon.differenceengines.com/.

[9] http://forum.mturk.com/.

researchers to create random numbers to display at the end of an externally hosted survey. A researcher can then ask workers to enter this same number into MTurk. To verify the responses, it becomes a matter of just matching the numbers in order to identify which results to reject. Alternatively, a researcher can program shared secrets based on an algorithm that can be verified. For instance, the algorithm might print out a 6-digit random number that is also a multiple of 39; multiple submissions that include identical numbers are likely to have colluded, whereas the researcher can also make sure that a worker did not simply enter a 6-digit number at random.

Managing Results

As results are submitted, the researcher can download real-time data files formatted as comma separated values (CSV). In addition to whatever data is explicitly collected as part of the HIT, MTurk also includes information such as unique worker identifiers and timestamps.

Once a worker submits a HIT, the requester then needs to decide whether or not to accept the worker's result. The API allows requesters to write scripts to automatically download newly submitted responses and then automatically decide whether or not to approve them. Likewise, requesters may also manually visit the website to view newly submitted results. If a HIT is not adjudicated within the specified time interval, it is automatically approved. If the worker did not follow the HIT's instructions or the requester has good reason to believe that the response is fraudulent (e.g., incomprehensible language, failure to correctly answer "gold standard" questions), the requester may reject the HIT. When a requester rejects a HIT, the worker does not receive compensation. Since MTurk uses worker approval rates as proxies for reputation, rejection also hurts a worker's reputation and may prevent that worker from completing future HITs that set a reputation threshold.

Closing the HIT

Finally, once a sufficient number of responses have been collected, the researcher will want to prevent additional workers from completing the HIT, as well as pay the workers who have completed it satisfactorily. When she receives either the target number of responses or the time limit passes, the HIT is said to have "expired" (i.e., it is no longer available for additional workers to complete).[10] Once the HIT is expired, one must make sure that all of the workers who completed the task satisfactorily have been paid (otherwise they will be paid automatically, regardless of the quality of their responses), which can also be done either from the web interface or the API. If the work is not satisfactory, requesters have the option of specifying a reason for the rejection.

[10] If for some reason the researcher wishes to expire the HIT early, this is possible to do from both the web interface and the API. Likewise, HITs can also be extended using either method.

Case Studies

In this section, we briefly describe our own experiences using crowdsourcing in research. In particular, we aim to give an informal account of difficulties we encountered and how they were addressed.

Case Study 1: Assessing Quality on Wikipedia

Ed H. Chi

In 2007, Aniket Kittur was an intern in my research group at PARC. One day early in the internship, we were exploring the question of how to assess the quality of Wikipedia articles. A huge debate was raging in the press about the quality of Wikipedia as compared with Encyclopedia Britannica articles (Giles, 2005). We became infatuated with the idea of using the crowd to assess the quality of the work of the crowd, and wanted to see if we could use Amazon MTurk to assess the quality of every Wikipedia article.

We knew that there was some limited ground truth data available on the quality of the articles from expert Wikipedians. In particular, one Wikipedia project systematically vetted a set of criteria for assessing the quality of Wikipedia articles, including metrics such as whether the articles were well-written, factually accurate, and used the required Neutral Point of View (NPOV). The project ranked some small set of articles with a letter grade from FA (Featured Article), A, GA (Good Article), B, C, and so on. By treating these ratings as ground truth, we embarked on a research project to find out if MTurk raters could reproduce expert ratings.

We asked workers to rate articles on a 7-point Likert scale on established metrics such as well-written, factually accurate, and of good quality. We also asked workers to give us free-text answers on how the articles could be improved. We paid workers $0.05 for each task. Within 2 days we had our data! Fifty-eight users made 210 ratings for 15 ratings per article, with a total cost of $10.50. We were thrilled!

However, the quality of the work was depressing. We obtained only a marginally significant correlation between the workers and the expert Wikipedians' consensus ratings ($r = 0.50$, $p = 0.07$). What was worse was that, by examining the rating data by hand, we saw that 59 % of the responses from workers appeared to be invalid. Forty-nine percent of the users did not enter any good suggestions on how to improve the articles, and 31 % of the responses were completed within 1 min, which is hardly enough time to actually read the article and form an opinion. What was worse was that 8 users appeared to have completed 75 % of the tasks! We felt frustrated and disappointed.

	Experiment 1	Experiment 2
Invalid comments	49%	3%
<1 min responses	31%	7%
Median time	1:30	4:06

Fig. 2 Dramatic Improvement in quality of the worker ratings on Wikipedia articles

We nearly gave up on the crowdsourcing approach at this point. We decided to try one more time. But in Experiment 2, we decided to completely change the design of the task:

- First, we decided that we would signal to the user that we were monitoring the results. We did this by asking some simple questions that were easy to answer just by glancing at the page. We used questions such as "How many images does this article have?" We could easily check these answers post hoc.
- Second, we decided to create questions where malicious answers were as hard to create as legitimate answers, such as "Provide 4–6 keywords that summarize this article." These questions make it hard for a worker to "fake" reading the article. Not only did these questions require some cognitive processing, they also allowed us to see the types of tags that users would generate.
- Third, we made sure that answering the above questions was somewhat useful to completing the main task. That is, knowing how many sections or images the article had required the worker to pay some attention to whether the article was well-organized, which in turn was useful in making a decision about its quality.
- Fourth, we put the verifiable tasks ahead of the main task, so that the workers had to perform these steps before assessing the overall quality of the article.

To our surprise, the 2nd experiment worked much better, with 124 users providing 277 ratings for 20 ratings per article. We obtained a significant correlation with the Wikipedia ratings this time ($r=0.66$, $p=0.01$), and there was a much smaller proportion of malicious responses (3 % invalid comments, 7 % <1 min responses). Moreover, the time on task improved dramatically (4:06 min instead of 1:30 min)! We were happy with this success. More details can be found in our CHI2008 conference paper (Kittur, Chi, & Suh, 2008) (Fig. 2).

Case Study 2: Shepherding the Crowd

Steven Dow

When I was a postdoc in the HCI Group at Stanford, we started using crowdsourcing to enable our research. When we needed quality and similarity ratings on a set of visual designs for our experimental work on prototyping practices, we turned to online crowds from Mechanical Turk and oDesk.com (Dow et al., 2010). Through these experiences, we realized that platforms like MTurk offered no real opportunity to communicate with workers or to provide feedback that would help them improve their work performance. Along with Bjoern Hartmann, Anand Kulkarni, and Scott Klemmer, we built a system called Shepherd to understand the effects of introducing real-time feedback into a crowdsourcing platform (Dow, Kulkarni, Klemmer, & Hartmann, 2012).

Our goal was to get unskilled crowds to produce better results on complex work. While other research efforts take a computational approach to this problem and focus on workflows that sequence and coordinate small individual contributions (Bernstein et al., 2010; Kittur, Smus, Khamkar, & Kraut, 2011; Kulkarni, Can, & Hartmann, 2012; Little, Chilton, Goldman, & Miller, 2010a), our work on Shepherd took a more human-centered stance. If we want crowdsourcing to become a viable part of the economy, we cannot be satisfied with paying workers $2–3 per hour on average. Our work examined how we can make crowd work—and the people doing the work—more valuable.

It was our belief that we could educate and motivate workers to do more complex work through process improvements. In particular, we hypothesized that shepherding the crowd—by providing workers meaningful real-time feedback—would lead to better work, learning, and perseverance. We built the Shepherd system to inject feedback into the crowdsourcing process. In our task, the worker writes a series of reviews for products they own. As the reviews start piling in from multiple workers, a requester monitors a work dashboard, reviews each piece of work, and fills in a feedback form. Workers then receive this feedback before they start on their next product review. In the feedback form, workers see what they wrote previously, a checklist of effective product review strategies, and a Likert rating for the product review.

To understand the effects of external feedback on crowdsourcing performance, we conducted a between-subjects study with three conditions. Participants in the *None* condition received no immediate feedback, consistent with most current crowdsourcing practices. Participants in the *Self-assessment* condition judged their own work. Participants in the *External* assessment condition received expert feedback. We found that *Self-assessment* alone yielded better overall work than the *None* condition and helped workers improve over time. *External* assessment also yielded these benefits, but it also resulted in more work. Participants who received external assessment made more revisions to their original reviews. More details about the experimental setup and results can be found in our 2012 CSCW paper (Dow et al., 2012).

Case Study 3: Scaling Up Recruitment and Diversity

Serge Egelman

My research mostly focuses on how humans make decisions concerning privacy and security. This means that at least half of my time is spent conducting experiments on people, both in the laboratory and in the field. Prior to being introduced to crowdsourcing, large-scale online surveys were seen as highly laborious. Back when I was a graduate student at Carnegie Mellon, many researchers would use dedicated participant pools that largely consisted of students and staff. In order to yield more diverse demographics, my group generally shied away from these participant lists in favor of recruiting participants online.

We would post to online forums, such as Craigslist, and ask people to fill out our surveys in exchange for a raffle incentive (e.g., we would give away gift cards to randomly selected survey respondents). Posting recruitment notices became a full-time task. For instance, in the case of Craigslist, I would post to as many different cities as possible in order to get a diverse sample. This involved slightly changing the wording on the posting for each city to which I posted, since the Craigslist spam filter would flag similar-looking postings from different cities. This also involved keeping track of when postings were expiring and needed to be reposted. All this effort—2 weeks of graduate student time—generally resulted in about 100–200 responses per week.

Because of the large time investment, it was not feasible to modify experimental designs. That is, if our data prompted new questions that could only be addressed by adding additional material to a survey, our results could be delayed by several weeks.

It was not until late 2009 that I read an article comparing both the demographics and efficiency of MTurk workers with survey respondents who had been recruited by a market research firm (Jakobsson, 2009). Jakobsson found similar results between the two samples. This, in addition to reading articles by Ipeirotis on the demographics of MTurk workers (2008, 2010a, 2010b), led me to investigate whether I could use MTurk to recruit a diverse sample of survey respondents in a much shorter amount of time than previously possible.

In my first experiment, we recruited workers to complete a survey regarding their workplace file-sharing habits. We offered participants $0.25 to participate, and the survey took roughly 10–20 min. We received over 350 legitimate responses in the course of 48 h. Even more interestingly, over 95 % of our respondents held white-collar jobs and were completing our survey midday. This indicated that they were not "professional" experimental subjects…they were instead amusing themselves at work, rather than participating solely for the compensation. When contrasting our results with previous studies using other recruitment methods, we found that not only did crowdsourcing cost much less while enabling quicker recruitment, but the number of obvious "cheaters" (i.e., those who submitted nonsensical responses) had not increased.

Since then, crowdsourcing has become my go-to recruitment mechanism for experiments that can be completed online. In addition to surveys, this has also included interactive tasks using embedded applets (Egelman et al., 2010), as well as workers downloading custom software (Christin et al., 2011). To give another example, my colleagues and I used crowdsourcing to study password creation habits by recruiting over 5,000 participants (Komanduri et al., 2011). The cost per participant was roughly a dollar, while the quality of the results did not suffer—the paper received the honorable mention award at CHI 2011.

Prior to crowdsourcing, the thought that a researcher could recruit over 1,000 subjects in under a week for under $1,000 was unheard of, but this is the new reality.

Crowdsourcing Research and Resources

Beyond using existing general-purpose crowdsourcing platforms to serve core HCI research activities, a growing number of researchers are creating new crowdsourcing platforms—either from scratch or on top of existing general-purpose platforms—to explore novel systems and applications.

Crowd-powered software: The Soylent project is perhaps the best-known harbinger of using crowds as a first-class entity in a software application. Bernstein et al. (2010) created a word processing interface that enables writers to "hire" MTurk workers to shorten, proofread, and edit documents on demand. Soylent pioneered a *Find-Fix-Verify* pattern to help manage micro-task crowds by splitting tasks into a series of generation and review stages. Since this project appeared, a number of other crowd-powered systems have emerged including PlateMate, which uses crowd workers to perform nutrition analysis from photographs of food (Noronha, Hysen, Zhang, & Gajos, 2011).

Real-time crowdsourcing: One significant thrust by developers of crowd-powered systems has been the goal of tapping the abilities of crowds in (near) real-time. For example, to help answer everyday questions from visually impaired users, the VizWiz application asks the same question of multiple people at the same time through crowdsourcing (Bigham et al., 2010). To achieve near real-time response rates (just over 2 min on average), VizWiz proactivity recruited and queued workers to work on a simple separate task and then pulled them into the VizWiz application on an as-needed basis. A number of other real-time crowdsourcing applications have since emerged, including Adrenaline, which gets crowd workers to quickly filter a short video down to the best single photo (Bernstein et al., 2011), and Legion, which employs crowds to control UIs, such as a remote control interface for a robot (Lasecki, Murray, White, Miller, & Bigham, 2011).

Complex tasks with constraints: A key characteristic of crowdsourcing is the ability to employ people to make small contributions to a larger and more complex problem. Zhang et al. (2012) explored the use of crowds for trip itinerary planning, where a requester has specified any number of high-level goals and constraints (e.g., "at least one fresh local food restaurant"). The researchers created a collaborative planning system called Mobi that allows crowd workers to view the solution context and make additional changes based on current problem needs. This approach enables requesters to iteratively add, subtract, or re-prioritize goals; workers can contribute a small amount or continue working on the list of needs.

Crowd toolkits: Managing crowds can be challenging, especially for complex workflows. A number of research efforts focus on creating worker visualizations and workflow management tools. Kittur et al. (2011) implemented the CrowdForge workflow tool based on the MapReduce programming paradigm where tasks are partitioned into subproblems, mapped to workers, and then combined back into one result. Kulkarni et al. (2012) took a similar approach with Turkomatic by asking workers, "can you finish this task in 1 min? If not, please break the task down into multiple smaller tasks." The authors also developed workflow visualizations for Turkomatic to help requesters better facilitate this process.

Crowd-specific studies: Other crowdsourcing research focuses on gathering empirical data about how particular workflows and conditions affect the work performance and attitudes of crowd workers. For example, Little, Chilton, Goldman, and Miller (2010b) explored the tradeoffs of iterative and parallel processes for human computation tasks. They reported that, in general, iteration improves work quality, except on more generative tasks like brainstorming, where showing a previous worker's ideas may limit the creativity of the next worker. In addition to specific workflow issues, researchers have examined crowd feedback (Dow et al., 2012), social transparency (Stuart, Dabbish, Kiesler, Kinnaird, & Kang, 2012), and labor concerns (Quinn & Bederson, 2011) with respect to crowdsourcing environments.

Conclusions

Crowdsourcing offers a technique for recruiting lots of people online to perform work, which has the potential to change HCI research. By utilizing both paid workers and unpaid volunteers, researchers can greatly expand the diversity and reduce the time it takes to conduct user studies and large-scale data analysis. While this is a powerful method, the technique presents a number of potential pitfalls. This chapter summarizes these common pitfalls and gives examples of how to avoid them. We also included a short summary of how to use Amazon's Mechanical Turk so that researchers can quickly get started with this technique. However, this is a relatively new technique that continues to evolve rapidly. As such, we expect certain aspects of this chapter to become outdated in the future. It is our hope that HCI researchers will use the tips in this chapter to further refine and expand on this valuable new research method.

References

André, P., Bernstein M., & Luther K. (2012). Who gives a tweet?: Evaluating microblog content value. In *Proceedings of the ACM 2012 Conference on Computer Supported Cooperative Work* (pp. 471–474). New York, NY: ACM

Bernstein, M. S., Brandt, J., Miller, R. C., & Karger, D. R. (2011). Crowds in two seconds: Enabling realtime crowd-powered interfaces. In *Proceedings of the 24th Annual ACM Symposium on User Interface Software and Technology* (pp. 33–42). New York, NY: ACM

Bernstein, M. S., Little G., Miller R. C., Hartmann B., Ackerman M. S., Karger D. R., et al. (2010). Soylent: A word processor with a crowd inside. In *Proceedings of the 23nd Annual ACM Symposium on User Interface Software and Technology* (pp. 313–322). New York, NY: ACM

Bigham, J. P., Jayant, C., Ji, H., Little, G., Miller, A., Miller, R. C., et al. (2010). VizWiz: Nearly real-time answers to visual questions. In *Proceedings of the 23nd Annual ACM Symposium on User Interface Software and Technology* (pp. 333–342). New York, NY: ACM

Christin, N., Egelman, S., Vidas, T., & Grossklags, J. (2011). It's all about the Benjamins: An empirical study on incentivizing users to ignore security advice. *Financial Cryptography and Data Security* 16–30

Dow, S. P., Glassco, A., Kass, J., Schwarz, M., Schwartz, D. L., & Klemmer, S. R. (2010). Parallel prototyping leads to better design results, more divergence, and increased self-efficacy. *ACM Transactions on Computer-Human Interaction (TOCHI), 17*(4), 18.

Dow, S., Kulkarni, A., Klemmer, S., & Hartmann, B. (2012). Shepherding the Crowd Yields better work. In *Proceedings of the ACM 2012 Conference on Computer Supported Cooperative Work* (pp. 1013–1022). New York, NY: ACM

Egelman, S., Molnar, D., Christin, N., Acquisti, A., Herley, C., & Krishnamurthi, S. (2010). Please continue to hold: An empirical study on user tolerance of security delays. In *Proceedings (Online) of the 9th Workshop on Economics of Information Security*

Fowler, F. J., Jr. (1995). *Improving survey questions: Design and evaluation* (Vol. 38). Thousand Oaks, CA: Sage. Incorporated.

Giles, J. (2005). Internet encyclopaedias go head to head. *Nature, 438*(7070), 900–901.

Grier, D. A. (2005). *When computers were human* (Vol. 316). Princeton, NJ: Princeton University Press.

Hand, E. (2010). Citizen science: People power. *Nature, 466*(7307), 685.

Heer, J., & Bostock, M. (2010). Crowdsourcing graphical perception: Using mechanical turk to assess visualization design. In *Proceedings of the 28th International Conference on Human Factors in Computing Systems* (pp. 203–212). New York, NY: ACM

Ipeirotis, P. (2008). *Mechanical turk: Demographics*. Retrieved September 15, 2009, from http://behind-the-enemy-lines.blogspot.com/2008/03/mechanical-turk-demographics.html

Ipeirotis, P. (2010a). Demographics of mechanical turk. Working Paper, CeDER-10-01. http://archive.nyu.edu/handle/2451/29585

Ipeirotis, P. (2010b). *The new demographics of mechanical turk*. Retrieved July 2, 2012, from http://www.behind-the-enemy-lines.com/2010/03/new-demographics-of-mechanical-turk.html

Jakobsson, M. (2009). Experimenting on mechanical turk: 5 How tos. Retrieved November 4, 2009, from http://blogs.parc.com/blog/2009/07/experimenting-on-mechanical-turk-5-how-tos/

Kittur, A., Chi, E. H., & Suh, B. (2008). Crowdsourcing user studies with mechanical turk. In *Proceedings of the Twenty-Sixth Annual SIGCHI Conference on Human Factors in Computing Systems* (pp. 453–456). New York, NY: ACM

Kittur, A., Smus, B., Khamkar, S., & Kraut, R. E. (2011). Crowdforge: Crowdsourcing complex work. In *Proceedings of the 24th Annual ACM Symposium on User Interface Software and Technology* (pp. 43–52). New York, NY: ACM

Kittur, A., Suh, B., & Chi, E. H. (2008). Can you ever trust a Wiki?: Impacting perceived trustworthiness in Wikipedia. In *Proceedings of the 2008 ACM Conference on Computer Supported Cooperative Work* (pp. 477–480). New York, NY: ACM

Komanduri, S., Shay, R., Kelley, P. G., Mazurek, M. L., Bauer, L., Christin, N., et al. (2011). Of passwords and people: Measuring the effect of password-composition policies. In *Proceedings of the 2011 Annual Conference on Human Factors in Computing Systems* (pp. 2595–2604). New York, NY: ACM

Kulkarni, A., Can, M., & Hartmann, B. (2012). Collaboratively crowdsourcing workflows with turkomatic. In *Proceedings of the ACM 2012 Conference on Computer Supported Cooperative Work* (pp. 1003–1012). New York, NY: ACM

Kumar, R., Kim, J., & Klemmer, S. R. (2009). Automatic retargeting of web page content. In *Proceedings of the 27th International Conference (Extended Abstracts) on Human Factors in Computing Systems* (pp. 4237–4242). New York, NY: ACM

Landsberger, H. A. (1958). *Hawthorne revisited: Management and the worker, its critics, and developments in human relations in industry*. Ithaca, NY: Cornell University.

Lasecki, W. S., Murray, K. I., White, S., Miller, R. C., & Bigham, J. P. (2011). Real-time crowd control of existing interfaces. In *Proceedings of the 24th Annual ACM Symposium on User Interface Software and Technology* (pp. 23–32). New York, NY: ACM

Lewis, S., Dontcheva, M., & Gerber, E. (2011). Affective computational priming and creativity. In *Proceedings of the 2011 Annual Conference on Human Factors in Computing Systems* (pp. 735–744). New York, NY: ACM

Little, G., Chilton, L. B., Goldman, M., & Miller, R. C. (2010a). TurKit: Human computation algorithms on mechanical turk. In *Proceedings of the 23nd Annual ACM Symposium on User Interface Software and Technology* (pp. 57–66). New York, NY: ACM

Little, G., Chilton, L. B., Goldman, M., & Miller, R. C. (2010b). Exploring iterative and parallel human computation processes. In *Proceedings of the ACM SIGKDD Workshop on Human Computation* (pp. 68–76). New York, NY: ACM

Noronha, J., Hysen, E., Zhang, H., & Gajos, K. Z. (2011). Platemate: Crowdsourcing nutritional analysis from food photographs. In *Proceedings of the 24th Annual ACM Symposium on User Interface Software and Technology* (pp. 1–12). New York, NY: ACM

Priedhorsky, R., & Terveen, L. (2008). The computational geowiki: What, why, and how. In *Proceedings of the 2008 ACM Conference on Computer Supported Cooperative Work* (pp. 267–276). New York, NY: ACM

Quinn, A. J., & Bederson, B. B. (2011). Human computation: A survey and taxonomy of a growing field. In *Proceedings of the 2011 Annual Conference on Human Factors in Computing Systems* (pp. 1403–1412). New York, NY: ACM

Ross, J., Irani, L., Silberman, M., Zaldivar, A., & Tomlinson, B. (2010). Who are the crowdworkers?: Shifting demographics in mechanical turk. In *Proceedings of the 28th International Conference (Extended Abstracts) on Human Factors in Computing Systems* (pp. 2863–2872). New York, NY: ACM

Stuart, H. C., Dabbish, L., Kiesler, S., Kinnaird, P., & Kang, R. (2012). Social transparency in networked information exchange: A theoretical framework. In *Proceedings of the ACM 2012 Conference on Computer Supported Cooperative Work* (pp. 451–460). New York, NY: ACM

Toomim, M., Kriplean, T., Pörtner, C., & Landay, J. (2011). Utility of human-computer interactions: Toward a science of preference measurement. In *Proceedings of the 2011 Annual Conference on Human Factors in Computing Systems* (pp. 2275–2284). New York, NY: ACM

Tuite, K., Snavely, N., Hsiao, D. -Y., Smith, A. M., & Popović, Z. (2010). Reconstructing the world in 3D: Bringing games with a purpose outdoors. In *Proceedings of the Fifth International Conference on the Foundations of Digital Games* (pp. 232–239). New York, NY: ACM

von Ahn, L., & Dabbish, L. (2004). Labeling images with a computer game. In *Proceedings of the SIGCHI Conference on Human Factors in Computing Systems* (pp. 319–326). New York: ACM

von Ahn, L., Maurer, B., McMillen, C., Abraham, D., & Blum, M. (2008). reCAPTCHA: Human-based character recognition via web security measures. *Science, 321*(5895), 1465–1468.

Zhang, H., Law, E., Miller, R., Gajos, K., Parkes, D., & Horvitz, E. (2012). Human computation tasks with global constraints. In *Proceedings of the 2012 ACM Annual Conference on Human Factors in Computing Systems* (pp. 217–226). New York, NY: ACM

Zimmerman, J., Tomasic, A., Garrod, C., Yoo, D., Hiruncharoenvate, C., Aziz, R., et al. (2011). Field trial of Tiramisu: Crowd-sourcing bus arrival times to spur co-design. In *Proceedings of the 2011 Annual Conference on Human Factors in Computing Systems* (pp. 1677–1686). New York, NY: ACM

Sensor Data Streams

Stephen Voida, Donald J. Patterson, and Shwetak N. Patel

A Short Description of the Method

The increased availability of inexpensive sensors, tremendous processing capabilities (even in mobile devices), high-bandwidth wireless networks, and vast quantities of data storage have made it much more practical to continuously collect streams of low-level data about people and their environments. This approach enables researchers to compile detailed records of various contextual factors surrounding people's interactions with their world. Their locations, physiological states, contact with other people, situated uses of devices, and other digital traces can potentially be recorded and analyzed. Each of these kinds of data can be collected at virtually any frequency, with or without participant knowledge or intervention, and for extended periods of time. Due to the automated nature of the method, a large number of samples can be gathered quickly and with relatively low overhead by the researchers during sessions in the field. However, the degree of automation involved in this method requires a number of pragmatic and analytic considerations, beginning with careful experimental design to ensure appropriate sensor design, and including how the study is deployed, participant training, privacy safeguards, and data storage requirements.

S. Voida (✉)
Indiana University School of Informatics and Computing, Indianapolis (IUPUI),
535 W. Michigan St., IT 475, Indianapolis, IN 46205, USA
e-mail: svoida@iupui.edu

D.J. Patterson
Donald Bren School of Information and Computer Sciences, University of California,
Irvine, 5084 Donald Bren Hall, Irvine, CA 92797-3440, USA
e-mail: djp3@ics.uci.edu

S.N. Patel
Department of Computer Science and Engineering, University of Washington,
Box 352350, Seattle, WA 98195-2350, USA
e-mail: shwetak@cs.washington.edu

J.S. Olson and W.A. Kellogg (eds.), *Ways of Knowing in HCI*,
DOI 10.1007/978-1-4939-0378-8_12,

While the large amounts of recorded data captured using this technique lend themselves to quantitative analysis—for example, counting the frequency, duration, or variance of signals within an event stream—analyzing and interpreting the data often requires the application of additional qualitative methods.

Its History, Intellectual Tradition, Evolution (in Brief)

While sensors of various kinds have been used to collect data about people and their activities for a long time across a wide variety of scientific domains, the combination of data collection and analysis techniques that we present here primarily evolved from various quantitative data collection methods employed within the domain of experimental psychology. There is a parallel here with psychological research. Psychologists frequently define objective measures for various human or group behaviors and develop various technical or nontechnical instruments to collect data to characterize, quantify, or investigate variance in measurable phenomena based on a specific set of research questions. Similarly, streams of data are a modern equivalent frequently employed in the computer and information sciences. In order to study people, groups, or environments using streams of data, one or more sensing devices are used to automatically collect data on researchers' behalf. The decreasing cost and increasing availability of powerful computational devices and sensors (Weiser, 1991) make it possible to bring a wide array of various sensors to bear on observing a single phenomenon, environment, or individual. Innovations in data fusion, user modeling, and inferencing are then used to aggregate, filter, and interpret the data to answer research questions.

Automated capture of sensed participants' behavior also complements in situ survey techniques like the *experience sampling method* (*ESM*), pioneered by Larson and Csikszentmihalyi (1983). ESM is also known as *ecological momentary assessment or EMA* (Stone, Shiffman, & DeVries, 1999). Like sensor data streams, ESM involves collecting data from a participant while they go about their everyday business; typically, they are asked to answer a question or complete a very short survey when they receive an alarm, alert, or telephone notification. ESM is a sociological data collection technique that elicits information about participants' actions and beliefs *as they happen* and *in the context* of everyday activity, overcoming the recall/recollection limitations of post-interviews or diary studies. Although technology-oriented adaptations of ESM (e.g., Consolvo & Walker, 2003, Intille, Rondoni, Kukla, Ancona, & Bao, 2003) have been utilized for some time within the ubiquitous computing research community, this approach still relies on active responses from research participants during data collection, making the approach less than ideal for longitudinal studies or for research in which introduction of interruptions (e.g., studies of multitasking) or foregrounding of the research study would influence or alter the behavior(s) of interest. Sensor data streams can provide an alternative, or at least complementary, means of collecting contextually situated data as well, but typically through the capture of implicit participant actions or behaviors.

What Questions the Method Can Answer

Sensor data stream collection may be used to understand people's activities, behaviors, or practices by instrumenting:

- a person (Bao & Intille, 2004; Choudhury et al., 2008; Choudhury & Pentland, 2003; Consolvo, Klasnja et al., 2008; Consolvo, McDonald et al., 2008; Liao, Patterson, Fox, & Kautz, 2006; Marmasse & Schmandt, 2000; Olguín & Pentland, 2008; Olguín et al., 2009; Patterson, Fox, Kautz, & Philipose, 2005; Patterson, Liao, Fox, & Kautz, 2003; Philipose et al., 2004),
- an environment, such as an office (Begole & Tang, 2007; Fogarty et al., 2005; MacIntyre et al., 2001; Mark, Voida, & Cardello, 2012), or
- the home (Brumitt, Meyers, Krumm, Kern, & Shafer, 2000; Cohn, Gupta, Froehlich, Larson, & Patel, 2010; Cohn, Stuntebeck et al. 2010; Froehlich et al., 2009; Froehlich et al., 2011; Gupta, Reynolds, & Patel, 2010; Intille, 2002; Intille et al., 2006; Kidd et al., 1999; Kientz et al., 2008; Orr & Abowd, 1999; Patel, Robertson, Kientz, Reynolds, & Abowd, 2007, Tapia, Intille, & Larson, 2004).

The resulting streams of sensor data can be used to help answer a wide variety of questions about the people instrumented with sensors or occupying the instrumented spaces, for example:

- Where do people travel over the course of a day?
- With whom do they normally communicate or collaborate?
- What tools or information resources do they use at various points during the day? When, where, and with whom?
- What routines help to define a "typical" or "atypical" day?
- How healthy are a person's daily behaviors? Is he or she making good health choices?

HCI researchers often use information gathered in this way to learn about people's current behaviors—for example, assessing the duration or frequency of an established technology's use throughout the day. The same data can also form a baseline in studies of how people's behaviors change over time, such the changes affected by the deployment of a new technology among a representative set of end users (e.g., Hutchinson et al., 2003). The ability to make such a comparison is important and enables researchers to draw conclusions about the impact of changing people's tools or environment.

Sensor data often begins as a stream of very fine-grained events. These events include timestamps and durations that are much more accurate than most human observers could possibly record in real time (without having to code a video-taped record of the session after the fact). The kinds of actions that can be recognized depends greatly on how tightly coupled the action is to a specific sensor. For example, a pressure sensor on a chair might suggest an individual's presence in the office, but is less definitive as an indicator of, say, task focus or interruptibility (after Horvitz, Koch, & Apacible, 2004; Mark et al., 2012). The more abstract the action or the more the action veers from being physically descriptive to being descriptive of intent, the more uncertainty is introduced in the interpretation, even with the availability of relevant sensing. Nonetheless, these kinds of data can answer questions about what people do, relevant stimuli, and physical or social constraints [e.g., the frequency with which members of a social group check in on one another's social network status (Miluzzo et al., 2008)]. Currently, inexpensive and widely available sensors like accelerometers and barometers can quantify things as disparate as motion signatures and changes in air–pressure, providing insight into questions about participants' activity levels, gestures, and other bodily movement. Capacitive, pressure, and acoustic sensors can detect the intensity and number of physical contacts made between people and objects or among objects. Video- and depth-based camera systems can count the number of people or objects in a space, their orientation with respect to one another and gestures or motion taking place, enabling a range of studies about participants' interpersonal interactions and proxemics.

The low-level data collected by various sensing devices to answer these questions can be analyzed to determine higher-level behavior such as activity (Bao & Intille, 2004; Tapia et al., 2004), routing and place-visiting behavior (Patterson et al., 2003), gesture performance (Fan et al., 2012; Westyn, Brashear, Atrash, & Starner, 2003), or communication roles (Wyatt, Choudhury, Bilmes, & Kitts, 2011). Statistical summaries can then be created which may help with the triangulation of other methods around phenomena of interest or to generate predictive models. (See Looking Back: Retrospective Study Methods for HCI.)

Historically, questions of *where*?—that is, questions that demanded location-awareness capabilities—were some of the earliest research questions thoroughly explored by the ubiquitous computing community. Many projects have been extremely successful in answering questions about the locations or places that a participant has visited (Abowd et al., 1997; Liao et al., 2006, Patel, Truong, & Abowd, 2006, Patterson, 2009), other people with whom they have been in close proximity (Want, Hopper, Falcão, & Gibbons, 1992; Choudhury & Pentland, 2003; Choudhury et al., 2008), and the participant's routines and patterns of mobility (Consolvo, McDonald et al., 2008; Liao et al., 2006). The level of fidelity of these streams of information varies greatly, depending both on the geo-location technology being used to collect the data and the level of fidelity at which the data are represented and stored; Hightower and Borriello explore many of these considerations in their overview article in *IEEE Computer* (2001b) and Varshavsky and Patel provide an updated perspective (Ubiquitous Computing Fundamentals 2009).

Research Questions with Various Units of Analysis

One advantage of using streams of data as a means of understanding people's activities and behaviors is that the technique can be used to answer research questions across a range of units of analysis. The main trade-off in realizing this analytic flexibility is that the "right" number, type, and combination of sensor data sources used to collect the data do vary based on the specific scope of research and the research questions that are being asked.

Egocentric Sensor Data Streams

Sensors focused on monitoring the movements, activities, and interactions of a single individual can answer questions at an *egocentric* unit of analysis. Studies that collect log data from an individual's use of a computer, a smartphone, or a different piece of technology; measure a person's physiological state; monitor their movements, location, or interactions using the sensors on a mobile phone; or collect information about activity within a private or semiprivate space could all be described as *egocentric* in this way. This is a typical study design as derived from the experimental and engineering psychology tradition and can be replicated over a population sample to look at patterns or trends. Egocentric data streams are often used to answer research questions like *How do people allocate their time or attention? How is a person's mental state or mood affected by real-world stimuli? How do electronic communications or mobile computing interactions affect daily routines? How do people's own understanding or interpretations of their activities, colleagues, or environment differ from what a ubiquitous computing application or tool is able to automatically sense?*

Group-Centric Sensor Data Streams

Sensors can also be used to instrument—and answer research questions about—groups of people at the same time. This *group-centric* approach can involve simply capturing the same signals as for a single person, but across a group over the same window of time, or it might involve deploying a broader set of environmental or infrastructural sensors in a shared/community space or collecting data about more interpersonal types of interactions (e.g., sociometer-style data, after Choudhury & Pentland, 2003). This style of sensor data stream-based research is often utilized for understanding research questions about organizational and group dynamics, such as *How often do members of this group interact with one another? What do these interactions entail? How do power relations manifest in different kinds of work environments or work teams?*

Space-Centric Sensor Data Streams

Finally, researchers can use streams of sensor data to answer questions about how spaces are used, irrespective of their particular inhabitants, given appropriate instrumentation of a space. This kind of *space-centric* research is often carried out in semipublic or public spaces, such as museums (Hornecker & Nicol, 2012; Sparacino, 2003). A standard approach to *space-centric* sensing is to employ a wide variety of sensors to detect signals from a place of interest. Commonly used devices include high information density sensors, such as cameras and microphones, as well as low density sensors such as pressure sensitive floor tiles (Orr & Abowd, 1999), passive infrared (PIR) motion detectors, or RFID readers. If an experiment requires a large number of sensors to be distributed throughout the environment, special networking support (either wired or wireless) may additionally need to be installed in the space to move the data from the site of collection to the storage and analysis servers. Maximizing these sensors' area of coverage often necessitates that they be mounted on a room's walls, ceiling, or doors so that the sensors have unobstructed views of a wide area; the aesthetic impact of these placements can lead to reduced adoption by homeowners or altered behavior on the part of the participants because they are reminded that they are being observed. The use of high information density sensors in certain spaces often raises concerns about the balance between the value of the proposed use of the data stream and impact of the resulting surveillance, particularly in home settings.

The alternative, which is low density sensing, includes the use of many simple, low-cost sensors, such as motion detectors, pressure mats, break beam sensors, and contact switches, to determine activity and movement. The principal advantages of the low-density approach are lower bandwidth and processing needs, and potentially reduced privacy concerns. In some cases, limitations imposed by the need to take into account participants' privacy risks can prohibit the use of sensors that might be beneficial for answering specific research questions, such as putting cameras in bathrooms. In such a case, low-density sensors may be the only alternative. Notable examples of space-centric sensing that were designed for multiple deployments include Georgia Tech's Aware Home (Kidd et al., 1999; Kientz et al., 2008) and MIT's House *n* (Intille, 2002; Intille et al., 2006; Tapia et al., 2004).

Researchers have addressed some of the limitations of *space-centric* research, by inventing a technique called "Infrastructure Mediated Sensing" or IMS. IMS involves using existing home infrastructure to detect activity within a home (Patel et al., 2007; Patel, Reynolds, & Abowd, 2008; Froehlich et al., 2009; Gupta et al., 2010). Electrical, plumbing, and HVAC systems, as well as natural gas piping and computer networks, are already widely deployed in much of the world. This existing infrastructure can be instrumented to sense when the home occupants engage in activities that utilize those infrastructures (e.g., monitoring the flow of electricity throughout a home to detect when individual appliances are being used). In addition, those infrastructures can be used to communicate the detection signals through a home. This approach reduces the need for researchers to install many sensors throughout a space, and enables researchers to answer questions like *How are the occupants of a home spending their time throughout the day and night? Is a senior adult living by herself continuing to maintain healthy levels of physical activity? What is the impact of ambient feedback promoting environmental awareness on cooking, cleaning, and hygiene activities within different types of families?*

There is a parallel between the ways that researchers use IMS to conduct infrastructure-based sensing with the ways that egocentric sensing is carried out using smartphones. In both cases, researchers have piggybacked sensing onto infrastructures that were already deployed but that were not originally meant for data collection. In a hybrid example, Isaacman and colleagues (2012) developed a technique that bridged between egocentric data streams and IMS by using cell phone billing records to analyze large-scale commuting patterns.

The attractiveness of IMS is that millions of locations and people, in effect, have been instrumented by highly refined and well-understood technologies. Additionally, this instrumentation occurs across a wide (but not exhaustive) variety of places and people. The amount of additional effort that must be done to leverage these infrastructures, however, varies greatly. Newer approaches, such as IMS, require more effort than location-based services, for example. Depending on the effort required to enable data collection through these infrastructures, these techniques can enable scaling that would not be possible with traditional sensor deployment approaches.

What Data Are Captured by Sensor Data Streams

Sensor data streams are similar to other kinds of approaches that are commonly used to capture and analyze user behavior, but are distinguished by their source (the physical world) and their level of interpretive fidelity. Sensor data streams are time-stamped data that flow from the physical environment surrounding and permeating user interactions. The range of sensors that might be employed in these kinds of studies are virtually limitless; sensing toolkits often facilitate data collection using a variety electrical switches, motion sensors, pressure sensors, voltmeters, photometers, thermometers, moisture sensors, proximity sensors, RFID tags and beacons, microphones, or cameras (e.g., Greenberg & Fitchett, 2001; Villar, Scott, & Hodges, 2011). New sensors are continually being developed and incorporated into consumer devices and research-oriented toolkits all the time.

Data streams might also come from log files or usage statistics from instrumented user interfaces (see also Looking Back: Retrospective Study Methods for HCI). These logs can be thought of as "virtual sensor" data streams, as these data *are* being collected in situ during an interaction but are the result of observations of the *digital* world as opposed to the physical world. Another source of "virtual sensor" data streams can come from the nonautomated human observation of interactions in which time stamped annotations of events are also produced—albeit with considerably more overhead and somewhat less precision (e.g., Mark et al., 2012). This approach can include observations of recorded audio or video data and differs from sensor data streams in that the quantity of information is typically much smaller, although the kinds of data collected can be much richer. For example, a log file analysis might present a user clicking a mouse on a "save" button as a "virtual sensor" event. A human observer might record very similar data from a video recording by annotating that the person "saved a document" at a particular moment in the video. A typical sensor data stream, however, would record the sound of a mouse click, some evidence that the person and the computer were collocated and perhaps some fluctuation in the power used by the computer at the same time. The boundaries between log file analysis, human observation, and sensor data stream-based data collection are not clearly delineated, however, and many of the same data storage, processing, and analytic techniques could be used to make sense of the information gathered using any of these data sources without much difficulty. Clearly, these data sources could strongly reinforce and inform one another, as well.

This style of quantitative data collection and analysis to understand human activity has been utilized within the HCI community for some time, most commonly in the analysis of software interactions by treating existing data logging capabilities as virtual sensors or by explicitly adding the ability to log new events, such as those that occur in desktop computing systems (Brdiczka, Su, & Begole, 2010; Hutchings, Smith, Meyers, Czerwinski, & Robertson, 2004; Kaptelinin, 2003; MacIntyre et al., 2001; Nair, Voida, & Mynatt, 2005; Stumpf et al., 2005), in awareness and social networking tools (Begole, Tang, Smith, & Yankelovich, 2002; Monibi & Patterson, 2009; Patterson, Ding, Kaufman, Liu, & Zaldivar, 2009; Tang & Patterson, 2010), or on the World Wide Web (e.g., Perkowitz, Philipose, Fishkin, & Patterson, 2004).

It is often helpful to augment automatically collected physical and virtual sensor data streams with other data collection techniques. This can be done to measure phenomena that span both the physical and virtual worlds, such as workplace interruptions (Bailey & Iqbal, 2008; Bailey & Konstan, 2006; Bailey, Konstan, & Carlis, 2001; Horvitz et al., 2004). It can also be done to provide clearly indexed points in the mountains of collected data that can be used for subsequent qualitative interviews, somewhat reducing the burden on an interviewee of recalling what was happening as the data was being collected (see Chapter on Looking Back: Retrospective Study Methods for HCI).

For the balance of this chapter, we will focus on the use of physical sensors as a way of studying people and their behaviors. While many of the techniques that we discuss generalize to the user of virtual sensors or log analysis to a greater or lesser extent, we refer the readers to other chapters in this volume for a more in-depth treatment of these specific topics.

Sensor Data Streams and Context-Aware Computing

The use of data streams as a research method also shares a number of similarities with *context-aware computing* research efforts, including the data sources and analytic techniques. While a complete overview of context-aware computing is beyond the scope of this chapter, several good surveys have been published, including Baldauf, Dustdar, and Rosenberg (2007); Bolchini and colleagues (2007); and Hong, Suh, and Kim (2009). Generally speaking, context-aware computing is a form of interactive computing in which a user's implicit behavior—that is, their location, their physical activity, or their interactions with other people—or the environment in which a system is being used can both serve as alternative or auxiliary inputs to the system (Dey & Abowd, 2001; Salber et al., 1999). The central differences between collecting sensor data streams to understand user behavior and leveraging streams of data as inputs for a context-aware computing system are as follows:

- the research goals (studying user behaviors versus developing interactive systems),
- when the collected data are processed and analyzed (as part of the analysis versus in real time), and
- whether the supporting technologies are primarily intended to *generate* a user model versus *predict* user behavior based on a preexisting model.

Generally, when collecting sensor data streams, there are no real-time processing or inference-generating requirements—such processing can usually be handled offline once collection is complete. This decouples the analytic concerns from data collection, relaxing constraints that are often imposed on most context-aware computing systems. In context-aware computing, data collection and analysis often need to be tightly coupled so that the results of the data stream analysis are available as soon as possible following the physical interactions that generated them. Interactive systems require that researchers minimize latency in analyzing sensed data, which often necessitates aggressive data pruning, matching against coarse, preexisting user models, or the use of heuristic approximations; data collection for understanding users does not necessarily need to pursue these kinds of optimizations. Even given this distinction, many of the inferencing techniques that have been developed within the ubiquitous computing and context-aware computing research communities may still be of interest to researchers who are more interested in using streams of data to learn about their participants, their actions, and their surroundings, since they represent a useful set of tools for transforming raw data into higher-level constructs.

Limitations of the Approach

While sensor data streams are useful for gathering large volumes of continuous, high-fidelity data about the way that people behave and interact in the real world, there are a number of important considerations and limitations associated with capturing, processing, storing, and analyzing sensor data streams:

The primary weakness of using streams of sensor data as a technique for learning about participants is that **sensor data generally does a poor job of answering questions of *why*** things have happened in the real world. This limitation goes above and beyond the self-evident challenges in detecting *what* is happening, since a single sensor reading can often be interpreted in many different ways. Even accurately sensed participant actions (or interactions or contexts) can only record the physical impact of what happened; these kinds of readings provide little insight about the intention behind the actions or the broader aims, goals, or internal, mental states of a participant. One way that this limitation can be minimized is by *triangulation* or *data fusion*, that is, combining many data streams and methodologies together.

The **phenomena to be measured or observed must be well understood** at the outset of the study, and an appropriate sensor or sensors must be acquired (or, in many cases, built from scratch) and then deployed to capture the "right" set of data to measure or observe those phenomena. Even with a well-designed sensor deployment, additional techniques may be required to automatically process the raw data so that the phenomena of interest can be isolated, contextualized, and measured. This gap between what can be captured in a sensor data stream and what actually took place in the real world may necessitate a probabilistic interpretation of whether or not an event has occurred and may impact the knowledge generated (e.g., understanding whether or not a gap in logged computer activity is due to an external interruption or simply because a participant is rereading a prompt or looking for additional information). Because the data collection instruments (the sensors) are such an integral part of the research design, when using this approach, it is often more difficult to make changes to a sensor-based data collection protocol than it is for a corresponding, human-administered study (e.g., fielding interviews or surveys).

Sensors (and their associated processing and recording technologies) **have limitations in the quality of data they can collect** during a study. They can sometimes be expensive (limiting the amount of sensing redundancy that can be employed), are sometimes unreliable, are often prone to generating noisy output, and are often limited in the scope and range of actions they can record. Sensor-based research protocols should employ a combination or suite of sensing technologies to balance these drawbacks; however, this does increase the overhead of setting up, storing, maintaining and analyzing the data that are produced, in addition to the design and implementation work necessary when sensors need to be created or modified. However, one of the central tenets of using sensor data streams to conduct user research is that the results of a sensor data stream study can be no better than the sensed data.

If the goal is to capture *everyday* activity or behavior, **care must be taken to select sensors that can effectively capture the right kind of data at the right fidelity and minimize intrusiveness and discomfort** for the study participants. Creating a study design that requires participants to dramatize, enlarge, or vary behaviors that are part of their normal routine in order to detect the activity with sufficient accuracy (e.g., walking through a particular area of a room to indicate presence) undermines the ecological validity gained by using this method. Furthermore, there is a balance to be struck between collecting data unobtrusively and collecting data without a participant's awareness or consent. Care must be taken to clearly communicate at least a high-level overview of the sensors' data collection capabilities to participants and to be up front about *what* data are being collected; *when* or *how often* they are collected; *how* they will be stored, shared with researchers, and analyzed; *whether data can be excluded or removed* from the study; and *what confidentiality or anonymity protections* will be in place to minimize risk to participants' privacy.

Many sensing technologies will produce very large streams of (typically low-level) **data** over the course of a moderate-length deployment. Although this can be a useful characteristic of this style of research in that it quickly produces a large corpus of data, managing these large data sets can be challenging. Transporting the data across networks can take significant battery power and time and can expose the data to privacy threats. Streams of sensor data may need to be aggressively filtered, aggregated, and timestamp-synchronized before any significant analysis of the output can be carried out.

Sensors also introduce a level of technical complexity to a research project that is not always present when other methodologies are used. These kinds of studies often require a moderate level of technical expertise in order to select appropriate sensors; to configure their recording frequency, fidelity, and output representations; to manage data storage, either locally to the sensor or on a network-connected server; to provide technical support when sensors or recording media inevitably fail; to parse or convert the recorded data into a format suitable for analysis; to protect the data from inadvertent release; and, ultimately, to analyze the results.

How to Do It: What Constitutes Good Work

Several steps are necessary for designing a sensor data stream collection study to ensure that it will effectively answer the desired research question(s). Issues specifically related to data sensor streams include:

- Generating the research questions and planning how to analyze the data streams
- Building, acquiring, or provisioning the sensors
- Determining how frequently and at what level of fidelity to collect data samples
- Installing the sensors
- Storing the data representation
- Making sense of the collected corpus of data

At each of these steps, researchers must make specific decisions about their research design, based on the kinds of research questions that they wish to answer and the limitations inherent in using automated systems to conduct data collection.

Generating the Research Questions and Identifying the Data to Collect

Research based on sensor data streams shares many characteristics with other kinds of in situ empirical studies. Depending on the research questions that the study is designed to answer, which unit of analysis (egocentric, group-centric, or space-centric) is chosen, the recording capabilities of the specific sensor devices used, and the frequency and fidelity at which samples are collected, there may be more or less significant risks to participants' privacy and concerns raised about the intrusiveness of the research.

One major difference between data collection by sensor versus in-person observation or "shadowing" is that with a human observer, regardless of their skill level or subtlety, it is almost always obvious to the study participants when they are being observed and (roughly) how much detail can be perceived about their activities by the researcher. With sensors, particularly ones that are small in size, worn for long periods of time, or "invisibly" embedded into a space or its supporting infrastructure, it is very possible for both consented participants and incidental passers-by to have their actions recorded without their knowledge. Having sensors "disappear" into the background does have some advantages in terms of reducing the chance of collecting data that are distorted or amplified as part of an intentional or unintentional performance for the researchers by the study participants. However, there are also very real ethical issues with capturing data from participants without their knowledge.

As with any other observational study, it is essential to be up front with participants about the scope, duration, frequency, and fidelity of the data that will be captured during the study. It might be advisable to have sample data sets available to share with the participants during solicitation of informed consent, so that participants will have a grounded sense about what the researchers will be able to "see." It may also be advantageous to provide participants some mechanism for revoking their consent to be recorded—for example, a button that will suspend data collection for some period of time or delete the data collected by the sensor for some number of minutes immediately preceding the button press.

Many institutional review boards have guidelines about getting participants' informed consent when automated sensing or recording devices are used for research purposes. Learning what expectations your institution has for carrying out this style of research at the outset of the study design can be an invaluable asset and prevent administrative delays later in the process of data collection and analysis (see also Chapter on Research and Ethics in HCI).

Building, Selecting and Acquiring, or Provisioning Sensors

Selecting the data sensor(s) used to acquire the data can be a function of cost, availability, technological capability, intrusiveness, or methodological needs. Good research requires that researchers either choose a sensor(s) that can reliably and accurately sense the desired phenomena or they must construct (and validate) their own custom sensor for this purpose. Broadly speaking, there are two categories of sensors that can be used in this type of research: sensors that are worn or carried by the participants, and sensors that are deployed in a particular space and record the activities of that space's occupants (Figs. 1 and 2).

The downside in instrumenting a space, as opposed to instrumenting a participant, is that data can only be collected within the instrumented environment. When the research questions require egocentric data collection, wearable sensors might be

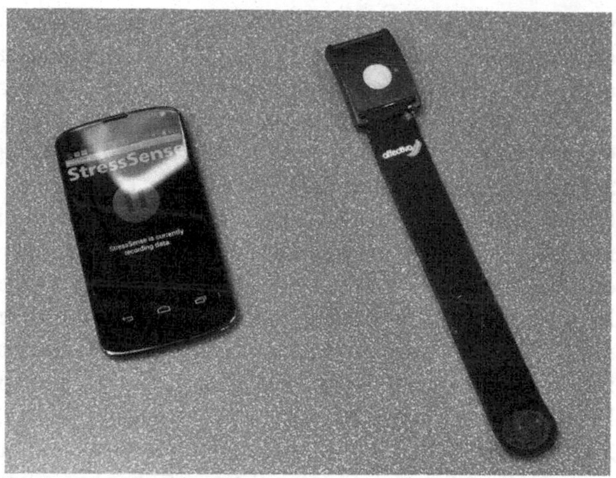

Fig. 1 A pair of sensors that are worn/carried to collect data about a person's stress level throughout the day. The smartphone captures audio of a person's voice to look for indications of increased stress levels; the wristwatch-style sensor is a commercial electrodermal analysis device used to identify episodes of stress and arousal based on variations in a person's skin conductivity (Poh, Swenson, & Picard, 2010)

Fig. 2 Examples of sensors that have been deployed in physical spaces to collect data about the spaces' occupants. (**a**) and (**c**), two different versions of an infrastructure-based system for identifying activities related to water use in a home. (**b**), Sensing suites that collect data about multitasking activity in an office, including interruptions at the door, activity with physical artifacts on a desk, use of the telephone, and presence of a person (and their posture) seated in the office chair

required in order to collect data samples over the full range of activities, behaviors, and interactions that a person experiences throughout a day (e.g., at home, in the car, at the café, and at work), regardless of the other drawbacks to this approach.

Some technologies collect data more effectively in some environments than in others, so another consideration in the selection or design of sensing technologies are the contexts in which data collection is envisioned to take place. For example, GPS can provide good outdoor positioning information in open areas, but becomes much less useful when location is needed in situations with no power, in "urban canyons," indoors, or when outdoor episodes are too short to acquire satellite lock. In these situations, it might be more appropriate to utilize alternate positioning technologies (e.g., Wi-Fi or Bluetooth signal strength triangulation, deployment of infrared or ultrasonic location beacons) in order to capture desired movement or location information (Hightower & Borriello, 2001a; Hightower & Borriello, 2001b). In some cases, context-aware computing middleware exists that can select from the best data source given the sensor hardware limitations and current conditions. Intel's Place Lab is one such platform that was developed for capturing location information (LaMarca et al., 2005); more recently, this kind of location-sensing data fusion functionality has been integrated into mobile computing operating systems, including Google's Android and Apple's iOS.

The degree of intrusiveness of the sensing devices can also have a substantive effect on the success of the study or the comfort level of the participants (Klasnja, Consolvo, Choudhury, Beckwith, & Hightower, 2009). Some physiological sensors (e.g., heart rate monitors, galvanic skin response meters, pupil dilation detectors) can be uncomfortable for participants to wear for extended periods of time, or they may make it more difficult for the participants to carry out their everyday tasks due to physical limitations or social concerns. On the other hand, installation of a few, relatively "invisible" sensors in the participants' home or work environments, such as the kinds of infrastructure-mediated sensing developed by Patel and colleagues (Cohn, Gupta et al., 2010; Cohn, Stuntebeck et al. 2010; Froehlich et al., 2009; Froehlich et al., 2011; Gupta et al., 2010; Patel et al., 2008, Patel et al., 2007) or the pervasive sensors used in the MIT House *n* project (Tapia et al., 2004), can help to improve the ecological validity of the study, as the sensors' presence is less likely to influence or affect the participants' behavior.

Determining How Frequently and at What Level of Fidelity to Collect Data Samples

Once sensors have been acquired, constructed, or provisioned, there are a number of key considerations about how those sensors are configured to collect data during a study, namely how a balance is maintained between the sampling rate of the data streams and storage/bandwidth, processing, and power requirements. These concerns are pushed to the forefront when using sensors incorporated into platforms like mobile phones. Aggressive sensing can:

- quickly exhaust the phones' limited on-board storage;
- lead to excessive data plan usage if the collected data is continuously transmitted to a server via the phone's cellular radios; and
- exhaust the phone's battery rapidly, leading to a loss of potential data and annoyance for an participant who expects to be able to use to phone for other purposes throughout the day.

The question of how much detail is captured by a sensor is bound up in the notions of data collection fidelity. In many cases, a sensor may be able to collect very accurate readings at very fast sampling rates, but this may lead to the collection of more data than is desired or necessary, given a particular research question(s). Usually, it is possible to generalize, abstract, or blur the data to reduce the processing and storage requirements or to protect the privacy of participants while still achieving a study's goals. Protecting participants' privacy may require a more nuanced evaluation, however, as behaviors may be able to be completely reconstructed even from low-fidelity data. Even if data collection with reduced fidelity is utilized, participants may be uncomfortable because they know that the sensor *can* collect higher fidelity data and researchers' assurances that this is not being done may be met with skepticism. This can be the case, for example, if video cameras are deployed as surrogates for motion detectors. These issues may create a negative reaction to being a subject of surveillance (Klasnja et al., 2009) and impact the naturalistic collection of data. Choudhury and colleagues explicitly discuss this tradeoff with regard to their collection of audio data for analysis of speech prosody using their *sociometric badges* (2003), a device worn around the neck to automatically activity—specifically, face-to-face conversations—during the work day.

Furthermore, a single unit of contextual data can have multiple interpretations, and the researchers' choice of which interpretation is modeled can have a significant impact on how participants perceive their privacy to be protected, as well as the kinds of analyses that can be carried out with the data. Location data, for example, can be represented in multiple possible ways (Liao et al., 2006): it could be represented as a latitude/longitude/altitude tuple, it could be represented as a geo-located address (e.g., "Donald Bren Hall"), it could be represented as being inside or outside of a particular municipal locality (e.g., "in the city of Irvine" or "on the UC Irvine campus"), or it could be represented as a semantically defined location (e.g., "at work"). Fine-grained location data can provide valuable insights about a participant's behavior, including their daily routines, paths of travel, or whether or not they cross paths with or are close enough to communicate with other individuals. But these data might also serve to identify an individual participant when analyzed or presented. The data may also reveal details about a person's activities that the participant would rather not—or does not intend to—disclose. Hightower and colleagues provide a good introduction to this problem (Hightower, 2003), and some data about user concerns in this regard has been collected, as well (Patterson et al., 2008).

Generally, however, the problem of interpreting contextual data applies to many different domains, including activity recognition, gesture recognition, etc. One solution that has been developed to navigate these concerns is to vary the fidelity of data collection based on input from other sensors. For example, by only collecting GPS location readings in an egocentric study when accelerometer readings indicate participant motion, the amount of data collected can be reduced with little to no impact and battery life can be extended. Another common solution is to save data on smartphones until a participant is near a Wi-Fi access point at which point the data is transferred to a storage server. This eliminates the need to use expensive and battery draining cellular connections, and provided there is sufficient on board storage to hold data between participants' encounters with Wi-Fi, has little impact on the data collection.

Installing the Sensors

The number of sensors required for coverage of a large space presents an inherent complexity hurdle. Installation and maintenance of (typically) tens of sensors in a home, or hundreds to thousands of sensors in a larger building such as a hotel, hospital, or assisted-living facility, results in high labor costs during installation, and an ongoing sensor network management challenge during routine operation. Furthermore, these sensors will require some type of power source, such as batteries, which create an additional maintenance schedule and costs. It is also difficult to balance the value of in-home sensing and the complexity of the sensing infrastructure. One example that illustrates this is the Digital Family Portrait (Mynatt, Rowan, Craighill, & Jacobs, 2001; Rowan & Mynatt, 2005), which communicates activity information from an elderly person's home to a remote caregiver. In the system's study, movement data was gathered from a collection of strain sensors attached to the underside of the first floor of an elder's home. The installation of these sensors was difficult, time-consuming, and required direct access to the underside of the floor. Though the value of the application was proven, the sensors' complexity limited the number of homes in which the system could be easily deployed. If the value of the proposed sensor deployment is not clear in advance, a Wizard-of-Oz method (Dahlbäck, Jönsson, & Ahrenberg, 1993) can be used to mitigate the deployment risk. This approach tests the outcome of a perfect sensor system by using a researcher to simulate the sensing infrastructure—that is, to visit the field site in person and manually record data of a corresponding type and level of fidelity as would be expected from an automated sensor. If the results are promising, then actual sensor deployments can proceed in earnest.

Storing the Collected Data

One of the more pragmatic concerns about conducting a study using streams of data are decisions about how—and where—the collected streams are stored between the time at which they are collected and when they are aggregated and analyzed by the research team. The options available to researchers here are often directly impacted by the decisions made about how frequently and at what level of fidelity data are collected.

The two most common options for capturing a data stream are (1) to record the sequence of observed events on a device connected to (or nearby) the sensor data source and to collect a copy of the stream at various points of time during and/or at the conclusion of the study or (2) to continually transmit collected data back to a central server via a network connection. Both of these options have pros and cons that must be considered. Although creating a local cache of study data is often more technically straightforward, this approach can be problematic when large volumes of data (e.g., video or audio) need to be stored, and relying on multiple storage media to capture study data can increase the complexity of data collection. Transmitting data to a research fileserver that is connected to the Internet can make it easier for researchers to monitor the volume of data in the stream, dynamically add storage capacity, and maintain continuous backups of the collected data. However, this approach requires that each of the sensors be able to connect to the network and necessitates additional technical support to ensure that gaps in connectivity do not result in skipped data and that participants' data are reasonably secured while in transit. This technique also requires that researchers consider whether/when data might be aggregated, filtered, or pruned to protect participants' privacy, to weigh demands on sensors' batteries, and to think about whether data transmission costs might come into play (e.g., if large volumes of data need to be uploaded over a cellular data connection).

Given the falling cost and rising capacity of digital storage, as well as the increased processing capabilities of tiny mobile computing devices, storing as much data as possible at the point of sensing (e.g., on a smartphone) would, in many cases, be beneficial to researchers. However, this is not the only option. Especially when using a large number or wide variety of sensor data sources, aggregating data from multiple sources into a single "stream" at the time of collection can dramatically simplify the subsequent analysis. Creating a single event stream eliminates (or at least minimizes) the overhead involved in temporally synchronizing observations from different sources. (Synchronization can also be accomplished by planting various markers in the data, similar to the way that a movie production clapboard leaves a visual trace in the film and an audible sound in the audio recording. However, these independent data streams still need to be aligned using these markers before they can be manipulated; this process can be significantly burdensome when many streams need to be synchronized.)

Making Sense of the Collected Corpus of Data

Researchers can conduct various post hoc analyses of data streams using standard statistical software packages. A common example would be comparing the frequency or distribution of sensed events under different conditions, revealing correlations between these conditions and the observed behaviors. However, due to the volume of sensor data that are collected by automated systems and the noise inherent in sensor data sources, it is often more valuable to use machine-learning techniques to classify sensor data into higher-level representations of participants' behaviors or to perform hypothesis testing. Examples of such classification include identifying when particular events occurred, classifying the mood of participants at any given time, or figuring out where a participant is. Some of the machine-learning techniques that are valuable for making sense of sensor data streams include data filtering and smoothing [e.g., particle or Kalman filtering (Krumm, 2010)], activity detection (Philipose et al., 2004), and sensor fusion [e.g., using Hidden Markov Models (Patterson et al., 2005), naïve or dynamic Bayesian Networks (Fox, Hightower, Kautz, Liao, & Patterson, 2003), or time series data analysis (Liao, Patterson, Fox, & Kautz, 2007)].

Oftentimes, the first step in analyzing the data collected during a sensor data stream study is to classify the data stream into segments or to identify particular events in the data (a special case of classification). This would be true of activity recognition and gesture recognition studies, for example. These classifications, in turn, may become the focus of analysis by conducting studies of their frequency or purpose. In the case of context-aware computing, they may serve as input to another system component, for example, presenting search results or map directions filtered by the user's location. In either case, it is important to carefully evaluate the classification algorithm based on machine learning techniques before advocating its broader use.

User Modeling and Event Detection

The first step in making sense of streams of sensor data is to clearly identify which data will serve as inputs to the classifier. This is equivalent to identifying the dependent variables or the "features" of a machine-learning task. Then, it is necessary to establish the classification categories or, alternatively, the metric(s) that is going to be detected; these become the independent variables or output of the classifier. A classification task would be one in which a reading of dependent variables is mapped to one of *N* categories (e.g., mapping accelerometer readings on a smartphone to one of {sitting, standing, lying down}). A continuous variant of classification, regression, would map the reading to a real-valued metric (e.g., mapping a skin galvanic response sensor and a heart rate monitor to a number from −1.0 to 1.0, representing emotional valence).

If the researcher has access to several (and, hopefully, many) examples of validated, true mappings from dependent variables to the independent variable—sometimes called a "gold standard"—then it is possible to use supervised machine learning to construct a classifier. In this case, the mapped values form a training set from which a classifier can be automatically trained. If the researcher does not have any examples from which to train a classifier, unsupervised machine learning, or clustering, is an option.

In both cases, it is important that the researcher consider how the training data generalizes, and if that is the right generalization for the study. For example, a gesture classifier that is trained on one person's gold standard data should do a good job identifying gestures in un-annotated data from the same person. However, because the training data only came from one person, it may not work well when applied to a different person's unannotated data. This would be true if gesture performance varied greatly among people even for the same gesture. Another example of when generalization might fail is if accelerometer data collected from one type of hardware is used to train a classifier. That classifier may be ineffective if it is given unannotated data from different hardware as an input. This would be true if due to calibration, sensitivity or something else, different devices produced different readings for the same phenomenon. A closely related research concern is to be sure that the classifier is being trained on data that is generated by the correct phenomena; although this may seem obvious, this principle can be violated in subtle ways. For example, if a person-tracking application is trained to classify motion detector activation as being indicative of a person's presence in a particular room, this classifier will not result in correct behavior if the training data contains examples of motion detector activation that was caused, instead, by the family pet.

Unfortunately, gold-standard data is often available in limited quantities; it is time-consuming and can be expensive to capture and accurately label sensor data streams, especially across a diverse or distributed population. Nonetheless, the goal of user modeling is to create an algorithm that can both be shown to classify the available gold-standard data correctly and that will work just as well with real-world sensor data. One problem that can arise in user modeling is known as "overfitting." That is, it is possible to create or tune a classifier that works very well on the gold-standard data, but will not work for as-yet-unseen data. Overfitting is typically addressed through a process of cross-validation, described below.

Validating User Models

Cross-validation refers to the process of evaluating a classifier by dividing a single set of gold-standard data into a number of subsets and then using different permutations of these subsets as training and testing data across a set of validation experiments. Each permutation is called a "fold" and corresponds to a single sample experiment; multiple experiments are carried out, with the number of experiments matching the number of folds used. The folds are often created to exhaustively cover all available annotated data—testing each data point one time as part of the process and using it as training data the rest of the time. An example would be tenfold cross-validation. In tenfold cross-validation, the gold-standard data is divided into ten subsets of equal size. A total of ten experiments are run, such that in any one experiment, nine of the subsets are used to train the classifier. Once the training phase is complete, the tenth, "held-out" subset is classified, but with the preexisting annotations on this part of the data hidden from the classifier. Finally, the annotations that are computed by the classifier for the training subset are compared to the actual, hidden annotations, providing a way to assess the degree to which the classifier was correct. Typically, the ten experiments will produce accuracy metrics, which can then be statistically analyzed for mean, variance, etc.

Generally speaking, it is possible to choose any number of folds from two through the number of data samples that have been collected. Broadly speaking, this is referred to as "*n*-fold" cross-validation. Regardless of the value of *n* that is chosen, each data sample is only tested (classified) once. A special case is "Leave One Out Cross Validation" (LOOCV). This is when *n* is chosen to be equal to the number of data samples, *N*. This requires the most computational power, as the classifier must be learned *N* times, and this process of training the classifier is typically far more computationally intensive than testing. Many software frameworks exist to automate the process of creating and testing using cross-fold validation, including the popular WEKA machine learning toolkit (Hall et al., 2009).

When using *n*-fold cross-validation to train and test classifier, it is important that no information "leaks" from the training set to the testing set during cross-validation; that is, that samples assigned to the training set are selected to be as independent as possible from the samples used for training. Even information leaks that appear to be very minor can create impressively accurate classifiers, which would completely fail during a real application of the same technique. How might such a subtle leak occur during experimentation? One example would be if sensor readings (e.g., temperature readings from an instrumented apartment space) were collected and stored as a difference from a running average of all previously collected data points. In this case, independence is broken; when the gold standard is separated into *n* folds, any given testing fold will be related to the training fold through the common mean implicitly used to encode the data. In this way, the training data "knows" something about the data to be tested. One way to solve this problem is to carefully consider the representations that are to be used for stored sensor data and to separate raw data samples into folds for cross-validation *before* calculating any aggregate statistics across the fold or subsets of the fold.

Some special cases require separating the data in unique ways in order to enforce the right kinds of independence. For example, in a gesture recognition problem, the most appropriate way to create folds might be one in which samples collected from different individuals are placed in their own fold. While this may create folds that vary in the number of samples that they contain, it will ensure that the classifier is learning gestures that will generalize to unseen *people* instead of to unseen samples from known people. This would help to support a claim that this technique would ultimately work when applied to a person for whom training has not been individually conducted. Another special case relates to the temporal nature of data streams: a sample collected from a sensor at a particular point in time is often explicitly or implicitly related to the samples collected just before or after. As a result, separating data into folds where the testing samples have a consistent temporal relationship to the samples in one or more folds might produce abnormally good results. Perhaps, in this case, creating folds based on day is more appropriate, or perhaps creating training sets of the last hour's worth of data applied to the next data sample would be more appropriate. Analysis tools like WEKA offer a variety of capabilities for partitioning a data set into folds based on various criteria and can be particularly helpful when taking advantage of these techniques.

The nature of cross-validation assumes that all data is available for use in an offline context in which it can be repeatedly accessed in different sequences as required. A common application of this approach is to use extensive cross-validation testing to create one classifier whose parameters are fixed and then to apply it in an online setting in a testing-only configuration. Such a classifier might then be used for a context-aware application, for example. The machine-learning literature includes extensive variations to the above approaches, which relax, in various ways, nearly every possible dimension of the described approach to achieve different effects, maintain sample independence, or account for different kinds of input data. (A good introduction to issues related to creating classifiers in the machine-learning style appears in Langley, 2000 and Domingos, 2012).

The metrics for evaluating classifiers are also influenced by the goals of the research and the questions that the researchers are interested in answering. Accuracy in matching the gold standard may seem to be a natural evaluation, but, in fact, this may be inappropriate given the ultimate use of the system. For example, should a researcher care about whether a classifier can accurately assess whether a doctor has entered an exam room or not? Such a classifier may identify such an event correctly over 99 % of the time, but if the goal is to figure out when a doctor doesn't see a patient, then false positives may be a more important metric. Similarly, it may not be important to detect the exact *moment* when you, for example, walk outside, get angry, speak to your child or eat something, provided that you *do* detect each of those events within some reasonable timeframe; an excellent discussion of evaluation metrics was written by Ward, Lukowicz, and Gellersen (2011). Finally, with regard to metrics, it is important to consider the statistical significance of the findings in addition to the accuracy of the predictions. These significance values help researchers to understand whether an algorithmic improvement on a small data set is as important as a small improvement on a big data set, and this is an often-overlooked component of classifier-based data analysis (see also Demšar, 2006).

What to Report in a Study Using Sensors

Although collecting streams of data from participants can be a powerful technique for observing existing behaviors (or in situ adaptations to novel technologies), incorporating sensor data into an experimental protocol requires careful planning. Additionally, reporting study findings based on streams of sensor data necessitates careful quantitative analysis that takes into account noisy sensor sources, ambiguities in sensed data, or technical problems with the sensors, which can lead to missing, corrupt, or misleading data. A well-described study will report the experiment in sufficient detail that another researcher would be able to reproduce the study and achieve the same results. This would include describing:

- Hardware: What type and quantity of hardware was utilized? What mode was the hardware placed in? In what ways were the hardware components configured? What mode were they utilized in?
- Experimental setup: Where were the sensors placed? Why were those locations chosen? Who installed the sensors? What did the environment look like with the hardware installed? Did the setup change? At what point? How did this impact the participants? How did the experiment end? What was the final disposition of the hardware?
- Participant knowledge: Exactly what did the participants know about the hardware and data collection? How were they told? Were they asked to do anything different from—or in addition to—their normal activities during the study? Did participants' interactions with the study hardware change during the course of the study? How were participants compensated?
- Experimental execution: During the course of the study, what transpired with respect to the experimental setup? Were they outages? How much and what type of maintenance was conducted?
- Software infrastructure: How was the data collected? Was there specialized software used in collection? What configuration was the software put into? How was data transmitted, protected, and integrity assured?
- Analysis: What analysis was conducted on the data? At what point in the running of the experiment was the analysis conducted? What was an example of the data that was collected? Were there any changes made during the course of the analysis? Was particular software used to conduct analysis? What were the parameters of the algorithms used? How accurate was the analysis based on the gold-standard data? What other metrics were evaluated to justify the approach used? Was the analysis statistically significant?
- Why were these choices, and not others, made?

Personal Story About How the Authors Got into Using this Method, What Attracted Them to It, What Set of Methods They Use with It

In a study of the effect that e-mail has on both individual multitasking and information flow within small groups in an information work site (Mark et al., 2012), Voida and colleagues used a combination of:

- semi-structured interviews;
- surveys;
- in-person "shadowing";
- logging of electronic activities (i.e., desktop window switches); and
- environmental, social, and physiological sensors

to collect a large corpus of data about the pace of work, pace of interpersonal communication, and communication channels used in the workplace. We compared behavior under typical circumstances to when targeted individuals within the organization were asked to shut off their e-mail for 1 week. (During the e-mail "holidays" that we instituted, our participants' managers and coworkers were told about our study, and informed that the participant was in the office and working but could not be reached via e-mail.)

Collecting streams of sensor data was particularly important in the course of this research for three reasons. First, we wanted to collect as many different kinds of data about these individuals' work practices during the study as possible. We were able to examine many different facets of work and to observe the variations in the ways that these work practices were carried out by comparing the differences among repeated measurements using the same set of sensors. Second, one of the more interesting—and less well-understood—measures that we could collect about workplace collaboration and multitasking was the level of stress experienced by our participants throughout the workday. To understand changes in stress levels, we used commercial heart rate monitors to collect readings about heart rate hundreds of times each day. Finally, the field site for this study happened to be located nearly 3,000 miles from our home institution. In order to collect the volume of data that we needed to draw conclusions about work practices in light of the per-day and per-participant variation in information work activities, we set up the sensors once and then used them to collect data continuously from each participant for multiple weeks. This approach substantially minimized the cost—both in terms of travel and time on-site—for running the study.

Based on the data streams that we collected, we were able to quantify the amount of time that our participants spent each day conducting various kinds of activities (based on a "virtual sensor" observing window changes on the desktop computer), examine the composition and strength of social ties in the workplace (based on face-to-face conversations sensed using sociometric badges), and measure changes in stress levels when e-mail was present and when it was not (based on readings provided by a commercial heart rate monitor). In order to make sense of these streams of data, we conducted manual statistical analyses of the aggregated event streams after data collection was complete. Generally speaking, we looked at daily averages of these events, using ANOVAs, *t*-tests, and parametric statistics to determine whether our e-mail availability intervention, individual differences, or other factors were more likely to explain any variation that we observed in event frequencies or durations during different phases of the study. Even with a relatively small group of fully instrumented participants ($n = 13$), the environmental, social, and physiological sensors collected millions of data points over the course of the study.

Another key aspect of our study was that we used a combination of log data (window changes recorded on each participant's primary desktop computer) and in-person observations to help label the sensor data. These discrete (and less ambiguous) data points enabled us to triangulate events of interest and helped us to make sense of a very large, complex, and messy corpus of real-world sensor readings.

In ongoing work, we are investigating the degree to which the log and sensor data that we collected can be used to train a classifier to predict the multitasking observations that we were able to collect from in-person observation—the traditional "gold standard" for this kind of research. The outcomes of this research will help to clarify which data streams are most useful for understanding information work as well as the circumstances in which sensors can (and cannot) serve as a substitute for in-person, ethnographically inspired fieldwork.

In another study, Patel and colleagues wanted to conduct an in-depth, empirical investigation of the proximity of the mobile phone to its owner over several weeks of continual observation (Patel, Kientz, Hayes, Bhat, & Abowd, 2006). The overall aims of this study were to determine if a mobile phone is a suitable proxy for its owner (an assumption that had never been empirically studied in the literature), to understand the reasons behind separation between a user and his or her mobile phone, and to offer guidelines for building applications. This study relied on a mixed-methods approach that required collecting data from some sort of proximity sensing technology. The sensors used were small, custom-built, battery-powered Bluetooth tags that the user would wear throughout the day. These tags would transmit a beacon signal continuously and the phone would record its distance to each tag based on the amount of time that it took the radio signal to propagate. This approach allowed us to continuously record the user's distance from his or her phone and to gather quantitative data not otherwise possible with other investigational means. Because the experience sampling method (ESM) or self-report would have created artificial changes to the user's behavior with their phone (i.e., picking it up to answer when the very thing that we were measuring was proximity to the phone), automatic sensing was a necessity. Additionally, the quantitative data allowed us to explore whether it was possible to apply machine-learning techniques to predict proximity. In the end, since the user's behavior was modified only by wearing a Bluetooth tag, and because no qualitative evidence to the contrary emerged in follow-up interviews, it was reasonable to argue that there was little modification to the user's natural behavior with respect to the proximity of the phone during the investigation. The resulting quantitative proximity traces proved valuable during the mixed-method interview process and the final analysis.

Because the logging application resided on the user's phone, the researchers took care to design it so that it would not impact the user's normal phone use. We also created a tool that would produce daily visualizations of people's proximity to their phone at 1-min intervals. Because the sensing was not perfect (e.g., it could be impacted by clothing, multiple people between the tag and the phone, and noisy RF environments), we used the visualization during our interviews with the participants as a guide for them to think through the day, an example of a retrospective analysis technique (see Looking Back: Retrospective Study Methods for HCI). They were allowed to bring their calendars or schedules to help them remember reasons for being separated from the phone. It turned out that the sensing did not have to be perfect. Even high-level activity information captured by the sensors was sufficient to help participants bridge the gaps in their memory, reconstruct their day, and articulate how they were using the phone. In many cases, we found that it was not until participants actually saw their activity data that they recalled the details. Combining automatically collected sensor data and prompted interviews that were grounded in the sensor data served as a powerful tool for studying detailed behavior and the reasons behind it. It provided a balance between quantitative and qualitative results that might be difficult to achieve with a single method. A few years later another research group replicated this study on smartphone proximity with similar success (Dey et al., 2011).

In a final study, Patterson et al. used a variety of sensors available in a laptop in order to measure, and, subsequently, to predict, when users were in a particular, semantically defined place (Patterson et al., 2009). This study required users to install a piece of software on their laptops that periodically asked them to provide information about where they were to establish an annotated training set, in the style of the ESM (Larson & Csikszentmihalyi, 1983). Since place is often a key component of personal context, this information was used to set custom Instant Messaging status messages, populated with the user's choice of place names. The focus of the work was on collecting accurate place data information along with sensor data so that, eventually, well-informed context-aware services could be developed. While the software likely impacted the participants' use of their laptop, it was unlikely to have affected their mobility, which was a key focus of the study. This study required a lot of engineering work on the software that was installed on participants' computers in order to support a wide variety of software and hardware configurations.

Conclusion and Further Reading

Sensor data streams can provide large quantities of fine-grained data for observational studies and can illuminate behaviors or phenomena that would otherwise be impossible to study at a comparable level of detail—if at all. Preparing a research study based on this technique requires careful planning to ensure that the correct hardware is used, configured, and managed. Additionally, some degree of technical sophistication is required to store, process, and analyze the data at scale and with appropriate methodological and statistical rigor. Finally, participants need to be appropriately informed about the scope and nature of the data collection, since sensors can be difficult to see (or be forgotten over the course of a long study) and since the data collected by sensors can introduce privacy-related risks. When these issues are managed thoughtfully, however, sensor data streams can form the basis for rigorous and replicable research, and they can be of great value when used as a complement to other types of data-collection methodologies.

For further reading about these topics, consider looking into the following publications, which are also emphasized with a bold typeface in the chapter reference listing. Although we have found these articles to be useful for the reasons enumerated below, they are also informative in their motivations for using sensor data streams, their methodological and analytical approaches, and the ways in which they use this research technique to draw conclusions for the research community:

- For more information about machine learning, classification and event-detection methodology, refer to Langley (2000) and Domingos (2012). For good examples of this technique being applied to answer specific research questions, see Patterson et al. (2005), Liao et al. (2007), and Horvitz et al. (2004).
- For more information on data processing and filtering, see Krumm (2010).
- For more information on managing participant privacy, see Langheinrich (2010) and Klasnja et al. (2009).
- A good example of egocentric research can be found in Fogarty et al. (2005).
- A good example of group-centric research is Choudhury and Pentland's paper on the *sociometer* (2003).
- A good example of infrastructure-centric research appears in Cohn, Stuntebeck et al. (2010).
- A good example of sensor triangulation can be found in Mark et al. (2012).

Exercises

1. What gold standards of behavior could you use to compare your interpretation of sensor streams against? Do you have to have a gold standard?
2. Compare the kinds of behaviors you can sense with a wearable sensor vs. a fixed sensor (e.g., on a wall)? What situations are appropriate for each?

References

Abowd, G. D., Atkeson, C. G., Hong, J., Long, S., Kooper, R., & Pinkerton, M. (1997, October). CyberGuide: A mobile context-aware tour guide. *Wireless Networks, 3*(5), 421–433.

Bailey, B. P., & Iqbal, S. T. (2008). Understanding changes in mental workload during execution of goal-directed tasks and its application for interruption management. *ACM Transactions on Computer-Human Interaction, 14*(4), 1–28.

Bailey, B. P., & Konstan, J. A. (2006). On the need for attention-aware systems: Measuring effects of interruption on task performance, error rate, and affective state. *Computers in Human Behavior, 22*(4), 685–708.

Bailey, B. P., Konstan, J. A., & Carlis, J. V. (2001). The effects of interruptions on task performance, annoyance, and anxiety in the user interface. In *Proceedings of the IFIP TC.13 international conference on human-computer interaction (INTERACT '01)* (pp. 593–601). Amsterdam: Ios Press.

Baldauf, M., Dustdar, S., & Rosenberg, F. (2007). A survey on context-aware systems. *International Journal of Ad Hoc and Ubiquitous Computing, 2*(4), 263–277.

Bao, L., & Intille, S. (2004). Activity recognition from user-annotated acceleration data. In *Proceedings of the second international conference on pervasive computing (PERVASIVE 2004)* (pp. 1–17). Berlin: Springer.

Begole, J., & Tang, J. C. (2007). Incorporating human and machine interpretation of unavailability and rhythm awareness into the design of collaborative applications. *Human–Computer Interaction, 22*(1), 7–45.

Begole, J., Tang, J. C., Smith, R. B., & Yankelovich, N. (2002). Work rhythms: Analyzing visualizations of awareness histories of distributed groups. In *Proceedings of the 2002 ACM conference on computer supported cooperative work (CSCW '02)* (pp. 334–343). New Orleans, LA: ACM Press.

Bolchini, C., Curino, C., Quintarelli, E., Schreiber, F. A., & Tanca, L. (2007). A data-oriented survey of context models. *SIGMOD Record, 36*(4), 19–26.

Brdiczka, O., Su, N. M., & Begole, J. B. (2010). Temporal task footprinting: Identifying routine tasks by their temporal patterns. In *Proceedings of the 14th international conference on intelligent user interfaces (IUI '10)* (pp. 281–284). New York, NY: ACM Press.

Brumitt, B., Meyers, B., Krumm, J., Kern, A., & Shafer, S. (2000). EasyLiving: Technologies for intelligent environments. In *Proceedings of the second international symposium on handheld and ubiquitous computing (HUC '00)* (pp. 97–119). Berlin: Springer.

Choudhury, T., Consolvo, S., Harrison, B., Hightower, J., LaMarca, A., LeGrand, L., et al. (2008). The mobile sensing platform: An embedded activity recognition system. *IEEE Pervasive Computing, 7*(2), 32–41.

Choudhury, T., & Pentland, A. (2003). Sensing and modeling human networks using the sociometer. In *Proceedings of the seventh IEEE international symposium on wearable computers (ISWC '03)* (pp. 216–222). Los Alamitos, CA: IEEE Computer Society.

Cohn, G., Gupta, S., Froehlich, J., Larson, E., & Patel, S. N. (2010). GasSense: Appliance-level, single-point sensing of gas activity in the home. In *Proceedings of 8th international conference on pervasive computing (PERVASIVE 2010)* (pp. 265–282). Berlin: Springer.

Cohn, G., Stuntebeck, E., Pandey, J., Otis, B., Abowd, G. D., & Patel, S. N. (2010). SNUPI: Sensor nodes utilizing powerline infrastructure. In *Proceedings of the 12th international conference on ubiquitous computing (UbiComp 2010)* (pp. 159–168). New York, NY: ACM Press.

Consolvo, S., Klasnja, P., McDonald, D. W., Avrahami, D., Froehlich, J., LeGrand, L., et al. (2008). Flowers or a robot army?: Encouraging awareness & activity with personal, mobile displays. In *Proceedings of the 10th international conference on ubiquitous computing (UbiComp '08)* (pp. 54–63). New York, NY: ACM Press.

Consolvo, S., McDonald, D. W., Toscos, T., Chen, M. Y., Froehlich, J., Harrison, B., et al. (2008). Activity sensing in the wild: A field trial of ubifit garden. In *Proceedings of the SIGCHI conference on human factors in computing systems (CHI '08)* (pp. 1797–1806). New York, NY: ACM Press.

Consolvo, S., & Walker, M. (2003). Using the experience sampling method to evaluate ubicomp applications. *IEEE Pervasive Computing, 2*(2), 24–31.

Dahlbäck, N., Jönsson, A., & Ahrenberg, L. (1993). Wizard of Oz studies: Why and how. In *Proceedings of the 1st international conference on intelligent user interfaces (IUI '93)* (pp. 193–200). New York, NY: ACM Press.

Demšar, J. (2006). Statistical comparisons of classifiers over multiple data sets. *The Journal of Machine Learning Research, 7*, 1–30.

Dey, A. K., & Abowd, G. D. (2001). A conceptual framework and a toolkit for supporting rapid prototyping of context-aware applications. *Human-Computer Interaction, 16*(2–4), 7–166.

Dey, A., Wac, K., Ferreira, D., Tassini, K., Hong, J., & Ramos, J. (2011). Getting closer: An empirical investigation of the proximity of user to their smart phones. In *Proceedings of the 13th international conference on ubiquitous computing (UbiComp '11)* (pp. 163–172). New York, NY: ACM Press.

Domingos, P. (2012, October). A few useful things to know about machine learning. *Communications of the ACM, 55*(10), 78–87.

Fan, M., Gravem, D., Cooper, D. M., & Patterson, D. J. (2012). Augmenting gesture recognition with erlang-cox models to identify neurological disorders in premature babies. In *Proceedings of the 2012 ACM conference on ubiquitous computing (UbiComp '12)* (pp. 411–420). New York, NY: ACM Press.

Fogarty, J., Hudson, S. E., Atkeson, C. G., Avrahami, D., Forlizzi, J., Kiesler, S., et al. (2005). Predicting human interruptibility with sensors. *ACM Transactions on Computer-Human Interaction, 12*(1), 119–146.

Fox, D., Hightower, J., Kautz, H., Liao, L., & Patterson, D. J. (2003). Bayesian techniques for location estimation. In M. Hazas, J. Scott, & J. Krumm (Eds.), Research paper presented at the *2003 Workshop on Location-Aware Computing*, held in conjunction with the *Fifth International Conference on Ubiquitous Computing (UbiComp 2003)*, Seattle, WA, USA, October 12, 2003.

Froehlich, J. E., Larson, E., Campbell, T., Haggerty, C., Fogarty, J., & Patel, S. N. (2009). HydroSense: Infrastructure-mediated single-point sensing of whole-home water activity. In *Proceedings of the 11th international conference on ubiquitous computing (Ubicomp 2009)* (pp. 235–244). New York, NY: ACM Press.

Froehlich, J., Larson, E., Saba, E., Campbell, T., Atlas, L., Fogarty, J., et al. (2011). A longitudinal study of pressure sensing to infer real-world water usage events in the home. In *Proceedings of the 9th international conference on pervasive computing (PERVASIVE 2011)* (pp. 50–69). Berlin: Springer.

Greenberg, S., & Fitchett, C. (2001). Phidgets: Each development of physical interfaces through physical widgets. In *Proceedings of the 14th annual ACM symposium on user interface software and technology (UIST 2001)* (pp. 209–218). New York, NY: ACM Press.

Gupta, S., Reynolds, M. S., & Patel, S. N. (2010). ElectriSense: Single-point sensing using EMI for electrical event detection and classification in the home. In *Proceedings of the 12th international conference on ubiquitous computing (UbiComp 2010)* (pp. 139–148). New York, NY: ACM Press.

Hall, M., Frank, E., Holmes, G., Pfahringer, B., Reutemann, P., & Witten, I. H. (2009). The WEKA data mining software: An update. *SIGKDD Explorations, 11*(1), 10–18.

Hightower, J., (2003). From position to place. Research paper presented at the 2003 Workshop on Location-Aware Computing, held in conjunction with the *Fifth International Conference on Ubiquitous Computing (UbiComp 2003)*, Seattle, WA, USA, October 12, 2003.

Hightower, J. & Borriello, G. (2001). *A survey and taxonomy of location systems for ubiquitous computing*. Technical Report UW-CSE-01-08-03, Intel Research Seattle and the University of Washington, Seattle, WA.

Hightower, J. & Borriello, G. (2001, August). Location systems for ubiquitous computing. *IEEE Computer, 34*(8), 57–66.

Hong, J.-H., Suh, E., & Kim, S.-J. (2009). Context-aware systems: A literature review and classification. *Expert Systems with Applications, 36*(4), 8509–8522.

Hornecker, E., & Nicol, E. (2012). What do lab-based user studies tell us about in-the-wild behavior? Insights from a study of museum interactives. In *Proceedings of the ACM conference on designing interactive systems (DIS '12)* (pp. 358–367). New York, NY: ACM Press.

Horvitz, E., Koch, P., & Apacible, J. (2004). BusyBody: Creating and fielding personalized models of the cost of interruption. In *Proceedings of the ACM conference on computer supported cooperative work (CSCW '04)* (pp. 507–510). New York, NY: ACM Press.

Hutchings, D. R., Smith, G., Meyers, B., Czerwinski, M., & Robertson, G. (2004). Display space usage and window management operation comparisons between single monitor and multiple monitor users. In *Proceedings of the working conference on advanced visual interfaces (AVI '04)* (pp. 32–39). New York, NY: ACM Press.

Hutchinson, H., Mackay, W., Westerlund, B., Bederson, B. B., Druin, A., Plaisant, C., et al. (2003). Technology probes: Inspiring design for and with families. In *Proceedings of the SIGCHI conference on human factors in computing systems (CHI '03)* (pp. 17–24). New York, NY: ACM Press.

Intille, S. S. (2002). Designing a home of the future. *IEEE Pervasive Computing, 1*(2), 76–82.

Intille, S., Larson, K., Tapia, E., Beaudin, J., Kaushik, P., Nawyn, J., et al. (2006). Using a live-in laboratory for ubiquitous computing research. In *Proceedings of the 4th international conference on pervasive computing (PERVASIVE 2006)* (pp. 349–365). Berlin: Springer.

Intille, S. S., Rondoni, J., Kukla, C., Ancona, I., & Bao, L. (2003). A context-aware experience sampling tool. In *Extended abstracts of the SIGCHI conference on human factors in computing systems (CHI '03)* (pp. 972–973). New York, NY: ACM Press.

Isaacman, S., Becker, R., Cáceres, R., Martonosi, M., Rowland, J., Varshavsky, A., et al. (2012). Human mobility modeling at metropolitan scales. In *Proceedings of the 10th international conference on mobile systems, applications, and services (MobiSys '12)* (pp. 239–252). New York, NY: ACM Press.

Kaptelinin, V. (2003). UMEA: Translating interaction histories into project contexts. In *Proceedings of the SIGCHI conference on human factors in computing systems (CHI '03)* (pp. 353–360). New York, NY: ACM Press.

Kidd, C., Orr, R., Abowd, G., Atkeson, C., Essa, I., MacIntyre, B., et al. (1999). The aware home: A living laboratory for ubiquitous computing research. In *Proceedings of the second international workshop on cooperative buildings (CoBuild '99)* (pp. 191–198). Berlin: Springer.

Kientz, J. A., Patel, S. N., Jones, B., Price, E., Mynatt, E. D., & Abowd, G. D. (2008). The Georgia Tech aware home. In *Extended abstracts of the SIGCHI conference on human factors in computing systems (CHI 2008)* (pp. 3675–3680). New York, NY: ACM Press.

Klasnja, P., Consolvo, S., Choudhury, T., Beckwith, R., & Hightower, J. (2009). Exploring privacy concerns about personal sensing. In *Proceedings of the 7th international conference on pervasive computing (PERVASIVE 2009)* (pp. 176–183). Berlin: Springer.

Krumm, J. (2010). Processing sequential sensor data. In J. Krumm (Ed.), *Ubiquitous computing fundamentals* (pp. 286–319). Boca Raton, FL: CRC Press.

LaMarca, A., Chawathe, Y., Consolvo, S., Hightower, J., Smith, I., Scott, J., et al. (2005). Place Lab: Device positioning using radio beacons in the wild. In *Proceedings of the third international conference on pervasive computing (PERVASIVE 2005)* (pp. 301–306). Berlin: Springer.

Langheinrich, M. (2010). Privacy in ubiquitous. In J. Krumm (Ed.), *Ubiquitous computing fundamentals* (pp. 286–319). Boca Raton, FL: CRC Press.

Langley, P. (2000). Crafting papers on machine learning. In *Proceedings of the seventeenth international conference on machine learning (ML '00)* (pp. 1207–1211). Stanford, CA: Morgan Kaufmann.

Larson, R., & Csikszentmihalyi, M. (1983). The experience sampling method. *New Directions for Methodology of Social and Behavioral Science, 15*, 41–56.

Liao, L., Patterson, D. J., Fox, D., & Kautz, H. (2006, December). Building personal maps from GPS data. *Annals of the New York Academy of Sciences, 1093*, 249–265.

Liao, L., Patterson, D. J., Fox, D., & Kautz, H. (2007, January). Learning and inferring transportation routines. *Artificial Intelligence, 171*, 311–331.

MacIntyre, B., Mynatt, E. D., Voida, S., Hansen, K. M., Tullio, J., & Corso, G. M. (2001). Support for multitasking and background awareness using interactive peripheral displays. In *Proceedings of the 14th annual ACM symposium on user interface software and technology (UIST '01)* (pp. 41–50). New York, NY: ACM Press.

Mark, G. J., Voida, S., & Cardello, A. V. (2012). "A pace not dictated by electrons": An empirical study of work without email. In *Proceedings of the SIGCHI conference on human factors in computing systems (CHI 2012)* (pp. 555–564). New York, NY: ACM Press.

Marmasse, N., & Schmandt, C. (2000). Location-aware information delivery with ComMotion. In *Proceedings of the 2nd international symposium on handheld and ubiquitous computing (HUC '00)* (pp. 157–171). London: Springer.

Miluzzo, E., Lane, N. D., Fodor, K., Peterson, R., Lu, H., Musolesi, M., et al. (2008). Sensing meets mobile social networks: The design, implementation and evaluation of the CenseMe application. In *Proceedings of the 6th ACM conference on embedded network sensor systems (SenSys '08)* (pp. 337–350). New York, NY: ACM Press.

Monibi, M., & Patterson, D. (2009). Getting places: Collaborative predictions from status. In *Proceedings of the European conference on ambient intelligence (AmI 2009)* (pp. 60–65). Berlin: Springer.

Mynatt, E. D., Rowan, J., Craighill, S., & Jacobs, A. (2001). Digital family portraits: Supporting peace of mind for extended family members. In *Proceedings of the SIGCHI conference on human factors in computing systems (CHI 2001)* (pp. 333–340). New York, NY: ACM Press.

Nair, R., Voida, S., & Mynatt, E. D. (2005). Frequency-based detection of task switches. In *Proceedings of the 19th annual conference of the British HCI group (HCI 2005)* (pp. 94–99). Berlin: Springer.

Olguín, D. O., & Pentland, A. (2008). Social sensors for automatic data collection. In *Proceedings of the 14th annual Americas conference on information systems (AMCIS 2008)* (p. Paper 171). Atlanta, GA: Association for Information Systems.

Olguín, D. O., Waber, B. N., Kim, T., Mohan, A., Ara, K., & Pentland, A. (2009). Sensible organizations: Technology and methodology for automatically measuring organizational behavior. *IEEE Trans. On Systems, Man, and Cybernetics – Part B: Cybernetics, 39*(1), 12 pages.

Orr, R. J., & Abowd, G. D. (1999). The smart floor: A mechanism for natural user identification and tracking. In *Extended abstracts of the SIGCHI conference on human factors in computing systems (CHI 1999)* (pp. 275–276). New York, NY: ACM Press.

Patel, S. N., Kientz, J. A., Hayes, G. R., Bhat, S., & Abowd, G. D. (2006). Farther than you may think: An empirical investigation of the proximity of users to their mobile phones. In *Proceedings of the 8th international conference on ubiquitous computing (UbiComp 2006)* (pp. 123–140). Berlin: Springer.

Patel, S. N., Reynolds, M. S., & Abowd, G. D. (2008). Detecting human movement by differential air pressure sensing in HVAC system ductwork: An exploration in infrastructure mediated sensing. In *Proceedings of the 7th international conference on pervasive computing (PERVASIVE 2008)* (pp. 1–18). Berlin: Springer.

Patel, S. N., Robertson, T., Kientz, J. A., Reynolds, M. S., & Abowd, G. D. (2007). At the flick of a switch: Detecting and classifying unique electrical events on the residential power line. In *Proceedings of the 9th international conference on ubiquitous computing (UbiComp 2007)* (pp. 271–288). Berlin: Springer.

Patel, S. N., Truong, K. N., & Abowd, G. D. (2006). PowerLine positioning: A practical sub-room-level indoor location system for domestic use. In *Proceedings of the 8th international conference on ubiquitous computing (UbiComp 2006)* (pp. 441–458). Berlin: Springer.

Patterson, D. J. (2009). Global priors of place and activity tags. In *Proceedings of the AAAI spring symposium on human behavior modeling* (pp. 75–79). Palo Alto, CA: AAAI Press.

Patterson, D. J., Baker, C., Ding, X., Kaufman, S. J., Liu, K., & Zaldivar, A. (2008). Online everywhere: Evolving mobile instant messaging practices. In *Proceedings of the 10th international conference on ubiquitous computing (UbiComp '08)* (pp. 64–73). New York, NY: ACM Press.

Patterson, D. J., Ding, X., Kaufman, S. J., Liu, K., & Zaldivar, A. (2009). An ecosystem for learning and using sensor-driven IM status messages. *IEEE Pervasive Computing, 8*(4), 42–49.

Patterson, D. J., Fox, D., Kautz, H., & Philipose, M. (2005). Fine-grained activity recognition by aggregating abstract object usage. In *Proceedings of the ninth IEEE international symposium on wearable computers (ISWC '05)* (pp. 44–51). Los Alamitos, CA: IEEE Computer Society.

Patterson, D., Liao, L., Fox, D., & Kautz, H. (2003). Inferring high-level behavior from low-level sensors. In *Proceedings of the 5th international conference on ubiquitous computing (UbiComp 2003)* (pp. 73–89). Berlin: Springer.

Perkowitz, M., Philipose, M., Fishkin, K., & Patterson, D. J. (2004). Mining models of human activities from the web. In *Proceedings of the 13th international conference on World Wide Web (WWW '04)* (pp. 573–582). New York, NY: ACM Press.

Philipose, M., Fishkin, K. P., Perkowitz, M., Patterson, D. J., Fox, D., Kautz, H., et al. (2004). Inferring activities from interactions with objects. *IEEE Pervasive Computing, 3*(4), 50–57.

Poh, M.-Z., Swenson, N. C., & Picard, R. W. (2010). A wearable sensor for unobtrusive, long-term assessment of electrodermal activity. *IEEE Transactions on Biomedical Engineering, 57*(5), 1243–1252.

Rowan, J., & Mynatt, E. D. (2005). Digital family portrait field trial: Support for aging in place. In *Proceedings of the SIGCHI conference on human factors in computing systems (CHI 2005)* (pp. 521–530). New York, NY: ACM Press.

Salber, D., Dey, A. K., & Abowd, G. D. (1999). The context toolkit: Aiding the development of context-aware applications. In *Proceedings of the SIGCHI conference on human factors in computing systems (CHI 1999)* (pp. 434–441). New York, NY: ACM Press.

Sparacino, F. (2003). Sto(ry)chastics: A Bayesian network architecture for user modeling and computational storytelling for interactive spaces. In *Proceedings of the 5th international conference on ubiquitous computing (UbiComp 2003)* (pp. 54–72). Berlin: Springer.

Stone, A. A., Shiffman, S. S., & DeVries, M. W. (1999). Ecological momentary assessment. In D. Kahneman, E. Diener, & N. Schwarz (Eds.), *Well-being: The foundations of hedonic psychology* (pp. 26–39). New York, NY: Russell Sage.

Stumpf, S., Bao, X., Dragunov, A., Dietterich, T. G., Herlocker, J., Johnsrude, K., Li, L., & Shen, J. (2005). Predicting user tasks: I know what you're doing! Research paper presented at the 20th national conference on artificial intelligence (AAAI-05) workshop on human comprehensible machine learning, Pittsburgh, PA, USA, July 9–13.

Tang, J., & Patterson, D. (2010). Twitter, sensors and UI: Robust context modeling for interruption management. In *Proceedings of the 18th international conference on user modeling, adaptation, and personalization (UMAP 2010)* (pp. 123–134). Berlin: Springer.

Tapia, E., Intille, S., & Larson, K. (2004). Activity recognition in the home using simple and ubiquitous sensors. In *Proceedings of the second international conference on pervasive computing (PERVASIVE 2004)* (pp. 158–175). Berlin: Springer.

Varshavsky, A., & Patel, S. (2009). Location in ubiquitous computing. In J. Krumm (Ed.), *Ubiquitous computing fundamentals* (pp. 286–319). Boca Raton, FL: CRC Press.

Villar, N., Scott, J., & Hodges, S. (2011). Prototyping with .NET Gadgeteer. In *Proceedings of the 5th international conference on tangible, embedded, and embodied interaction (TEI 2011)* (pp. 377–380). New York, NY: ACM Press.

Want, R., Hopper, A., Falcão, V., & Gibbons, J. (1992). The active badge location system. *ACM Transactions on Information Systems, 10*(1), 91–102.

Ward, J. A., Lukowicz, P., & Gellersen, H. W. (2011). Performance metrics for activity recognition. *ACM Transactions on Intelligent Systems and Technology, 2*(1), Article 6.

Weiser, M. (1991). The computer for the 21st century. *Scientific American, 265*(3), 94–104.

Westyn, T., Brashear, H., Atrash, A., & Starner, T. (2003). Georgia tech gesture toolkit: Supporting experiments in gesture recognition. In *Proceedings of the 5th international conference on multimodal interfaces (ICMI '03)* (pp. 85–92). New York, NY: ACM Press.

Wyatt, D., Choudhury, T., Bilmes, J., & Kitts, J. A. (2011). Inferring colocation and conversation networks from privacy-sensitive audio with implications for computational social science. *ACM Transactions on Intelligent Systems and Technology, 2*(1), Article 7.

Eye Tracking: A Brief Introduction

Vidhya Navalpakkam and Elizabeth F. Churchill

The countenance is the portrait of the soul, and the eyes mark its intentions.

Markus Tullius Cicero

When the arm is stretched in front of one's face, the size of the thumb is approximately what we see in high resolution. Visual acuity drops as we move toward the periphery. It is remarkable that despite this drop in acuity, we perceive, scan, recognize, and navigate visual information in the world around us—apparently effortlessly. This is largely due to eye movements. To counter limitations in peripheral resolution, our eyes rapidly shift from one position to another about three to four times per second, to sample visual information from the interesting areas of the world. The brain stitches together these different pieces of information in real time to present a picture of the world around us in good visual resolution. These sudden jumps in eye position that occur through fast eye movements are known as saccades.

Information processing is thought to occur during fixations,[1] when the eye position is relatively static. Therefore, eye fixations are taken to be a good proxy for cognitive

[1] Although saccades and fixations are most commonly analyzed for information processing tasks, there exist other types of eye movements such as pursuit, vergence, and vestibular eye movements. Pursuit eye movements have lower velocity than saccades and occur when the eyes follow a moving object. Vergence eye movements occur when the eyes move toward each other, to fixate on a nearby object. Vestibular eye movements occur when the eyes rotate to compensate for head and body movements in order to maintain the same direction of vision. Other smaller movements of the eyes include drifts and microsaccades.

V. Navalpakkam
Google, Inc., 1600 Amphitheatre Parkway, Mountain View, CA 94043, USA
e-mail: vidhyan@google.com

E.F. Churchill (✉)
eBay Research Labs, eBay Inc., 2065 Hamilton Ave, San Jose, CA 95125, USA
e-mail: churchill@acm.org

J.S. Olson and W.A. Kellogg (eds.), *Ways of Knowing in HCI*,
DOI 10.1007/978-1-4939-0378-8_13,

attention[2] and focused problem solving, and have been of interest to communities studying the perceptual, cognitive, and social processing of information.

In this chapter we briefly introduce eye tracking as a way of knowing in Human Computer Interaction. Specifically, we look at how eye tracking is used as a method for assessing how people perceive, process, and interact with images and interfaces to digital, computer-based technologies. In addition, we consider how eye tracking can facilitate understanding and supporting human-to-human communication and collaboration in technologically mediated environments. We offer a brief grounding with some examples and ask how can eye tracking help us understand the way humans interact with each other, and with displays.

We briefly consider the anatomy of the eye, and list eye tracking measurements using popular and contemporary technologies. We also discuss the benefits of eye tracking (including strengths and limitations), and illustrate with examples on how to effectively apply this method for research in Human Computer Interaction.

What Is Eye Tracking?

Eye tracking is the process of measuring either the point of gaze ("where we are looking") or the movement of the eye relative to the head. Although eye tracking as a method has gained a lot of press in recent years, eye tracking has been a method for understanding conscious and unconscious information processing since the 1800s (for example, Javal, 1990). Much of the early work into eye tracking was conducted through direct observation of people's gaze. However, today an array of sophisticated eye tracking technologies is readily available from trusted vendors who offer services as well as software and hardware products. An *eye tracker* is a device for measuring eye positions and eye movement. Eye trackers are used for research on the human (primate) visual system, in a number of research areas including psychology, cognitive science, marketing, and product design. There are a number of methods for measuring eye movement using eye trackers. Before we discuss the details of how eye tracking data is measured, gathered, and interpreted, we briefly discuss the anatomy of the eye and the various ways to obtain eye tracking data.

[2] Attention can be of two types: overt (the focus of attention matches where the eyes look) and covert (the focus of attention is different from where the eyes look). For example, when one is looking up to concentrate, where their eyes look is not correlated with they are thinking. This is a case of covert attention. It has been argued that for most natural viewing conditions, the focus of attention correlates with where the eyes look. In the rest of this article, we refer to overt attention as simply attention.

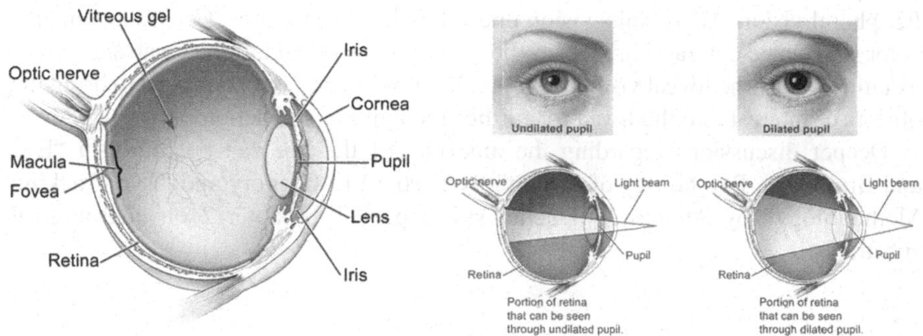

Fig. 1 Anatomy of the eye. Courtesy: National Eye Institute, National Institutes of Health (NEI/NIH)

The Anatomy of the Eye

The human eye, a slightly asymmetrical globe, is filled with a clear gel called the vitreous humor. There are seven parts of the eye that are worth knowing about for the purposes of understanding how eye tracking works (see Fig. 1). These are the iris (the pigmented part which gives us our eye color), the cornea (a clear dome over the iris), the pupil (the open circle at the center of the iris where light comes into the eye, appearing black), the sclera (the white part of the eye), the conjunctiva (a clear layer of tissue that is not visible, but which covers all but the cornea of the eye), the lens (which lies behind the pupil and the iris and which helps to focus incoming light on the back of the eye), and the retina (which is comprised of light-sensing cells the inside lining of the eye).

In the center of the retina is an area called the macula; at *its* center is the fovea, a slight depression which is responsible for high-resolution vision. Light waves enter the eye through the cornea, and pass through the pupil. As light intensity changes, so does the size of the pupil: Brighter light constricts the pupil; less light causes the pupil to dilate. Light is converted into electrical impulses by the retina, and the optic nerve transmits these impulses to the brain via the visual pathway, to the occipital cortex at the back of the brain.

There are two kinds of light receptor cells found in the retina: the cone cells and the rod cells. Foveal vision is created by tightly packed cone cells; these only account for 6 % of the total retinal light receptors. Cone cells require the most light for creating a clear, detailed image. Rod cells account for the other 94 % of light receptors in the retina. They require less light but create the blurry, less colorful qualities of peripheral vision.

Once the electrical signals get to the brain they are interpreted or "seen" by the brain as a visual image, sometimes called the "visual field." The visual field is a combination of the two primary types of vision mentioned in the introduction: *foveal* (high resolution and colorful) and *peripheral* vision (blurry and less colorful).

Peripheral vision also permits vision under low light conditions. Gross movements, color, and shape contrasts in the periphery are processed and, if they warrant further examination by the foveal vision we typically move our eyes and/or head to bring the objects of interest into the fovea for further recognition and action.

Deeper discussions regarding the anatomy of the eye can be found in "Eye and Brain: The Psychology of Seeing" by Richard L. Gregory and "Eye Tracking Methodology" by Andrew T. Duchowski, especially chapter 2 on neurological substrates.

Eye Tracking Methods

Since eye tracking is considered to provide a window onto the user's attention, we begin by addressing the strengths and current limitations of current methods. Eye gaze reveals a lot about the user that is otherwise hard to know: It tells us where the user looked, for how long, and in what order. Eye trackers are easy to use; most commercial eye trackers come with user-friendly data collection capabilities and data management interfaces, and some offer built-in data analysis software.[3] Eye tracking as a method enables tracking user eye gaze at a fine temporal resolution (~2–20 ms/sample) and high spatial resolution (<0.5° error in accuracy). The high temporal and spatial resolution can be valuable for a variety of applications. The uses range from diagnosis of medical disorders to determining user examination strategies on web search pages. In the latter, they examine the order in which people examine the search results, how much time do they spend on titles, urls, snippets, that are critical for inferring document relevance, and for applications such as ranking and search optimization. The current limitations of the method are that eye trackers are currently expensive (commercial equipment ranges from $10 K upwards), that studies tend to be small scale (involving 10–30 users), and that studies are usually conducted in controlled lab settings (raising questions about ecological validity—that is, whether results generalize to natural settings). As we discuss toward the end of this chapter, recent work has started addressing some of these limitations.

Eye tracking methods have come a long way since the method was first proposed. Early studies of eye gaze—"looking behavior"—involved simply filming subjects while they looked at a picture or watched a video clip. Researchers hand scored

[3] There are many companies offering hardware and software for eye tracking studies both in the laboratory or in controlled desktop settings and also for mobile contexts. Well-known companies include SMI (SensoMotoric Instruments) a spin-off from led by Dr. Winfried Teiwes and his academic mentors in 1991 (http://www.smivision.com/), Tobii Technology established in 2001 by John Elvesjö, Henrik Eskilsson, and Mårten Skogö (http://www.tobii.com/), and Arrington Research which was founded in 1995 by Dr. Karl Frederick Arrington as part of a technology transfer initiative at the Massachusetts Institute of Technology (http://www.arringtonresearch.com/). Other companies include Applied Science Laboratories (ASL), EyeTech, Mirametrix, Seeing Machines and SR. Webcam-based eye tracking solutions include GazeHawk and eye-trackShop.

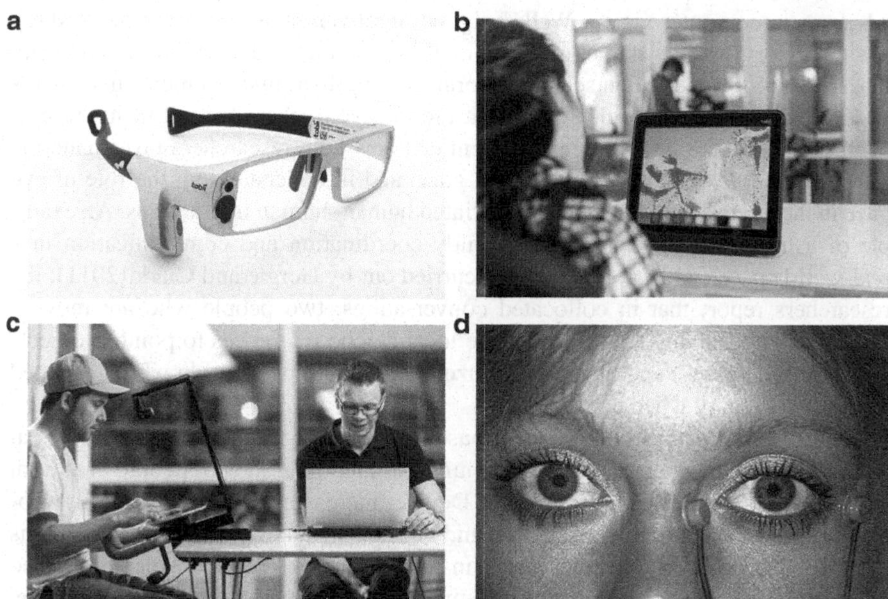

Fig. 2 Some examples of contemporary eye tracking equipment. From *top* to *bottom*: (**a**) Head-mounted mobile eye tracker in the form of eye tracking glasses (Courtesy: Tobii Technology); (**b**) table-based remote eye tracker (Courtesy: Tobii Technology); (**c**) eye tracking setup for mobile and personal devices (Courtesy: Tobii Technology); and (**d**) EOG (*Source*: Utah Medical School, http://webvision.med.utah.edu/)

recorded material to obtain a crude indication of their gaze direction. Since these early beginnings, a variety of eye tracking methods have evolved to determine the direction of gaze more accurately. Techniques include:

(a) Surface electrodes, electrooculogram (EOG)
(b) Infrared corneal reflections
(c) Video-based pupil monitoring
(d) Scleral search coils

These methods differ in their utility and in their invasiveness: EOG techniques are helpful in measuring saccade latency, but not good at measuring location; while scleral coils offer high spatial resolution (0.01°) and high temporal resolution (1,000 Hz), they are invasive and uncomfortable for participants, hence less preferred, except in clinical settings. These methods also differ depending on whether the head is free to move or not. For some applications, the head position is fixed using a forehead support or a bite bar or some other restraining mechanism that holds the eye position steady. In other cases, the head is free to move; here, head movement is accounted for with magnetic or video-based head trackers. Examples of these are illustrated in Fig. 2.

Methods are constantly evolving thanks to new technologies that are appearing—lighter weight, mobile, high resolution, infrared-enabled webcams—and due to advances in computer vision. Well-designed, lightweight and comfortable desktop and laboratory-based eye tracking equipment is nowadays standard fare for usability labs, psychology and vision science laboratories. Further, improvements in cameras and in recording technologies mean that mobile eye trackers for following gaze as people navigate the physical environment can now help researchers investigate the effect of complex environments on eye gaze, and in understanding the role of eye gaze in face to face and technology-mediated human–human interactions. An example of using mobile eye tracking to study coordination and communication in a real-world space can be found in work carried out by Gergle and Clark (2011); the researchers report that in collocated conversations, two people who are moving around and conversing tend to use more local deictic references to point to objects (e.g., "this," "these") and have lower gaze overlap than two people who are seated and thus stationary when conversing.

Finally, the possibility of webcam-based eye tracking offers hope of studies at large scale, where eye gaze patterns of hundreds and thousands of people in natural settings can be tracked using webcams. Devices increasingly come fitted with webcams, and software for data collection and analysis is readily available. In fact, this offers the possibility of eye tracking as an easy-to-use web service, where the video of the eye captured by the webcam is sent to the server in a cloud infrastructure, which then extracts eye positions from the video, analyzes eye tracking data, and sends the results back to the user's device, thereby enabling the use of eye tracking on mobile devices and phones with low computing power. Eye tracking as a cheap web service could lead to several interesting opt-in applications such as hands-free, eye-controlled scrolling, swiping, navigating, typing, and gaming on any web and webcam-enabled device (big or small) including smart phones, tablets, laptops, and desktop computers, and more importantly, can enable patients with motor disorders to interact with computers and mobile devices. A current limitation is that the accuracy of these methods is still low and a rigorous comparison against high accuracy commercial eye trackers is lacking. Thus, there exists a tradeoff between accuracy and scalability of eye tracking. However, it remains the case that the increasing availability and accuracy of high-resolution, inexpensive, lightweight, and highly configurable sensors like cameras means that eye tracking equipment is becoming readily available for researchers cheaply.

The resolution in terms of temporal sampling and point of gaze in the collection of eye tracking data varies according to the type and model of eye tracker used. Today, reduced price means reduced accuracy. Boraston and Blakemore (2007) offer examples of the variations in temporal sampling for various methods, reporting that technologies are improving all the time. At the time of their paper, pupil-only and pupil-CR eye trackers typically operated at sampling rates of between 50 Hz and 2 kHz (i.e., sampled eye position at 0.5–20 ms/sample). Direct tracking of the fovea was being accomplished at speeds of up to 200 Hz (5 ms/sample). Spatial resolution varies from 0.005° of visual angle (Clarke, Ditterich, Druen, Schonfeld, & Steineke, 2002) to 0.5°, or approximately 0.1° for methods that involve direct detection of the fovea (Gramatikov, Zalloutm, Wu, Hunter, & Guyton, 2007). Indeed, more recently, for the purposes of evaluating equipment for usability studies where considerable accuracy at the pixel-on-screen level can be needed, Johansen et al. (2011) conducted a series of tests of eye trackers, comparing an open source remote eye tracking system with a state-of-the-art commercial eye tracker. While both devices were fairly stable over time, the commercial tracker was clearly more accurate at the pixel level. They concluded that low cost eye tracking is a viable alternative to expensive equipment only when usability studies do not need to distinguish between particular words or menu items that participants are looking at. If the research is focused on larger areas of interest, e.g., whether a person is looking at an object or another person in the room, cheaper solutions are adequate.

Video-based eye trackers are the most commonly used method today; indeed, when people talk about eye tracking, they usually mean video-based methods. Most commercial eye trackers use infrared cameras with high zoom to capture high-resolution images of the eye. Points of interest such as the center of the pupil and corneal reflection are extracted from these images to determine the point of regard of the user, or simply the eye position (Goldberg & Wichansky, 2003). In order to learn how the eye position (on the image of eye) maps to what the user is looking at (on a screen for example), a short procedure known as "calibration" is performed where the user is asked to look at various points (usually in a 3×3 grid) on the display, and the relationship between the two coordinate systems (pupil-center/corneal-reflection on the image of the eye, and the x,y coordinate on the display being viewed) is established. Once good calibration is achieved (high accuracy is $<1/2°$ error) the study can commence.

Below are a few general practices for accurate and reliable calibration. One must ensure a good initial view of the eye that is robust to wide-angle glances (especially for participants wearing glasses), and use a calibration grid that is at the approximate distance of the testing stimuli (e.g., don't calibrate on a wall then test on a nearby display or vice versa). Next, one must ensure that the calibration grid covers just outside the boundaries that will be used by the participants. Participants must be requested to move their eyes then their head when performing calibration and during the study.

Once the raw eye tracking data are obtained using one of the methods above, the data can be parsed to obtain various measures such as eye fixations (brief pauses in eye position lasting around 200–250 ms each on average) and saccades (fast eye movements). Most commercial eye trackers come with built-in software for extraction of fixations and saccades, and provide output in the form of a sequence of fixations (with timestamp, x,y position, duration, and link to the display viewed). Because these are now provided automatically, we skip the discussion of these computations and refer the reader to Salvucci and Goldberg for an overview of algorithms to extract these measures (Salvucci & Goldberg, 2000).

For the remainder of the chapter, we focus on the process of inferring useful information from eye movement recordings, which involves the researcher defining "areas of interest" over certain parts of a display or interface under evaluation, and analyzing the eye movements that fall within such areas. Commonly used measures for areas of interest are fixation duration (how long do users notice as measured by dwell-time on a part of the visual scene), number of fixations (how often do users

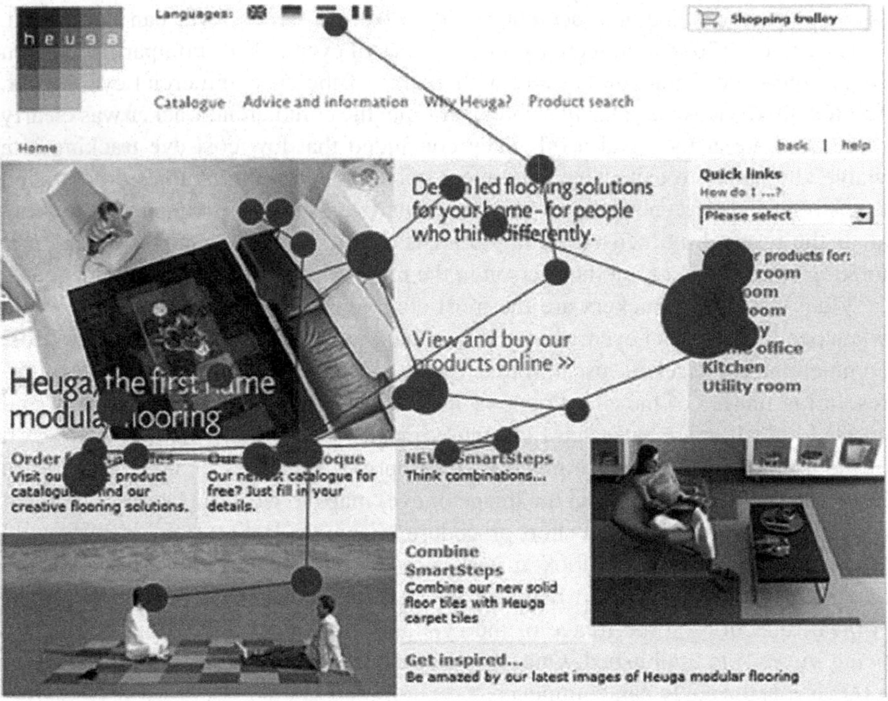

Fig. 3 An example of a sequence of eye fixations obtained from a single user. The *lines* indicate saccades and the *circles* indicate fixations. The size of the circle is proportional to the duration of the fixation. *Source*: Nielsen & Pernice, 2010

notice a part of the visual scene), sequence of fixations (the order in which users notice different parts of the visual scene), and transitions between pairs of areas of interest (how frequently users visit one area of interest from another). Figure 3 shows an example of a sequence of eye fixations interspersed by saccades.

What Is Measured with Eye Tracking? What Questions the Method Can Answer?

Eye fixations are known to be driven by perceptual salience and relevance as determined from prior experience to be important or informative (Loftus & Mackworth, 1978). A strong hypothesis is the "eye-mind" hypothesis (Just & Carpenter, 1976), according to which the eye provides a window to the user's mind, i.e., it provides a "dynamic trace of where a person's attention is being directed in relation to a visual scene." Although several exceptions to this hypothesis have been reported (for example, in covert attention, where the focus of user attention is different from where the eye is looking) for most natural viewing scenarios, eye fixations are thought to reflect

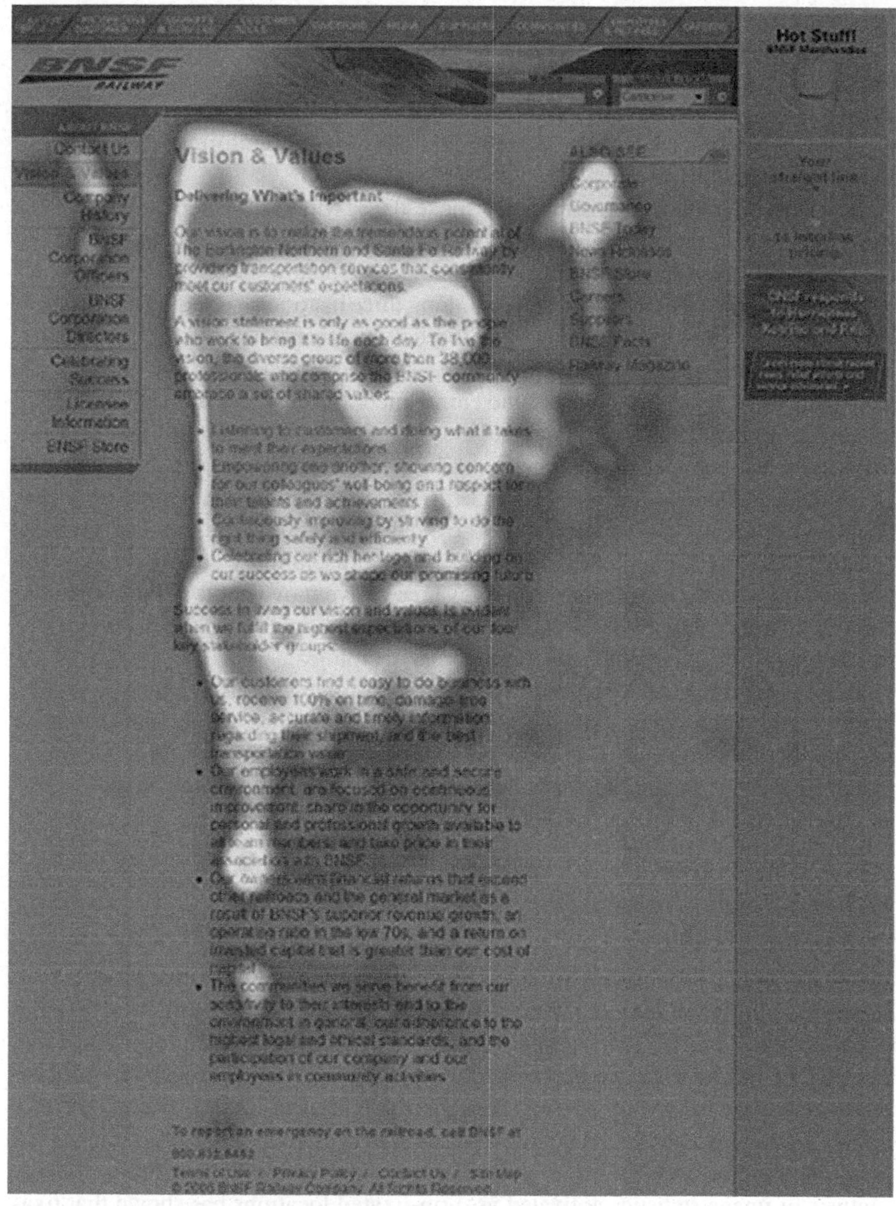

Fig. 4 Example F-pattern heatmap (*Source*: Nielsen Norman Group, http://www.nngroup.com/articles/f-shaped-pattern-reading-web-content/)

the current focus of the user's attention and the amount of cognitive processing on the fixated object(s).

In contemporary studies, a typical representation of results from an eye tracking study aggregated over several users is the "heatmap." Heatmaps use different colors to visualize the distribution and intensity of user attention on the display (see Fig. 4).

This contrasts with Fig. 3, which shows the data from a single user. Areas where users looked the most are colored red, yellow areas indicate fewer fixations, followed by the least-viewed areas in blue. Gray areas did not attract any fixations. The example in Fig. 4 is from a website's "About Us" pages. The heatmap clearly shows users' tendency to read in an "F" pattern, and their focus on information that is presented in bulleted lists.

While heatmaps are in common use these days, with colors depicting levels of attention/interest, there have been a number of different ways of representing what holds peoples' gaze and how to interpret that gaze. Duchowski (2002) characterizes eye tracking research as having developed along three historical periods. The first from 1879 to 1929 focused on psychophysiological characteristics of eye movements. Movement characteristics of the eyes, such as latency of saccadic eye movements, were studied in this period. The second came under the auspices of the behaviorism movement in understanding human behavior (1930–1958). The third era has been focused on technological developments and the production of increasingly accurate and reliable systems. Mele and Federici (2012) offer a thorough review of papers about eye tracking and discuss what they call the "fourth era" of eye tracking research: here, we see a greater emphasis on multidisciplinary contributions to the understanding of the significance and relevance of eye movements to the situational context in which the observed participant is engaged, including the disciplines of neuro-, cognitive-, and social-psychology and sociology.

Below we offer a further breakdown of the kinds of questions to which eye tracking as a method has been applied across different disciplines. While many of these may not be directly reported in the Human Computer Interaction (HCI) literature, they lead to new possibilities for understanding the power of this technique.

Vision Science (Neuroscience/Psychology)

Eye tracking has been widely used to study perceptual and cognitive processes in attention (e.g., visual search, memory, scene perception). We can think of fixation as either being "pulled" to a particular scene location by the visual properties at that location, or "pushed" to a particular location by cognitive factors related to what we know and what we are trying to accomplish (Henderson, 2003). For example, a bright or colorful area of a scene might attract the eyes simply because of its visual properties, where gaze reflects low-level processing of the human brain. Indeed, analysis of image statistics at fixated vs. non-fixated locations has shown that fixations tend to have higher density of edges, higher brightness contrast, and more generally, higher image saliency (Reinagel & Zador, 1999; Parkhurst, Law, & Niebur, 2002; reviewed in Itti & Koch, 2001). However, at a higher processing level, a viewer might want to look at scene regions that are relevant given current tasks and goals, whether or not those regions are visually prominent. Fixated regions differed in both their image statistics and their semantic content compared to regions that were not fixated (Henderson, Brockmole, Castelhano, & Mack, 2007). An ongoing focus of research is understanding the extent to which fixations are pulled by the stimulus or pushed by cognitive processes. Hayhoe and Ballard (2005) provide a good review of eye movements in natural behavior.

Computer Vision: Perceptual Models of Eye Gaze

Based on insights from eye tracking, several computational models have been developed to predict eye gaze on images and videos. These models compute visual saliency or eye-catchiness of image regions based on differences in visual properties such as color, orientation, size, brightness, motion, etc. A popular such model, which is inspired by the functioning of primate visual cortex, is the saliency model of Itti and Koch (2000). Other approaches include:

- Computing visual "surprise" in a Bayesian sense, the difference between prior and posterior belief in distribution of visual features in the world (Itti & Baldi, 2009)
- Use information theoretic measures (Bruce & Tsotsos, 2009; Zhang, Tong, Marks, Shan, & Cottrell, 2008)
- Machine learning classifiers to differentiate between fixated and non-fixated locations in images

These models perform reasonably accurately in predicting eye gaze on images and videos, especially for the first few seconds of viewing. They have potential applications in evaluating the visual catchiness of web page designs and advertisements.

In addition to visual attention models that are solely driven by image properties, there have also been attempts to model the role of user knowledge about the world. For example, when searching for cars or pedestrians, people tend to look at salient objects in the bottom half of the image where the street is most likely to appear. Thus, eye gaze is driven as a combination of saliency and knowledge of scene context. Torralba and colleagues have developed models of eye gaze that take both factors into account, and provide better prediction of user eye gaze on natural scenes (Torralba, Oliva, Castelhano, & Henderson, 2006).

An ongoing challenge is to develop predictive models of eye gaze that combine low-level image saliency with high-level semantics of the image and user intent. Examples of initial attempts in this direction and further refinements in the context of visual search task can be found in Navalpakkam and Itti (2002, 2005, 2007). In addition, there is an increasing number of models based on the emerging notion that attention and eye movement strategies serve to optimize human visual task performance (Najemnik & Geisler, 2005; Navalpakkam & Itti, 2007; Renninger, Verghese, & Coughlan, 2007; Stritzke, Trommershäuser, & Gegenfurtner, 2009).

Psychology: Reading Behavior

The study of eye movements during reading has a long and rich history dating back to the latter part of the nineteenth century (see Rayner (1998) for a good review). Eye tracking has been used for a critical examination of the cognitive processes underlying reading. For example, when reading English, eye fixations last about 200–250 ms and the mean saccade size is 7–9 letter spaces. Interestingly, many words are skipped so that foveal processing of each word is not necessary. As word length increases, the probability of fixating a word increases (Rayner & McConkie, 1976). As text becomes conceptually more difficult, fixation duration increases, saccade length decreases, and the frequency of backward saccades increases (Jacobson & Dodwell, 1979; Rayner & Pollatsek, 1994).

Language Processing

Methodologically, the drive to look at objects as they are mentioned provides an important tool for studying online language processing. For example, viewers will typically look to a scene area that contains an object when that object is mentioned by a speaker. Thus, eye fixation provides a pointer or index (Ballard, Hayhoe, Pook, & Rao, 1997) that anchors cognitive processes such as language understanding to entities in the world (Henderson & Ferreira, 2004; Tanenhaus, Spivey-Knowlton, Eberhard, & Sedivy, 1995).

Neuroscience: Medical Conditions/Disorders

Eye tracking has also been used as a method for detecting medical conditions and/or disorders such as autism and attention deficit disorders. Kanner's original description of autism highlighted the social and emotional aspects of this disorder and demonstrated it with eye tracking data (Kanner, 1943). The most commonly used stimuli are pictures of human faces, but videotapes of social interactions, human voices, and abstract animations have also been employed. Normal adults show a very specific pattern of gaze when viewing faces, fixating mainly on the eyes, but also on the nose and mouth, the so-called "core features" (Walker-Smith et al., 1977). People with autism spend less time examining the eyes (Dalton et al., 2005; Pelphrey et al., 2002) and look more frequently at the mouths and bodies, and at other objects in the scene. Eye tracking could therefore be used to diagnose and understand cognitive processing in individuals with autism (Klin, Jones, Schultz, Volkmar, & Cohen, 2002).

Market Research

Eye tracking has been used extensively in Market Research to assess product designs and also the impact of advertisements on salience and memorability of brands, logos, and products. Common use cases include comparative experiments that determine which advertisement designs attract more attention (e.g., Lohse, 1997), as are experiments focused on determining whether Internet users look at banner advertising on websites (they do not) (Burke, Hornof, Nilsen, & Gorman, 2005). An example of a recent study using eye tracking to examine the relative impact of visual salience of the design and its perceived value on user choice can be found in Milosavljevic, Navalpakkam, Koch, and Rangel (2012).

Human Computer Interaction (HCI)

Turning specifically to HCI, eye tracking has been used as a method for several purposes:

- Understanding the perceptual aspects of user attention on displays (what do users notice)
- Cognitive aspects of attention (what do users focus on, or spend time processing)
- Social aspects of attention (e.g., mutual gaze in human–human interactions, explained later)
- As an input method, using gaze as an alternative to the keyboard and mouse

These various use cases are outlined in the next section.

The Variety of Uses of Eye Tracking in Human Computer Interaction

Research over the years has investigated eye gaze patterns while driving, flying, and reviewing X-ray images. More recently, researchers are increasingly using eye tracking as a method to understand *information seeking*, *searching, and browsing* with desktop and handheld devices. The process of inferring useful information from eye movement recordings involves defining "areas of interest" over certain parts of a display or interface under evaluation, and analyzing the eye movements that fall within such areas. Using heatmaps and measures such as fixation duration described in the methods section, the visibility, meaningfulness, and placement of specific interface elements can be objectively evaluated. The findings can be used to improve the design of the interface (Goldberg & Kotval, 1999). For example, usability studies in HCI routinely use eye gaze heatmaps and the sequence of eye fixations to evaluate websites and designs (Nielsen & Pernice, 2010). Dabbish and Kraut

(2004) use it to assess attentional distribution to better understand awareness on collaborative tasks.

Eye tracking has also been used to study people's *web search behaviors* (Cutrell & Guan, 2007; Granka & Rodden, 2006) and the relationship between mouse tracking and eye tracking (Rodden & Fu, 2007; Rodden, Fu, Aula, & Spiro, 2008). For example, Rodden et al. identified different types of eye-mouse coordination behavior, including the mouse moving randomly without any correlation to eye movements, or being parked at some spot on the page, or marking something as important, or following the eye vertically and, to a lesser extent, horizontally. A more recent study by Huang, White, and Buscher (2012) identified how correlations between eye tracks and mouse tracks vary with time from page load, and attempted to model eye position from mouse position. In one of our studies (Navalpakkam & Churchill, 2012), we identified eye gaze and mouse markers that are predictive of when users struggle to read content. Because mouse tracking is scalable (and unlike current forms of eye tracking, it doesn't require the user to wear special equipment), understanding user attention and other behaviors through mouse tracking and in relation to eye tracking is becoming a hot topic of research. Certainly there are newly emerging eye tracking techniques that are more scalable (e.g., using embedded laptop cameras and webcams); however, as mentioned in the method section, these have low accuracy and may have particular challenges if successful, webcam-based eye tracking could have a big impact on HCI and other fields (advertising, web page optimization, marketing research).

Eye tracking can be useful in studying *social interactions*, and *human–human conversations*. Within sociological studies, eye gaze has been used to uncover the role of mutual gaze in the ongoing conduct of social organization and social interactions (Argyle & Cook, 1976; Goffman, 1964; Goodwin, 1984; Kendon, 1967). Goffman, in particular, points out that eye gaze plays a crucial role in the initiation and maintenance of social encounters, describing what he calls an "eye-to-eye ecological huddle which tends to be carefully maintained, maximizing the opportunity to monitor one another's mutual perceivings" (Goffman, 1964, page 95). Goodwin and Kendon both offer detailed accounts with careful notations of the ways in which eye gaze is used as part of conversational turn-taking and to manage embodied mutual orientation within a conversation or toward a shared resource. In the context of human–human communication, eye gaze has been used in two ways:

- A "measure" of interpretation (e.g., to resolve ambiguity in utterances in face-to-face conversations, Henderson & Ferreira, 2004; Tanenhaus et al., 1995)
- To place "constraints" upon interpretation (Hanna & Brennan, 2007) that make eye gaze a powerful cue about attention and intention in face-to-face communication.

Eye tracking has been used to determine mutual gaze achievement in video conferencing and also with synthetic interface characters and on-screen avatars (Roberts et al., 2009; Steptoe et al., 2009).

Eye tracking has also been used as a method in Computer Supported Cooperative Work (CSCW) to understand and support human-to-human communication in mediated environments—both to better understand the process as well as to design and develop systems to support coordination and collaboration. For example, work by Fussell and colleagues show how the transfer of gaze in remote collaboration scenarios (i.e., sharing a speaker's gaze with listener by transferring or projecting it on the listener's system) can be incredibly useful to support communication and coordination (e.g., Fussell, Setlock, Parker, & Yang, 2003; Ou, Oh, Yang, & Fussell, 2005). Further, Ou et al. (2005), Ou, Oh, Fussell, Blum, and Yang (2008), and Ou, Shi, Wong, Fussell, and Yang (2006) use eye gaze to predict focus of attention and attentional distribution in a visual setting.

Researchers have recently started using eye tracking to track two people (dyadic eye tracking) at once, and also in mobile settings to understand things like *initiative*, *lead and follow patterns*, *attention*, *ambiguity resolution*, *coordination failures*, and so on (Brennan, Chen, Dickinson, Neider, & Zelinsky, 2008; Cherubini, Nüssli, & Dillenbourg, 2008; Gergle & Clark, 2011; Jermann & Nüssli, 2012). An ongoing topic of research, some of the methodological challenges that arise for dyadic (more generally, multi-person) eye tracking are discussed in Richardson, Dale, and Kirkham (2007).

Finally, although oriented to more traditional web page interaction, recent work has used eye tracking as an investigative method to uncover patterns in human–device interaction from the perspective of a *social engagement*, rather than purely as cognitive information processing (Moore & Churchill, 2011; Moore et al., 2011).

In addition to being used as a method to evaluate people's interaction with devices and to study mediated communication (e.g., human–human communication through video conferencing), eye tracking has also been used as an input method for adaptive interfaces (Jacob & Karn, 2003a, 2003b). Experiments suggest that selection with eye gaze can be a robust input/selection technique, and in fact, eye selection can be faster than using a mouse. In 2007, Oyekoya demonstrated experimentally that eye gaze can be a viable selection method for visual search tasks, and that prior experience on visual tasks with a mouse can create a "training effect" (Oyekoya, 2007). Focusing on assistive technologies and what they call "psychotechnologies," Mele and Federici (2012) outline several opportunities for eye tracking technologies as the underlying technology for eye gaze to be more fully realized as an input technique.

How to Do It: What Constitutes Good Work?

Like any other method, a clean experiment design makes eye tracking a viable method for understanding user behavior.[4] We describe the main aspects of a good eye tracking study (described in detail in Duchowski, 2007) with an example from our work below (Navalpakkam & Churchill, 2012; Navalpakkam, Rao, & Slaney, 2011).

[4] For more details on experimental design, please see the chapter on Experimental Research in HCI in this volume.

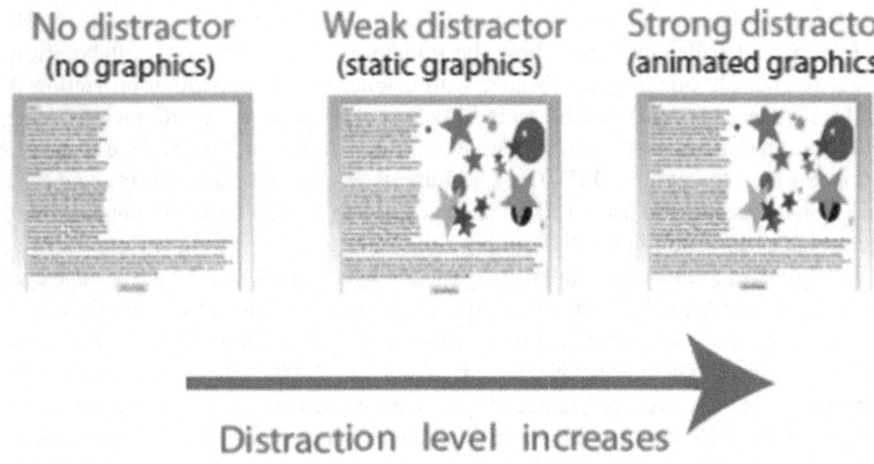

Fig. 5 This figure illustrates the experiment design with control group (no distractor), treatment group 1 (weak distractor), and treatment group 2 (strong distractor). *Source*: Navalpakkam et al., 2011

1. *Hypothesis*: Formulate clear null and alternate hypotheses, motivate them, and state the underlying assumptions.

   For example, in our case study, for *Motivation*, we wished to understand how the presence of distracting elements on a web page affects eye gaze patterns. Our *Null hypothesis H0* was that eye gaze patterns on a web page do not change in the presence of distracting elements; our *alternate hypothesis H1* was that users spend more time looking at the distracting element; our second *alternate hypothesis H2* was that users spend more time looking at the page content.
2. *Design*: Determine whether it is an observational study or an experimental study. If the latter, design control and treatment groups to test the hypotheses. Identify independent and dependent variables; change the independent variable's values, while keeping everything else more or less a constant (thus avoiding confounding factors); and measure the impact on the dependent variable. Also determine whether it is a within-subjects design or between-subjects design. There can be differences in eye movements between participants on identical tasks, thus it may be prudent to use a within-participants design in order to make valid performance comparisons (Goldberg & Wichansky, 2003).

   In the case study, the control group did not see any distracting elements on the web page. Treatment group 1 saw a moderately distracting element in the form of a static, irrelevant graphic on the top right of the page. Treatment group 2 saw a highly distracting element in the form of an animated, irrelevant graphic on the top right of the page, shown in Fig. 5. A within-subject design was used, and the order of experimental conditions was randomized to balance familiarity and fatigue. Each participant saw three essays (from the Test of English Language Fluency) that were randomly paired with one of the graphic types. Each essay consisted of 300–400 words, followed by 5 factual/theme-based multiple-choice questions (to confirm that users indeed read the essays and performed the task as told), and 2 subjective questions where subjects were asked to rate their user experience on a scale of 1–5 for pleasantness.

In this study, the independent variable was the level of distraction (varied from "none" in the control group, to "medium" in treatment 1 to "high" in treatment 2). Dependent variables included the amount of time users spent looking at the page content and distracting element, the corresponding number of eye gaze fixations, the time to first eye fixation, and the user-reported levels of pleasantness of experience.

3. *Task description*: What task is assigned to the participants? Are they freely viewing the displays, or are they performing a task such as searching for a particular object in the display. The task description is a critical part of an eye tracking study. Its importance is highlighted in a classic study by Yarbus (1967) that shows how eye movements on a painting are influenced by the task given to the user, with other things kept constant. The eye tracking data shows that when asked to determine the ages of the people in the painting, eye gaze focused on the faces of people, whereas when asked to assess material circumstances of the family, eye gaze focused on the clothes that people wore, the furniture, and other features in the visual scene (see Fig. 6). In our example study, the task assigned to participants was a reading comprehension task—"Read the article on this web page and answer the questions that follow." Unknown to the participants, the goal of our study was to test how the presence of distracting elements on the page affects eye gaze. Thus we avoided any potential biasing of participants' behavior that may result from them knowing the study's goal.
4. *Participants, Apparatus, and Procedure*: A good eye tracking study should include a description of the participants (e.g., number, age group, gender, demographics); compensation or incentive structure (is there a performance bonus?); apparatus used (eye tracker model, monitor resolution, display viewing angle, calibration accuracy); and procedure of the study (instructions before the study, flow of the study, posttask feedback).

In the case study, there were 20 participants (8 female, 12 males; residents of United States), aged 19–60, with normal or corrected vision. Participants were fluent in English (spoken and written), and had either completed or were pursuing undergraduate education. We recorded participants' gaze patterns during task performance using a Tobii 1750 eye tracker (50 Hz sampling frequency), with a 17″ LCD monitor, set at resolution 1,024 × 768, at roughly 85 cm viewing distance. We collected a log of eye and mouse movement.

Participants were compensated as follows: a flat payment rate for participation in the study, and in addition, $1 for every correct answer (5 questions per essay × 3 essays). During data cleaning, three participants were excluded for the following reasons: poor calibration (two did not maintain their head in the correct position), or outliers in fixation duration or number of fixations (3 standard deviations, 1 participant).

The study began with a 5-point calibration procedure followed by the task-instruction screen. This was followed by one practice essay paired with animated

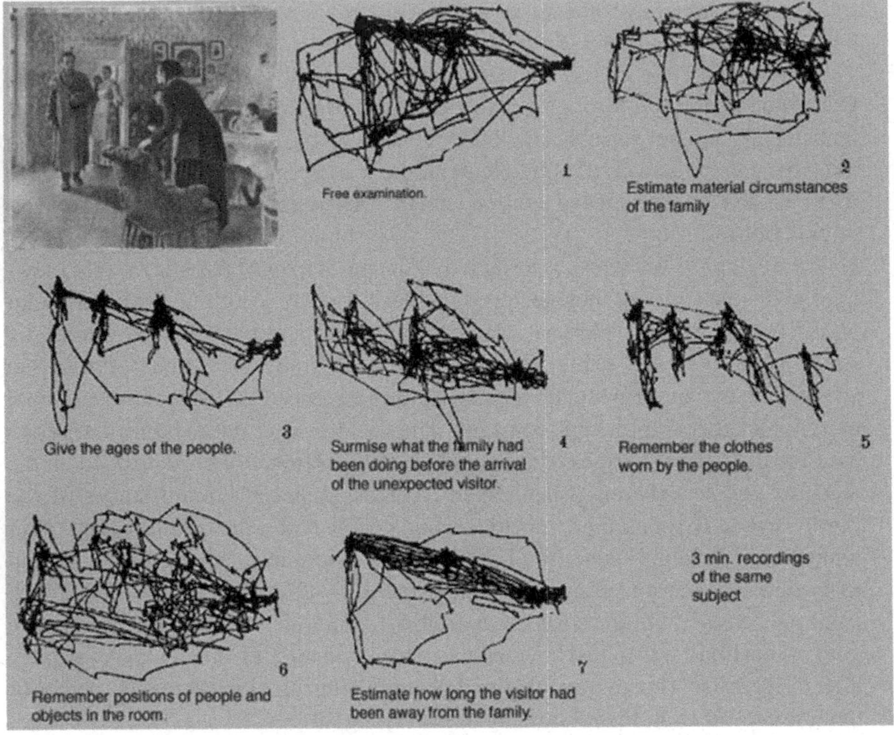

Fig. 6 Eye movements given different tasks—From Yarbus (1967)

graphics to help participants familiarize themselves with the task, types of graphics, and the format of the questions in the reading comprehension test. Following the practice trial, participants saw three essays randomly paired with no, static, or animated graphics (i.e., control, treatment 1 and 2 groups in randomized order). At the conclusion of the study, participants were paid based on task performance.

5. *Analysis*: A critical part of a good eye tracking study is conducting careful and appropriate analysis. Eye tracking data can be analyzed in qualitative ways (e.g., heatmap visualization, observing where people look) and using rigorous quantitative methods. The latter consists of defining areas of interest and extracting measures for each area of interest, such as the number of eye fixations, duration of eye fixations, number of saccades, time to first fixation, and number of backward saccades (called "regressions," suggesting confusion or distraction). As mentioned later under challenges, defining and determining areas of interest (AOIs) for analysis can be highly complex as one moves toward dynamic and/or longer-term tasks. It is much easier for a stable, 2D image, with predefined areas of interest where we want to analyze gaze. If interested in questions such as "where does the user look next," one could also look at the temporal ordering of eye fixations by extracting transition probabilities P(x,y), which describe the

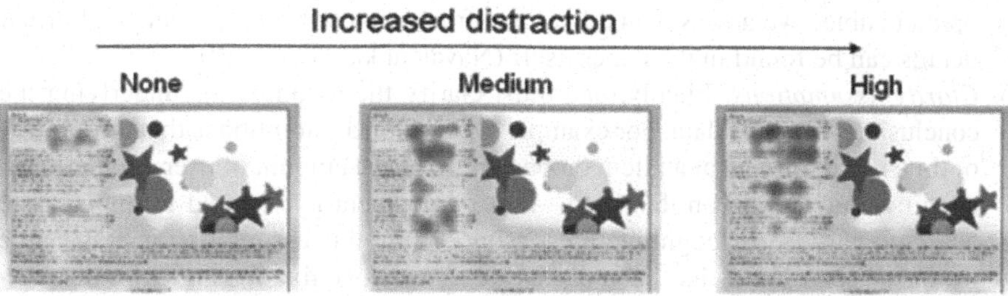

Fig. 7 Example of heatmap from Navalpakkam et al., CHI 2011

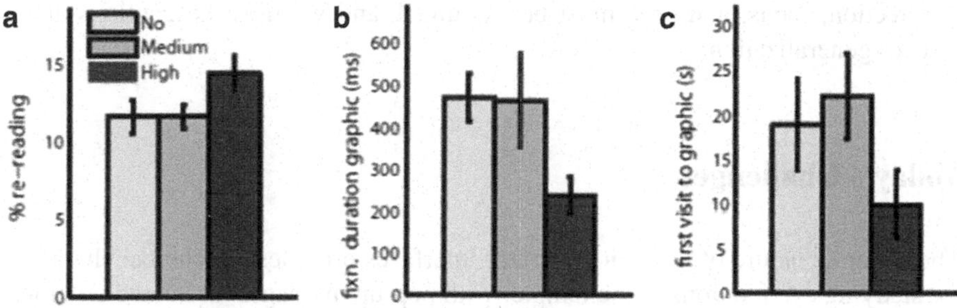

Fig. 8 Example eye tracking measures for the distraction study

probability that the user will look next at display item or location "y," given that she is currently looking at item or location "x." Navalpakkam et al. (2011) offer an example of using the above eye tracking metrics and analysis in the context of understanding and modeling how image presence, position, and user interest drive the way users attend to and select online news content. More common types of analyses such as heatmaps and fixation metrics are illustrated below with an example from the case study.

Figures 7 and 8 show that users spend more time processing the page content as the amount of distraction on the page increases. Similar effects were observed with mouse tracks as well. For easy comparison, we overlay the heatmaps on the same image; however, note that in the "no" distraction condition, there was no graphic on the right-hand side of the page.

A particularly interesting finding in the study was that more time on page is not always good. We found that users spent more time on page in the high distraction condition, but they reported being more annoyed (very low pleasantness scores). Analysis of eye tracking data revealed that the increased time was due to struggle in reading, and increased cognitive effort in processing the content in the presence of a highly distracting graphic. For example, in Fig. 8, although the highly (animated) distracting element was noticed earlier in time (panel C), it was rejected faster (panel B), and users spent more time processing and re-reading the page content (panel A), which is an indication of struggles with reading. The authors further show in the paper that further, these eye tracking patterns can predict subjective assessments, like user frustration with high accuracy. Further details can be found in the paper itself (Navalpakkam et al., 2011).

6. *Clarify assumptions*: Finally, one must clarify the assumptions underlying the conclusions from the data. For example, a common assumption is that the amount of time spent looking at an item on page reflects the amount of user attention and cognitive processing on that item—more time spent is assumed to imply more attention and deeper cognitive processing. While this assumption is reasonable for most scenarios, as has been discussed already eye fixation and gaze duration are sometimes not correlated with the depth of the user's cognitive processing. For example, sometimes people look up when they concentrate or are conducting hard mental operations. Assumptions about the correlations between eye gaze direction, focus, and time must be examined, and we must be cautious about over-generalization.

Today's Challenges

The dynamic nature of modern computer interfaces provides a technical challenge for studying eye fixations. For example, with pop-up messages, animated graphics, and user-initiated object movement and navigation, objects can move around a screen, or move off of 2D screens. As a result, the definition of areas of interest becomes a challenging problem. Knowing that a person was fixating 10° above and 5° to the left of the display's center does not allow us to know what object the person was looking at in the computer interface unless we keep track of the changes in the computer display. Analysts must bear this in mind while considering dynamic displays. Recent software packages sold by the various eye tracking vendors now have definable areas of interests that can change over time, but even those often must be generated post hoc. Gergle and Clark (2011) suggest another solution that couples eye tracking with vision tracking techniques by using "objects of interest" as opposed to static areas of interest.

Conclusions

As modern computer interfaces continue to shrink in size (from desktops to laptops to tablets to phones) and become more mobile, understanding how users process information as they move around in the world becomes important. Upcoming technologies like webcam-enabled eye tracking are exciting and offer hope, but need to deal with challenges in calibration and accuracy (that are rendered difficult due to varying distance between user and device, varying head pose, and lighting conditions).

Finally, as Chi and colleagues (2009) have discussed, ideally, eye tracking methods should possess the following factors:

1. Accuracy
2. Reliability
3. Robustness
4. Non-intrusiveness
5. The possibility for free head movements
6. No prior calibration
7. Real-time response

We add to this list that eye tracking methods should be

8. Work for Dynamic displays
9. Allow for study participants' mobility
10. Be Scalable

Achieving all of these factors in one system is not yet possible as systems still require calibration, and because there are accuracy-intrusiveness and accuracy-scalability trade-offs. However, it is possible to imagine, as cameras improve and mounts for mobile eye trackers become increasingly lightweight, that we will see the emergence of more powerful eye tracking opportunities in the future. With these we will be able to more accurately discern and assess degree of attention, level of interest, and management of cognitive and social interaction.

Exercises

1. Unlike regular experiments, what do experiments with eye tracking have to begin with? How is this done?
2. When is the eye's direction not a good indicator of what the person is looking at or thinking about? How would you separate those from "real" perception?

References and Further Reading[5]

References for getting more expert in this method

Duchowski, A. T. (2007). *Eye tracking methodology: Theory and practice*. Secaucus, NJ: Springer.

Duchowski, A. T. (2002). A breadth-first survey of eye tracking applications. *Behavior Research Methods, Instruments, and Computers, 34*, 455–470.

Goldberg, J. H., & Wichansky, A. M. (2003). Eye tracking in usability evaluation: A practitioner's guide. In J. Hyönä, R. Radach, & H. Deubel (Eds.), *The mind's eye: Cognitive and applied aspects of eye movements* (pp. 493–516). Oxford, UK: Elsevier Science.

Holmqvist, K., Nystrom, M., Andersson, R., Dewhurst, R., Jarodzka, H., & Van De Weijer, J. (2011). *Eye tracking. A comprehensive guide to methods and measures*. Oxford: Oxford University Press.

[5] There are many texts that can be drawn on to learn more about eye tracking. We list some below.

Rayner, K. (1998). Eye movements in reading and information processing: 20 years of research. *Psychological Bulletin, 124*(3), 372–422.

Rayner, K. (2009). Eye movements and attention in reading, scene perception, and visual search. *The Quarterly Journal of Experimental Psychology, 62*(8), 1457–1506.

Henderson, J. (2003). Human gaze control during real-world scene perception. *Trends in Cognitive Sciences, 7*(11), 498–504.

Hayhoe, M., & Ballard, D. (2005). Eye movements in natural behavior. *Trends in Cognitive Sciences, 9*(4), 188–194.

Jacob, R. J. K., & Karn, K. S. (2003a). Eye tracking in human-computer interaction and usability research: Ready to deliver the promises. In J. Hyönä, R. Radach, & H. Deubel (Eds.), *The mind's eye: Cognitive and applied aspects of eye movement research* (pp. 573–603). Oxford, England: Elsevier.

Poole, A., & Ball, L. J. (2005). Eye tracking in human-computer interaction and usability research: Current status and future prospects. *Psychology, 10*(5), 211–219.

References to example papers that have used it well, a short commentary about each reference and what questions they answer can be found in the footnotes

Goldberg, J. H. & Kotval, X. P. (1999). Computer interface evaluation using eye movements: Methods and constructs. *International Journal of Industrial Ergonomics, 24*, 631–645[6]

Torralba, A., Oliva, A., Castelhano, M. S., & Henderson, J. M. (2006). Contextual guidance of eye movements and attention in real-world scenes: The role of global features in object search. *Psychological Review, 113*, 766–786[7]

Reinagel, P. & Zador, A. M. (1999). Natural scene statistics at the centre of gaze. *Network: Computation in Neural Systems, 10*, 341–350[8]

Milosavljevic, M., Navalpakkam, V., Koch, C., & Rangel, A. (2012). Relative visual saliency differences induce sizeable bias in consumer choice. *Journal of Consumer Psychology, 22(1)*, 67–74[9]

Navalpakkam, V., Rao, J. M., & Slaney, M. (2011). Using gaze patterns to study and predict reading struggles due to distraction. *CHI 2011*, 7–12 May, 2011, Vancouver, BC, Canada[10]

[6] Comparative assessment of measures of eye movement locations and scanpaths used for evaluation of interface quality. Revealed that well-organized functional grouping of icons result in shorter scanpaths, covering smaller areas. Less well organized interfaces produce less efficient search behavior. However, poorly organized icon groupings do not affect users' ability to interpret/understand icons.

[7] While searching for objects in scenes, eye gaze is affected by the scene context, e.g., one searches for pedestrians on the lower half of a scene, where the street is most likely to be; with this one searches for birds in the upper half of the scene, where the sky is most likely to be.

[8] While freely viewing images, eye gaze is biased towards image locations that have high spatial contrast.

[9] Fast eye movements and choices between food items are driven by perceptual factors such as the visual catchiness or saliency of items, while slower choices are driven by high-level factors such as the value of items.

[10] Eye movements differ when users are distracted compared to when they are not. This study identifies eye gaze markers that are predictive of user reading struggle, or frustration. This study and its companion paper (Navalpakkam & Churchill CHI 2012, focusing mainly on the relationship between eye and mouse tracking) have been used as examples of designing, conducting and analyzing an eye tracking study.

Moore, R. J. & Churchill, E. F. (2011). Computer Interaction Analysis: Toward an Empirical Approach to Understanding User Practice and Eye Gaze in GUI-Based Interaction. *Computer Supported Cooperative Work, 20 (6)*, 497–528[11]

Other references

Argyle, M., & Cook, M. (1976). *Gaze and mutual gaze*. Cambridge: Cambridge University Press.

Ballard, D. H., Hayhoe, M. M., Pook, P. K., & Rao, R. P. (1997). Deictic codes for the embodiment of cognition. *Behavioral and Brain Sciences, 20*(04), 723–742.

Brennan, S. E., Chen, X., Dickinson, C., Neider, M., & Zelinsky, G. (2008). Coordinating cognition: The costs and benefits of shared gaze during collaborative search. *Cognition, 106*, 1465–1477.

Boraston, Z., & Blakemore, S. (2007). The application of eye tracking technology in the study of autism. *The Journal of Physiology, 581*, 893–898.

Burke, M., Hornof, A., Nilsen, E., & Gorman, N. (2005). High-cost banner blindness: Ads increase perceived workload, hinder visual search, and are forgotten. *ACM Transactions on Computer-Human Interaction, 12*(4), 423–445.

Buswell, G. T. (1935). *How people look at pictures*. Chicago, IL: University of Chicago Press.

Bruce, N. D., & Tsotsos, J. K. (2009). Saliency, attention, and visual search: An information theoretic approach. *Journal of Vision, 9*(3), 5. 1–24.

Cherubini, M., Nüssli, A. M., & Dillenbourg, P. (2008). Deixis and gaze in collaborative work at a distance: A computational model to detect misunderstandings. In *ETRA '08: Proceedings of the 2008 symposium on eye tracking research and applications* (pp. 173–180). New York, NY: ACM.

Chi, J.-n., Zhang, P.-y., Zheng, S.-y., Zhang, C., & Huang, Y. (2009). Key techniques of eye gaze tracking based on pupil corneal reflection. In *WRI global congress on intelligent systems, 2009: GCIS '09, 19–21 May, 2009, Xiamen, CN* (pp. 133–138). Washington, DC: IEEE.

Clarke, A. H., Ditterich, J., Druen, K., Schonfeld, U., & Steineke, C. (2002). Using high frame rate CMOS sensors for three-dimensional eye tracking. *Behavior Research Methods, Instruments, and Computers, 34*, 549–560.

Cooke, L. (2006). Is eye tracking the next step in usability testing? In *IEEE international professional communication conference: IPCC '06, 23–25 October, 2006, Saratoga Springs, NY* (pp. 236–242). Washington, DC: IEEE. doi:10.1109/IPCC.2006.320355.

Cutrell, E., & Guan, Z. (2007). What are you looking for?: An eye tracking study of information usage in web search. In *Proceedings of the SIGCHI conference on human factors in computing systems, Montréal, QC, Canada, 22–27 April, 2006* (pp. 407–416). New York, NY: ACM.

Dabbish, L., & Kraut, R. (2004). Controlling interruptions: Awareness displays and social motivation for coordination. In *Proceedings of the 2004 ACM conference on computer supported cooperative work (CSCW '04), Chicago, IL, 6–10 November, 2004* (pp. 182–191). New York, NY: ACM.

Dalton, K. M., Nacewicz, B. M., Johnstone, T., Schaefer, H. S., Gernsbacher, M. A., Goldsmith, H. H., et al. (2005). Gaze fixation and the neural circuitry of face processing in autism. *Nature Neuroscience, 8*, 519–526.

[11] The authors outline a novel methodological approach to understanding how humans interact with interactive applications and services. They call their approach "computer interaction analysis". Computer interaction analysis extends traditional eye tracking approaches by interweaving and drawing out the relationship between people's input actions, system display events and people's eye movements.

Ellis, S., Candrea, R., Misner, J., Craig, C. S., & Lankford, C. P. (1998). Windows to the soul? What eye movements tell us about software usability. *7th annual conference of the Usability Professionals' Association conference: UPA '98*, June 22–26, 1998, Washington, DC

Exline, R. V. (1974). Visual interaction: The glances of power and preference. In S. Weitz (Ed.), *Nonverbal communication* (pp. 65–92). New York, NY: The Oxford University Press.

Exline, R. V., & Winters, L. G. (1965). Affective relations and mutual glances in dyads. In S. Tomkins & C. Izzard (Eds.), *Affect, cognition and personality*. New York, NY: Springer.

Fussell, S. R., Setlock, L. D., Parker, E. M., & Yang, J. (2003). Assessing the value of a cursor pointing device for remote collaboration on physical tasks. In *CHI '03 extended abstracts on human factors in computing systems* (pp. 788–789). New York, NY: ACM.

Garrett, J. J. (2003). *The elements of user experience: User-centered design for the web*. New York, NY: New Riders Press.

Gergle, D., & Clark, A. T. (2011). See what I'm saying? Using dyadic mobile eye tracking to study collaborative reference. In *Proceedings of the ACM 2011 conference on computer supported cooperative work (CSCW '11)* (pp. 435–444). New York, NY: ACM.

Gibson, J. J., & Pick, A. D. (1963). Perception of another person's looking behavior. *American Journal of Psychology, 76*, 386–394.

Gramatikov, B. I., Zalloutm, O. H., Wu, Y. K., Hunter, D. G., & Guyton, D. L. (2007). Directional eye fixation sensor using birefringence-based foveal detection. *Applied Optics, 46*, 1809–1818.

Goffman, E. (1964). *Behavior in public places*. Glencoe: The Free Press.

Goodwin, C. (1984). Notes on story structure and the organization of participation. In M. Atkinson & J. Heritage (Eds.), *Structures of social action* (pp. 225–246). Cambridge: Cambridge University Press.

Granka, L. & Rodden, K. (2006). Incorporating eye tracking into user studies at Google. In *Workshop Position paper presented at CHI*.

Hanna, J., & Brennan, S. (2007). Speakers' eye gaze disambiguates referring expressions early during face-to-face conversation. *Journal of Memory and Language, 57*(4), 596–615.

Henderson, J. M., Brockmole, J. R., Castelhano, M. S., & Mack, M. (2007). Visual saliency does not account for eye movements during search in real-world scenes. In R. van Gompel, M. Fischer, W. Murray, & R. Hill (Eds.), *Eye movements: A window on mind and brain* (pp. 537–562). Oxford: Elsevier.

Henderson, J. M., & Ferreira, F. (Eds.). (2004). *The interface of language, vision, and action: Eye movements and the visual world*. New York, NY: Psychology Press.

Huang, J., White, R., & Buscher, G. (2012). User see, user point: Gaze and cursor alignment in web search. In *Proceedings of the 2012 ACM annual conference on human factors in computing systems* (pp. 1341–1350). New York, NY: ACM.

Humphrey, K., & Underwood, G. (2009). Domain knowledge moderates the influence of visual saliency in scene recognition. *British Journal of Psychology, 100*, 377–398.

Itti, L., & Baldi, P. (2009). Bayesian surprise attracts human attention. *Vision Research, 49*(10), 1295–1306.

Itti, L., & Koch, C. (2000). A saliency-based search mechanism for overt and covert shifts of visual attention. *Vision Research, 40*(10–12), 1489–1506.

Itti, L., & Koch, C. (2001). Computational modeling of visual attention. *Nature Reviews Neuroscience, 2*, 194–203.

Jacob, R. J., & Karn, K. S. (2003b). Eye tracking in human-computer interaction and usability research: Ready to deliver the promises. *Mind, 2*(3), 4.

Jacobson, J., & Dodwell, P. C. (1979). Saccadic eye movements during reading. *Brain and Language, 8*(3), 303–314.

Javal, E. (1990). Essay on the physiology of reading. *Ophthalmic and Physiological Optics, 10*, 381–384.

Jermann, P., & Nüssli, M. A. (2012). Effects of sharing text selections on gaze cross-recurrence and interaction quality in a pair programming task. In *Proceedings of the ACM 2012 conference on computer supported cooperative work (CSCW '12)* (pp. 1125–1134). New York, NY: ACM.

Johansen, S. A., Agustin, J. S., Skovsgaard, H., Hansen, J. P., & Tall, M. (2011). Low cost vs. high-end eye tracking for usability testing. In *CHI '11 Extended Abstracts on Human Factors in Computing Systems*. ACM, New York, NY, USA, 1177–1182.

Just, M. A., & Carpenter, P. A. (1976). Eye fixations and cognitive processes. *Cognitive Psychology, 8*, 441–480.

Kanner, L. (1943). Autistic disturbances of affective contact. *Nervous Child, 2*(3), 217–250.

Kendon, A. (1967). Some functions of gaze-direction in social interaction. *Acta Psycholoigica, 26*, 22–63.

Klin, A., Jones, W., Schultz, R., Volkmar, F., & Cohen, D. (2002). Visual fixation patterns during viewing of naturalistic social situations as predictors of social competence in individuals with autism. *Archives of General Psychiatry, 59*, 809–816.

Land, M. F., & Hayhoe, M. (2001). In what ways do eye movements contribute to everyday activities? *Vision Research, 41*, 3559–3565.

Latimer, C. R. (1988). Eye-movement data: Cumulative fixation time and cluster analysis. *Research Methods, Instruments and Computers, 20*(5), 437–470.

Loftus, G. P., & Mackworth, N. H. (1978). Cognitive determinants of fixation location during picture viewing. *Journal of Experimental Psychology: Human Perception and Performance, 4*(4), 565–572.

Lohse, G. L. (1997). Consumer eye movement patterns on yellow pages advertising. *Journal of Advertising, 26*(1), 61–73.

Mele, M. L., & Federici, S. (2012). A psychotechnical review on eye tracking systems: Toward user experience. *Disability and Rehabilitative Technology, 7*(4), 261–281.

Moore, R. J., Churchill, E. F., & Kantamneni, R. G. P. (2011). Three sequential positions of query repair in interactions with internet search engines. *In Proceedings of CSCW 2011* (pp. 415–424). New York: ACM

Najemnik, J., & Geisler, W. S. (2005). Optimal eye movement strategies in visual search. *Nature, 434*(7031), 387–391.

Navalpakkam, V., & Churchill, E. (2012). Mouse tracking: Measuring and predicting users' experience of web-based content. In *Proceedings of the 2012 ACM annual conference on human factors in computing systems* (pp. 2963–2972). New York, NY: ACM.

Navalpakkam, V., & Itti, L. (2002). A goal oriented attention guidance model. In *Biologically motivated computer vision* (pp. 81–118). Berlin: Springer.

Navalpakkam, V., & Itti, L. (2005). Modeling the influence of task on attention. *Vision Research, 45*(2), 205–231.

Navalpakkam, V., & Itti, L. (2007). Search goal tunes visual features optimally. *Neuron, 53*(4), 605–617.

Nielsen, J., & Pernice, K. (2010). *Eye tracking web usability*. Berkeley, CA: New Riders.

Ou, J., Oh, L. M., Yang, J., & Fussell, S. R. (2005). Effects of task properties, partner actions, and message content on eye gaze patterns in a collaborative task. In *Proceedings of the SIGCHI conference on human factors in computing systems* (pp. 231–240). New York, NY: ACM.

Ou, J., Shi, Y., Wong, J., Fussell, S. R., & Yang, J. (2006). Combining audio and video to predict helpers' focus of attention in multiparty remote collaboration on physical tasks. In *Proceedings of the 8th international conference on multimodal interfaces* (pp. 217–224). New York, NY: ACM.

Ou, J., Oh, L. M., Fussell, S. R., Blum, T., & Yang, J. (2008). Predicting visual focus of attention from intention in remote collaborative tasks. *IEEE Transactions on Multimedia, 10*(6), 1034–1045.

Oyekoya, O. (2007). Eye tracking: A perceptual interface for content based image retrieval. Unpublished Ph.D. thesis, Department of Electronic & Electrical Engineering, Adastral Park Campus, University College London.

Parkhurst, D., Law, K., & Niebur, E. (2002). Modeling the role of salience in the allocation of overt visual attention. *Vision Research, 42*(1), 107–124.

Poole, A., & Ball, L. (2006). Eye tracking in human-computer interaction and usability research: Current status and future prospects. In C. Ghaoui (Ed.), *Encyclopedia of human computer interaction* (pp. 211–219). London, UK: Idea Group Reference.

Pelphrey, K. A., Sasson, N. J., Reznick, J. S., Paul, G., Goldman, B. D., & Piven, J. (2002). Visual scanning of faces in autism. *Journal of Autism and Developmental Disorders, 32*, 249–261.

Rayner, K., & McConkie, G. W. (1976). What guides a reader's eye movements? *Vision Research, 16*(8), 829–837.

Rayner, K., & Pollatsek, A. (1994). *The psychology of reading*. Hillsdale, NJ: Lawrence Erlbaum.

Reimer, M. D. (1955). Abnormalities of the gaze – A classification. *Psychiatric Quarterly, 29*, 659–672.

Renninger, L. W., Verghese, P., & Coughlan, J. (2007). Where to look next? Eye movements reduce local uncertainty. *Journal of Vision, 7*(3), 6. 1–17.

Rensink, R. A., O'Regan, J. K., & Clark, J. J. (1997). To see or not to see: The need for attention to perceive changes in scenes. *Psychological Science, 8*, 368–373.

Richardson, D. C., Dale, R., & Kirkham, N. Z. (2007). The art of conversation is coordination: Common ground and the coupling of eye movements during dialogue. *Psychological Science, 18*(5), 407–413.

Roberts, D., Wolff, R., Rae, J., Steed, A., Aspin, R., McIntyre, M., et al. (2009). Communicating eye-gaze across a distance: Comparing an eye-gaze enabled immersive collaborative virtual environment, aligned video conferencing, and being together. In *Virtual reality conference, 2009. VR 2009. IEEE* (pp. 135–142). Washington, DC: IEEE.

Robinson, G. H. (1979). Dynamics of the eye and head during movement between displays: A qualitative and quantitative guide for designers. *Human Factors, 21*(3), 343–352.

Rodden, K. & Fu, X. (2007). Exploring how mouse movements relate to eye movements on web search results pages. In *SIGIR 2007 workshop on web information seeking and interaction (WISI)*, July 27, 2007, Amsterdam, The Netherlands, pp. 29–32.

Rodden, K., Fu, X., Aula, A., & Spiro, I. (2008). Eye-mouse coordination patterns on web search results pages. In *CHI '08 extended abstracts on human factors in computing systems* (pp. 2997–3002). New York, NY: ACM.

Salvucci, D. D., & Goldberg, J. H. (2000). Identifying fixations and saccades in eye-tracking protocols. In Proceedings of the Eye Tracking Research and Applications Symposium (pp. 71–78). New York: ACM Press.

Steptoe, W., Oyekoya, O., Murgia, A., Wolff, R., Rae, J., Guimaraes, E., et al. (2009). Eye tracking for avatar eye gaze control during object-focused multiparty interaction in immersive collaborative virtual environments. In *Virtual reality conference, 2009. VR 2009. IEEE* (pp. 83–90). Washington, DC: IEEE.

Stritzke, M., Trommershäuser, J., & Gegenfurtner, K. R. (2009). Effects of salience and reward information during saccadic decisions under risk. *JOSA, A26*(11), B1–B13.

Tanenhaus, M. K., Spivey-Knowlton, M. J., Eberhard, K. M., & Sedivy, J. C. (1995). Integration of visual and linguistic information in spoken language comprehension. *Science, 268*, 632–634.

Torralba, A., Oliva, A., Castelhano, M. S., & Henderson, J. M. (2006). Contextual guidance of eye movements and attention in real-world scenes: The role of global features in object search. *Psychological Review, 113*, 766–786.

Walker-Smith, G. J., Gale, A. G., & Findlay, J. M. (1977). Eye movement strategies in face perception. *Perception, 6*, 313–326.

Yarbus, A. L. (1967). *Eye movements and vision*. New York, NY: Plenum Press.

Zhang, L., Tong, M. H., Marks, T. K., Shan, H., & Cottrell, G. W. (2008). SUN: A Bayesian framework for saliency using natural statistics. *Journal of Vision, 8*(7), 32.

Understanding User Behavior Through Log Data and Analysis

Susan Dumais, Robin Jeffries, Daniel M. Russell, Diane Tang, and Jaime Teevan

Overview of Log Data and Analysis in HCI

Behavioral logs are traces of human behavior seen through the lenses of sensors that capture and record user activity. They include behavior ranging from low-level keystrokes to rich audio and video recordings. Traces of behavior have been gathered in psychology studies since the 1930s (Skinner, 1938), and with the advent of computer-based applications it became common practice to capture a variety of interaction behaviors and save them to log files for later analysis. In recent years, the rise of centralized, web-based computing has made it possible to capture human interactions with web services on a scale previously unimaginable. Large-scale log data has enabled HCI researchers to observe how information diffuses through social networks in near real-time during crisis situations (Starbird & Palen, 2010), characterize how people revisit web pages over time (Adar, Teevan, & Dumais, 2008), and compare how different interfaces for supporting email organization influence initial uptake and sustained use (Dumais, Cutrell, Cadiz, Jancke, Sarin, & Robbins, 2003; Rodden & Leggett, 2010).

In this chapter we provide an overview of behavioral log use in HCI. We highlight what can be learned from logs that capture people's interactions with existing computer systems and from experiments that compare new, alternative systems. We describe how to design and analyze web experiments, and how to collect, clean and use log data responsibly. The goal of this chapter is to enable the reader to design log studies and to understand results from log studies that they read about.

S. Dumais (✉) • J. Teevan
Microsoft Research One Microsoft Way, Redmond, WA 98005, USA
e-mail: sdumais@microsoft.com; teevan@microsoft.com

R. Jeffries • D.M. Russell • D. Tang
Google, Inc., 1600 Amphitheatre Parkway, Mountain View, CA 94043, USA
e-mail: robin@jeffries.org; drussell@google.com; diane@google.com

J.S. Olson and W.A. Kellogg (eds.), *Ways of Knowing in HCI*,
DOI 10.1007/978-1-4939-0378-8_14,

Table 1 Different types of user data in HCI research

	Observational	Experimental
Lab Studies		
Controlled interpretation of behavior with detailed instrumentation	In-lab behavior observations	In-lab controlled tasks, comparison of systems
Field Studies		
In the wild, ability to probe for detail	Ethnography, case studies, panels (e.g., Nielsen)	Clinical trials and field tests
Log Studies		
In the wild, little explicit feedback but lots of implicit signals	Logs from a single system	A/B testing of alternative systems or algorithms

What Are Behavioral Logs?

In HCI research behavioral logs arise from the activities recorded when people interact with computer systems and services. Behaviors of interest can include: low-level actions such as the keystrokes used when interacting with a productivity application; the content viewed in web browsers or e-readers; the search queries and result clicks captured by web search engines; browsing patterns and purchases on e-commerce sites; content generated and shared via social media; the history of edits to wikis or other documents; detailed traces of eye gaze position when playing a computer game; physiological responses when driving; etc.

An important characteristic of log data is that it captures actual user behavior and not recalled behaviors or subjective impressions of interactions. While logs can be captured in laboratory settings, they are increasingly captured at a much larger scale in situ as people interact with applications, systems, and services. Behavioral observations can be collected on a client machine or on remote servers. Client-side logging can be included in operating systems, applications such as browsers or e-readers, or special purpose logging software or hardware. Server-side logging is commonly used by service providers such as web search engines, e-commerce sites, or online courses. Some behavioral logs are publically available (e.g., Wikipedia content and edit history, Twitter posts, Facebook public feeds, Flickr photos, and Pinterest collections), but many are private, available only to individuals or service providers.

What Can We Learn from Behavioral Logs?

To understand what HCI researchers and practitioners can learn from behavioral logs, it is useful to compare them with other types of data. This book summarizes many different HCI methods for understanding behavior and improving design. In Table 1 we highlight a simplified view of a few approaches, described in more detail in other chapters, which are useful to contrast with log studies. The two dimensions represented in the table are: (1) whether the studies are observational or experimental, and (2) the naturalness, depth and scale of the resulting data.

Lab Studies represent the most controlled approach. In lab studies participants are brought into the laboratory and asked to perform certain tasks of interest. Demographic and other data can easily be collected about participants. The lab setting affords control of variables that are not of interest and enables detailed instrumentation of novel systems that could not be easily deployed more broadly. Researchers can learn a good deal about participants and their motivations in this way, but the observed behavior happens in a controlled and artificial setting and may not be representative of behavior that would be observed "in the wild." For example, a person may invest more time to complete a task in the lab than they might otherwise to please the investigator (Dell, Vaidyanathan, Medhi, Cutrell, & Thies, 2012). In addition, laboratory studies are often expensive in terms of the time required to collect the data which limits the number of different people and systems that can be studied. Lab studies can be either observational, examining interactions with a specific system of interest, or experimental, comparing two or more variables or systems.

Field Studies collect data from participants in their natural environments conducting their own activities, and, commonly, periodically ask them for additional information. Data collected in this manner tends to be less artificial than in lab studies but also less controlled. As with lab studies, demographic and other data can be collected about participants, but the researcher may still interfere with people's interactions by asking them about what they are doing. Field studies can be observational (e.g., where TV watching behavior is recorded for Nielsen panelists) or experimental (e.g., in clinical trials where new medical treatments are compared with a control).

Log Studies collect the most natural observations of people as they use systems in whatever ways they typically do, uninfluenced by experimenters or observers. As the amount of log data that can be collected increases, log studies include many different kinds of people, from all over the world, doing many different kinds of tasks. However, because of the way log data is gathered, much less is known about the people being observed, their intentions or goals, or the contexts in which the observed behaviors occur. Observational log studies allow researchers to form an abstract picture of behavior with an existing system, whereas experimental log studies enable comparisons of two or more systems.

Log studies are a valuable complement to other kinds of studies for several reasons. They represent traces of naturalistic human behavior uninfluenced by observers. Because log data provide a portrait of uncensored behavior, they give a more complete, accurate picture of all behaviors, including ones people are unlikely to talk about or remember accurately. Early analyses of web search logs, for example, found that searches for porn were common and were associated with different interaction patterns than other types of search (Silverstein, Henzinger, Marais, & Moricz, 1998). Similarly (Teevan, Adar, Jones, & Potts, 2007) observed that many queries were repeated, a behavior that would probably not be seen in the lab.

Logs also have the benefit of being easy to capture at scale. While laboratory and field studies typically include tens or hundreds of people, log studies can easily include data from tens or hundreds of millions of people. This large sample size means that even small differences that exist between populations can be observed. In particular, large-scale logs give a good picture of unusual but important behavior that is hard to capture in smaller studies. For example, if fewer than 1 in 100 people click on banner advertisements, a lot of effort would be required to collect a reliable number of such clicks in a laboratory setting. In contrast, the behavior can be significantly and reliably understood in web browser logs where there are millions of clicks on advertisements. As another example, student behavior logged during massively online courses can provide detailed insight into individual learning strategies and how they relate to educational success.

In spite of the benefits, logs have disadvantages, including non-random sampling (people must choose to use the system), uncontrolled tasks, and the absence of annotations to indicate motivations, success, or satisfaction. Logs provide a good deal of information about *what* people are doing but much less about *why* they are doing so and whether they are satisfied. This must be taken into account in analyses and complemented by other techniques to provide a more complete understanding of behavior.

In the remainder of this chapter, we describe in more detail large-scale log studies, providing examples of observational (section "Observational Log Studies: Understanding Behavior Through Log Analysis") and experimental (section "Experimental Log Studies: Comparing Alternative Systems Through Log Analysis") approaches, and discussing how to collect, clean (section "Collecting, Cleaning, and Using Log Data"), and share log (section "Using Log Data Responsibly") data. While publically available log data such as Wikipedia edit history or Twitter posts can support interesting observational analyses, there are fewer reports of large-scale experimental design and analysis, so we cover this aspect in particular detail. Most of the examples that we present in this chapter come from our own experiences in collecting and analyzing large-scale behavioral logs from web search engines, but the methods are more broadly applicable.

Observational Log Studies: Understanding Behavior Through Log Analysis

Most analyses of log data collected through observational studies provide a descriptive overview of human behavior. Simply observing behavior at scale provides insights about how people interact with existing systems and services, often revealing surprises. For example, an early analysis of more than a million web searches found that queries were short, averaging only 2.35 terms, and that over 80 % of all queries did not include advanced operators (Silverstein et al., 1998). Although these findings are consistent with what we expect for web search today, they were initially quite surprising because they differed from previous observations of search conducted in other contexts, such as in libraries. Librarians tended to issue queries that were much longer and included many more advanced operators. Another important observation from early web search engine logs was that query frequency was not uniform. Some queries were asked frequently (the 25 most common queries in the Silverstein et al. study accounted for 1.5 % of all queries), while others occurred very infrequently (64 % of the queries occurred only once). For additional results from observational studies of web search logs, we recommend (Jansen, 2006; Silverstein et al., 1998; Spink, Ozmutlu, Ozmutlu, & Jansen, 2002).

When analyzing log data, researchers extract a variety of metrics. *Metrics* are measurable quantities that matter to the users or system stakeholders. Metrics can emerge directly from the data, such as, in the case of search, query length or frequency. However, other metrics can be computed that allow researchers to infer information that is not directly represented in the raw data. For example, researchers have developed behavioral proxies for search success based on clicking and dwelling behavior (Fox, Karnawat, Mydland, Dumais, & White, 2005). While inferences based on such analyses can be noisy and imperfect, the large scale of log data can overcome distortions due to randomness. Noise due to factors that are not systematically related to the phenomenon of interest tends to even out with a large number of observations.

In order to place descriptive metrics in context, it is necessary to compare them to similar metrics for different contexts. For this reason, the process of learning about user behavior from log data typically involves partitioning the data into meaningful subsets, called *partitions*, and comparing across the different partitions. There are many different ways behavioral log data can be partitioned, including by language (Ghorab, Leveling, Zhou, Jones, & Wade, 2009), geography (Efthimiadis, 2008), device (Baeza-Yates, Dupret, & Velasco, 2007), time (Beitzel, Jensen, Chowdhury, Grossman, & Frieder, 2004), and user (Kotov, Bennett, White, Dumais, & Teevan, 2011; Teevan et al., 2007). Log data can also be partitioned by system variant, where behavior is observed under two different system conditions, such as when comparing two different user interfaces. How to successfully partition, collect, and analyze experimental data from system variants is discussed in section "Experimental Log Studies: Comparing Alternative Systems Through Log Analysis."

Two common ways to partition log data are by time and by user. Partitioning *by time* is interesting because log data often contains significant temporal features, such as periodicities (including consistent daily, weekly, and yearly patterns) and spikes in behavior during important events. It is often possible to get an up-to-the-minute picture of how people are behaving with a system from log data by comparing past and current behavior. For example, researchers can accurately predict the strength of seasonal flus based on search engine log data with a lag of only 1 day (Ginsberg, Mohebbi, Patel, Brammer, Smolinski, & Brilliant, 2009). In contrast, the Center for Disease Control and Prevention (CDC) typically reports such information with a 1–2 week lag. Care must be taken when partitioning log data by time because logs contain observations from many different time zones. This is discussed in greater detail in section "Collecting, Cleaning, and Using Log Data."

It is also interesting to partition log data *by user characteristics*. For example, researchers have looked at how advanced search users compare with novices (White & Morris, 2007), and how domain experts use different vocabulary, resources, and strategies than people who do not know about a domain (White, Dumais, & Teevan, 2009). In addition to comparing across users, it is also possible to look for patterns of behavior within an individual. In this way, researchers have discovered that people often return to common topics when they search (Kotov et al., 2011) and even repeat the same query over and over again (Teevan et al., 2007; Tyler & Teevan, 2010). One challenge with partitioning data by user is that it can be hard to accurately identify a user from log traces, and we discuss common ways this is done in section "Collecting, Cleaning, and Using Log Data."

Ideally the log partitions will be similar in all aspects other than what is being studied, to control for other factors potentially influencing the observed differences. It is useful to run a sanity check across partitions to confirm that metrics that should be consistent actually are. For example, White et al. (2009) examined differences in search strategies and outcomes for domain experts as compared to novices. Within the domain of interest, experts used different vocabulary, visited different sites, used different strategies, and achieved higher success. But, outside their domain of expertise, there were no differences in search performance, indicating that differences were isolated to the variable of interest and not to more general differences between the groups.

Although many useful things can be learned from observational log analysis, there are drawbacks to the approach. For one, rarely is there much information about the people who generate the data; not their age, gender, or even whether events observed at different times or on different machines are from the same person. Even less is known about user motivations. Logs cannot tell us people's intent, success, experience, attention, or beliefs about what is (or is not) happening. For example, when a person leaves a search page without clicking anything, it could be because they could not find what they were looking for, or it could be because the content presented on the result page was sufficient to satisfy their information need. As another example, click entropy, or the variation in what people click following a query, is often used as a proxy for how ambiguous a query is. However, while variation in clicks can arise due to variation in intent, it can also be caused by changing results or an ambiguous need that requires synthesis across multiple pages to meet (Teevan, Dumais, & Liebling, 2008). We also do not know when people are confused. For example, a person may inadvertently switch from general web search to image search, yet still believe they are searching the entire web when they issue a query.

Additionally, when a system uses log data to drive its own performance, people may have ulterior (often adversarial) motives to create artificial traces. For example, if a search engine boosts results that are consistently clicked on for a query towards the top of the ranked list, then spammers may game the system by repeatedly clicking a result solely for the purpose of boosting its rank (Fetterly, Manasse, & Najork, 2004).

One way to mitigate such limitations is by controlling for as many external factors as possible. In the case of query ambiguity, for example, appropriate metrics will consider not just click entropy but also the entropy of the results returned and the average number of clicks per user. This kind of deep analysis is best done by examining a sample of the user traces directly (often by a hand examination of a representative sample) and not just computing metrics over them, so the researchers can be sure that important insights have not been lost in the averaging of millions of data points. We describe more about how to look at and clean data is discussed in section "Collecting, Cleaning, and Using Log Data."

Another way to better understand what is going on within log data is to supplement the logs by capturing context that may not be directly represented. Capturing as much contextual data as possible up front, including the version of the system being interacted with and what users of the system actually see, can be critical to understanding log data or comparing data collected at different points in time. Additionally, log data can be enhanced with insight about what the user is doing via field trials or critical incident studies. For example, Broder (2002) conducted a critical incident study in which users were occasionally interrupted with a pop up window asking the motivation behind the query they had just issued. This led to the classification of queries as navigational (i.e., targeted at getting to a particular web page), informational (i.e., intended to find information that is present in one or more pages), or transactional (i.e., intended to perform a web-mediated activity).

Even when the log data is very rich, researchers should not rely solely on logs to understand user behavior. Converging methods can help confirm and provide insight into what is learned from log data. Methods that complement log analysis include usability studies, eye tracking studies (see Eye Tracking in HCI: A Brief Introduction, this volume), field studies, diary studies, retrospective analysis (see Looking Back: Retrospective Study Methods for HCI, this volume), and surveys (see Survey Research in CHI, this volume). For example, Teevan and Hehmeyer (2013) analyzed the logs of a popular enterprise communication system that infers and projects availability state of users. They found that people were significantly more likely to answer the phone when their status indicated that they were busy, but were unable to tell from the logs why this was the case. By conducting a complementary survey study they discovered that busy people perceived incoming phone calls as particularly important because they knew the caller chose to call as opposed to email. The survey helped to explain the rich, real-world picture provided by the log files, and gave a view into communication behavior that would not have been possible otherwise.

In addition to inspiring complementary studies and suggesting interesting areas for further research, the results of observational log studies make it possible to design computer systems that support people's actual behavior rather than their presumed behavior. For example, search engines were designed to cache search result pages because query log analysis revealed that only a handful of unique queries represent a significant portion of search engine traffic (Silverstein et al., 1998).

Log data can also be used to test hypotheses that researchers develop about user behavior. Lau and Horvitz (1999) used log data to learn a probabilistic model of how users refine their queries over the course of a session, and then evaluated how well they could predict the next action in a sequence using log data. Likewise, Kotov et al. (2011) used log data to learn a predictive model of whether web searchers were likely to return to their current search task sometime in the future. To avoid over-fitting, models should be learned from a subset of log data and evaluated on data that is similar in characteristic to (but not the same as) the data used to construct the model.

All of the studies described above rely on logs recorded from existing systems. As such, the findings are limited to understanding existing interactions. It is impossible to learn from observational log analysis how people might interact with a different system. For example, people may want to use facets to navigate web search results, filtering results by how recent the content is or the quality of the web site. But search logs cannot reveal this because people do not currently have the option to use facets. Similarly, people might want to search for old posts on Twitter, but this behavior is rarely observed (Teevan, Ramage, & Morris, 2011). This could be because old posts are not interesting, or it could be an artifact of the fact that the Twitter search interfaces currently only returns the most recent tweets. Observational log analysis can only reveal what people do with the tools they have. A richer way to test hypotheses that allows researchers to explore new interaction paradigms is to vary the system users interact with in an experimental framework, and compare how the logged behavior differs across system variants. This is discussed in detail in the next section.

Experimental Log Studies: Comparing Alternative Systems Through Log Analysis

To understand how people react to different user experiences, a typical approach is to run an in situ experiment designed to compare behavior across different system variants. Web experiments are commonly used to understand and improve a variety of services (Kohavi, Longbotham, Sommerfield, & Henne, 2009; Tang, Agarwal, O'Brien, & Meyer, 2010) and may be colloquially called *A/B tests* (meaning comparing system A with system B) or *bucket tests* (because users are "bucketed" into different user experiences, from the term "hash bucket").

Log-based experiments are often the only way to evaluate small changes to a system (e.g., a change in font; a change in the text label on a button or in a status message) and can also be useful to evaluate larger changes (e.g., complete redesign of a site's page layout or a change in the workflow to complete a task). The nature of an experiment and its analysis differ as a function of the change being studied. Large changes tend to produce noisier data and thus require more data points to get a statistically reliable signal. For small changes, the analyst will usually be able to identify a small number of metrics that should improve for the design change to be considered a success. In the case of a major change, the analyst may have metrics specific to the goals of the particular experiment (e.g., how many people completed the task flow; how much time people spent on the newly redesigned site), but may also need to look at a broad suite of metrics, particularly since a priori hypotheses about how behavior will change are harder to arrive at. Even if tools that automatically compute typically useful metrics are available [e.g., Google Analytics (Google,

2012)], the researcher may need to conduct a custom analysis for a large change. If the goal of a redesign is to move clicks from one category of UI element to a different one (e.g., changing from clicks on elements that move one forward to clicks on elements that invite deeper exploration), then a custom analysis that looks at exactly those clicks will be needed.

Sometimes it feels that doing a formal experiment to compare two approaches is unnecessary because the answer is obvious (typically that the new design is "better"). Experience says otherwise: those of us accustomed to doing log analysis have many times seen the outcome of an experiment differ from the supposedly obvious answer. For example, changing a text button's font to make it more visible and more likely to be clicked may backfire if instead it makes the button seem not to be a clickable object. Moreover, even when the intended outcome does occur, there can be unanticipated side effects. For example, people may use additional information available in one of the conditions, such as automatic spell correction in search, but are slowed down because they feel compelled to click through to the results for their misspelled query. Side effects are typically not easily predicted, but may be important enough to overpower the positive aspects of a change. Without an experiment and examination of a broad set of metrics, a team may not discover tradeoffs inherent in a proposed change.

The basics of log data collection are covered in section "Data Collection." We next cover how to design an experiment that will give valid results (meaning if it were repeated by a different set of experimenters, similar results would be expected) and the basics and common pitfalls of analyzing experiments.

Definitions

In order to talk about experiments we need to have a common language. Here are a few important definitions that are important in understanding server-based web experiments:

Request: A user action or user request for information. For web applications, a new web page (or change in the current page) is requested via a set of parameters (this might be a query, a submit button at the end of a form, etc.). The result of the request is typically the unit of analysis that the researcher is interested in (e.g., a query, an email message displayed, a program being debugged).

Cookies: A way of identifying a specific session in a browser. We typically equate a cookie with a user, but this is an important source of bias in log studies as described in more detail in section "Collecting, Cleaning, and Using Log Data."

Diversion: How traffic is selected to be in a particular experimental condition. It might be random at the request level; it might be by user id, by cookie, or as a hybrid approach, by cookie-day (all requests from a given cookie are either in or out of the experiment each day). Typically experiments for user experience changes are user-id or cookie based.

Triggering: Even if a request or cookie is in the experiment, the experimental change may not occur for all requests. For example, if the change is to show the current weather at the user's location on weather-related queries, this will only trigger on the small fraction of queries that are weather related; all conditions in the experiment will provide identical experiences for other queries. On the other hand, for an experiment that changes the size of the logo shown in the upper left of all pages, all requests diverted into non-control conditions will trigger the change.

Experiment Design

A web-based experimental condition diverts some subset of the incoming requests to an alternate processing path and potentially changes what is shown to the user. A control condition diverts some subset of incoming requests, but does not change what is shown to the user. For example, in an experiment that explores the effect of the size of the logo in the upper left corner of the page, the control would leave the logo unchanged, one experimental condition would increase the size of the logo, and another condition would decrease the size of the logo.

The design of an experiment begins with a set of questions to be answered. In HCI experiments, the question will be a variant of, "Will this user experience be better for users (in some way that needs to be defined and quantified) than another experience (which might be the existing experience)?" "Better" may be defined, for example, as getting results faster, or it may be defined as being more fun, resulting in people spending more time on the site. A log experiment only makes sense with specific, concrete research questions. For a small change, this is likely to be easy; for example, if a UI manipulation changes the font of the text on the page, the associated hypotheses will most likely have to do with engagement with the page. But with a larger change, such as changing the location and wording of links on the homepage of a site, it is harder to attribute changes in behavior to specific changes made. For example, one result of the page layout change might be that more people click on the "Contact Us" link. First, are more clicks on "Contact Us" good? Second, if the primary metrics for this experiment are site-wide—e.g., how much time do people spend on the site? Do they end up engaging with the site in a way that represents the core purpose of the site?—how does the improved access to the Contact Us page contribute to that metric, and is it a positive or negative contribution?

Questions such as these still needs to be turned into testable hypotheses about the qualitative and quantitative behavioral changes expected. For example, the experimenter may hypothesize that the condition with a large red button reading "sign up" will result in significantly more signups. There may be several testable hypotheses that relate to different metrics. Alternatively, if the goal is to explore a space of design choices (e.g., whether a button 1, 2, or 3 in. tall results in the most signups), there may not be specific predictions beyond the expectation that behavior will differ across the conditions.

Testable hypotheses will lead to a set of conditions to test. A condition may or may not be "user visible" (one might experiment with different delays in delivering the page, which changes the user experience but not in a visible way), or even "user noticed" (a change in the order of search results on a page will not be noticed unless a user can see results from more than one condition in the experiment).

A special condition is one designated as the *control*. The control is the "baseline" user experience to which the "new, improved" experience is being compared. Typical controls in web experiments include the existing treatment of a feature, or to not show a feature shown in another condition. If there are two novel treatments, neither can be considered the control—there needs to be a third condition to serve as the control. In some experiments, there may be multiple controls.

In an ideal world one would test a set of parameters (the simplest case being one two-level parameter: feature X is "on" or "off") and all the factorial combinations of them. In practice, only a subset of the possible combinations is tested. This may be:

A practical constraint: if the parameters are the background color of the page and the text color, setting both to the same value produces an unreadable page,

A logical constraint: neither a large nor a small image makes sense for the condition where no image is shown, or

A resource constraint: there may not be enough traffic to make statistically meaningful comparisons among a large number of different conditions unless the experiment is run for an unrealistically long time.

Because of this log experiments are seldom factorial experiments and are typically analyzed as a set of pairwise comparisons with the appropriate control.

We have covered the aspects of experimental design that are determined by the study's research questions; next we discuss aspects that influence the analysis phase or have pragmatic significance.

Making Conditions Comparable

A good deal of experiment analysis depends on having a control condition that is directly comparable to the experimental conditions. However, this is often more complex than imagined. There are a variety of ways that comparisons between conditions go wrong.

It is important to run all conditions in the experiment concurrently, if possible. It may be tempting to run the control and the experimental conditions at different times, because the traffic to the site being experimented on is low, and it is less work to change the user experience for the entire site than it is to randomly divert traffic to several different versions of the site. However, all sorts of things can happen that make those time periods different: a major event that relates to the site (e.g., a sports event for a sport related site); a significant shopping period, like Valentine's Day; a situation that causes people to be on-line more (or less). If the site attracts visitors from many different countries, the experimenter may not even be aware of all the relevant holidays or important news events.

Users should be assigned to conditions by some random process, not by, for example, an opt-in process. Opt-in can be useful for getting feedback on how well a particular user experience works, but this approach does not constitute a valid experiment. Participants in the experimental condition will be people who are early adopters, interested in technology, and perhaps need the feature being provided. The control will be full of people who do not like change, could care less about the latest technology and might find the new feature gets in the way of what they want to get done. These two groups will likely behave differently even with identical user interfaces, and we cannot accurately attribute the differences we see to the new user experience. In addition, the experiment is only measuring the reactions of a fraction of the intended population.

There are, of course, many other ways the various conditions may differ subtly. All conditions should be diverting on people from the same countries, speaking the same language, on the same types of devices (e.g., tablets are different from desktops). Similarly, all requests need to be assigned to conditions by the same method: request (every incoming request is randomly and independently assigned to a condition), cookie (all requests from a single cookie are in a particular condition throughout the entire experiment), or cookie-day (within a day, requests from a cookie are in a specific condition, but each day which condition the cookie is in is randomly selected). The decision about which method to use for assignment is a tradeoff among the need for consistency in user experience, possible changes due to learning, and independence of observations (e.g., those based on cookies or cookie-day assignments are not independent).

The idea of *counterfactuals* is important in designing and analyzing log experiments. Often the experimental change in user experience does not occur for every request. As described earlier, if the feature is to show the current weather whenever the user searches for "weather," people in the experimental condition will only see the weather information for a small fraction of searches. The counterfactuals are the requests where the person *would* have seen weather information if they had been in the experimental condition, but did not, because they were in the control. Thus it is very important to mark the counterfactual events in the control logs. This enables the analyst to easily identify the comparable subset of requests. Otherwise, any effects on the small subset of searches which change will be diluted in the larger body of unchanged behavior.

Experiment Sizing

Experiment sizing is about determining the power the experiment requires in order to detect differences of interest (Huck, 2011). A power calculation (Wikipedia: Power) is the number of observations needed to see a statistically reliable difference, assuming one exists. This determines the minimum number of data points needed in each condition of the experiment. One reason experimenters want to control for size is that log experiments are typically run with real-world customers and carry the risk of a negative user experience, so efficiently discovering whether the experience is positive or negative is important.

When sizing log experiments it is important to recognize that the kinds of changes we typically want to measure (e.g., a 5 % increase in the number of visitors who book a hotel room, a 3 % increase in the number of visitors who sign up for the newsletter or sign the petition) require large numbers of observations. This is in part because the changes are often small. But the larger issue is that log data is extremely noisy. It is an amalgam of people doing what appear to be the "same" actions, but with very different tasks and intents behind them. And in many cases, they come from multiple countries, with different languages, different cultural assumptions, etc. Thus, we often need tens of thousands of observations or more to get significant results, sometimes a lot more.

Given the need for very large numbers of observations to get statistically reliable results, it can be frustrating to run an experiment and discover that while the control and experiment conditions differ by an amount that would be considered practically meaningful, those differences are not statistically meaningful.

To estimate the number of observations needed to detect differences that are "interesting," the analyst needs to determine:

- The metric(s) of interest,
- For each metric, the minimum effect size change that the experiment should be able to detect statistically, e.g., a 2 % change in click-through rate,
- For each metric, the standard error.

Each of these is explained further below.

Deciding what effect size matters can be a challenge until an analyst or group has carried out enough experiments to know what level of change is practically important. But a larger problem is how to estimate the standard error, especially for metrics that are a ratio of two quantities (e.g., CTR (click-through-rate)—the number of clicks/number of queries). The most common problem arises when the unit of analysis is different than the experimental unit. For example, for CTR, the unit of analysis is a query, but for cookie-based experiments (as most user experience experiments will be), the experimental unit is a cookie (a sequence of queries) and we cannot assume that queries from the same cookie are independent. There are a variety of ways to calculate the standard error when the observations are not independent—two common ones are the delta method (Wikipedia: Delta method) and using "uniformity trials" (Tang et al., 2010).

Sizing is also impacted by the triggering rate, i.e., the fraction of the traffic that the experimental change actually impacts (see definition in section "Definitions"). If a particular experimental change impacts only 5 % of the traffic, then it will take 20 times as long to see an effect than if the change happens on all the traffic. Table 2 shows the effect that different triggering rates have on the number of observations (queries) needed to see an effect. Column 5 shows that more queries are required for lower triggering rates, and Column 7 shows that if counterfactuals are not logged

Table 2 The number of observations needed to achieve a given standard error as a function of the triggering rate (the fraction of traffic the experimental change impacts) and the effect size (the size of a change that would be meaningful in the experiment)

Metric standard error	Trigger rate (%)	Effect size on affected traffic (%)	Needed queries (affected)	Queries needed in expt. (counterfactuals logged)	Effect size if no counterfactuals (measured on all traffic)	Queries needed in expt. (no counterfactuals logged)
5	1	10	52,500	5,250,000	0.1 % (10 % * 1 %)	525,000,000
5	5	10	52,500	1,050,000	0.5 % (10 % * 5 %)	21,000,000
5	20	10	52,500	262,500	2 % (10 % * 20 %)	1,312,500
5	50	10	52,500	105,000	5 % (10 % * 50 %)	210,000

The required number of observations grows inversely with triggering rate and effect size

even more queries are required since experimental differences are diluted by all the observations that are the same across conditions. In practice, it helps to have some historical information to make an educated guess about what the triggered fraction will be in calculating the number of (diverted) observations needed for the given power level.

Interpreting Results

Once the experiment has been designed, users have been exposed to the experimental variants, and information has been logged about their behavior, the researcher needs to perform the analyses that will answer the original experimental questions.

Sanity Checks: The analyst's first task is to make sure that the data makes sense. An initial step is to calculate means, standard deviations, and confidence intervals for the metrics of interest. This may be done via a dashboard, such as in Google Analytics, via a script written in a general purpose language such as Python, or in a special purpose language specifically optimized for log analysis [e.g., map-reduce languages, such as Hadoop (Wikipedia: Hadoop)]. It is particularly important to look at overall traffic in all the conditions, which should be the same with random assignment. If there are differences in overall traffic across conditions, be sure to rule out the many artifacts (including bugs in logging) before assuming that an observed difference is a real effect. It is also important to break down the data in as many ways as possible—by browser, by country, etc. It may be that some differences are actually caused by a small subset of the population instead of the experimental manipulation.

Interpreting the metrics: In log analyses, it is standard practice to use confidence intervals (Huck, 2011) rather than analysis of variance significance testing, because the conditions are not organized into a factorial design with interaction terms, and because a confidence interval gives useful information about the size of the effect and its practical significance that is not as easily visible in a significance table.

It is conventional to use a 95 % confidence interval in comparing each of the experimental conditions to the control.

As described earlier, it is common to consider many different metrics, e.g., click-through rate (on results, ads, whole page), time to first click, and time on page. Given a large number of metrics, some of those supposedly significant differences are spurious (about 1 in 20, to be precise). How does one decide which to trust? Look for converging evidence; are there other metrics that ought to increase/decrease when this one does, and do they move in the appropriate direction? Might a logging error account for the effect? Is this a difference seen before in other experiments? How this is done depends on the domain and on previous experience—for example, in search, clicks per ad-shown and clicks per page are very likely to be correlated, but these click metrics are unlikely to be correlated with conversions (when someone purchased something on a page that they got to from an ad). Recognizing which changes may be artifactual comes from experience, and is somewhat of an art. It is most important to have converging evidence for the metrics that will directly impact decision-making. An important source of artifacts to be aware of are those that lead to Simpson's Paradox, which arises when ratios that have different denominators are compared (Wikipedia: Simpson's Paradox). These situations are very common in log experiments, and all experimenters should be aware of them, how to identify them, and how they change the analysis.

Now that the believable significant metrics have been identified, what is the broader interpretation of the results? Generally, there will be agreement of whether a metric's "good" direction is an increase (number of clicks) or a decrease (latency, time to click). Most likely some metrics will move in the "good" direction and some in the "bad" direction. If this is not a logging error, it is a tradeoff. People may be clicking more (because the experimental UI gives them more things to click on), but it takes them longer (because there are more things to decide among). Is the net effect positive or negative? This is always a judgment call, but a broad set of metrics may lead to a more holistic answer. For example, look for a measure of the total interaction with the site, rather than just the clicks on an individual UI element. Otherwise, go back to the original goals of the study. If the goal is to get people to the best possible information/outcome, then what they click on matters most. If the goal is to get them there quickly, then latency matters (possibly at the cost of more clicks). If the goal is to maximize the number of people who find useful information (or sign up or are able to send a message), then the fraction of people who click through to an "end result" page is the most critical measure. Sometimes it is not possible to make a single statement about what constitutes "good," there are a classic set of tradeoffs to make among speed, efficiency, usability, and design consistency. It is the analyst's job to decide and justify the decision to stakeholders (or conference paper reviewers).

Practical Significance: While we have been emphasizing the importance of statistical significance, it is also important to take practical significance into account. The experiment may show a statistically reliable difference when, say, the number of "undo" commands goes from 0.1 to 0.12 %, but that small a number might not have any practical significance (it, of course, depends on the application and what people use undo for). Do not get so blinded by the statistics (which are easy to compute) that the practical importance (which can be harder to determine) gets lost from the discussion.

Whenever possible, it is useful to have an understanding of the range of values the metrics being used take in a steady state—this is obviously not possible when just starting out with a product or prototype that has not previously been exposed to users. But knowing what typical values are and how hard it is to move those values (based on previous experiments) will help assess meaningful changes in the metrics.

So far in this chapter we have taken the existence of logs for granted. In the next two sections we consider how to collect log data, focusing on three important practical challenges: data collection, data cleaning, and using data responsibly.

Collecting, Cleaning, and Using Log Data

A typical log for a web service requires infrastructure that creates data (at the server end) about users' interactions with the service, recording all relevant information needed for later analysis. If the infrastructure also supports experiments, it must enable different users, chosen according to some appropriately random scheme, to be shown pages associated with different conditions in the experiment.

This recording infrastructure might be web server logs (Ogbuji, 2009) or logs created by special analytics packages such as Google Analytics (Google, 2012). There will be ways to configure the log-recording program to record parameters specific to the experiment or the activities the researcher is interested in (Brown, 2012).

Logs can also be created with code at the client end, but this requires users to download some sort of logging program and for that program to send data back to the server (Capra, 2011; Fox et al., 2005). Both client and server logging are lossy, but each type tends to lose different kinds of data: the server may not receive full information about aborted or timed-out operations; the client will not be aware of data that is not reported to the application that contains the logging code. Both the kind of data loss that is acceptable and the challenges of users needing to install a software plugin are important considerations for deciding between server- and client-side logging.

Data Collection

Consider a simple example of how to understand the success and strategies of searchers using a web search engine. At a minimum, a useful log would capture what queries people issue, which (if any) search results they click, and a timestamp for each of these actions. An ideal log would additionally allow the experimenter to reconstruct exactly what the user saw at the moment of behavior. But at a minimum, a web service should log:

- The time an event happened,
- The user session it was part of (often via a cookie),
- The experiment condition, if any,
- The event type and any associated parameters (e.g., on login page, user selected "create new account").

Logging a simple query is straightforward. But practitioners need to understand the different challenges that need to be addressed when collecting data for deeper analysis.

Recording accurate and consistent time is often a challenge. Web log files record many different timestamps during a search interaction: the time the query was sent from the client, the time it was received by the server, the time results were returned from the server, and the time results were received on the client. Server data is more robust but includes unknown network latencies. In both cases the researcher needs to normalize times and synchronize times across multiple machines. It is common to divide the log data up into "days," but what counts as a day? Is it all the data from midnight to midnight at some common time reference point or is it all the data from midnight to midnight in the user's local time zone? Is it important to know if people behave differently in the morning than in the evening? Then local time is important. Is it important to know everything that is happening at a given time? Then all the records should be converted to a common time zone.

The language of the interaction is often an important variable in studying behavior from people in multiple countries. If grouping by language is necessary to analyze the data, one has to determine what does "language" mean? Be careful not to confuse the user's country with their language, or the language the UI is presented in with the language of the words people type in their interactions or queries. They will often differ, especially if the experiment runs in countries where people speak multiple languages. Depending on the question of interest, it may be appropriate to partition by the language of the query or UI.

UserIDs are another challenge for identifying distinct users accurately. HTTP cookies, IP addresses, and temporary IDs are broadly applicable and easy to use, but there is a great deal of churn in such IDs (Jupiter Research Corporation, 2005). Further, cookies and temporary IDs are not uniquely associated with individuals—several people can use the same browser instance, and the same person may use multiple devices. A closer correspondence between IDs and individuals can be achieved via logins or client code, but this requires that people sign in or download client code, raising other issues (e.g., of a biased sample). Either way there is a bias in data that is captured, which can have significant impact on the results and is often overlooked by analysts.

In web logs, knowing where a page request, such as a query, came from can be important in understanding unique behavioral patterns. Queries can be generated from many different entry points—from the home page of search engines, the search results page, a browser search box or address bar, an installed toolbar, by clicking on query suggestions or other links, etc. Other applications will also have large numbers of entry points. Metadata of this kind (e.g., about the point of request origin) may be useful in later analysis, but involves additional planning and effort to collect at this stage. Without this kind of contextual information, partitioning and interpreting the data is much harder. Ultimately, all of the data and metadata collected needs to be in service of the overall goals of analysis—this defines what needs to be logged.

Data can also be distorted by exogenous factors that might also need to be considered. The site might become unexpectedly viral, perhaps being picked up by SlashDot, Reddit, or the New York Times. Virality can cause a huge swing in logged behavior. This usually occurs when a blogger shares a link that has an implicit parameter assigning all those visitors to the same condition. Even if the virality does not cause a change in how users are allocated to conditions, the behavior of people who visit out of curiosity is different from that of regular users. And if the onslaught of new users causes the site to slow down or go down, the data is even less realistic. In addition, real world events, such as the death of a major sports figure or a political event can often cause people to interact with a site differently. Again, be vigilant in sanity checking (e.g., look for an unusual number of visitors) and exclude data until things are back to normal.

Data Cleaning

A basic axiom of log analysis is that the raw data cannot be assumed to correctly and completely represent the data being recorded. Validation is really the point of data cleaning: to understand any errors that might have entered into the data and to transform the data in a way that preserves the meaning while removing noise. Although we discuss web log cleaning in this section, it is important to note that these principles apply more broadly to all kinds of log analysis; small datasets often have similar cleaning issues as massive collections. In this section, we discuss the issues and how they can be addressed.

How can logs possibly go wrong? Logs suffer from a variety of data errors and distortions. The common sources of errors we have seen in practice include:

- *Missing Events*: Sometimes client applications make optimizations that (in effect) drop events that should have been recorded. One example of this is the web browser that uses a locally cached copy of a web page to implement a "go back" action. While three pages might be visited, only two events may be logged because the visit to the cached page is not seen on the server.
- *Dropped Data*: As logs grow in size, they are frequently collected and aggregated by programs that may suffer instabilities. Gaps in logs are commonplace, and while easily spotted with visualization software, logs still need to be checked for completeness.

- *Misplaced Semantics*: For a variety of reasons, logs often encode a series of events with short (and sometimes cryptic) tags and data. Without careful, continual curation, the meaning of a log event or its interpretation can be lost. Even more subtle, small changes in the ways logging occurs can change the semantics of the logged data. (For instance, the first version of logging code might measure time-of-event from the first click; while later versions might measure time-of-event from the time the page finishes rendering—a small change that can have a substantial impact.) Since data logging and interpretation often take place at different times and with different teams of people, keeping semantics aligned is an ongoing challenge.

Data transformations: The goal of data-cleaning is to preserve the meaning with respect to an intended analysis. A concomitant lesson is that the data-cleaner *must track all transformations performed on the data*. As data is modified (e.g., removing spurious events, combining duplicates, or eliminating certain kinds of "non-signal" events), the data-cleaner must annotate the metadata for the log file with each transformation performed. Ideally, the entire chain of cleaning transformations should be maintained and tightly associated with progressive copies of the log. Not all data transformations can be reversed, but they should be recreatable from the original data set, given the log of actions taken.

The metadata associated with a dataset is necessary for other analysts to understand what the log files include. The metadata should have enough information so that the "chain of change" can be tracked back from the original file to the one that is used in the final analysis. If metadata about the cleaning of the log file is missing, analysts cannot know the semantics of the data they are analyzing. No matter how complete the record may appear, without the metadata the researcher can never be sure, and confidence in conclusions based on analysis is undermined.

Understanding the structure of the data: In order to clean log data properly, the researcher must understand the meaning of each record, its associated fields, and the interpretation of values. Contextual information about the system that produced the log should be associated with the file directly (e.g., "Logging system 3.2.33.2 recorded this file on 12-3-2012") so that if necessary the specific code that generated the log can be examined to answer questions about the meaning of the record before executing cleaning operations. The potential misinterpretations take many forms, which we illustrate with encoding of missing data and capped data values.

A common data cleaning challenge comes from the practice of encoding missing data with a value of "0" (or worse; "zero" or "−1"). Only with knowledge of what is being captured in the log files, along with the analyst's judgment of the meaning of missing data, can reasonable decisions be made about how to treat such data. Ideally, data logging systems represent missing data as NIL, ø, or some other non-confusable data value. But if the logger does not and uses a value that is potentially valid as a behavioral data value, the analyst will need to distinguish valid "0"s from missing data "0"s (for example), and manually replace the missing data with a non-confusable value.

Capped data values, usually expressing some value on a scale that has an arbitrary max (or min) value, can cause both cleaning and analysis problems. Unless the analyst knows that a particular data value being captured varies between integer values 0…9, validating data as well as making decisions about cleaning are compromised. For example, if the log is capturing data whose value is capped at 9 (because we "know the value can never go higher"), this can lead to an insight when a long string of 9's suddenly appears in the log stream.

Our point is that while the logs might not lie—they represent the values actually written out by the logging system—the interpretation of those values relies on knowledge and expertise about the data being captured on the part of the person doing the interpretation. It is not enough to know that a given action took place at a given time, the analyst also needs to know things like the possible range of values the data may take, whether measurements can be capped, and how missing data is encoded, detected, and interpreted.

Outliers: All data sets have an expected range of values, and any actual data set also has outliers that fall below or above the expected range. (Space precludes a detailed discussion of how to handle outliers for statistical analysis purposes, see: Barnett & Lewis, 1994 for details.) How to clean outliers strongly depends on the goals of the analysis and the nature of the data.

Outliers often indicate the range of human possibility, frequently in ways that experimenters (and system designers) do not anticipate. It is not uncommon in log studies to find outliers that are many standard deviations away from the mean. For example, while web search query sessions typically have a mean of around two queries/user-session, the upper end of the distribution can be in the high hundreds. People's behavior is widely variable.

If a system needs to perform correctly over a wide behavioral distribution, then outliers give valuable information about what the boundaries are, and how the system will need to respond. On the other hand, outliers can also often signal underlying exceptional cases in the logging system, the system/application being logged, or spurious signals that have been added to the log stream. Outliers should always be checked for signal integrity. That is, are outlier data points *actual* data, or are they due to some kind of system or logging error along the way? Verifying what causes the outliers to appear will dictate the approach to data cleaning that should be taken.

For a more complete description of data cleaning practices, see (Osborne, 2012).

In summary, it is important for analysts to understand the data they are analyzing. Publications that summarize a log analysis study need to include a careful description of what data cleaning steps were taken, and why they were undertaken. When a log dataset is handed from one researcher to another, data cleaning metadata must be included, along with descriptions of how the cleaning was done (preferably with pointers to the tools and settings used so that reanalysis can be done if needed).

Using Log Data Responsibly

Companies and universities may have data collection, retention, access, and use policies, and it is the researcher's responsibility to seek these out, be aware of their implications, and to make sure they comply with them in the course of carrying out research involving log data. These policies arise from internal business practices, usage agreements with users of a service, and sometimes from government regulations regarding particular types of data or privacy protections. It is fair to say that standards and best practices continue to evolve in an effort to balance privacy concerns with the potential benefits that derive from a richer understanding of user behavior. For additional perspectives on these policies and issues, see Research and Ethics in HCI, this volume.

Additional concerns arise when researchers wish to share data more broadly, often to support academic research. Risks associated with personally identifiable or sensitive information must be addressed before data is shared. Such risks may not be obvious, so researchers must proceed with due caution. Two recent examples serve to highlight the subtlety of risks associated with indirectly identifiable information—in one case involving the use of locations and names; and another case involving linking multiple sources of information.

AOL. On Aug 4, 2006, AOL search data was released to the academic community on. The data consisted of <AnonID, Query text, Query time, Rank of clicked item, URL of clicked item>. Two days later a New York Times story revealed that AnonID 4417749 was Thelma Arnold, a 62 year old woman from Lilburn GA. (Barbaro & Zeller, 2006). Two weeks later two AOL employees were fired and the CTO resigned. How did the anonymized data lead to Thelma Arnold? She issued multiple queries for businesses in Lilburn GA (a town of ~11 k people). She also issued multiple queries for people named Arnold (only 14 people with the name Arnold in Lilburn). A reporter from the NY Times contacted all 14 of these people and Thelma acknowledged to the reporter that the queries were hers. Truly anonymizing log data can be extraordinarily difficult (Wikipedia: AOL Search).

Netflix. On Oct 2, 2006 Netflix announced the Netflix Prize, an open competition for developing new collaborative filtering algorithms for predicting a user's ratings for films, along with a $1 million prize for the first team to beat the current Netflix algorithm by 10 %. The data consisted of: <MovieID, UserID, Rating, Date of Rating> <MovieID, Title, Year>. Care was taken in the data released to the public, including the introduction of random noise. The prize was awarded on Sept 21, 2009, and another competition was announced. Narayanan and Shmatikov (2008) published a paper in which they described how to de-anonymize Netflix IDs in a way that was robust to noise perturbations in the data using background knowledge of reviews in IMDB. On Dec 17, 2009 a suit was filed by one of the people identified in this manner. On Mar 12, 2010 the second Netflix competition was cancelled and no new data released (Wikipedia: Netflix).

When data is shared beyond the set of people bound by the privacy protection policies associated with the original data collection, there may be benefits to the research community, but there are also risks to the privacy of the individuals whose behavior has been logged that are extremely difficult to predict in advance. For this reason, releasing data, even when anonymized, has serious ethical risks.

Summary

In this chapter we have discussed an increasingly important way of knowing how people interact with computer systems: log analysis. As it becomes easier for computer systems to record people's interactions with technology, it is also becoming vital for HCI researchers and practitioners to know how to understand this information. We started the chapter with an eye toward understanding what we can learn about how humans interact with online systems using both observational and experimental log data. Observational log studies enable HCI researchers to summarize patterns of user interactions with existing systems. Experimental log studies enable researchers to compare behaviors in two or more systems.

Large-scale log studies give rise to a wide range of practical problems that need to be addressed to produce reliable results. From setting up the logging system, through experiment design, data collection, data cleaning and interpretation, log analysis rewards careful tracking of what is being done at every step along the way. As we have seen, it is a mistake to think that an incomplete or unreliable logging system will actually reveal deep truths about human behavior. Constant sanity checking and validation of sample sets of log data is essential to developing confidence that what is being logged and interpreted, accurately reflects human use and behaviors.

Although web logs are commonly used to understand how people interact with web services, there are few resources for becoming more expert in the method. Kohavi et al. (2009) and Tang et al. (2010) provide examples of web experimentation in practice and describe the underlying experimental infrastructure and associated analysis tools needed to carry out such experiments. Crook et al. (2009) and Kohavi et al. (2012) present interesting examples of common pitfalls, highlighting that producing reliable results requires ongoing vigilance about every aspect of experimental design, logging, and analysis. There are some available software tools for parsing web server logs and summarizing metrics (Ogbuji, 2009; Google Analytics).

Despite the challenges of dealing with massive data sets, the use of logs to extract useful insights about people's interaction with technology is becoming more common and more useful in HCI research.

Exercises

1. Name all the digital things you can think of to log. What inferences might you draw from each? What would you like to infer but can't?
2. What are the differences between logs and sensor data streams? What analysis method details are the same for both?

References

Adar, E., Teevan, J., & Dumais, S. T. (2008). Large scale analysis of web revisitation patterns. In *Proceedings of CHI 2008* (pp. 1197–1206). New York: ACM.

Baeza-Yates, R., Dupret, G., & Velasco, J. (2007). A study of mobile search queries in Japan. In *Proceedings of WWW 2007 workshop on query log analysis: Social and technical challenges.* New York, NY: ACM.

Barbaro, M. & Zeller, T. (2006). A face is exposed for AOL searcher No. 4417749, *New York Times,* Retrieved on August 9, 2006, from http://www.nytimes.com/2006/08/09/technology/09aol.html?_r=1

Barnett, V., & Lewis, S. (1994). *Outliers in statistical data.* New York, NY: Wiley & Sons.

Beitzel, S. M., Jensen, E. C., Chowdhury, A., Grossman, D. A., & Frieder, O. (2004). Hourly analysis of a very large topically categorized web query log. In *Proceedings of SIGIR 2004* (pp. 321–328). New York, NY: ACM.

Broder, A. (2002). A taxonomy of web search. *SIGIR Forum, 36*(2), 3–10.

Brown, C. (2012). Split testing with Google analytics experiments. Retrieved on December 16, 2012, from http://webdesign.tutsplus.com/tutorials/applications/split-testing-with-google-analytics-experiments/

Capra, R. (2011). HCI browser: A tool for administration and data collection for studies of web search behavior. In *Proceedings of HCIHCI 2011* (pp. 259–268). New York, NY: Springer.

Crook, T., Frasca, B., Kohavi, R., & Longbotham, R. (2009). Seven pitfalls to avoid when running controlled experiments on the web. In *Proceedings of KDD 2009* (pp. 1105–1114). New York, NY: ACM.

Dell, N., Vaidyanathan, V., Medhi, I., Cutrell, E., & Thies, W. (2012). "Yours is better!": Participant response bias in HCI. In *Proceedings of CHI 2012* (pp. 1321–1330). New York, NY: ACM.

Dumais, S. T., Cutrell, E., Cadiz, J. J., Jancke, G., Sarin, R., & Robbins, D. C. (2003). Stuff I've seen: A system for personal information retrieval and re-use. In *Proceedings of SIGIR 2003* (pp. 72–79). New York, NY: ACM.

Efthimiadis, E. N. (2008). How do Greeks search the web?: A query log analysis study. In *Proceedings iNews 2008* (pp. 81–84). New York, NY: ACM.

Fetterly, D., Manasse, M., & Najork, M. (2004). Spam, damn spam, and statistics: Using statistical analysis to locate spam web pages. In *Proceedings WebDB 2004* (pp. 1–6). New York, NY: ACM.

Fox, S., Karnawat, K., Mydland, M., Dumais, S. T., & White, T. (2005). Evaluating implicit measures to improve web search. *ACM: Transactions on Information Systems (TOIS), 23*(2), 147–168.

Ghorab, M. R., Leveling, J., Zhou, D., Jones, G. J. F., & Wade, V. (2009). Identifying common user behaviour in multilingual search logs. In *Proceedings of CLEF 2009*, pp. 518–525.

Ginsberg, J., Mohebbi, M. H., Patel, R. S., Brammer, L., Smolinski, M. S., & Brilliant, L. (2009). Detecting influenza epidemics using search engine query data. *Nature, 457*, 1012–1014.

Google. (2012). Google analytics. Retrieved on December 16, 2012, from http://www.google.com/analytics/

Huck, S. (2011). *Reading statistics and research* (6th ed.). Boston, MA: Pearson.

Jansen, B. J. (2006). Search log analysis: What it is, what's been done, how to do it. *Library and Information Science Research, 28*(3), 407–432.

Jupiter Research Corporation. (2005, March 9). *Measuring unique visitors: Addressing the dramatic decline in the accuracy of cookie-based measurement*

Kohavi, R., Deng, A., Frasca, B., Longbotham, R., Walker, T., & Xu, Y. (2012). Trustworthy online controlled experiments: Five puzzling outcomes explained. In *Proceedings of KDD 2012* (pp. 786–794). New York, NY: ACM.

Kohavi, R., Longbotham, R., Sommerfield, D., & Henne, R. M. (2009). Controlled experiments on the web: Survey and practical guide. *Data Mining and Knowledge Discovery, 18*(1), 140–181.

Kotov, A., Bennett, P., White, R. W., Dumais, S. T., & Teevan, J. (2011). Modeling and analysis of cross-session search tasks. In *Proceedings of SIGIR 2011* (pp. 5–14). New York, NY: ACM.

Lau, T., & Horvitz, E. (1999). Patterns of search: Analyzing and modeling web query refinement. In *Proceedings of user modeling 1999* (pp. 119–128). New York, NY: ACM.

Narayanan, A., & Shmatikov, V. (2008). Robust de-anonymization of large sparse datasets. In *Proceedings of IEEE symposium on security and privacy 2008* (pp. 111–125). Washington, DC: IEEE.

Ogbuji, U. (2009). Working with web server logs. Retrieved on December 16, 2012, fromhttp://www.ibm.com/developerworks/web/library/wa-apachelogs/

Osborne, J. W. (2012). *Best practices in data cleaning: Everything you need to know before and after collecting your data*. Thousand Oak, CA: Sage Publications.

Rodden, K., & Leggett, M. (2010). Best of both worlds: Improving Gmail labels with the affordance of folders. In *Proceedings of CHI 2010* (pp. 4587–4596). New York, NY: ACM.

Silverstein, C., Henzinger, M., Marais, H., & Moricz, M. (1998). *Analysis of a very large web search engine query log*. Technical Report 1998-014. Digital SRC.

Skinner, B. F. (1938). *The behavior of organisms: An experimental analysis*. Oxford, England: Appleton-Century.

Spink, A., Ozmutlu, S., Ozmutlu, H. C., & Jansen, B. J. (2002). U.S. versus European web searching trends. *ACM SIGIR Forum, 36*(2), 32–38.

Starbird, K. & Palen, L. (2010). Pass it on? Retweeting in mass emergencies. In *Proceedings of ISCRAM 2010*, pp. 1–10.

Tang, D., Agarwal, A., O'Brien, D., & Meyer, M. (2010). Overlapping experiment infrastructure: More, better, faster experimentation. In *Proceedings KDD 2010* (pp. 17–26). New York, NY: ACM.

Teevan, J., Adar, E., Jones, R., & Potts, M. (2007). Information re-retrieval: Repeat queries in Yahoo's logs. In *Proceedings of SIGIR 2007* (pp. 151–158). New York, NY: ACM.

Teevan, J., Dumais, S. T., & Liebling, D. J. (2008). To personalize or not to personalize: Modeling queries with variation in user intent. In *Proceedings of SIGIR 2008* (pp. 163–170). New York, NY: ACM.

Teevan, J., & Hehmeyer, A. (2013). Understanding how the projection of availability state impacts the reception of incoming communication. In *Proceedings of CSCW 2013* (pp. 753–758). New York, NY: ACM.

Teevan, J., Ramage, D., & Morris, M. R. (2011). #TwitterSearch: A comparison of microblog search and web search. In *Proceedings of WSDM 2011* (pp. 35–44). New York, NY: ACM.

Tyler, S. K., & Teevan, J. (2010). Large scale query log analysis of re-finding. In *Proceedings of WSDM 2010* (pp. 191–200). New York, NY: ACM.

White, R., Dumais, S. T., & Teevan, J. (2009). Characterizing the influence of domains expertise on web search behavior. In *Proceedings of WSDM 2009* (pp. 132–141). New York, NY: ACM.

White, R., & Morris, D. (2007). Investigating the querying and browsing behavior of advanced search engine users. In *Proceedings of SIGIR 2007* (pp. 255–262). New York, NY: ACM.

Wikipedia: AOL search. Retrieved on December 16, 2012, from http://en.wikipedia.org/wiki/AOL_search_data_scandal

Wikipedia: Delta method. Retrieved on December 16, 2012, from http://en.wikipedia.org/wiki/Delta_method

Wikipedia: Hadoop. Retrieved on December 16, 2012, from http://en.wikipedia.org/wiki/Apache_Hadoop

Wikipedia: Netflix. Retrieved on December 16, 2012, from http://en.wikipedia.org/wiki/Netflix_Prize

Wikipedia: Power. Retrieved on December 16, 2012, from http://en.wikipedia.org/wiki/Statistical_power

Wikipedia: Simpson's Paradox. Retrieved on December 16, 2012, from http://wikipedia.org/Simpsons_Paradox

Looking Back: Retrospective Study Methods for HCI

Daniel M. Russell and Ed H. Chi

The think-aloud protocol (Ericsson & Simon, 1985) has participants talk *while* doing the behavior of interest. While this approach is often used, speaking aloud during the activity can introduce social, cognitive load, and attention aberrations, creating a somewhat unnatural behavioral response (Dickson, McLennan, & Omodei, 2000; Wilson, 1994). On the other hand, as Ericsson and Simon points out (1985), Ericsson (2006), in the retrospective cued recall (RCR) approach, the amount of time that passes between mental action and recollection of that action necessarily introduces artifacts of memory and post-event processing that interfere with accurate recall. Neither is perfect.

With all this in mind, retrospective analysis is a methodology for conducting studies where the participant does their normal behavior without taking any disruptive action such as writing a diary entry, talking about their behavior, or responding to an interruption. RCR methods can be used to reconstruct participants' behaviors, rationales, affective reactions, and responses for events that have been recorded. In essence, a RCR method is whenever the participant is later asked to recall (or explain) their earlier behavior when prompted by cues such as images taken during their behavior, videos of the event, eye tracking showing what they were looking at during the task, etc. The central element of a study is that have important recollection-aiding cues have been captured during the experience and then used in post-event discussion and analysis. This method of gathering user behaviors is remarkably accurate when recollection cues and interview methods are well designed, even when there are fairly lengthy delays between action and recall.

D.M. Russell (✉) • E.H. Chi
Google, Inc., 1600 Amphitheatre Parkway, Mountain View, CA 94043, USA
e-mail: drussell@google.com; edchi@google.com

J.S. Olson and W.A. Kellogg (eds.), *Ways of Knowing in HCI*,
DOI 10.1007/978-1-4939-0378-8_15, © Springer Science+Business Media New York 2014

Retrospective Methods: Introduction

Traditional HCI study techniques tend to be very active, in the moment, and event driven, using many of the methods described in the chapters of this book. The analysis of events that have taken place during the course of an HCI study or experiment over a longer period have largely been diary studies (Czerwinski, Horvitz, & Wilhite, 2004; Rieman, 1993) experience sampling (Hektner, Schmidt, & Csikszentmihalyi, 2007; Larson & Csikszentmihalyi, 1983), or log data analysis [see Chap. "Understanding Behavior Through Log Data and Analysis"].

By contrast, while in-lab usability studies are effective at discovering some kinds of UI use patterns, it is notoriously difficult to track the more naturalistic, longer-term behavior of users in the lab or in the wild (Russell & Grimes, 2007). Specialized tracking devices, such as diaries (physical or online) or interruption studies (such as "experience sampling methods") (Brandt et al., 2007; Kuniavsky, 2003) can all materially affect user behavior by continually reminding the user that their behavior is being tracked and monitored. The very fact that the participants are in a laboratory setting (see, the Hawthorne effect (McCarney et al., 2007)), or that they are consciously updating a diary, can lead to important changes in their behavior through observer-expectancy effects (Steele-Johnson, 2000).

Similarly, log data analysis alone does not provide insights into user motivations, nor are logs capable of providing much context about what a user did before, during, or after, particularly when the behaviors of interest take place outside of the system under study.

Diary studies and interviews are difficult to sustain for a long period of time since they rely on participant motivation, which tends to decline as the study continues (Blackwell, Jones, Milic-Frayling, & Rodden, 2005). Furthermore, diaries and interviews tend to focus on specific events and tasks and therefore limit observations to the specific tasks or events, which is not ideal for studying behaviors where users are passively watching or infrequently interacting, instead of actively interacting.

For cases where observations of normal, non-lab user behavior is desired, and where researchers are interested in the context and motivations of participants in a study, retrospective methods are often useful techniques to consider. These methods are particularly useful when it comes to the need for understanding user perceptions and the ability to observe the context of user activity that is not triggered by an event or task.

Let us begin with a definition of these methods:

Definition: a *retrospective* study is one that records data about the behavior of the participant(s) over some period of time. This study-period data is reviewed by the participant afterwards, with the participant providing context and commentary on their behavior as prompted by examining the data that was collected during the course of the study.

A retrospective study is defined by several important experimental design dimensions:

1. *Data collection*: the way in which data (and what type of data) is collected for later review by the participant.
2. *Study duration* (varying from minutes to days).
3. *Review instruments, interview methods, and process* used by the participant to elicit recall of prior events.
4. *Sampling frequency* of data collection (samples/time-unit).
5. *Delay of review after collection*: how much time has elapsed since the data was collected and the samples reviewed.

In this chapter, we provide an analysis of retrospective methods in HCI, first discussing the nature of human memory vis-à-vis retrospective recall, then outlining the methodologies used for conducting retrospective studies, presenting a sample retrospective study analyzed along the dimensions given above, concluding with a review of the features and challenges that come with the methods.

Human Memories and How They Work

The retrospective analysis approach takes advantage of the well-known human ability to visually recognize images of earlier situations they had been in, and comment on them. Images, particularly images of an environment (e.g., the computer screen) that have been created as part of the normal course of work, are particularly powerful at bringing about recall of the situation at the time (Brewer, 1986, 1988). Images, especially when used during post-study interviews with the participants, can give an improved view into what was happening in the actual setting of use with nearly imperceptible intrusion. This human ability to recall situational information when retrospectively cued by some data, sound, or visual imagery obtained at the time is the key to these methods.

Yet, at the same time, human memory is notoriously fragile and subject to many kinds of recall errors; memories are often imbued with a sense of accuracy and quality they actually lack (Roediger & McDermott, 1995). A good understanding of the ways in which memory is subject to alteration can help us design retrospective methods that yield useful data for HCI studies. Important factors from these memory studies are the tendency:

1. To reconstruct a memory of a prior event according to a widely held, prototypical pattern of this event category (Van Boven et al., 2009; Schacter, 2001) rather than by accurate recall of the actual events.
2. To follow the researcher's lead in answering questions about the event (Steele-Johnson, 2000; Weisberg, Krosnick, & Bowen, 1996).
3. To make associations about events based on perceived similarities between the recalled event and other, similar experiences that influence memory of the event to be similar to those previous events (Underwood, 1965).

One of the key techniques for avoiding false memories and improving accuracy in recall is to use *cueing* techniques. Cues are used in retrospective studies (usually images or videos) from the record of the participant's behavior. It has become clear that even highly meaningful events will be inaccurately recalled if there is no cueing to orient and remind the participant about the event and that giving cues improves the accuracy of recall (Lamming et al., 1994). Shiffman et al. (1997) demonstrated the poor quality of retrospectively recalling behaviors from 12 weeks ago, even though when the events in question were actively logged by the participant on a personal handheld device, the act of simply *recording* an event seemed to have little impact on quality of recollection. By contrast, actually *seeing* contextually appropriate cues that capture salient cues of the time, place, and activity is an important piece of the method.

Given that the accuracy of uncued memory rapidly deteriorates after about 1 day, there is good reason to wonder about the accuracy of retrospective recall. Some critiques of retrospective methods question their accuracy when the recall is from more than 1 week in the past and when the recollection is performed without the use of cues to support recollection (such as with post-event interviews and surveys) (Novick, Santaella, Cervantes, & Andrade, 2012). However, the careful use of recall cues derived from the participant's own history has been shown to lead to more accurate and useful recollections from some time in the past (see Sect. "A Sample Retrospective Analysis Method" below).

Consequently, setting up the method of a retrospective study requires attention to the details of data collection and event review with the participant to avoid introducing false memories, or asking the participant to recall more than they can accurately report upon (Loftus, 1996).

Earlier Retrospective Work

There is a long tradition of using photos of key events to cue retrospective memories. Collier (1967) is mostly closely associated with the photographic technique in anthropological settings, when photos are used as both prompts and foils to elicit memories and context around some circumstance. Van Gog, Paas, and Van Merriënboer (2005) reports on the tradeoffs involved in using concurrent versus retrospective reporting of problem solving behaviors, ending with the observation that a retrospective recall is often preferred to avoid interfering with the problem-solving process as it occurs.

The idea that images can also be used for HCI recollection purposes can be seen in the work of Van House (2006) and Intille, Kukla, and Ma (2002), although these (and other similar systems) capture images of the world context, and do not provide the detailed internal tracking of events in the user's experience of the online world as a logging system could.

There have been a variety of retrospective methods developed to understand behavior by looking back at their performance. Here we discuss logging tools to track user behavior for later analysis by the participant, video recording and playback, eye tracking with post-task commentary, the Day Reconstruction Method (DRM), and the Experience Sample method.

Logging: Many systems have been built to unobtrusively log user events over time for later analysis. These range from the obvious logging analysis systems of Web behavior to client-side tooling that records user behavior in great detail, allowing the user to look at their behavior afterwards and comment accurately on what (and why) they were acting in a particular way.

LogViewer (Blackwell et al., 2005) is a tool that logs user events and screen images in Web behavior for later analysis. LogViewer also creates a tree analysis visualization of user behavior to track which clicks generate which subsequent Web page views. Their data was primarily intended to facilitate the tracing of user behavior—how many times was the back button used to return to earlier pages, and how user navigation is organized in terms of landmarks pages, etc. They also interviewed their participants with the screen captures as cues, but with a focus on gathering contextual information to aid in their application redesign purposes.

Kellar, Watters, and Shepherd (2006) built a logging system is attached to a customized browser (a modification of Internet Explorer) that allowed the user to label their own behavior as they completed tasks. As with all systems that ask for manual labeling in near real time, the presence of the logging system is hard to ignore, and Kellar points out that there is good evidence that users modified their behavior because of its presence. This system was also used as an object of discussion in the retrospective style, but again, the focus was on accurately labeling sequences of behaviors, rather than using the event log as a cue for recollection of overall behavior.

Other loggers such as (Al-Qaimari & McRostie, 1999; Chi, Pirolli, & Pitkow, 2000; Jones, Milic-Frayling, Rodden, & Blackwell, 2007; Siochi & Hid, 1991) log events for later analysis and are intended to support the understanding of user behavior in search and information browsing tasks, often coordinating the log data with other kinds of user data (e.g., field observations). See (Ivory & Hearst, 2001) for a summary of many such tools developed for tracking and logging Web behavior to improve usability analysis.

Video: (Capra, 2002) developed and evaluated a retrospective analysis version of the self-reported critical incident technique. In this study, researchers showed participants a video replay of their entire working session, asking them to detect and describe critical incidents as they observed them in the video.

To speed up the process and simplify the interaction from the participants perspective (Akers, Simpson, Jeffries, & Winograd, 2009), logged critical events in a participant's use of the CAD solid modeling system SketchUp. Each critical event was then automatically extracted from the video 20 s around each incident. After the entire task was completed, the participants answered a series of questions about their performance while watching the video clips of each incident. This approach gave the participants enough visual context, and enough perspective (coming after the task was completed), to be able to explain why this moment was a crucial incident for them.

In these uses of video as the cue stimulus, the events in question were freshly in mind (having just been completed) and the participants were able to comment on their performance in accurate and useful ways.

Eye tracking: Tracking a participants' eyes as they perform a task gives another kind of video trace that can be used to elicit information [See Chap. "Eye Tracking in HCI: A Brief Introduction"]. When gathered shortly after the completion of the task, a participant can narrate what they were spending their attention on, and why. Hyrskykari, Ovaska, Räihä, Majaranta, and Lehtinen (2008) and Guan, Lee, Cuddihy, and Ramey (2006) both use eye tracking video as cues to help participants describe their visual motions in broad-brush terms. By comparing subjects' retrospective narration with their eye movements, they found the post-event accounts to be valid and reliable, providing a useful account of what people paid attention to while completing tasks. They also found that this has a low risk of introducing fabrications, and is unaffected by overall task complexity.

Eye tracking can also be used to have people retrospectively comment on what was noticed, or not, in a user-interface. Muralidharan, Gyongyi, and Chi (2012) describe a retrospective think-aloud protocol (RTA) where, immediately after the tasks, participants were asked to take the researcher through what they were doing while watching a replay of a screen capture (with eye tracks) of their tasks. The researcher would ask probing questions for clarification, and then move on to talk about another task.

If the participant never mentioned the features being tested (even if the feature was always visible in all of the tasks), the researcher would return to a screen capture, point out the UI feature explicitly and ask a series of question: "What is that? Did you notice it? What does this element of the user interface suggest to you? Tell me what you think about this." The goal was to learn what the participant thought the feature was, whether it had been noticed at all, to understand whether or not they perceived it as useful, and why.

Day Reconstruction Method: Another approach that has been used extensively in psychology studies is the "Day Reconstruction Method" (DRM). As introduced by Kahneman, Krueger, Schkade, Schwarz, and Stone (2004), the DRM combines the advantages of an offline method with the accuracy of introspective approaches such as Experience Sampling (described below).

A DRM study asks participants to reconstruct their daily experiences as a continuous series of episodes, writing a brief name for each one. Experiential episodes are recalled in relation to preceding ones, which lets participants draw on episodic memory when reporting on the experience (Schwarz et al., 2009). To minimize retrospection biases, the DRM is typically conducted at the end of a reported day or at the beginning of the next day. Hence, participants are better able to capture the properties of a single experiential episode, avoiding inferences from their global beliefs about the experience. The DRM method works well when the participants understand the nature of what the study is about (for instance, if it is trying to capture their hedonic experience of a particular system), but less well when the research questions are about their experience over multiple days.

Experience Sampling: The "experience sampling method" (ESM) can be thought of as a diary study where the diary entries are driven by some external signal, typically a beeper, phone message, text message, or another way to remind the participant to fill out a questionnaire about their experience at that moment in time (Kuniavsky, 2003; Larson & Csikszentmihalyi, 1983). A modification of the ESM method that makes it much more like a RCR method is "Image-based experience sampling and reflection" (Intille et al., 2002), where a still image (or short video clip) of the participants environment is captured at the sample moment, and then later reflected upon for subsequent analysis.

A Sample Retrospective Analysis Method

Our use of the tool IE-Capture (short for Internet Explorer Capture) (Russell & Oren, 2009) illustrates a retrospective method in HCI. This was a browser add-on that captured not just moments in a user's behavior of a Web browser but also, crucially, complete screen images (the entire screen extent—more than just the Internet Explorer (IE) window being used to work on the Web). Having the complete-screen proved valuable for helping the participants recall their behavior accurately.

In terms of the methodology design dimensions mentioned above, IE-Capture had:

- *Data collection*: complete screen captures, URL, time-stamp; triggered by the completion of the loading of a Web page in the browser.
- *Study duration*: varying from 1 to 6 weeks (mostly 2 weeks in length).
- *Review instruments*: the collected screenshots were reviewed with a custom-built data viewer (see Fig. 3) that allowed the participant to browse forward and backward in time through the collection. Each screen capture occurred whenever a Web page completed loading, so the sampling frequency varied depending on the participant's use of the Web. Typically, this would measure in the many hundreds over the course of the study.
- *Sampling frequency*: samples were collected on-event (at document-load time) whenever the participant was using Internet Explorer as their browser.
- *Delay of review*: 1–6 weeks (most often 2 weeks) after data collected.

IE-Capture was designed to help us to understand how search-engine users would approach questions that required a long effort over time. By their nature, such tasks are difficult to capture in laboratory settings; an unobtrusive data capture method was needed, and hence the creation of the logging system that could be left in place without any intervention on the part of the participant.

However, a key to this research study was to understand *how* participants thought about and framed questions as they went through their research process over hours, days, and weeks. In this study, IE-Capture logged screenshots for a period of 2 weeks (sometimes longer); then a retrospective review was held with the participant in their home or workplace. As seen in Fig. 1, a series of whole-screen captures were collected for the interview. An individual frame of that sequence can be seen

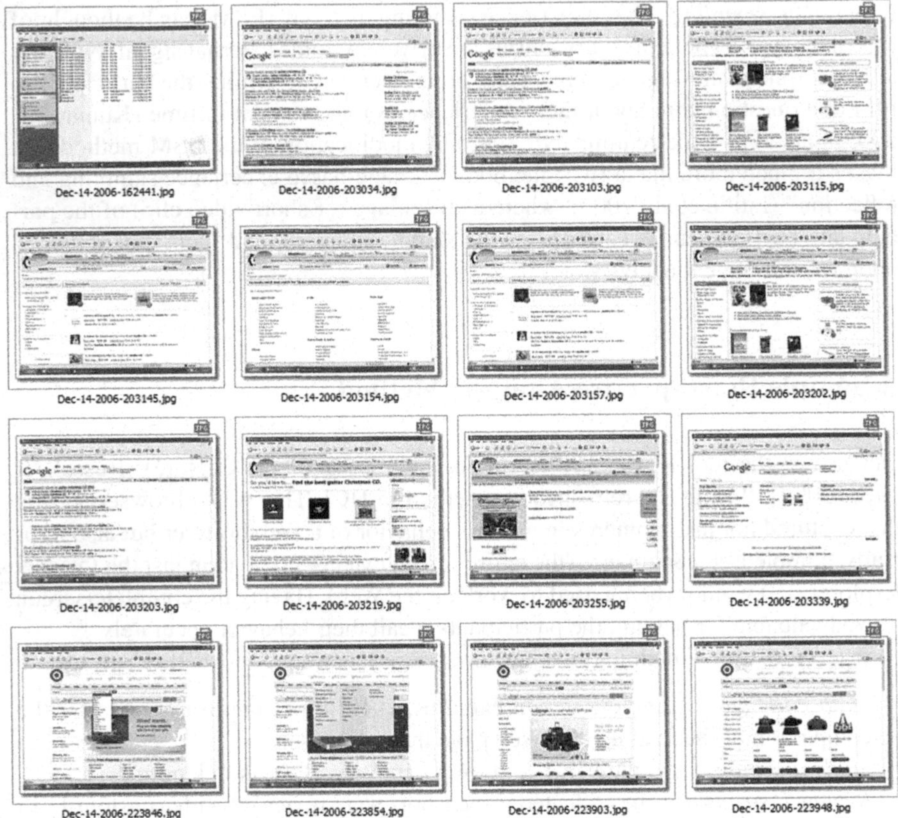

Fig. 1 A series of screen snapshots used in a post-event retrospective interview. These kinds of captured cues support accurate recall by providing a great deal of context to the participant, allowing them to reconstruct what was happening at the time of the events in question (In this case, the browser is maximized to full screen size)

in Fig. 2. With all of the additional visual information available (such as which applications are open, which documents are visible at the same time, the state of work in progress), the participant can recollect what was going on at the time from many different cues.

During the interview, the participant would review the collected series of screen captures, providing the backstory in response to questions asked by the interviewer. Since the number of screen images and logged events could be in the large hundreds, the researcher would select a sample of the events to review in detail.

As is typical for RCR studies, the interviewer began by acquainting the participant with the review instrument operation (how to move forward and backward in the sequence of captured screen images) and then setting the context by reviewing the very earliest data collected in the study. Then, a series of semi-structured review questions were asked, determining the properties of the experience during the study period, elaborated in the next section.

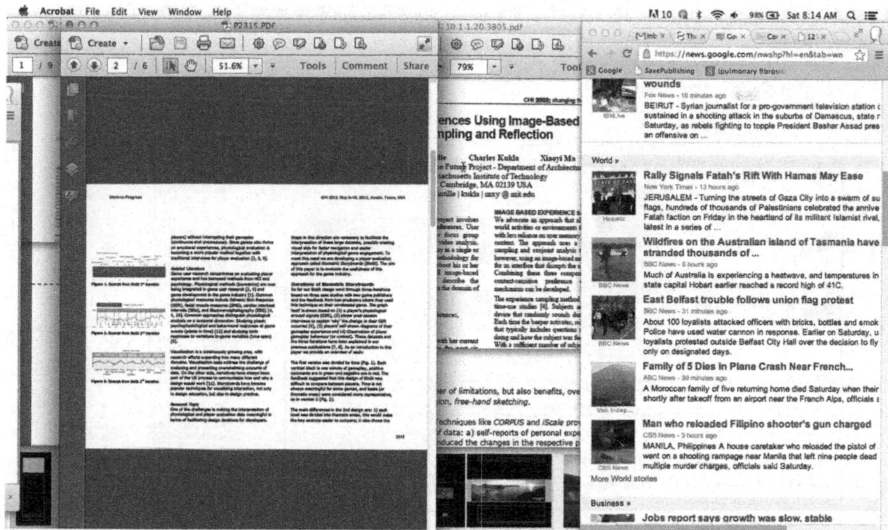

Fig. 2 The choice of cue stimulus is crucial in being able to elicit high quality retrospective recollections. In (Russell & Oren, 2009) the user's entire desktop image was used as a cue to ask questions about overall task intent and search behavior. Cue stimuli with less contextually useful information (e.g., only the browser screen image) were not as successful, and led to lower quality retrospection data

Interview questioning procedure: At the beginning of the interview, after introducing the review tool and reviewing the first day of collected data, the experimenter would view selected visits each day in the log with IECaptureViewer, our tool to view and scroll through the data logs and screenshots (see Fig. 3 for IECaptureViewer). Since this was a study investigating how well people could remember their search tasks, nearly all participants had searches on days that we probed. In the few missing cases, we used the next prior search event (e.g., substituting day 3 for day 4).

For each of the days in question, the experimenter would jump to the first search query made on that day and show it to the participant. (Note that "jumping" to the screen image in question was important, as to avoid showing the participant later screen images that would have shown them the sequence of events.)

The experimenter then asked the participant to describe "what happened next in the search process." The participant was instructed to describe the next event if they felt "reasonably confident" that they knew what happened, in particular, focusing on what search terms were used, and whether or not that particular next search was successful.

While the participant was not prompted for a particular kind of answer, we noted possible variations on their answer. Was the search successful with this query alone? Did the participant have to continue searching after this point in time? If they continued, did they have to continue refining the current query or do something else entirely? This free form question made it easy to assess whether or not the

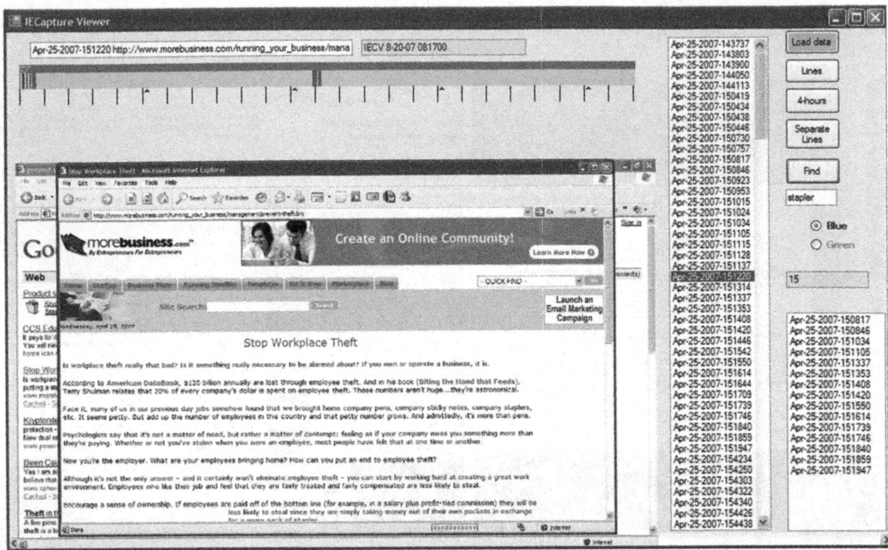

Fig. 3 IE-Capture Viewer—a tool for reviewing the participant's log and screen images for discussion and retrospective cueing. The participant's screen image is visible in the center of the display, with the stack of windows present at the time of screen capture, an essential part of cueing for long-term recall. The lists on the *right hand side* are for quickly moving among the log events and captured images for discussion purposes with the participant

participant could recall the situation at all since we had data on what actually did happen next.

If the participant could not recall, then the experimenter would go forward in time, showing them one event image after another, pushing forward in time, until the participant could recollect what was going on and was able to predict what the next search event would be.

We were interested in measuring the participant's ability to speak accurately about the next major event in their search process. That is, having cued their memory of an event, we measured their ability to recall the next step in the process. (For example: looking at a screen image from 6 days ago, the participant would be asked "What's the next search you did after this point?") In nearly all cases, the assessment by the researcher of the participant's memory was clear and evident: either the participant could accurately predict what was coming up in the log, or they just could not say. Only rarely did a participant guess and feel confident; when they guessed, they would say so and express a lack of confidence in their prediction.

Results: For each participant we had two measures—the number of correct predictions based on a cued recall, and the number of times they had to go to a previous page before they could recollect what was going on in the search (see Fig. 5). A good recollection happens when the participant can accurately recall the next search event after just one or two "cue" screen images.

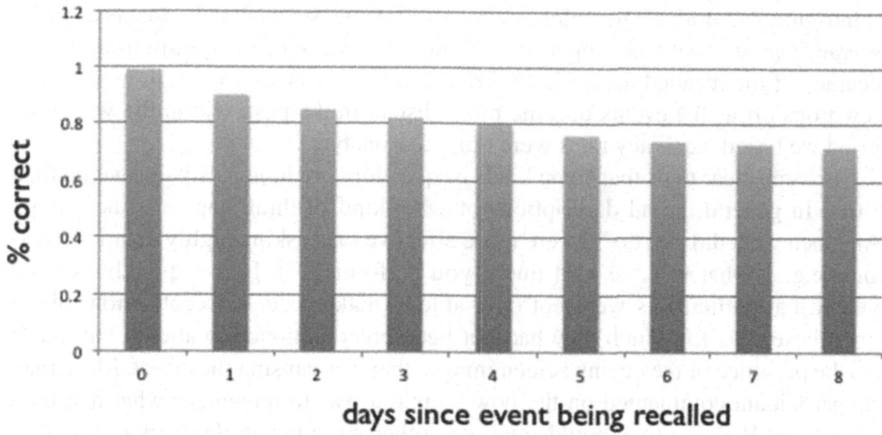

Fig. 4 The number of correct next event predictions drops below 86 % for events 2 days in the past, but is still at 75 % correct for events that are 7 days in the past

As can be seen in Fig. 4, the majority of participants could accurately recall the next search event after the probe within the past 4 days. This is not terribly surprising, given that searches are relatively infrequent. In this participant pool, the average number of searches per week was 11. A search done 3 or 4 days ago is relatively recent and is memorable (and recallable) by its relative rarity among the total number of Web interactions in that time, and its distinctness (a search is an event that requires explicit interaction to achieve some goal).

However, as we tested farther into the past (6 days and 7 days out), participant recall was still quite good. Even after nearly a week, participants were able to recall the next search event correctly around 75 % of the time. With the additional prompting of advancing to the next page in their cached screenshots using the IE-Capture viewer, participants could recall their next search event accurately after seeing only three additional screen images taken from the log/screen files. (Remember that the cue screen images are usually *not* search events, but usually just the next Web page that the participant visited.)

It was clear during the interviews that participants really could recollect not just the next event, but also how this search fit into the larger story of what was going on at the time. Even for events 7 days in the past, participants were able to not just make a prediction about the next event, but also complete the story and say whether or not the entire task (of which search was just a part) was successful or not. We viewed this "story ability" as suggestive that the entire sequence of behaviors was being recollected, and not just the single search event in isolation. The relatively high accurate recall rate after cueing also suggested that more than just one frame of the sequence was necessary for context restoration—a few images in sequence seemed to work the best for accurate recall. As is shown in the context capture method of (Akers et al., 2009), truly effective retrospective analysis means capturing the *context* of use over time, as well as memorable instances with visual context.

Intriguingly, during the interviews, participants seemed able to speak with assurance about what had happened even quite a while ago. But questions about the accuracy of the recalled memories worried us. As we can see in Fig. 4, while accuracy drops off as the events become more distant in the past, within the weeklong period we tested, accuracy rates were quite reasonable.

It became clear to us that some kinds of questions are more easily answered than others. In general, broad descriptions of what kind of thing happened next (e.g., "And then what did you do?") were more effective than asking highly detailed questions (e.g., "What was the next query you performed?"). It also quickly became evident that participants were not only able to make accurate recollections about particular events for which they had not been preconditioned to attend, but that it was the presence of the cueing screen images that was causing the effect. More than one participant commented on the how simple it was to remember what had happened then. Because they could often see other windows in the background (the corner of the Excel spreadsheet, say) those small peripheral cues would give them a distinct sense of time, activity, and place (Wilson, Evans, Emslie, & Malinek, 1997).

Retrospective HCI Methods: Three Time Spans

In HCI, retrospective studies have been used to elicit reflections from study participants on time scales varying from minutes to weeks. Because retrospective memories (and the reflections elicited) vary so much by the amount of time from the original event, it is useful to divide retrospective studies into three categories: Each time period has a distinctive character, with particular challenges and properties.

1. *Short-term studies* (study period <2 h; the retrospective is gathered immediately after task) are typically performed in usability labs, often with the retrospective gathered by a think-aloud protocol as the participant observes a playback of the actual study as captured by video recording of the participant, their screen behavior, or their eye movement behavior (Guan et al., 2006; Hyrskykari et al., 2008). While such studies can be valuable for understanding the instantaneous motivations and reasons for making the choices they do, the temptation is to ask the participant to tell "more than they can know" about their performance. By asking for motivational responses to behavior that might be not open to conscious understanding (such as "why did you choose to read that particular passage"), the participant might easily fall into rationalization about prior behavior that is actually only inadequately remembered. On the other hand, different attributes of the interaction (e.g., why a particular behavior strategy was followed) that are explicitly informational (rather than motivational) can still be commented on accurately (Kuusela & Paul, 2000).
2. *Intermediate-term studies* (study period ≥ 2 h, <2 days; retrospective gathered 1 or 2 days after completion). These studies are currently somewhat rare in the HCI literature, but strike a nice balance between the accuracy of immediate, short-term labs studies versus the long term studies required to gather enough

rare events, like errors. Studies that run for 1 or 2 days can be naturalistic in ways that the short-term studies are not, because they are conducted in lab settings under tight time constraints.

3. *Very long term* studies (study period >1 day; retrospective gathered 1 or 2 days after the end of the study). For retrospective studies over a long term, the participant cannot ignore the memories and experiences that have happened since the study period. The participant knows how it all turns out, so every recollection will be informed by that knowledge. This effect can be useful (by giving a report about the outcome of actions taken and decisions made), but it can also lead to the "irresistible tendency for subjects to clean up their act and to describe a more coherent and well-thought-out strategy than is normal" (Kuusela & Paul, 2000). Longer-term studies often use daily debriefs of the participant by the researcher. Remote-usability studies sometimes follow this daily check-in protocol as a way to keep in touch with their participants, growing a rapport with them and learning additional information that is nominally outside the scope of the study (Brush, Ames, & Davis, 2004). Furthermore, retrospection from several days in the past is also subject to bias effects—forgetting of the options *not* explored (and in particular, options that we considered at the time, but that left no trace in the cueing record), current mood, and beliefs acquired since the study period (Schacter, 1999).

Evaluating Retrospective Methods in HCI

Many HCI studies have a retrospective aspect to them. At any time a study that has a performance component followed by an evaluation that occurs a significant amount of time after the performance is effectively a retrospective study, even though it may not be labeled as such. (And, in particular, most studies of this sort are not *cued* retrospective studies.) However, many research works in HCI have some aspect of retrospection, for example, when a survey is given to a user population that asks them to reflect on their experiences with experimental software, or when a longitudinal study asks questions about earlier uses of the system under study (both of these are retrospective analyses) (Jain & Boyce 2012).

What about retrospective studies is broadly useful to know for HCI practitioners? We believe that there are two answers. First, what kinds of biases and response effects occur as a result of the passage of time over the course of a study? And second, how do the experimental methods used influence the validity of the retrospection?

It is useful to consider a retrospective study in terms of the important experimental design choices (briefly described in Sect. "Earlier Retrospective Work", above),

- *Data collection:* how will the data be collected and what kinds of data will be collected? Automatically? Or will it be collected by manual intervention (as in the DRM and manual labeling methods)? To what extent will manual annotation interfere with the actual behaviors under study?

- *Study duration*: how long will the study run? Longer runs have the advantage of collecting larger amounts of data, and thus have a higher chance of observing events of real interest, but this interacts with longer term biasing effects.
- *Review instruments*: how will the participant and the researcher review the data? Usually some kind of playback system is needed to select salient episodes or events from the data stream. Such playback systems need to have the ability to "jump to" the prompt of interest without revealing any of the interstitial events. (This avoids giving the participant unanticipated cues which would then degrade the value of the recall).
- *Sampling frequency*: what data sampling rate should be expected and what events cause the data to be collected? Will it be random time sampling (à la ESM), event driven (e.g., by a user action being taken), or periodic (e.g., every hour or at the end of the day).
- *Delay of review*: when will the data be reviewed with the participant? Periodic reviews are useful for longer term experiments, but it becomes difficult to avoid giving the participant subtle clues about what kind of behaviors are the "right" ones (or the opposite—it is difficult to not reveal with responses are surprising from the researcher's perspective).

Can retrospection bias be useful? One may wonder how much retrospection biases influence the accuracy of recall, even during RCR studies. As we have seen, bias is inevitable over the passage of time. But there is an important way to consider this bias: *The memory is what matters*. The veridicality of reconstructed experiences can be of minimal importance as these memories will guide future behavior of the individual and will be communicated to others (Karapanos, Zimmerman, Forlizzi, & Martens, 2009; Norman, 2009). In other words, while what participants remember might be different from what they experienced at the time, memories that are consistent over multiple recalls provide valuable information about future actions. In effect, the memory (no matter how inaccurate), and not the actuality of what happened, becomes the basis on which future decisions are made (Karapanos et al. 2009, 2010).

The influence of retrospective experimental methods. As is true with most psychological or HCI experiment designs, the experimental methods used during retrospective studies can have a profound influence on the results. Even time-honored experimental design patterns can be influential. It is well-known that even something as simple as *assigning* tasks to participants (versus having them use their own, ecologically valid and personally important tasks) can heavily influence outcomes (Russell & Grimes, 2007). Likewise, choices in retrospective experiment designs can be highly influential as well.

We found, for instance, that the choice of cues gathered for recall purposes can spell the difference between no useful results and highly reliable results (Russell & Oren, 2010; van den Haak, De Jong, & Schellens, 2003). In an early (and naïve) version of our study, we tried cueing previous behavior recollection with the search queries presented as strings and associated dates (e.g., "You searched for {Vancouver hotel OR B&B} on Nov 7, 2007. What was your next search for?"). We quickly

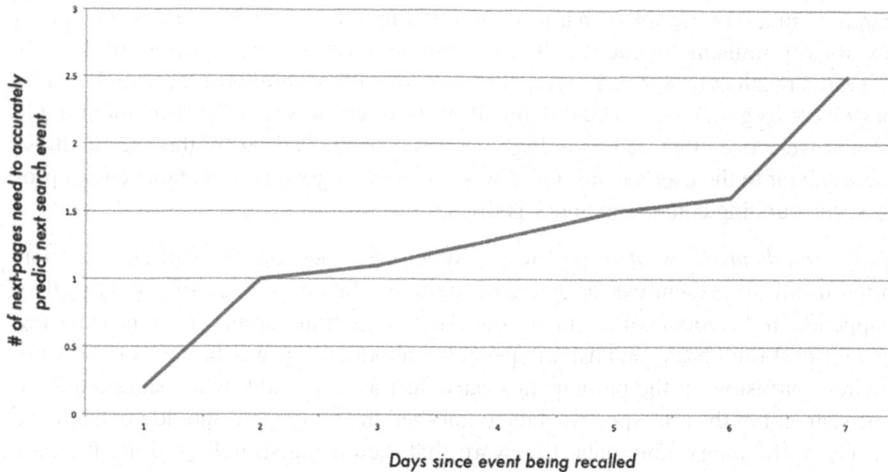

Fig. 5 Events farther in the past required more and more pages to accurately recall the next search event. After about 6 days out, a participant usually needed around 2 or 3 pages to remember what happened next (The numbers are non-integers as they represent the average number of pages required by all study participants)

found that such cues are effectively useless as memory prompts—people simply cannot remember their Internet searching behaviors with probes of this kind.

When we switched to capturing just the browser window (as seen in Fig. 5), recall clearly improved, but the recall error rates were still fairly high. The fairly small switch of capturing the *entire screen* (rather than just the browser window) ended up also triggering memory for a good deal of additional task context information.

The experimental protocol must also include methods to validate that the behavior recalled by the participant is actually the behavior that was performed. As is well known from the design of surveys [cf. Chap. "Survey Research in HCI"] (Holbrook, Green, & Krosnick, 2003), biases in recollection also occur as a consequence of trying to conform to social expectations, simple satisficing, pleasing the experimenter, or to rationalize behavior that seems awkward in after-the-fact review.

Pragmatics of Using Retrospective Methods

While an entire book could be written just on good experimental practices for retrospective methods, a few pragmatic guidelines will be useful for the practitioner.

Choosing good cues: When designing a retrospective study, it is important to capture data that will provide useful memory cues. In general, memories are cued by images or data that encapsulate a good deal of rapidly recognizable context. Thus, images such as screen captures or videos of user performance can be used as cues. Reconstructions of a situation (for example a simulation whose state can be captured in a few variables) that lose memorable or recognizable contextual details are not as promising for cued recall. For example, in an un-cued memorability study of search results (Teevan & Karger, 2005), many of the features of the search results page were forgotten within 60–90 min after the query was run. The only memorable results were ones that had been highly ranked or clicked-on by the user as these were salient to the user's goals. Otherwise, without a good recognizable cue to provide surrounding context, memory is difficult.

Walkthrough methods of interviewing: When interviewing the participants, it is often useful to present earlier data and maintain the chronology of events as they happened in the course of the interview. That is, jumping around from near-past to distant-past (and back) and asking questions about each segment out of order only invites confusion on the participant's part. Just as important, when skipping from one segment of the retrospective data to another, the participant should not be aware of any of the intervening data. Be aware that the cueing stimuli used are the *only* stimuli being tested for recall. Seeing additional cues may significantly alter the answers (and improve!) to later interview questions.

Asking for predictions: While there are many ways to validate the accuracy of the recollections being elicited by the cues, one particularly useful approach is to ask for predictions from the participant. (Roughly, "After you saw this screen, what was your next action?") While not always applicable in exactly this form, the general idea of looking for inter-response consistency in retrospective reports is a valuable thing to measure. This is similar to a method often used in survey design is to ask slight variations on the same question at different points in the survey, testing for consistent replies in responses across variations (Weisberg et al., 1996).

Face-to-face interviews: For retrospective interviews, a face-to-face connection between the participant and the researcher is more effective than distance methods (e.g., telephone surveys). Holbrook et al. (2003) have shown that satisficing and social desirability response biases are more likely to take place in telephone interviews than in face-to-face interviews.

Ways to avoid the false memories effect: As is well-known (Loftus, 2005), interviewers can easily (and often accidentally) introduce false memories by the way they frame their questions. While there is an entire literature on interviewing techniques to avoid introducing spurious information (Loftus, 2005; Memon, Wark, Holley, Bull, & Koehnken, 1997), in an HCI context the challenges are far simpler. Typically, the behavior under question is not emotionally laden (thereby avoiding the effects of eye-witness testimony when charged events occur), and the cueing stimuli are usually data gathered from the participant's own behavior. Good advice to follow when asking questions of past behaviors include:

1. *Avoid direction* about what parts of the behavior should be noticed. That is, avoid cueing the participant to pay special attention to the behaviors that are the focus of the study. If they skip over the important parts, the researcher can ask follow-up questions, noting that they are replies to direct questions.

2. *Avoid value statements* about the behaviors in question, e.g., "When did it stop acting badly?" or "When did you start liking that awesome new interaction widget?" Introducing affectively laden terms can easily alter people's responses.
3. *Avoid asking for global affective responses from experiences in the past.* As (Schwarz et al., 2009) shows, asking for accurate evaluations of emotional perceptions from the past *cannot help* but be influenced by subsequent events and especially the perception of the entire experience at the end. No amount of rationality can apparently overcome this strong cognitive self-perception bias effect. The participant might intellectually understand that they enjoyed using a system at the beginning of their use experience. But if a later experience turned out to be highly negative, it is tremendously difficult to evaluate the entire experience as positive, even though the average experience might be highly positive. Although factual information about specific events in the past can be accurate, the reconstructive nature of emotional memories makes accuracy difficult.

Avoiding testing children: The age of the participants can be another factor for caution in using RCR. van Kesteren, Bekker, Vermeeren, and Lloyd (2003) found that children between the ages of 6 and 7 often have difficulty holding multiple concepts in memory at once, limiting their ability to both watch a retrospective video of their behavior *and* comment on what they were doing at that time, although it was clear that they could correctly report on changes in their understanding that occurred during the study. (See also (Höysniemi, Hämäläinen, & Turkki, 2003) who found similar cognitive limits on younger children's ability to reflect on previous performances.) However, Baauw and Markopoulos (2004) found that post-task interviews for usability problems worked about as well as in-lab, real-time usability analysis for children between the ages of 9 and 11.

Another age-related issue that appears with younger children is that reviewing videos is not always an exciting prospect, leading to a certain amount of attentional drift during the retrospective review part of the study. Retrospective interviewing of children is often a researcher's most challenging task.

Summary

With all this in mind, retrospective studies are a set of methods to gain insight into behavior that is otherwise very difficult to learn. As we have seen, RCR methods can be used to reconstruct participants' behaviors, rationales, affective reactions, and responses for events that have been recorded. However, there are many challenges to creating a carefully design retrospective study. Such studies must be designed with care, paying particular attention to capturing cues that are useful and engaging for recall, asking questions that do not ask the participant to over-infer what they can accurately recall, and continually validating the responses with the record of actual behavior.

We find this method of gathering user behaviors to be remarkably accurate when recollection cues and interview methods are well-designed, even when there are fairly lengthy delays between original action and recall.

Further Reading and Resources

For us, the development of the RCR technique grew out of a frustration with not being able to see normal user behavior over an extended period of time. Logs analysis is a splendid technique [See Chap "Understanding Behavior Through Log Data and Analysis"], but it does not allow for any particular insight into attitudinal data or an understanding of individual responses over a longer period of time.

To deal with this issue, we built IE-Capture (see above) as a tool to allow our users to "tell their own story" and give us those additional insights into their use of our system. As we interviewed more and more participants, it became clear that the RCR method was both powerful and sensitive. The concern for not over-interpreting the data became evident when we found our participants rephrasing things they had said earlier in our interviews. This in turn led us to study the accuracy of recalled behavior, and to develop our own skills in asking questions that would not bias the participant.

For additional information about the pragmatics of asking questions in retrospective interview settings, please see (Beatty & Willis, 2007) and (Willis, 2005).

For guidance in using the Experience Sampling Method (reconstructing the day's events at the end of each day), see (Hektner et al., 2007).

Exercises

1. Which of the other methods in this book work well with retrospective study methods?
2. What kinds of reports are not generally accurate when people are reviewing a record and/or visualization of their past behavior?

References

Akers, D., Simpson, M., Jeffries, R., & Winograd, T. (2009, April). Undo and erase events as indicators of usability problems. In *Proceedings of the 27th International Conference on Human Factors in Computing Systems* (pp. 659–668). New York, NY: ACM.

Al-Qaimari, G., & McRostie, D. (1999). KALDI: A computer-aided usability engineering tool for supporting testing and analysis of human computer interaction. In J. Vanderdonckt & A. Puerta (Eds.), *Proceedings of the 3rd International Conference on Computer-Aided Design of User Interfaces (CADUI'99), Dordrecht, October, 1999*. Louvain-la-Neuve: Kluwer.

Baauw, E., & Markopoulous, P. (2004, June). A comparison of think-aloud and post-task interview for usability testing with children. In *Proceedings of the 2004 Conference on Interaction Design and Children: Building a Community* (pp. 115–116). New York, NY: ACM.

Beatty, P. C., & Willis, G. B. (2007). Research synthesis: The practice of cognitive interviewing. *Public Opinion Quarterly, 71*(2), 287–311.

Blackwell, A., Jones, R., Milic-Frayling, N., & Rodden, K. (2005, April). *Combining logging with interviews to investigate web browser usage in the workplace.* Position paper for Workshop Usage Analysis: Combining Logging and Qualitative Methods, ACM Conference on Human Factors in Computing Systems (ACM CHI 2005).

Brandt, J., Weiss, H., & Klemmer, S. (2007). txt 4l8r: Lowering the burden for diary studies under mobile conditions. In *CHI '07*, April 23–May 3, 2007, San Jose, California.

Brewer, W. F. (1986). What is autobiographical memory? In D. Rubin (Ed.), *Autobiographical memory* (pp. 25–49). Cambridge, UK: Cambridge University Press.

Brewer, W. F. (1988). Qualitative analysis of the recalls of randomly sampled autobiographical events. In M. M. Gruneberg, P. E. Morris, & R. N. Sykes (Eds.), *Practical aspects of memory: Current research and issues* (Vol. 1, pp. 263–268). Chichester, UK: Wiley.

Brush, A. J., Ames, M., & Davis, J. (2004, April). A comparison of synchronous remote and local usability studies for an expert interface. In *CHI'04 Extended Abstracts on Human Factors in Computing Systems* (pp. 1179–1182). New York, NY: ACM.

Capra, M. (2002). Contemporaneous versus retrospective user-reported critical incidents in usability evaluation. In *Proc. Human Factors 2002,* HFES, 1973–1977.

Chi, E. H., Pirolli, P., & Pitkow, J. (2000). The scent of a site: A system for analyzing and predicting information scent, usage, and usability of a web site. In *Proc. CHI 2000* (pp. 161–167). New York, NY: ACM Press.

Collier, J. (1967). *Visual anthropology: Photography as a research method.* New York, NY: Holt, Rinehart and Winston.

Czerwinski, M., Horvitz, E., & Wilhite, S. (2004). A diary study of task switching and interruptions. In *CHI* (pp. 175–182). New York, NY: ACM Press.

Dickson, J., McLennan, J., & Omodei, M. M. (2000). Effects of concurrent verbalization on a time-critical, dynamic decision-making task. *The Journal of General Psychology, 127*(2), 217–228.

Ericsson, K. A. (2006). Protocol analysis and expert thought: Concurrent verbalizations of thinking during experts' performance on representative tasks. In K. A. Ericsson, N. Charness, P. Feltovich, & R. R. Hoffman (Eds.), *Cambridge handbook of expertise and expert performance* (pp. 223–242). Cambridge, UK: Cambridge University Press.

Ericsson, K. A., & Simon, H. A. (1985). *Protocol analysis.* Cambridge, MA: MIT Press.

Guan, Z., Lee, S., Cuddihy, E., & Ramey, J. (2006). The validity of the stimulated retrospective think-aloud method as measured by eye tracking. In *Proc. CHI 2006* (pp. 1253–1262). New York, NY: ACM Press.

Hektner, J. M., Schmidt, J. A., & Csikszentmihalyi, M. (2007). *Experience sampling method: Measuring the quality of everyday life.* Thousand Oaks, CA: Sage Publications.

Holbrook, A. L., Green, M. C., & Krosnick, J. A. (2003). Telephone versus face-to-face interviewing of national probability samples with long questionnaires: Comparisons of respondent satisficing and social desirability response bias. *Public Opinion Quarterly, 67*(1), 79–125.

Höysniemi, J., Hämäläinen, P., & Turkki, L. (2003). Using peer tutoring in evaluating the usability of a physically interactive computer game with children. *Interacting with Computers, 15*(2), 203–225.

Hyrskykari, A., Ovaska, S., Räihä, K., Majaranta, P., & Lehtinen, M. (2008). Gaze path stimulation in retrospective think-aloud. *Journal of Eye Movement Research, 2*(4), 1–18.

Intille, S. S., Kukla, C., & Ma, X. (2002). Eliciting user preferences using image-based experience sampling and reflection. In *Proc. CHI '02 Extended Abstracts on Human Factors in Computing Systems* (pp. 738–739). New York, NY: ACM Press.

Ivory, M. Y., & Hearst, M. A. (2001). The state of the art in automated usability evaluation of user interfaces. *ACM Computing Surveys, 33*(4), 470–516.

Jain, J., & Boyce, S. (2012). Case study: Longitudinal comparative analysis for analyzing user behavior. In *Proceedings of the 2012 ACM Annual Conference Extended Abstracts on Human Factors in Computing Systems Extended Abstracts*. New York, NY: ACM.

Jones, R., Milic-Frayling, N., Rodden, K., & Blackwell, A. (2007). Contextual method for the redesign of existing software products. *International Journal of Human-Computer Interaction, 22*(1–2), 81–101.

Kahneman, D., Krueger, A. B., Schkade, D. A., Schwarz, N., & Stone, A. A. (2004). A survey method for characterizing daily life experience: The day reconstruction method. *Science, 306*(5702), 1776.

Karapanos, E., Martens, J.-B., & Hassenzahl, M. (2009). Reconstructing experiences through sketching. Arxiv preprint, arXiv:0912.5343.

Karapanos, E., Martens, J., & Hassenzahl, M. (2010). On the retrospective assessment of users' experiences over time: Memory or actuality?. In *Proceedings of the 28th of the International Conference Extended Abstracts on Human Factors in Computing systems*. New York, NY: ACM.

Karapanos, E., Zimmerman, J., Forlizzi, J., & Martens, J. B. (2009). User experience over time: An initial framework. In *Proceedings of the 27th International Conference on Human Factors in Computing Systems* (pp. 729–738). New York, NY: ACM.

Kellar, M., Watters, C., & Shepherd, M. (2006). A goal-based classification of web information tasks. In *Proceedings of the Annual Meeting of the American Society for Information Science and Technology*, Austin, TX (ASIS&T).

Kuniavsky, M. (2003). *Observing the user experience: A practioner's guide to user research*. New York, NY: Morgan Kaufman.

Kuusela, H., & Paul, P. (2000). A comparison of concurrent and retrospective verbal protocol analysis. *American Journal of Psychology, 113*(3), 387–404.

Lamming, M., Brown, P., Carter, K., Eldridge, M., Flynn, M., Louie, G., et al. (1994). The design of a human memory prosthesis. *Computer Journal, 37*(3), 153–163.

Larson, R., & Csikszentmihalyi, M. (1983). The experience sampling method. In H. T. Reis (Ed.), *Naturalistic approaches to studying social interaction: New directions for methodology of social and behavioral science*. San Francisco, CA: Jossey-Bass.

Loftus, E. F. (1996). Memory distortion and false memory creation. *Journal of the American Academy of Psychiatry and the Law Online, 24*(3), 281–295.

Loftus, E. F. (2005). Planting misinformation in the human mind: A 30-year investigation of the malleability of memory. *Learning & Memory, 12*(4), 361–366.

McCarney, R., Warner, J., Iliffe, S., van Haselen, R., Griffin, M., & Fisher, P. (2007). The Hawthorne effect: A randomised, controlled trial. *BMC Medical Research Methodology, 7*, 30.

Memon, A., Wark, L., Holley, A., Bull, R., & Koehnken, G. (1997). Eyewitness performance in cognitive and structured interviews. *Memory, 5*(5), 639.

Muralidharan, A., Gyongyi, Z., & Chi, E. (2012). Social annotations in web search. In *Proceedings of the 2012 ACM Annual Conference on Human Factors in Computing Systems*. New York, NY: ACM.

Norman, D. (2009). Memory is more important that actuality. *Interactions, 16*(2), 24–26.

Novick, D. G., Santaella, B., Cervantes, A., & Andrade, C. (2012, October). Short-term methodology for long-term usability. In *Proceedings of the 30th ACM International Conference on Design of Communication* (pp. 205–212). New York, NY: ACM.

Rieman, J. (1993). The diary study: A workplace-oriented research tool to guide laboratory efforts. In *Proceedings of CHI: ACM Conference on Human Factors in Computing Systems* (pp. 321–326).

Roediger, H. L., & McDermott, K. B. (1995). Creating false memories: Remembering words that were not presented in lists. *Journal of Experimental Psychology: Learning, Memory, and Cognition, 21*, 803–814.

Russell, D. M., & Grimes, C. (2007). Assigned tasks are not the same as self-chosen Web search tasks. In System Sciences, 2007. *HICSS 2007. 40th Annual Hawaii International Conference on* (pp. 83-83). IEEE.

Russell, D. M., & Oren, M. (2009, January). Retrospective cued recall: A method for accurately recalling previous user behaviors. In *System Sciences, 2009, HICSS'09. 42nd Hawaii International Conference on* (pp. 1–9). IEEE.

Schacter, D. L. (1999). The seven sins of memory: Insights from psychology and cognitive neuroscience. *American Psychologist, 54*(3), 182–203.

Schacter, D. L. (2001). *The seven sins of memory: How the mind forgets and remembers*. Boston, MA: Houghton Mifflin.

Schwarz, N., Kahneman, D., Xu, J., Belli, R., Stafford, F., & Alwin, D. (2009). Global and episodic reports of hedonic experience. In R. Belli, D. Alwin, & F. Stafford (Eds.), *Using calendar and diary methods in life events research* (pp. 157–174). Newbury Park, CA: Sage Publishing.

Shiffman, S., Hufford, M., Hickcox, M., Paty, J. A., Gnys, M., & Kassel, J. D. (1997). Remember that? A comparison of real-time versus retrospective recall of smoking lapses. *Journal of Consulting and Clinical Psychology, 65*, 292.

Siochi, A. C., & Hid, D. (1991). A study of computer-supported user interface evaluation using maximal repeating pattern analysis. In *Proceedings of ACM CHI '91* (pp. 301–305).

Steele-Johnson, D. (2000). Goal orientation and task demand effects on motivation, affect, and performance. *The Journal of Applied Psychology, 85*(5), 724–738.

Teevan, J., & Karger, D. (2005). *The research engine: Helping people return to information on the Web*. Paper presented at the Proceedings of the ACM Symposium on User Interface Software and Technology (UIST'05), Seattle, WA.

Underwood, B. J. (1965). False recognition produced by implicit verbal responses. *Journal of Experimental Psychology, 70*, 122–129.

Van Boven, L., Kane, J., & McGraw, A. P. (2009). Temporally asymmetric constraints on mental simulation: Retrospection is more constrained than prospection. *The handbook of imagination and mental simulation* (131–147).

van den Haak, M., De Jong, M., & Schellens, P. J. (2003). Retrospective versus concurrent think-aloud protocols: Testing the usability of an online library catalogue. *Behaviour & Information Technology, 22*, 339–351.

Van Gog, T., Paas, F., & Van Merriënboer, J. J. G. (2005). Uncovering the problem-solving process: Cued retrospective reporting versus concurrent and retrospective reporting. *Journal of Experimental Psychology: Applied, 11*(4), 237–244.

Van House, N. (2006). Interview viz: Visualization-assisted photo elicitation. *Ext. Abstracts CHI 2006* (pp. 1463–1468). New York, NY: ACM Press.

van Kesteren, I. E., Bekker, M. M., Vermeeren, A. P., & Lloyd, P. A. (2003). Assessing usability evaluation methods on their effectiveness to elicit verbal comments from children subjects. In *Proceedings of the 2003 Conference on Interaction Design and Children* (pp. 41–49). New York, NY: ACM.

Weisberg, H., Krosnick, J. A., & Bowen, B. D. (1996). *An introduction to survey research, polling, and data analysis. Thousand Oaks*, CA: Sage Publications.

Willis, G. B. (2005). *Cognitive interviewing: A tool for improving questionnaire design*. Thousand Oaks, CA: Sage Publications.

Wilson, B. A., Evans, J. J., Emslie, H., & Malinek, V. (1997). Evaluation of NeuroPage: A new memory aid. *Journal of Neurology, Neurosurgery and Psychiatry, 63*, 113–115.

Wilson, T. D. (1994). The Proper Protocol: Validity and Completeness of Verbal Reports. Psychological Science, 5(5), 249–252.

Agent Based Modeling to Inform the Design of Multiuser Systems

Yuqing Ren and Robert E. Kraut

Why Agent-Based Modeling?

Decades have passed since the inception of the field of Human–Computer Interaction. The emergence of the Internet has shifted researchers' attention from understanding how individuals interact with computers to understanding how individuals interact with one another using computer technologies. A wide range of systems designed for multiple users have been labeled as groupware, collaborative computing, multiuser applications, and more recently social computing technologies. Designing these types of system is more challenging than designing single-user ones because other people and their behaviors are integral elements of the system as experienced by users (Grudin, 1994). The system itself, therefore, is non-deterministic and evolutionary because the experience of some users is partly the result of decisions that earlier users have made. Because the behavior of a multiuser system is not stable until a critical mass of users has developed a routine way of using it, it is difficult to predict how groups of users will respond to a particular design before the stable state is reached. As a result, interactive design and evaluation cycle, perhaps the most successful HCI technique for system design, is insufficient for the design of multiuser applications.

Consider the design of an online health support group like breastcancer.org and the decision about whether to employ moderators who ensure group members spend their time discussing cancer-related topics, channel off-topic content to sub-forums, or prevent users from posting advertisements. A member's decision to participate in

Y. Ren (✉)
Carlson School of Management, University of Minnesota,
321 19th Avenue South, Minneapolis, MN 55455, USA
e-mail: chingren@umn.edu

R.E. Kraut
Human–Computer Interaction Institute, Carnegie Mellon University,
5000 Forbes Ave., Pittsburgh, PA 15213, USA
e-mail: robert.kraut@cs.cmu.edu

J.S. Olson and W.A. Kellogg (eds.), *Ways of Knowing in HCI*,
DOI 10.1007/978-1-4939-0378-8_16,

the community depends in part upon the content that other members post and what moderators, if they are used, allow. But how should the designers go about deciding whether moderators will improve the site? Building different versions of the site with alternative design options would be impractical and costly. Another solution is to use computational modeling to simulate a system before building it. The simulation can be used to run virtual experiments to evaluate users' likely reaction to alternative design choices and to predict how it will actually be used under various scenarios. Assuming that the simulation can replicate known patterns of behavior in the phenomena it attempts to replicate, it also can be used to predict reactions to as yet undeveloped features.

A computer simulation is a program that embodies a partial theory of how some phenomenon operates. The method has been used for decades by social scientists to understand a wide variety of social dynamics and processes. For example, Schelling (1971) created a simple model to show how residential segregation can emerge even when most members of a community would tolerate living in an ethnically mixed environment. It is "runnable" in the sense that a scientist can turn on and off or vary input parameters (e.g., the initial sizes of the ethnic groups, the strength of members' preference for diversity or speed of housing turnover), and the simulation will generate output to predict e.g., the extent to which the society will become segregated.

A similar approach can be taken to study HCI phenomena characterized by bottom-up, self-organizing, and complex interactions among individual users. For example, the use of social media such as wikis, blogs, social networking, and social bookmarking has become very prevalent in many organizations (Treem & Leonardi, 2012) and has attracted great interest from HCI researchers (e.g., DiMicco et al., 2008; Shami, Ehrlich, Gay & Hancock, 2009; Thom-Santelli, Millen & Gergle, 2011; Wu, DiMicco & Millen, 2010). Simulation can answer questions such as the following: How does usage spread within an organization? What patterns will emerge in the use and adoption of these technologies by individual users? How will the adoption and use of such technologies change organizational hierarchy? How can an organization align system design, incentives, and its culture and policies to encourage effective use of the technologies?

Scientists and engineers have built several genres of simulation to simulate social systems, including statistical, causal models, mathematical models, system dynamics models, neural networks, cellular automata, multilevel simulations, evolutionary models, and agent-based models (Taber & Timpone, 1996). In this chapter we focus on agent-based modeling as a tool to inform the design of multiuser systems and to advance our knowledge of how these systems operate because of the isomorphism between the systems we are attempting to simulate and the simulation techniques. An agent-based model simulates a multiuser system by modeling the behaviors of and interactions among individual users who comprise the system. We start with a brief review of the method, followed by our key contribution, a seven-step roadmap that HCI researchers can follow to build or evaluate agent-based models. We then describe how we followed the steps and built an agent-based model that can inform the design of online communities. In the end, we share a personal account of how we encountered the method and include references for readers who would like to learn more about the method.

What is Agent-Based Modeling?

Agent-based modeling is a form of computational simulation that "enables a researcher to create, analyze, and experiment with models composed of agents that interact within an environment" (Gilbert, 2008). The agents can imitate a wide variety of physical and social entities such as human beings, animals, particles, or molecules. Agent-based modeling is similar to mathematical modeling in terms of rigor but better suited for situations when agents are autonomous and heterogeneous, when there are complex interactions between agents, and when lower-level actions and interactions can lead to the emergence of system-level structures. Compared with conventional methods of developing theories in social sciences, agent-based modeling is especially suitable for bottom-up theorizing (Kozlowski & Klein, 2000), and for understanding how individual agent behaviors interact over time and lead to emergent system-level patterns.

The system-level regularities are often the results of multiple forces working together. The tension among the forces may be temporal, structural, or spatial and often result in nonlinear relationships like tipping points (Davis, Eisenhardt & Bingham, 2007). A famous example is Reynolds' "boids" model (1987) that simulates the behaviors of flocks of birds. The agents in this model are birds with limited perception programmed with three simple rules as illustrated in Fig. 1: separation to avoid getting too close to other birds, velocity to travel at the speed of nearby

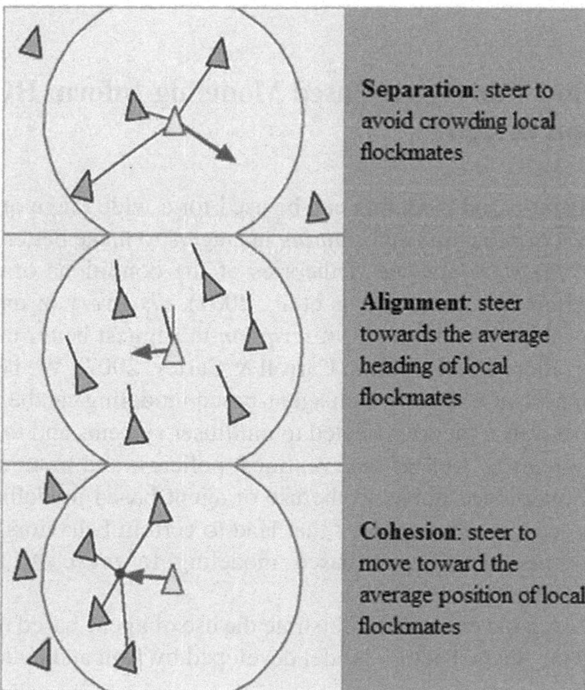

Separation: steer to avoid crowding local flockmates
Alignment: steer towards the average heading of local flockmates
Cohesion: steer to move toward the average position of local flockmates

Fig. 1 Illustration of Reynold's boids model (Reprinted with permission from http://www.red3d.com/cwr/boids/)

flockmates, and cohesion to head for the perceived center of nearby flockmates. The model does a remarkable job of replicating how flocks of birds fly together without bumping into each other. This tension of wanting to follow the crowd but not get too close applies to many human groups, too.

The history of agent-based models dates back to Von Neumann in the late 1940s, when he developed a machine that was capable of self-replicating (Gilbert, 2008). His creation of a self-replicating automaton without a computer eventually led to the creation of cellular automata, a popular technique for doing agent-based modeling by placing individual agents on a two-dimensional lattice or grid of cells and observing what patterns emerge as they interact with neighbors (Davis et al., 2007). The idea motivated the creation of Conway's Game of Life (Gardner, 1970) and gradually, the method morphed its way from mathematics into economics, social science, and other disciplines. The social science version of the game is called the Sugarscape model, created by Epstein & Axtell (1996) to simulate and study human societies.

In the past two to three decades, agent-based modeling has become much more widespread due to the exponential growth of computing power. A wide variety of models have been developed to simulate physical and social phenomena, such as flow in sand piles and the activities of animals such as birds and ants (Sawyer, 2003), social and organizational behaviors in cooperation and collective action (Macy, 1991), learning (March, 1991), social influence and norm formation (Axelrod, 1986; Axelrod, 1997a, 1997b), cultural dissemination (Harrison & Carroll, 1991), and innovation diffusion (Strang & Macy, 2001).

How Can Agent-Based Modeling Inform HCI Theory and Design?

Agent-based modeling can be used for a wide range of purposes such as *description* of behaviors and *training* managers to make better decisions (Burton & Obel, 1995), *development* of theories of the conditions or mechanisms that generate certain behaviors (Davis et al., 2007), *discovery* of unexpected consequences of local interactions, and *prescription* to suggest better modes of operation or organization (Harrison, Lin, Carroll & Carley, 2007). We believe there are at least two important ways in which agent-based modeling can be leveraged in HCI research: to advance theories related to multiuser systems and to inform the design of these systems as well as interventions, policies, and practices surrounding them. The former corresponds to the use of agent based modeling to *explain* mechanisms, processes, or conditions that lead to certain behaviors and the latter corresponds to the use of agent-based modeling to *prescribe* actions to obtain desired outcomes.

A good example to illustrate the use of agent-based modeling to *advance theory* is the Shape Factory model developed by Nan and colleagues (2005, 2008). Their research began with a laboratory experiment to investigate how geographic separation influences the performance of collocated and remote workers. Ten participants, five collocated and five remote, earn points by making and buying parts of different shapes to fill "customer" orders. A puzzling pattern emerged, in which collocated and remote players were equally successful even though collocated players had communication advantages. Two theoretically plausible mechanisms—in-group favoritism and communication delay—could have been at work, but they were confounded in the experiment, making it impossible to isolate their independent effects.

Agent-based modeling is well suited for addressing such challenges because it grants researchers the ability to computationally turn a mechanism on and off and observe how outcomes change as a result. Using behavioral patterns observed in the lab experiments as benchmarks, Nan and colleagues (2005, 2008) developed an agent-based model to separate the effects of in-group favoritism and communication delay. They implemented the two mechanisms as two behavioral rules: in-group favoritism meant collocated agents always transacted business with other collocated agents before contacting remote agents; communication delay meant a one-step time delay in communications with all remote agents. Their simulation results suggested in-group favoritism actually had a detrimental effect on the performance of collocated players (by limiting themselves to transact with only local agents) although (lack of) communication delay had a positive effect on their performance. The two effects cancelled out each other in the laboratory experiment and would be hard to disentangle without agent-based modeling.

The Nan study illustrates how agent-based modeling can be used to complement other empirical methods, in this case laboratory experiments, to enrich our theoretical understanding of the working of multiuser systems. As Davis et al. (2007) suggest, simulation occupies a "sweet spot" between theory-creating methods such as case studies and formal modeling, and theory-testing methods, such as survey and experiments. The model needs to be grounded in theoretical insights and empirical evidence, and in turn it can expand our understanding beyond the conditions that were observed in early research.

Because researchers in management, public policy, and sociology have already documented how to use simulation to develop and test theories (Axelrod, 2005; Davis et al., 2007; Harrison et al., 2007; Macy & Willer, 2002), in this chapter, we focus on the use of agent-based modeling in HCI research to inform the design of multiuser systems and policies and practices surrounding them. We use online communities as an example of multiuser systems. A key challenge in designing online communities is that designers must make numerous decisions about features, structures, and policies. Even experienced designers can be overwhelmed by the trade-offs involved in the decisions and fail to anticipate how users will respond. For instance, when launching an online community, if a community offers points for contributions and recognizes the most active contributors on a public "leader board," this feature may encourage the least active participants to increase their level of contribution, and heavy contributors to contribute less if the former perceive themselves as under-contributing and the latter perceive themselves as over-contributing; moreover, it could discourage most community members from contributing at all if they perceive that the leaders are providing sufficient content.

These contradictory predictions originate from two social science theories: social comparison theory (Festinger, 1954) and the Collective Effort Model (Karau & Williams, 1993). The former argues that people are motivated to match their performance to the performance of similar others, and thus increase their effort when being told that others have contributed more than they have (Harper et al., 2007). The latter argues that people exert less effort when working in groups than individually because they perceive their efforts are unnecessary to achieve group outcomes. Perhaps because of contradictory predictions like these, theories from social psychology, organizational behavior, sociology, and economics have been applied to describe behaviors in online communities, more than they have been applied prescriptively, to offer solutions for building successful communities (see Ling et al., 2005, for exception Kraut & Resnick 2012).

An important reason that these social science theories seem ill suited for design is that the logic of design, which manages trade-offs among tens or hundreds of parameters that can influence members' behaviors, is at odds with the logic of social science research, which examines the influence of a small set of variables while holding everything else equal. Agent-based models can bridge this gap, by synthesizing insights from multiple theories to identify the pathways through which particular design choices will affect the different outcomes that designers aim to achieve. In other words, agent-based models can be used to link and integrate what Davis et al. (2007) termed "simple theories" to infer "the combined implications of several theoretical assumptions or empirical results" (Taber & Timpone, 1996, p. 6).

What Constitutes Good Work: A Seven-Step Roadmap

In this section, we provide a roadmap with a set of guidelines HCI researchers can follow to build agent-based models, as shown in Table 1. To make the guidelines concrete and accessible, we use our personal experience to demonstrate how we followed these guidelines and built an agent-based model to inform the design of text-based online communities. We assume that you already have a research question and wonder if agent-based modeling is the appropriate method to study it.

Step 1: Evaluate the Appropriateness of Agent-Based Modeling for the Research Question

Whether agent-based modeling is appropriate for your research depends on several factors: the phenomenon of interest, the level of analysis of your research questions,

Table 1 Roadmap for using agent-based modeling (ABM) to inform HCI design

Steps	Activities/questions
Evaluate the appropriateness of ABM for your research question	– Can the overall system behavior be decomposed into decisions and actions by autonomous interacting agents? – Are their decisions and actions influenced by multiple forces? – Is the system likely to be multilevel, nonlinear, and dynamic? – Are there simple theories or empirical evidence available to ground the model?
Define boundary conditions and build a conceptual model	– Decide the scope of the model (types of agents, types of objectives, types of agent behaviors, the larger environment) – Identify theories to help construct the conceptual map – Identify key variables in the conceptual map – Start with a simple model and gradually expand
Translate the conceptual model to computational representations	– Operationalize three key elements: agents, environment, and timescale – Translate theories to behavioral rules governing agents' motion, communication, and action – Simulate time as forced parallel
Implement the model	– Decide whether to use an existing platform or build from scratch – Compare and choose a platform if needed – Program, debug, test the program
Validate the model	– Check program to make sure that it is an accurate translation of the conceptual model and is bug free – Calibrate the model by modifying the model to match theory predictions, stylized facts, or empirical training data – Test the external validity of the model by comparing simulation results with theory or empirical testing data
Experiment with the model	– Design virtual experiments (determine key factors and their values or range and number of runs) – Set parameters with theoretical or real-life values – Run experiments and gather output data
Publish the model and results	– Provide sufficient detail for others to replicate the model – Arrange to share the source code – Discuss practical as well as statistical significance

and the working body of knowledge from which you can borrow insights to ground the model. Phenomena well suited for agent-based modeling typically have the following characteristics:

1. They involve the actions and interactions of individual agents.
2. Individual agents have heterogeneous motivations, interests, or behaviors.
3. Individual agents form a large social system whose structure is determined by individual actions and the size and structure of the social system, in turn, shape individual behaviors.

4. The system dynamically evolves over time as individual agents interact with one another, and as a result, it can be characterized as multilevel, nonlinear, and dynamic.

   Here are some sample HCI problems that agent-based modeling may help tackle:

- *Attention management in communication exchanges.* Spontaneous, informal communication at work is important, yet, it often helps the initiator of the communication at the expense of the person being interrupted (Perlow, 1999). Interventions designed to balance the benefits and costs of spontaneous communication have often had unforeseen consequences. For example, pricing systems that increase the cost of interruptions can reduce the volume of communication below optimal levels (e.g., Kraut et al., 2002) or awareness displays that show when someone is interruptible can increase instead of reducing interruptions (e.g., Fogarty, Lai & Christensen, 2004). Agent-based models can help predict the long term impact of alternative interventions.
- *Feedback mechanisms in online contribution.* Online production communities like Wikipedia need high quality contributions. Interventions that aim to increase quality often have unintended consequences on the contributors. For example, making new members pass a quality test can increase their quality and motivation but reduce the number of members who join (Drenner, Sen & Terveen, 2008), and giving contributors corrective feedback may direct their attention away from the task and towards themselves and harm their performance (Kluger & DeNisi, 1996). Agent-based models can help community leaders manage these trade-offs.

To reiterate, several common themes run through these examples that make them appropriate for agent-based modeling. First, the phenomena are generated *bottom-up,* in the sense that individuals make autonomous decisions and the outcome—whether it is the success or failure of a system or communication patterns—are jointly determined by individual actions and interactions. Second, *multiple forces* drive individual behaviors, implying the model needs to combine multiple theories to be a valid representation of reality. Finally, the system-level regularities cannot be intuitively predicted based on rules for individual actions because the multiple forces affecting behaviors may work in opposite directions or cancel each other out.

These examples draw upon relatively mature theoretical and empirical understanding of the phenomena being studied. Such understandings should be past the exploratory stage, with sufficient literature available to ground the model. It is ideal to have multiple theoretical propositions or empirical results, none of which seems capable of explaining the observation alone but collectively have the potential to do so (Taber & Timpone, 1996). For example, an agent-based model to simulate how starting conditions influence a community's success can rely upon a rich literature on critical mass (Markus, 1987), network externalities (e.g., Shapiro & Varian, 1999), organizational ecology (Hannan & Freeman, 1989), and group commitment (e.g., Mathieu & Zajac, 1990). It is also desirable to have ways to gather new empirical data to fill in detail of the model where theories are silent or fail to provide detail to specify functions or parameters.

Step 2: Define Boundary Conditions and Build a Conceptual Model

Agent-based models, like mathematical and statistical models, are a simplified representation of reality. It is crucial to clearly define the boundaries of a model to capture the essence of the phenomenon being studied. Many multiuser systems are complex by nature, involving agents in different roles, artifacts of different types, and complicated connections between agents, artifacts, and their environment. For example, work in Wikipedia occurs in 270 different languages and depends upon the contribution of tens of thousands of volunteer editors who take on a variety of tasks from creating new articles to writing policies (Bao et al., 2012; Welser et al., 2011). The editors are organized into hundreds of subgroups known as WikiProjects, and they collaborate on a technical infrastructure run by a non-profit organization. The content ultimately becomes viewable to tens of millions of Internet users. If you were to build an agent-based model to understand the working of Wikipedia collaboration, where should you draw the boundaries? Besides editors and articles, should Wikipedia readers or other agents like the bots (automated programs) that repair vandalism be explicitly modeled? What about higher-level social entities like WikiProjects or the Wikimedia Foundation, which supports Wikipedia's infrastructure?

These are nontrivial decisions, and the answers are not straightforward. Trade-offs between simplicity and reality or between parsimony and accuracy plague agent-based modeling. An agent-based model, while needing to be sufficiently comprehensive and complete to be accurate, also needs to be a simplified representation of reality to be useful (Gilbert, 2008). Complex models can be more accurate in their predictions but are more difficult to debug and may become so incomprehensible that readers or even its developers have difficulty deciphering how variations in the model's features lead to its results. The right decision, therefore, requires a balance of capturing the central phenomenon while stripping away the nonessentials. To a large extent, this balancing act is a judgment call (Davis et al., 2007) or "the art of simulation" (Harrison et al., 2007). There are no universally correct answers; settling on one depends on individual researchers' preferences and style of research.

Some modelers lean toward simplicity. Simple models are especially good for theory development, exemplified by Schelling's (1971) racial segregation model. Simple models can be quite powerful if unexpected system-level patterns can be generated with simple rules at the agent level. On the other hand, simple models often fall short of making accurate predictions to guide practice because they fail to incorporate all the important forces or mechanisms driving a phenomenon. Agent-based modeling needs a reasonable level of complexity to provide useful guidance to design. Even when building complex models, a good practice is to start with a simple model and gradually expand to add more fidelity.

Once the boundaries are established, researchers can identify the important concepts they want to capture in the model and their relationships. Taber and

Timpone (1996) suggest the practice of creating "an inventory of concepts on paper when dealing with a complex model" (p. 15). This concept inventory should define the concepts in qualitative terms and propose a loose notion of how they might be operationalized. Researchers should consider theories from multiple disciplines to ground their models, because social behaviors and processes cannot be decomposed into separate subprocesses that neatly match the artificial divisions of different disciplines (Epstein, 1999). Individual behaviors can be driven by economic, psychological, political, and technological factors, and researchers should not let disciplinary boundaries prevent them from identifying important aspects of a phenomenon.

Step 3: Translate the Conceptual Model into Computational Representations

The next step is to operationalize the conceptual model by translating theoretical relationships to assumptions, agent attributes, and behavioral rules. Gilbert (2008) identified three key elements to specify an agent-based model: agents, environments, and timescales. An agent can be a person, animal, or object. Agents imitating humans can engage in the following activities:

- Perceive the environment including the presence of other agents or objects in their neighborhood.
- Perform a set of behaviors, such as moving within a space, communicating (sending messages to and receiving messages from other agents), acting or interacting with the environment (such as joining a group or contributing information to a corporate wiki).
- Remember their previous states, actions, or consequences (e.g., for learning purpose).
- Follow policies or adopt strategies that determine what actions to take next.

In the Shape Factory simulation, agents' perceptions of the environment include the awareness of collocated and remote players and the shapes they produce. Agents could not move but could communicate by sending and fulfilling shape requests. Agents did not have memories and could not learn from past behaviors. They did not engage in sophisticated strategies although they had a goal of maximizing the number of orders they filled.

In more complex models, agents could engage in more sophisticated behaviors. For instance, in an model developed to study transactive memory, agents possessed knowledge about both their own and other agents' areas of expertise (Ren, Carley & Argote, 2006). The transactive memory enabled agents to efficiently search for information and assign tasks to those with specialized knowledge.

Timescale is another key element in translating theories to agent behaviors. The order in which agent behaviors occur can significantly influence simulation results because output from early agents changes the environment that other agents experience later. For example, in simulating an online discussion group, researchers must consider whether a post will be broadcast to all other agents immediately after it is posted or kept in a repository until all agents finish the current round of activities. The choice arises because parallel processing is computationally costly, but less costly modeling procedures can approximate it. For example, researchers can buffer all interactions in the environment (e.g., recent messages posted in a community) and wait until all actions are completed before displaying the new interactions (i.e., new messages) to all agents. In the jargon of agent-based modeling, actions can be organized in staged episodes, with time simulated as "forced parallel." This technique is also called "simulated synchronous execution" (Gilbert, 2008). A useful tool in model design is a flowchart, which shows the sequence of actions together with the conditions under which a rule applies or an action is taken.

Step 4: Implement the Model

The implementation step is often mistaken as the core of agent-based modeling (all you need to do is to write computer code which will generate tons of numbers, right?). Implementation is important in the sense that the modeler must accurately translate the conceptual model to computer code so that the program runs efficiently and is bug free. Compared to the conceptual model, however, implementation is less important. Without a valid conceptual model, whatever data you get from the simulation will be worthless, i.e., "garbage in, garbage out."

There are typically two options for implementing an agent-based model: build on an existing simulation platform like NetLogo (Wilensky, 1999) or code it from scratch using a general-purpose computer language like Java, Python, or C++. This choice determines the user interface that the modeler will use to interact with the model. Myers and colleagues (2000) have identified criteria for evaluating user interface software. Criteria that are especially important in choosing tools for agent-based modeling include the software's threshold (i.e., the difficulty in learning the system and building initial software), ceiling (how much can be accomplished with the software), path of least resistance (i.e., whether the software helps the user produce appropriate models), and stability (i.e., whether the software is changing too rapidly for its users to gain significant experience with it).

Each approach has its pros and cons. Which to choose depends on the complexity of the model, the researchers' timeline and programming skills, and the extent to which a decent user interface and add-on features such as network analysis and visualization are needed. For beginners, we recommend building on a platform such as NetLogo (http://ccl.northwestern.edu/netlogo/), Repast (http://repast.sourceforge.net/), or Mason (http://cs.gmu.edu/~eclab/projects/mason/), especially if the model is simple, can be built using standard modules or if researchers have limited programming skills. Even experienced programmers can save a great deal of time and effort by building on a platform. On the other hand, if the model is complex and requires functions that are unavailable in established platforms, it can be difficult to coerce the platform to fit your purpose. If so, building from scratch may be the only option. The first author's experience of developing a model to simulate transactive memory systems fits this category. A key element of the model was the concept of transactive memory that stores information about other agents' areas of expertise (Ren et al., 2006). Existing platforms did not have a built-in module that can be modified to simulate the working of a transactive memory system. Therefore, we built our own model using Java. It was a multiyear effort (close to 3 years including model validation) but it allowed us to capture the core concept we wanted to study.

Detailed reviews and comparisons of agent-based modeling platforms are available from several sources. Gilbert (2008) compares four platforms—Swarm, Repast, Mason, and NetLogo—on user base, speed of execution, support for graphical interface and systematic experimentation, and ease of learning. He reports that "NetLogo stands out as the quickest to learn and the easiest to use, but may not be the most suitable for large and complex models. [...] Repast has the advantage of being the newest ... but also has a significantly smaller user base, meaning that there is less of a community that can provide advice and support" (p. 49). Our own experiences with the two platforms are consistent with Gilbert's assessment. In addition, we recommend Repast for building large, complex models that are computationally demanding.

Step 5: Demonstrate the Internal and External Validity of the Model

The next step is to ensure that the model is a valid representation of reality. Doing so consists of three processes: verification, calibration, and validation. Model *verification* involves checking that the agent-based model satisfies its specification, is correctly implemented and bug free (Gilbert, 2008). Model *calibration* involves tuning a model's rules or parameters to produce results that match real data or stylized facts (i.e., simplified representations of empirical findings) with reasonable accuracy (Carley, 1996). Model *validation* involves comparing model predictions to a holdout sample of data that was not used in the calibration process to see how well the two match (Gilbert, 2008). Verification ensures internal validity or the degree to which the implemented model corresponds with the conceptual model and calibration and validation ensure external validity or the degree to which the model corresponds with the real world (Taber & Timpone, 1996). All three processes can be time consuming so researchers must budget sufficient time when planning the project. No matter how carefully one has worked to develop a model, it will always contain bugs. Some bugs are obvious and easy to find because they prevent the model from running or they generate anomalous results. Other bugs are harder to find, because the model runs and produces results that appear superficially plausible. These bugs require more careful scrutiny and rigorous testing. Occasionally they survive the verification process and get caught in calibration or validation when researchers have difficulty producing results that match theoretical predictions or empirical data (although sometimes it is the theory that needs to be modified).

Once you are confident about a model's internal validity, you can move on to assess its external validity with calibration and validation techniques. The two processes are often confused as one. While both processes examine whether model output matches real world data, calibration involves the "tweaking" of the model iteratively so that its output matches (some of the) data. Validation involves running the model to assess its match to a new sample of data. To avoid overfitting, a good practice is to split data into two sets: one set used to calibrate the model and the other used to validate the model (similar to training and testing sets in machine learning).

For calibration and validation, researchers often focus on assessing outcome validity by comparing model predictions with real-world data or with the predictions of other competing models[1] (Taber & Timpone, 1996). The primary criterion is to show that the model can replicate the system-level regularities that the research seeks to explain (Gilbert, 2008). The replication can be assessed using multiple criteria such as correlations, analysis of variance, linear or nonlinear regression, or tests for comparison of means (Taber & Timpone, 1996). Carley (1996) describes four levels of assessing outcome validity: *pattern validity* requires the pattern of simulation results matches patterns of real data, *point validity* requires the output variables of the model, taken one at a time have the same mean as the real data, *distributional validity* requires the distribution of model output has the same distributional characteristics as the real data, and *value validity* requires the highest level of precision in matching, that is, the model output matches the real data on a point-by-point basis. Which level to choose is at the discretion of researchers depending on research purpose.

Step 6: Experiment with the Model

Once a model is validated, it can be used to run virtual experiments to generate simulated data for which no real data yet exist. This is the step where HCI researchers and practitioners will experience the value of agent-based modeling. They can vary parameters across a wide range and at great granularity—much beyond the level of control typical of field studies or laboratory experiments. Once the model is built, the costs of running a virtual experiment are minimal. More importantly, researchers can open the proverbial "black box" by observing and analyzing intermediate variables to reveal the mechanisms or processes that cause the resulting patterns.

[1] Another rare form of model validation is called model alignment or "docking" in short, under which researchers compare two or more models to see if they can produce the same results. A good example is Axtell and colleagues' work (1996) to align the cultural transmission model and the Sugarscape model. They call for wider practice of docking among modelers.

Similar to laboratory experiments, a virtual experiment generates data for each cell in an experimental design by running the model with a combination of parameters. Meaningful results require careful setting of parameters to match reality and to determine how many data points to simulate for each condition. A good practice is to use theory or empirical evidence to restrict the range of parameters, either qualitatively or quantitatively. Another approach is to sample the parameter space to cover a reasonable range (Gilbert, 2008). For instance, Nan (2011) built an agent-based model to simulate IT use in organizations and used data from a case study by Orlikowski (1996) to set initial conditions. It is also important to include counterfactual analysis or what is often referred to as "what-if" experiments to explore what might happen if parameters are to set to values different from existing empirical observations.

After the experiments, researchers should run sensitivity analysis or robustness checks (Davis et al., 2007) to assess how sensitive simulation results are to key assumptions and parameters built in the model. Sensitivity analysis is the process of relaxing assumptions or systematically changing functions and parameters to see how robust simulation results are or to understand the conditions under which the model yields the results (Gilbert, 2008). Researchers can be more confident about the results if they remain stable when key constraints are relaxed or key parameters are varied. Sensitivity analysis can also be used to facilitate model validation. This practice is especially valuable when little theory or empirical evidence is available to inform the specification of experimental parameters. One recommendation is to expand the parameter space to identify and report "boundary conditions" when simulation results no longer hold.

Step 7: Publish the Model and Results

Analogously to conducting a usability study in industry, if your only purpose of building an agent-based model to inform system design, you are done. However, if you are an academic working on peer-reviewed publications, more work still remains. Because many reviewers are unfamiliar with the method and because the details of a model are harder to describe than the details of an empirical study, it can be difficult to publish research using agent-based models. In this section, we share some of our experiences of reviewing and publishing simulation work.

Lesson 1—Write in plain English and provide enough detail about the model. This advice is easier said than done. Good writing is important for publishing all papers but especially crucial for simulation work because you must appeal to both domain and methodology experts and readers vary greatly in their familiarity with the method. Even a moderately complex model, like the one we describe below, might include dozens of rules, close to 100 variables and 1,300 lines of code built on a platform. You need to provide enough detail about how the model works, without making every reader read the original program. Some common mistakes are failing to include all rules that determine agent behaviors (e.g., saying an agent's opinion is influenced by close neighbors without specifying how the influence occurs), failing to specify the order in which behaviors occur (e.g., do agents express their opinions first and then get influenced or decide to switch groups or vice versa), or failing to clearly describe the initial conditions of virtual experiments (e.g., how many agents to begin with, the rate at which new agents enter, the number of runs for each condition). A rule of thumb to assess whether sufficient detail has been provided is that experienced modelers should be able to draft the pseudo code of the program based on the model description. If space permits, it is also a good idea to include pseudo-code, key functional forms, and a flow chart showing the sequence of actions.

Lesson 2—Prepare to share your code. Whether you are building on a platform or programming from scratch, write clear code with good documentation so that it can be easily read and understood by an average programmer. Some reviewers may request to see your code and other researchers may be interested in confirming or extending your model. There are multiple ways of sharing one's model, privately or publicly. One advantage of building on a platform is the ease of sharing models. For example, NetLogo hosts a Modeling Commons for its users to share models and search for others' models (http://modelingcommons.org/account/login). You can access our online community model at http://dl.dropbox.com/u/11116596/OnlineCommDesign.nlogo.

Lesson 3—Be mindful of sample size when reporting simulation results. Sample size is determined by the number of runs for each experimental condition. Because it is so easy to replicate an experiment once a model is developed, reporting statistical significance is insufficient. A reviewer comment we once received vividly illustrates the concern: "Could you have simulated 1,000 groups and got everything to be significant? How did you choose [the number we had chosen in the paper]?" Our advice is to report effect sizes (e.g., % increase in adoption rates or number of visitors for a day) in addition to statistical significance.

Following the Roadmap: Using ABM to Inform the Design of Online Communities

In this section, we show how we have followed the seven steps and built a model to inform the design of online communities. We began with the research question of how design choices such as topical breadth, message volume, and discussion moderation interact to influence the success of an online community. We believed agent-based modeling was appropriate to address the question. Online communities are bottom-up social structures whose success depends on the active participation and interaction of individual members. Members are heterogeneous in their attributes (e.g., interests, knowledge, experiences with the community) and motivations (e.g., seeking information, emotional support, reputation, entertainment, a sense of belonging) (Ridings & Gefen, 2004; Wasko & Faraj, 2005). When we were starting the project in 2006, there was a good body of knowledge from survey and interviews about what motivates users to participate in online communities (Bryant, Forte & Bruckman, 2005; Ridings & Gefen, 2004). We were also able to ground the model on well-established theories from economics and social psychology around individual decisions to join groups, participate in collective actions, and the influence of perceived benefits and costs. Agent-based modeling was a useful tool to integrate these theories to understand challenges of building successful online communities.

Based on our own empirical research and the literature, we decided the core concept to simulate was individuals' motivation to participate. We chose the Expectancy Theory of motivation (Vroom, Porter & Lawler, 2005) and one of its extensions, the Collective Effort Model (Karau & Williams, 1993) as our basis. These theories assume people contribute to a group to the extent they believe their efforts will lead to outcomes for themselves that they value. Neither theory, however, is specific about the types of benefits that motivate people. Research on online communities had identified six benefits that consistently drove participation: (1) information, (2) fulfillment of altruistic or expressive needs produced by helping others, (3) identification with the group, (4) relationships formed with group members, (5) entertainment, fun, and other forms of intrinsic motivation, and (6) reputation and other forms of extrinsic motivation (e.g., Ren, Kraut & Kiesler, 2007; Ridings & Gefen, 2004; Roberts, Hann & Slaughter, 2006; Wasko & Faraj, 2005).

We supplemented the Expectancy Theory with other theories to operationalize the six benefits. We drew insights from theories of group identity theory (Hogg, 1996) and interpersonal bonds (Berscheid, 1994) to calculate social benefits, and we drew insights from resource-based theory (Butler, 2001) and information overload theory (Jones, Ravid & Rafaeli, 2004) to calculate informational benefits. Theories of group identity and interpersonal bonds propose that members commit and contribute to a group if they feel psychologically attached to the group or its members (Prentice, Miller & Lightdale, 1994). Information overload theory proposes that human beings' information processing capacity is limited and too much information or irrelevant information is aversive (Rogers & Agarwala-Rogers, 1975).

This is where the value of agent-based modeling's ability to combine multiple theories becomes apparent. First, motivation has multiple causes, and each cause is typically treated by a separate social science theory. For example, information overload theory focuses on how informational benefits affect motivation while group identity theory focuses on the motivational influence of psychological attachment to the community. Therefore, multiple theories are needed to model motivation. Second, a single design choice, when routed through different theoretical lenses, can have divergent effects on motivation. One example is the effect of group size. When examined through resource-based theory of online social groups, large group size is a measure of resource availability and thus provides informational benefits. When examined through the Collective Effort Model (Karau & Williams, 1993), however, members of large groups tend to contribute less time and resources because of dilution of responsibility. When examined through the lens of interpersonal bonds (Frank & Anderson, 1971), large group size reduces motivation because it makes it difficult to form relationships with other members. Combining these effects

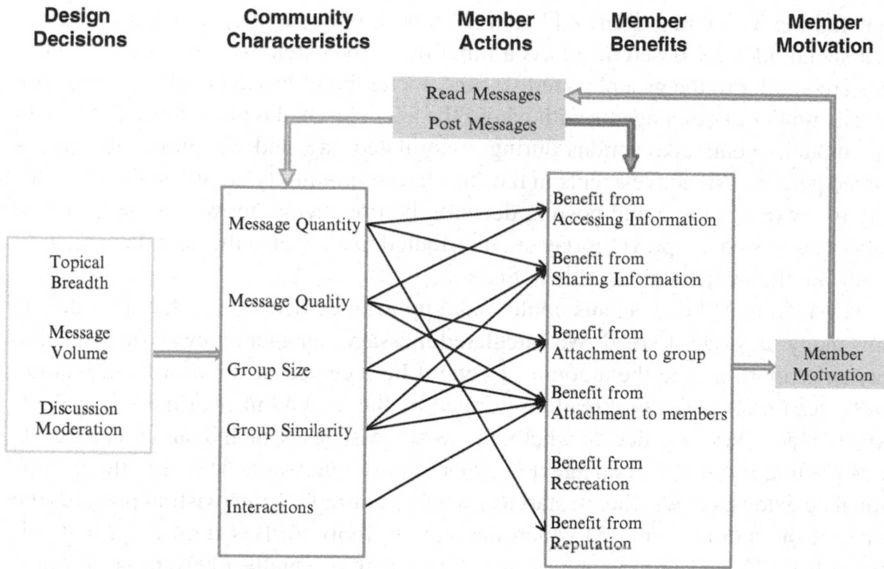

Fig. 2 The conceptual model

in an agent-based model allows us to better understand how potential design decisions affect member motivation and contribution through multiple routes.

In terms of boundary conditions, we decided to focus on members, their interactions, and how various design choices affect their experiences. Although theories from organizational ecology (Hannan & Freeman, 1989), which focus on intercommunity competition, could have been relevant, we ignored these to limit the model's scope and to make model development tractable. We also excluded, for example, the cost of implementing the design choices primarily because our goal is to assess the effectiveness of various designs and partially because it is complicated to model costs (e.g., due to different design contexts).

Figure 2 depicts our conceptual model. Member actions such as reading and posting messages are determined by benefits and costs associated with participation. Reading and posting behaviors change community dynamics such as the number and quality of messages, as well as the number of members and their relationships with one another; these, in turn, influence experienced benefits and motivation. Design interventions, such as the cost of posting messages, diversity of nominal topics, and moderation also influence community dynamics.

We then translated the conceptual model into agents' attributes and behavior rules. The two behaviors that agents engage in are reading and posting messages. Following the utility-like logic underlying the expectancy-value theories, we assumed an agent (1) logs in to read messages when expected benefit from participation exceeds expected cost, and (2) posts messages when expected benefit from contributing exceeds expected cost. Details about how we calculated member benefits can be found in Ren and Kraut (In press). For example, the model assumed that social, identity-based benefit is a function of the extent to which agents' interests are similar to the group's interests, and social, bond-based benefit is a function of the number of other agents with whom the agent has had repeated interactions. In this model, agents take actions during a simulated day, and we simulated time as forced parallel. All active agents in the simulated community are given the opportunity to make a reading and posting decision before anyone moves to the next day. Messages posted the previous day are distributed to all agents the next day and used to update their expectations of benefits.

For a simulated day, agents could make up to three decisions. They first decide how many messages to read. We calculated messages an agent viewed on a specific day as proportional to the amount of benefit he received in the past from reading messages minus the cost of reading, capped by the total number of messages available to read. They next decide whether to post messages, which incurs greater costs than reading messages. If an agent decides to post a message, he makes three additional decisions: (1) whether to start a new thread or reply to an existing post; (2) the topic of the message and (3) which message to reply to. Based on empirical evidence from Usenet groups, we assumed an agent is equally likely to start a new thread or to reply to one. The topic of the message is a joint function of the agent's interests, topics of the messages the agent has recently viewed, and the topic of the replied-to message if it is a reply. Theory and empirical evidence (Fisher, Smith & Welser, 2006; Faraj & Johnson, 2011) suggest three common patterns of interaction among community members: (1) preferential attachment, in which members respond to popular messages or posters; (2) reciprocity, in which members respond to others who have written to them in the past; and (3) interest matching, in which members respond to messages that match their interests. We thus assumed that agents in the model choose to reply to a message based on the average of (1) the number of replies the message has received; (2) the number of times the poster of the message has responded to the agent; and (3) the match between message topic and the agent's interests.

We first built our model in NetLogo, a cross-platform multi-agent modeling environment (Wilensky, 1999). Figure 3 shows a snapshot of the user interface. The buttons in the upper-left corner allow researchers to specify the initial members, messages, type of the community and run time. The window on the right shows members in the community. The plots track statistics such as member entry, exit, and the number of participants and contributors. It took us a year and half to design, build, and validate the model. The online communities we simulated grew to have thousands of members including both lurkers and active contributors and thousands of messages. We later re-implemented it using Repast to achieve greater speed. A virtual experiment with 540 runs that used to take three days to run in NetLogo took several hours in Repast.

We went through all three steps to ensure the validity of our model. Previous studies show that three statistics describing online communities—posts per member, replies per post, and communication partners (out-degrees) per member—demonstrate a power-law distribution (Fisher et al., 2006; Smith, 1999). We used these

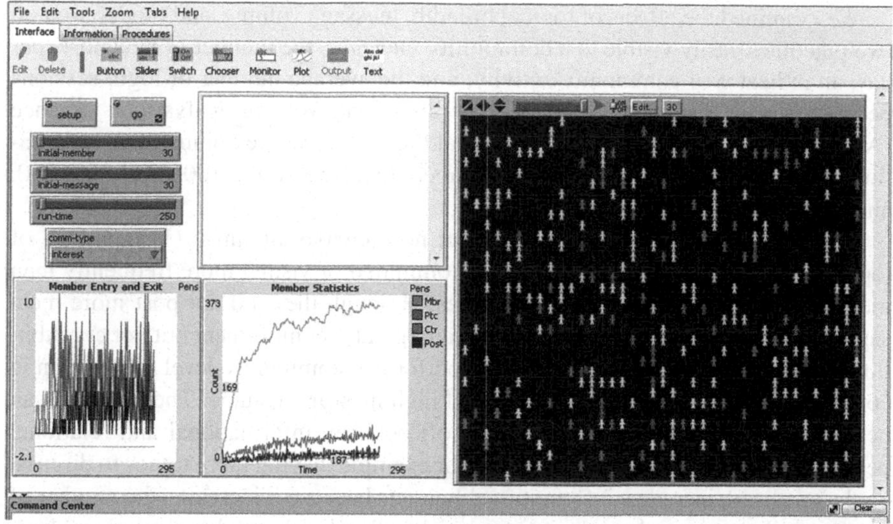

Fig. 3 Interface of the online community model in NetLogo

three stylized facts to calibrate the model. We constructed two data sets of *Usenet* groups and used a set of 12 groups to calibrate the model and a new set of 25 groups to validate it.

Calibration was an iterative process. After each run, we examined mismatches between the simulated and the real data, reexamined model assumptions, and made adjustments to the model in light of theoretical reasoning, empirical evidence, or knowledge about how the processes in the model operate. After ten iterations of tweaking, the model replicated the power-law distribution for all three statistics. We then simulated a new set of 25 Usenet groups. We used pattern validation and compared the pattern of three statistics from the model—posts per agent, replies per post, and out-degree ties per agent—with the pattern generated from real data. We also calculated Pearson correlations between the empirical data series and the simulated ones and the coefficients ranged between 0.90 and 0.96, confirming a good match between the two.

We used the model to explore three design decisions: How broad a set of discussion topics should the community encourage? What is an optimal level of message volume? What type of discussion moderation if any should the community adopt? We designed a full-factorial experiment to simulate three levels of topical breadth with one, five or nine topics, three levels of message volume, with an average of 10, 15, or 20 messages per day, and three types of moderation: no moderation, community-level moderation (under which off-topic messages are removed), and personalized moderation (under which a personalization algorithm presents a subset of messages that match a member's interests). We ran a 365-day simulation for each experimental condition on five randomly constructed groups. All groups began with 30 agents and 30 messages and evolved over time as newcomers joined and old-timers left.

We examined the effects of topical breadth, message volume, and moderation on two outcomes easily visible to a community manager: the number of new posts per day, an indicator of community activity, and the average number of login sessions per member, an indicator of member commitment. We ran analysis of variance (ANOVA) to examine the effects of topical breadth, message volume, and moderation. We also examined the benefits members received on the 100, 150, 200, 250, and 300th day of the experiment.

The model led to several plausible yet non-obvious findings: (1) members of topically broad communities were more committed or visited more frequently than members of topically narrow communities, although they did not post more messages, (2) community-level moderation led to greater commitment but not contribution, and (3) personalized moderation outperformed community-level moderation in communities with broad topical focus and high message volume. These results can be partially explained as a critical trade-off between informational and relational benefits, which the simulation revealed. For example, having more topics to discuss, on the one hand, increases informational benefits because it increases the number of messages likely to match one's interest; on the other hand, it reduces relational benefit because it reduces the chance of two members sharing a common interest.

To assure the robustness of our results, we ran a series of sensitivity analyses by relaxing key assumptions and varying key parameters. Results did not differ substantially. Some of the key parameters we varied were: the likelihood of posting a new message in a day (from 30 to 70 %), the criterion to be recognized as an active contributor (from top being in the top 5 % to the top 20 %), and the accuracy of personalization (from 60 to 100 %).

In terms of design implications, the simulation results call for reconsideration of well-established beliefs in the effectiveness of a narrow focus (Maloney-Krichmar & Preece, 2005) and community-level moderation (Preece, 2000). While these practices remain useful for some communities, our research suggests a contingency view of online community design. There is no universally optimal design for all online communities. The optimal choice depends on community characteristics (topical breadth and message volume) and the specific goals designers wish to accomplish (to make members loyal or to increase their contribution).

How We Discovered Agent-Based Modeling and Useful References

We were asked to also talk about our personal stories with agent-based modeling. For the first author, it could be traced to a belief she had since childhood that we could "simulate" (although at that time she did not know the word) and predict human society as accurately as we could predict the physical world. Serendipity also played a role when she started graduate school at Carnegie Mellon University working as a research assistant for Kathleen Carley, who is an expert in computational social and organizational theory. Later, she began conducting field studies and experiments because it was considered risky to do just simulation research (which may still be true in some disciplines). Years later, however, she enjoys and benefits greatly from being able to study a phenomenon using multiple methods including agent-based modeling.

The second author had been intrigued by the methodology for a long time and offered the first author a postdoctoral position, which started our years of collaboration to study online communities using various methods including agent-based modeling. So if you are foreign to the method, teaming up with someone who has done it can help you climb the learning curve. We should note that like other research skills, agent-based modeling is easy to learn but hard to do well. Experiences help and familiarity with the domain which you study helps as well. In addition, Axelrod (2005) has a book chapter in which he shares his experience of building the model to study the Prisoner's Dilemma game and his success as well as struggles of working with researchers from other disciplines and publishing his interdisciplinary work. It is a fun read.

Here is a list of papers that we have found useful and recommend as additional references:

- Schelling, T. C. (1969). Models of segregation. *American Economic Review*, 59(2), 488–493.
- Reynolds, C. W. (1987). Flocks, herds, and schools: A distributed behavioral model. *Computer Graphics*, 21(4), 25–34.
- Harrison, J. R., & Carrol, G. R. (1991). Keeping the faith: A model of cultural transmission in formal organizations. *Administrative Science Quarterly*, 36(4), 552–582.
- Carley, K. M. (1991). A theory of group stability. *American Sociological Review*, 56(3), 331–354.
- March, J. G. (1991). Exploration and exploitation in organizational learning. *Organization Science*, 2(1), 71–87.
- Carley, K. M. (1992). Organizational learning and personnel turnover. *Organizational Science*, 3(1), 20–46.
- Epstein, J. M., & Axtell, R. L. (1996). *Growing Artificial Societies: Social Science from the Bottom Up*. Boston, MA: MIT Press.
- Axelrod, R. (1997). *The complexity of cooperation: agent based models of competition and collaboration*. Princeton, NJ: Princeton University Press.
- Macy, M. W., & Willer, R. (2002). From factors to actors: computational sociology and agent-based modeling, Annual Review of Sociology, 28:143–166.

Concluding Remarks

To summarize, in this chapter, we presented a roadmap of how to use agent-based modeling to synthesize multiple social science theories to inform the design of multiuser systems, using our model on online community design as an example.

We encourage HCI researchers to consider either building your own agent-based models or adapting existing models as a new way of understanding and addressing challenges in designing multiuser systems. Researchers and designers can collaborate to perform "full-cycle research" (Chatman & Flynn, 2005) by alternating between agent-based modeling and field experiments and using the two to complement one another—the former to combine theories and generate new predictions and the latter to test the redesigns informed by simulation results. Once developed and validated, the agent-based model can be continuously extended to incorporate new theories or study new design choices. It can also serve a test bed to help designers navigate design spaces and choose features that fit their design goals. We also foresee the possibility of building agent-based models as collaboration platforms to allow researchers from different disciplines to collectively tackle formidable design challenges.

Exercises

1. Name some social behavior that might be amenable to agent based modeling, outside of the ones listed in this chapter.
2. Where do the rules come from that determine the agents' behaviors?

References

Axelrod, R. (1986). An evolutionary approach to norms. *American Political Science Review, 80*, 1095–1111.

Axelrod, R. (1997). *The complexity of cooperation: Agent based models of competition and collaboration*. Princeton, NJ: Princeton University Press.

Axelrod, R. (1997a). The dissemination of culture: A model with local convergence and global polarization. *Journal of Conflict Resolution, 41*, 203–226.

Axelrod, R. (1997b). *The complexity of cooperation: Agent-based models of competition and collaboration*. Princeton, NJ: Princeton University Press.

Axelrod, R. (2005). Agent-based modeling as a bridge between disciplines. In K. L. Judd, & L. Tesfatsion (Eds.), *Handbook of Computational Economics, Vol. 2: Agent-Based Computational Economics*. Amsterdam: Elsevier.

Bao, P., Hecht, B., Carton, S., Quaderi, M., Horn, M., & Gergle, D. (2012). Omnipedia: Bridging the Wikipedia language gap. *Proceedings of the ACM Conference on Human-Factors in Computing Systems* (pp. 1075–1084). New York: ACM Press.

Berscheid, E. (1994). Interpersonal relationships. *Annual Review of Psychology, 45*, 79–129.

Bryant, S. L., Forte, A., & Bruckman, A. (2005). Becoming Wikipedian: Transformation of participation in a collaborative online encyclopedia. *Proceedings of the 2005 International ACM SIGGROUP Conference on Supporting Group Work* (pp. 1–10). New York: ACM.

Burton, R. M., & Obel, B. (1995). The validity of computational models in organization science: From model realism to purpose of the model. *Computational and Mathematical Organization Theory, 1*(1), 57–71.

Butler, B. S. (2001). Membership size, communication activity, and sustainability: A resource-based model of online social structures. *Information Systems Research, 12*(4), 346–362.

Carley, K. M. (1991). A theory of group stability. *American Sociological Review, 56*(3), 331–354.

Carley, K. M. (1992). Organizational learning and personnel turnover. *Organizational Science, 3*(1), 20–46.

Carley, K.M. (1996). *Validating computational models*. Working paper, Pittsburgh, PA.

Chatman, J. A., & Flynn, F. J. (2005). Full-cycle micro-organizational behavior research. *Organization Science, 16*(4), 434–447.

Davis, J., Eisenhardt, K. M., & Bingham, C. B. (2007). Developing theory through simulation methods. *Academy of Management Review, 32*(2), 480–499.

DiMicco, J., Millen, D. R., Geyer, W., Dugan, C., Brownholtz, B., & Muller, M. (2008). Motivations for social networking at work. *Proceedings of the ACM Conference on Human-Factors in Computing Systems* (pp. 711–720). New York: ACM Press.

Drenner, S., Sen, S., & Terveen, L. (2008). Crafting the initial user experience to achieve community goals. *Proceedings of the 2008 ACM Conference on Recommender Systems.* (pp. 187–194). New York, NY: ACM

Epstein, J. M., & Axtell, R. L. (1996). *Growing artificial societies: Social science from the bottom up*. Boston, MA: MIT Press.

Epstein, J. M. (1999). Agent-based computational models and generative social science. *Complexity, 4*(5), 41–60.

Faraj, S., & Johnson, S. L. (2011). Network exchange patterns in online communities. *Organization Science, 22*(6), 1464–1480.

Festinger, L. (1954). *A theory of social comparison processes (Vol. 7)*. Indianapolis, IN: Bobbs-Merrill.

Fisher, D., Smith, M., & Welser, H.T. (2006). *You are who you talk to: Detecting roles in Usenet newsgroups*: *Proceedings of the 39th Hawaii International Conference on System Sciences in Waikoloa, Big Island, Hawaii.*

Fogarty, J., Lai, J., & Christensen, J. (2004). Presence versus availability: The design and evaluation of a context-aware communication client. *International Journal Human-Computer Studies, 61*(3), 299–317.

Frank, F., & Anderson, L. R. (1971). Effects of task and group size upon group productivity and member satisfaction. *Sociometry, 34*(1), 135–149.

Gardner, M. (1970). Mathematical games: The fantastic combinations of John Conway's new solitaire game "life". *Scientific American, 223*, 120–123.

Gilbert, N. (2008). Agent-based models. In T. F. Liao (Ed.), *Quantitative applications in the social sciences*. Los Angeles, CA: Sage.

Grudin, J. (1994). Groupware and social dynamics: Eight challenges for developers. *Communications of ACM, 37*(1), 92–105.

Hannan, M. T., & Freeman, J. (1989). *Organizational ecology*. Cambridge, MA: Harvard University Press.

Harper, F.M., Frankowski, D., Drenner, S., Ren, Y., Kiesler, S., Terveen, L., Kraut, R. E., & Riedl, J. T. (2007). Talk amongst yourselves: Inviting users to participate in online conversations. *Proceedings of the 12th International Conference on Intelligent User Interfaces* (pp. 62–71). Honolulu, Hawaii.

Harrison, J. R., & Carroll, G. R. (1991). Keeping the faith: A model of cultural transmission in formal organizations. *Administrative Science Quarterly, 36*(4), 552–582.

Harrison, J. R., Lin, Z., Carroll, G. R., & Carley, K. M. (2007). Simulation modeling in organizational and management research. *Academy of Management Review, 32*(4), 1229–1245.

Hogg, M. A. (1996). Social identity, self-categorization, and the small group. In E. H. Witte & J. H. Davis (Eds.), *Small group processes and interpersonal relations* (2nd ed., pp. 227–253). Hillsdale, NJ: Lawrence Erlbaum Associates.

Jones, Q., Ravid, G., & Rafaeli, S. (2004). Information overload and the message dynamics of online interaction spaces: A theoretical model and empirical exploration. *Information Systems Research, 15*(2), 194–210.

Karau, S. J., & Williams, K. D. (1993). Social loafing: A meta-analytic review and theoretical integration. *Journal of Personality and Social Psychology, 65*(4), 681–706.

Kluger, A., & DeNisi, A. (1996). Effects of feedback intervention on performance: A historical review, a meta-analysis, and a preliminary feedback intervention theory. *Psychological Bulletin, 119*(2), 254.

Kozlowski, K. J., & Klein, S. W. J. (2000). *Multilevel theory, research, and methods in organizations: foundations, extensions, and new directions* (Society for industrial and organizational psychology frontier series). San Francisco, CA: Jossey-Bass.

Kraut, R. E., & Resnick, P. (2012). *Building successful online communities: Evidence-based social design*. Cambridge, MA: MIT Press.

Ling, K., Beenen, G., Ludford, P. J., Wang, X., Chang, K., Li, X., et al. (2005). Using social psychology to motivate contributions to online communities. *Journal of Computer Mediated Communication,* 10(4), article 10.

March, J. G. (1991). Exploration and exploitation in organizational learning. *Organization Science, 2*(1), 71–87.

Macy, M. W. (1991). Chains of cooperation: Threshold effects in collective action. *American Sociological Review, 56*(6), 730–747.

Macy, M. W., & Willer, R. (2002). From factors to actors: Computational sociology and agent based modeling. *Annual Review of Sociology, 28*, 143–166.

Maloney-Krichmar, D., & Preece, J. (2005). A multilevel analysis of sociability, usability, and community dynamics in an online health community. *ACM Transactions on Computer-Human Interaction, 12*(2), 201–232.

Markus, M. L. (1987). Towards a "critical mass" theory of interactive media: Universal access, interdependence, and diffusion. *Communication Research, 14*(5), 491–511.

Mathieu, J., & Zajac, D. (1990). A review and meta-analysis of the antecedents, correlates, and consequences of organizational commitment. *Psychological Bulletin, 108*(2), 171–194.

Myers, B., Hudson, S. E., & Pausch, R. (2000). Past, present, and future of user interface software tools. *ACM Transactions on Computer-Human Interaction, 7*(1), 3–28.

Nan, N. (2011). Capturing bottom-up information technology use processes: A complex adaptive systems model. *MIS Quarterly, 35*(2), 505–532.

Nan, N., Johnston, E., & Olson, J. (2008). Unintended consequences in central-remote office arrangements: A study coupling laboratory experiment with multi-agent simulation. *Computational and Mathematical Organization Theory, 14*(2), 57–83.

Nan, N., Johnston, E. W., Olson, J. S., & Bos, N. (2005). Beyond being in the lab: Using multi-agent modeling to isolate competing hypotheses. *Proceedings of the ACM Conference on Human-Factors in Computing Systems* (pp. 1693–1696). New York: ACM Press.

Orlikowski, W. J. (1996). Improvising organizational transformation over time: A situated change perspective. *Information Systems Research, 7*(1), 63–92.

Perlow, L. A. (1999). The time famine: Toward a sociology of work time. *Administrative Science Quarterly, 44*(1), 57–81.

Preece, J. (2000). *Online communities: designing usability supporting sociability*. Chichester: Wiley.

Prentice, D. A., Miller, D. T., & Lightdale, J. R. (1994). Asymmetries in attachments to groups and to their members: Distinguishing between common-identity and common-bond groups. *Personality and Social Psychology Bulletin, 20*(5), 484–493.

Ren, Y., Carley, K. M., & Argote, L. (2006). The contingency effects of transactive memory: When is it more beneficial to know what others know? *Management Science, 52*(5), 671–682.

Ren, Y., Harper, F. M., Drenner, S., Terveen, L., Kiesler, S., Riedl, J., & Kraut, R. E. (2012). Building member attachment in online communities: Applying theories of group identity and interpersonal bonds. *MIS Quarterly, 36*(3), 841-864.

Ren, Y., & Kraut, R. E. (In press). Agent-based modeling to inform online community design: Impact of topical breadth, message volume, and discussion moderation on member commitment and contribution. *Human-Computer Interaction*.

Ren, Y., Kraut, R. E., & Kiesler, S. (2007). Applying common identity and bond theory to design of online communities. *Organization Studies, 28*(3), 377–408.

Reynolds, C. W. (1987). Flocks, herds, and schools: A distributed behavioral model. *Computer Graphics, 21*(4), 25–34.

Ridings, C. M., & Gefen, D. (2004). Virtual community attraction: Why people hang out online. *Journal of Computer Mediated Communication*, 10(1), article 4.

Roberts, J., Hann, I., & Slaughter, S. (2006). Understanding the motivations, participation, and performance of open source software development: A longitudinal study of the Apache projects. *Management Science, 52*(7), 984–1000.

Rogers, E. M., & Agarwala-Rogers, R. (1975). Organizational communication. In G. L. Hanneman & W. J. McEwen (Eds.), *Communication behavior* (pp. 218–236). Reading, MA: Addison-Wesley.

Sawyer, R. K. (2003). Artificial societies: Multiagent systems and the micro-macro link in sociological theory. *Sociological Methods & Research, 31*, 325–363.

Schelling, T. C. (1969). Models of segregation. *American Economic Review, 59*(2), 488–493.

Schelling, T. (1971). Dynamic models of segregation. *Journal of Mathematical Sociology, 1*, 143–186.

Shami, N. S., Ehrlich, K., Gay, G., & Hancock, J. T. (2009). Making sense of strangers' expertise from signals in digital artifacts. *Proceedings of the ACM Conference on Human-Factors in Computing Systems* (pp. 69–78). New York: ACM Press.

Shapiro, C., & Varian, H. R. (1999). *Information rules: A strategic guide to the network economy*. Boston, MA: Harvard Business School Press.

Smith, M. A. (1999). Invisible crowds in cyberspace: measuring and mapping the social structure of USENET. In M. A. Smith & P. Kollock (Eds.), *Communities in cyberspace: Perspective on new forms of social organization* (pp. 195–219). London: Routledge.

Strang, D., & Macy, M. (2001). "In search of excellence:" fads, success stories, and adaptive emulation. *American Journal of Sociology, 107*(1), 147–182.

Taber, C. S., & Timpone, R. J. (1996). *Computational modeling*. Thousand Oaks, CA: Sage.

Thom-Santelli, J., Millen, D. R., & Gergle, D. (2011). Organizational acculturation and social networking. *Proceedings of the ACM Conference on Computer Supported Cooperative Work* (pp. 313–316). New York: ACM Press.

Treem, J. W., & Leonardi, P. M. (2012). Social media use in organizations: Exploring the affordances of visibility, editability, persistence, and association. *Communication Yearbook, 36*, 142–189.

Vroom, V., Porter, L., & Lawler, E. (2005). Expectancy theories. In J. B. Miner (Ed.), *Organizational behavior 1: Essential theories of motivation and leadership* (pp. 94–113). New York, NY: M. E. Sharpe.

Wasko, M., & Faraj, S. (2005). Why should I share? Examining social capital and knowledge contribution in electronic networks of practice. *MIS Quarterly, 29*(1), 35–58.

Welser, H. T., Cosley, D., Kossinets, G., Lin, A., Dokshin, F., Gay, G., & Smith, M. (2011). Finding social roles in Wikipedia. *Proceedings of the 2011 iConference.*

Wilensky, U. (1999). *NetLogo*. Evanston, IL: Center for Connected Learning and Computer-Based Modeling, Northwestern University.

Wu, A., DiMicco, J. M., & Millen, D. R. (2010). Detecting professional versus personal closeness using an expertise social network site. *Proceedings of the ACM Conference on Human-Factors in Computing Systems* (pp. 1955–1964). New York: ACM Press.

Social Network Analysis in HCI

Derek L. Hansen and Marc A. Smith

Introduction

Social network analysis (SNA) is the systematic study of collections of social relationships, which consist of social actors implicitly or explicitly connected to one another. Social network analysts characterize the world as composed of *entities* (e.g., people, organizations, artifacts, nodes, vertices) that are joined together by *relationships* (e.g., ties, associations, exchanges, memberships, links, edges). SNA focuses on relational data about what transpires between entities in contrast to attribute data about individuals. Network analysts focus on the patterns generated within collections of many connections. For individuals, SNA is more about "who you know" than "what you know" or "who you are." At the group level, SNA illuminates how each person's individual connections aggregate to form emergent macrostructures like densely connected subgroups. Using the mathematics of graph theory, social network analysts calculate and visualize the properties of networks and the social actors that inhabit them.

HCI seeks to improve the ways people interact with information systems, many of which support interactions between people. SNA can be applied in many ways to HCI concerns, providing theory and methods for better understanding and evaluating the diffusion and impact of CSCW innovations like social media systems. Network analysis can be applied to capture the social structure of a user population before, during, and after new technologies are deployed. Network datasets can be used to measure changes in patterns of relationships and workflow that are not

D.L. Hansen (✉)
School of Technology, Brigham Young University, 265 Crabtree Bldg,
Provo, UT 84602, USA
e-mail: dlhansen@byu.edu

M.A. Smith
Social Media Research Foundation, 2617 Hallmark Drive, Belmont, CA 94002, USA
e-mail: marc@smrfoundation.org

J.S. Olson and W.A. Kellogg (eds.), *Ways of Knowing in HCI*,
DOI 10.1007/978-1-4939-0378-8_17,

visible in more common metrics like counts of users and rates of resource usage. A network perspective distinguishes between simple population growth and the development of important social structures within that population. The success of some systems may depend, for example, on attracting smaller populations of users who create a denser web of connections than systems that attract larger but more sparsely connected populations (see Ren & Kraut, in this volume). Attracting users in the first place is another HCI concern for which network methods can be useful. For example, SNA can help identify potential influencers who occupy strategic positions in existing networks who can recruit new users most effectively.

Social networks have formed for as long as people have interacted, traded, and engaged with one another. While social networks have existed long before the Internet, recent social networking services, such as Facebook and LinkedIn, support the creation of large, distributed, real-time social networks. When these services are used, they often generate data that is valuable for basic and applied research purposes. Prior to the widespread use of digital information systems, generating records of social interactions was challenging. In the era of pencil and paper data collection, datasets were often subjective, small, and time bound. Today, many legal, financial, educational, recreational, and personal communication systems generate the materials needed to analyze webs of human relations. Social networks are present in collections of e-mail, instant messaging, text messages, phone call logs, hyperlinks, message forum posts and replies, wiki page edits, tweets, "pins," video calls, multiplayer games, etc. These activities all generate network data that can be captured at a scale and pace never before possible, opening up new opportunities in computational social science (Kleinberg, 2008). Network analysis of online interactions is also proving to be a new source of actionable insights for community administrators, marketers, and designers of CSCW systems (Hansen, Shneiderman, & Smith, 2010).

Social media network maps can be a useful way to create a higher level understanding of collections of messages and the connections among authors that form in many information systems. Network maps can reveal divisions between subgroups of users that would otherwise be difficult to perceive. Network metrics can also be calculated for each participant to highlight the few people in key locations in a population, such as network hubs or bridge spanners. Visualization of networks along with calculated metrics can provide useful illustrations and summaries of the shape of a connected population. For example, Fig. 1 shows a network created from the connections among Twitter users discussing "global warming."

The graph represents a network of 415 Twitter users whose tweets contained "global warming." There is a green edge for each follows relationship. There is a blue edge for each "replies-to" or "mentions" relationship in a tweet. The tweets were made over the 4-h, 54-min period from Sunday, 11 November 2012, at 13:46 UTC to Sunday, 11 November 2012, at 18:41 UTC. The graph's vertices were grouped by cluster using the Clauset–Newman–Moore cluster algorithm. Each group is presented alone in a box, separated from all other clusters. The graph was laid out using the Harel–Koren Fast Multiscale layout algorithm. The vertex sizes are based on follower values. Visual attributes of this network map display multiple facets of each user and their connections. The size of nodes highlights important people,

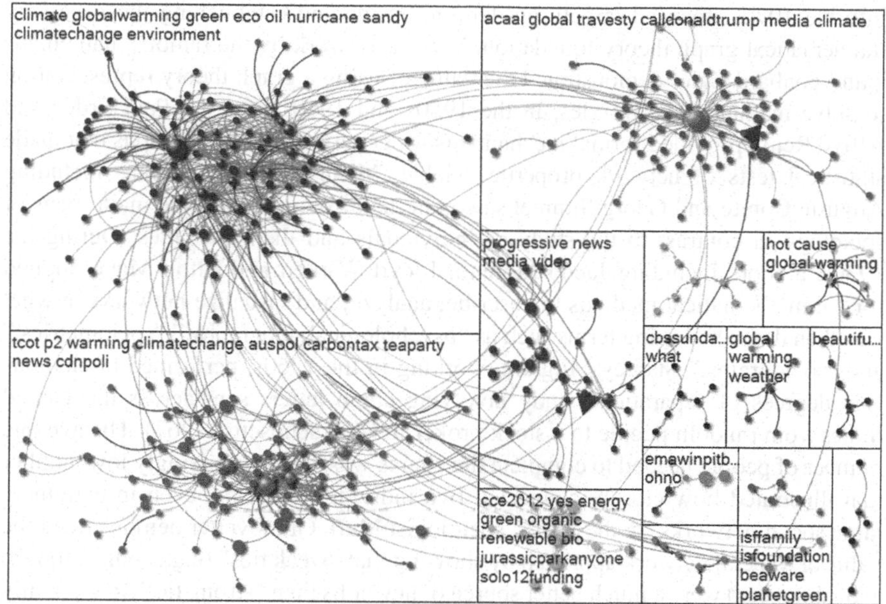

Fig. 1 A social network consisting of Twitter users (*nodes*) who have tweeted the word "global warming" connected to one another based on Follow, Reply, or Mention relationships (*edges*). Nodes are assigned different colors based on clusters, and hubs with many followers are indicated by size. Labels for each group are derived from frequently mentioned hashtags in the tweets from the users in each cluster. (Color figure online)

while color indicates membership in subgroups that are more densely connected to themselves than to other groups of users. The network is composed of a large connected component of people who are linked together by replying to or following one another. The connected group is subdivided into clusters or subgroups based on relative densities of connections. From analysis of this network and the content associated with it the groups can be labeled to indicate their focus or orientation. In this network climate change deniers are separate from people discussing climate science, sharing few follows, replying, or mentioning connections between the two groups.

A Brief History of Social Network Analysis

Though social networks are primordial, SNA is a relatively recently developed methodology whose history can be divided into roughly three phases: the foundational phase, the computational phase, and the network data deluge phase. See Linton Freeman's book on the development of SNA for a full treatment of the history of SNA (2004).

The early foundational phase, beginning in the eighteenth century and continuing into the 1970s, focused largely on defining terms and establishing the necessary mathematical graph theory foundation. Very early work by the famous mathematician Leonhard Euler demonstrated the value of using a graph theory representation to solve mathematical puzzles. In the 1950s and 1960s work by Paul Erdős and Alfréd Rényi provided formal mechanisms for generating random graphs that made statistical tests of network properties viable. Meanwhile, sociologists including Auguste Comte and Georg Simmel saw patterns of social ties as the main focus of sociology in contrast to the study of individuals and their attributes. During the 1930s, authors including Jacob Moreno, Lloyd Warner, and Elton Mayo applied formal mathematical methods to describe, analyze, and visualize networks in what was then described using terms such as "psychological geography," "sociometrics," and "sociograms." Stanley Milgram, working in the 1960s, performed his famous "six degrees of separation" study involving chain letters sent across the United States from random people to a stock broker in Massachusetts (1967). The average number of people needed to complete the chains was six, a surprisingly low number that illustrated how closely connected two individuals can be, even in extremely large social networks. In the 1970s, sociologist Mark Granovetter demonstrated the value of a social network approach by showing that "weak ties" (e.g., connections to acquaintances) were a much better source of new jobs than "strong ties" (e.g., family and very close friends) (1973). Later studies showed the "strength of weak ties" in other contexts including learning novel information, marketing, and politics.

The second major phase of the development of SNA, occurring largely in the 1970s through the mid-1990s, included the creation and systematic use of computational tools and methods. SNA as a methodological approach came into being during this phase, which leveraged the new capabilities of computers to analyze and visualize networks in novel ways. Lin Freeman built early tools for exploring networks (e.g., UCINet along with Borgatti and Everett) as well as identified core "centrality metrics" that provided objective measures of an individual's importance in a given network as described later in this chapter. George Homans developed new techniques for identifying subgroups (i.e., clusters) in networks, while Harrison White developed techniques for finding people that occupy similar network positions (via "structural equivalence"). Sociologist Barry Wellman founded the International Network for Social Network Analysis in 1976, which has served as a hub for social network researchers in a variety of fields ever since. Wellman has argued that SNA is not simply a method but is the core paradigm for explaining social action, particularly in our age of "networked individualism" where our work, community, and familial relationships no longer fit nicely within densely connected and bounded groups (2001). By the mid-1990s SNA was a well-respected approach in numerous fields ranging from organizational behavior (e.g., work by Ronald Burt and Rob Cross) to social psychology (e.g., Alex Bavelas' work) to communication networks (e.g., Noshir Contractor's work) to epidemiology. Perhaps the culminating work of this era is the "SNA bible" *Social Network Analysis: Methods and Applications* by Stanley Wasserman and Katherine Faust (1994), which rigorously summarized decades of research into a coherent mathematical framework, identifying the core metrics and techniques used by SNA tools and researchers today.

The current phase of SNA centers around the deluge of rich network data captured at Internet scale. A wealth of real-time social network data is captured by our everyday use of mobile phones, social networking sites, and commercial transactions. No longer is SNA a purely academic exercise, as corporations, governments, and nonprofit organizations utilize SNA techniques to find criminals, rank Web sites, recommend books, identify influencers, and restructure organizations. Authors such as Lada Adamic, Albert-László Barabási, Bernardo A. Huberman, Jon Kleinberg, Mark Newman, Steven Strogatz, and Duncan Watts have identified theoretical models that explain network generation and dynamics (e.g., see Newman, Barabási, & Watts, 2006; Newman, 2010), shown how information and influence propagate through them, and developed techniques for identifying clusters (i.e., communities) within them. Meanwhile, tools such as Pajek, developed by Vladimir Bagatteli, and the Stanford Network Analysis Platform (SNAP) by Jure Leskovec allow analysis of social networks at a scale never before possible. Other tools such as NodeXL and Gephi have focused on supporting SNA novices in their attempts to visualize small- to medium-sized networks. Computational social scientists have seized the moment by mining data from Facebook, Instant Messaging services, and other social media channels to more rigorously substantiate earlier work such as Milgram's 6 degrees of separation study, as described later in the chapter. Meanwhile, Nathan Eagle, Alex (Sandy) Pentland, and David Lazer have pioneered techniques for inferring friendship networks from data captured via mobile devices (2009). No doubt, this phase of SNA will continue to flourish as our social lives become increasingly mediated by technology.

Social Network Analysis and Human–Computer Interaction

Network analysis is a relatively new methodological and theoretical framework used within the HCI tradition. However, it has become prevalent in recent years, as social technologies have blossomed and tools for analyzing and visualizing networks have become more widely available. In this chapter we focus on how SNA can be used to design, evaluate, and understand CSCW and social media systems. We begin by describing five different goals that HCI researchers and practitioners can use SNA to achieve. We then move on to a discussion of specific questions that SNA can effectively address.

Goals of Social Network Analysis for HCI Researchers and Practitioners

1. Inform the design and implementation of new CSCW systems.

SNA can characterize the social structure of a population of intended users of a new CSCW system before the system is put in place. Understanding the social network properties of a target user population can help clarify requirements and challenges, leading to better initial designs and implementation strategies. Research has shown that mapping the social network of members of a large organization can help design social and technical strategies to facilitate more effective information flow (e.g., Cross, Parker, Prusak, & Borgatti, 2001). For example, tools may be needed that identify important bridge spanners or encourage the increased connection of groups that are too disconnected. Those implementing a new CSCW system could use SNA to identify, educate, and leverage those who will influence the maximal spread of adoption through the network to assure its rapid, effective use (Kempe, Kleinberg, & Tardos, É, 2003) or help others to know how to use a new technology (Eveland, Blanchard, Brown, & Mattocks, 1994).

Data for these analyses may come from network surveys (Marsden, 2005) or from existing data sources such as communication exchanges (e.g., e-mail, phone logs, IM, texts). Networks from these sources can characterize existing social structures and establish a baseline for measures of the impact of new CSCW systems (Goal #3). Furthermore, individuals with unique and important network positions can be identified and interviewed or observed as part of a comprehensive contextual inquiry process (Beyer & Holtzblatt, 1997).

2. Understand and improve current CSCW systems.

SNA of data from existing CSCW systems can illustrate the ways current features are utilized by users in different locations in the network. For example, the phenomenon of "unfollowing" someone on Twitter is partly explained by the social network structures of those involved (Kivran-Swaine, Govindan, & Naaman, 2011). Basic understanding of the pattern of user interactions can often inform the future design of social and technical improvements to CSCW systems. For example, network analysis of a technical support message board forum can help identify those who fill vital roles, such as "Answer Person" (Welser, Gleave, Fisher, & Smith, 2007). Community administrators can court these people to encourage them to remain active.

SNA may help community managers understand what is happening in large-scale communities where reading through even a meaningful sample of the content is not feasible. For example, a subgroup of users labeled "Theorists" was identified using network analysis techniques from among hundreds of thousands of Lostpedia wiki editors (Welser, Underwood, Cosley, Hansen, & Black, 2010). Knowing this subgroup exists could allow designers to develop tools that meet the particular needs of subpopulations, such as page templates that help systematically compare the competing theories. Similarly, unique social structures were found in Wikipedia's "breaking news" articles, which lead to insights about how people coordinate and potential designs to improve such work (Keegan, Gergle, & Contractor, 2012). Recently, several studies have developed recommendations for improving virtual reality games based on network analysis of guild networks and social interaction patterns (Ducheneaut, Yee, Nickell, & Moore, 2006; 2007). Other studies have shown variations in network structure by different users (e.g., teens and older adults) of the same discussion forum software (Zaphiris & Sarwar, 2006). Network methods that identify subpopulations can offer customized interfaces and services to different groups of users, using the history of other users in the same group as a guide. Education researchers have shown how students use different social features to interact within small groups and class-wide, with implications for system design and instructional strategies (Haythornthwaite, 2001). Work that shows separation between subgroups (e.g., conservative and liberal bloggers or readers) (Adamic & Glance, 2005) could be used to design tools that recommend posts that would increase cross-pollination of ideas (Munson & Resnick, 2010).

3. Evaluate the impact of CSCW system on social relationships.

SNA can be used to evaluate the impact of a CSCW system on the existing social structure of a population. Many CSCW systems are designed to, at least in part, influence the social relationships of those who use the system. Corporate intranets help employees find internal experts; online exchange markets match buyers and sellers; online community sites hope to develop sustainable communities around their niche topic; and collaboratories aim to facilitate scientific collaboration. Measuring the changes in aggregate and person-specific network metrics can help systematically evaluate the effectiveness of such systems. For example, the impact of CSCW systems designed to maintain weak ties between dispersed occupational communities could be measured (Pickering & King, 1992). Indeed, increased use of an internal, corporate social networking site has been shown to be positively associated with bonding relationships, sense of corporate citizenship, interest in connecting globally, and access to people and expertise (Steinfield, DiMicco, Ellison & Lampe, 2009). Evaluation can also be performed to assess the impact of a specific feature or social intervention. For example, the impact of an online "icebreaker" activity could be assessed by looking at changes in the network (e.g., network density) before and after. The majority of work in this arena relates to structuring social networks within organizations to improve knowledge creation, sharing, and innovation (e.g., Cross, Parker, & Borgatti, 2002; Borgatti & Foster, 2003; Müller-Prothmann, 2006). However, education researchers are also using network data to identify students using online course management systems that may be in need of extra support (Dawson, 2010).

Data for evaluation assessments may come from offline network surveys, existing communications (e.g., e-mail) captured over time, or system usage data (e.g., friendship or follow relationships). For large-scale evaluations, SNA can be used as part of a mixed method approach. For example, SNA can be used to identify individuals to interview based on their network positions (e.g., those with high, medium, and low network centrality; those from different subgroups).

4. Design novel CSCW systems and features using SNA methods.

SNA can be used as input to new CSCW systems and features. A growing number of research prototypes and innovative products leverage SNA metrics and methods to provide enhanced functionality. For example, a tool that recommends potential friends on a social networking site can use network properties to help identify likely candidates (Chen, Geyer, Dugan, Muller, & Guy, 2009). SNA has been used to help identify experts in technical support groups (Zhang, Ackerman, & Adamic, 2007) and organizations (Ehrlich, Lin, & Griffiths-Fisher, 2007; Perer & Guy, 2012), though early work showed that users often did not trust that their personal friends were the best experts (McDonald, 2003). Early work showed that social structure coupled with temporal patterns could be used to develop situated awareness tools (Fisher & Dourish, 2004). More recent work has used SNA to identify political tendencies of the followers of different news agencies on Twitter (Golbeck & Hansen, 2011), a technique that could be used for tools that personalize news or present alternative views. Tools have been developed that leverage network analysis and visualization to help gain insights into large datasets, such as published literature on a topic (Chau, Kittur, Hong, & Faloutsos, 2011). A novel feature that would show network diagrams of researchers who use similar queries in Citeseer has been proposed to help identify potential collaborators and research communities (Farooq, Ganoe, Carroll, & Giles, 2007). Recent work has explored the theoretical and practical design implications for promoting "social translucence" within directed social network systems, such as Twitter, where users can only see a portion of the social space, unlike chat rooms and discussion forums where everyone is visible to everyone else (Gilbert, 2012b). Related work has proposed novel information dissemination strategies that leverage social networking sites and semi-anonymity, such as "veiled viral marketing" (Hansen & Johnson, 2012). These examples give a flavor of the countless possible uses of SNA to enhance current CSCW systems, making this a particularly ripe area of research.

5. Answer fundamental social science questions.

Network analysis of data from CSCW system can be used to address fundamental questions about the nature of social relations. This research is part of the growing field of "computational social science," a set of techniques that use computational techniques to address core social science questions in novel ways. Because so much data is automatically captured via social media, they provide new opportunities to test hypotheses and theories at a much larger scale than previously possible. For example, Leskovec and Horvitz analyzed data from 180 million Microsoft Instant Messenger users finding an average path length of 6.6 between users, strikingly close to Milgram's original 6 degrees of separation work (2008). More recent work based on Facebook shows an average path length of just under five (Ugander, Karrer, Backstrom, & Marlow, 2011). Another example is a study of Facebook (Bakshy, Rosenn, Marlow & Adamic, 2012), which helped support and extend Granovetter's original work (1973) that showed the importance of weak ties. Other work predicts the strength of ties between individuals based on their social media interactions (Gilbert, 2012a; Gilbert & Karahalios, 2009) or mobile phone usage patterns (Eagle, Pentland, & Lazer, 2009). Such data can support further large-scale studies of social networks by reducing the need for raw data collection from users. Other studies are looking at the factors that lead to the sustained growth or death of online communities, such as the initial network structure (Kairam, Wang, & Leskovec, 2012).

Social Network Analysis Questions

SNA has been used to address a wide variety of questions in dozens of fields. While these questions vary considerably, they all share an emphasis on understanding social structures and how those structures influence outcomes of interest. SNA is designed to answer several types of specific questions as the categorized lists below illustrate.

Questions About Individual Social Actors

Often, network analysts are interested in identifying individuals who play an important, prominent, or unique role within a particular social network. Analysts use "centrality metrics" and "equivalence metrics" to address these questions. Some example questions include the following:

- Who are the most popular individuals in a network (e.g., network hubs)?
- Which individuals have the most influence?
- Who is a bridge spanner between different subgroups of users?
- If one is trying to disrupt a network, who should be removed?
- Are there different types of social actors that can be identified by unique network patterns? Who fills those social roles?

Questions About Overall Network Structure

Many questions relate to the overall structure of complete networks, such as the network of all Facebook users or all employees of an organization. Instead of focusing on the position of individuals within the network, these questions focus on the overall distribution. Analysts use "community detection algorithms" (i.e., network clustering algorithms) and a variety of "aggregate network metrics" to answer these questions. Some example questions include the following:

- How interconnected are a group of social actors (i.e., how dense is the network)?
- What is the distribution of individual network properties or social roles? For example, are there only a small percentage of "hubs" with a majority of "isolates"? Are there "enough" people that fill certain social roles?
- Are there subgroups of highly connected users (i.e., clusters, cliques)? If so, how many? And what is their relationship to one another? How do they differ from one another?
- What network properties or *motifs* (i.e., recurring network patterns) are related to social outcomes of interest? For example, what are the network structures of highly efficient groups, teams, businesses, and markets?

Questions About Network Dynamics and Flows

Other questions look at how networks change over time (i.e., network dynamics) or how information, objects, and attributes flow through networks (e.g., information diffusion, technology diffusion). Some example questions include the following:

- How do the structures of social relationship vary over time? For example, does the network become more interconnected or diffuse with use of a CSCW system?
- How does the importance of specific individuals, social roles, or clusters change over time? For example, does an intervention designed to bring separate subgroups together have the intended effect?
- How does information spread through a network (e.g., Twitter)? How can information propagation be catalyzed or minimized? What other attributes spread through a network?
- How does the use of new technologies spread through social networks? Who influences adoption of technology the most?

Performing Social Network Analysis

Despite the many types of analyses that can be performed, there is a common set of key steps including identifying the goals of the analysis, gathering data, and visualizing and analyzing the data using various network analysis software programs. This is a highly iterative process (Hansen et al., 2009). Analysts refine their goals after realizing the limitations of their datasets. Exploratory visualizations help identify the types of quantitative analyses that should be performed. And, additional data is often needed to validate or refute preliminary results.

Identify Goals and Research Questions

HCI researchers use SNA to accomplish a variety of high-level goals, each of which includes a large number of potential subgoals and research questions. It is essential that analysts hone in on a few critical goals and turn them into specific research questions, lest they spend unreasonable amounts of time aimlessly meandering around the data. Having said that, within HCI, SNA is often exploratory in nature and as with some types of qualitative research, analysts may only recognize what they are looking for once they see it. Often, after a preliminary analysis of initial data the questions are refined, another round of data collection is completed, and a final analysis is performed.

Table 1 Key sources of network data

Data source	Comments	Effort level
Raw data from system usage (e.g., database or XML files)	If you have access to the source data for a CSCW system (e.g., you are hosting it), you can query the data directly in highly customizable ways	Medium–high
Network survey	Network surveys ask people to manually characterize their relationships with other people (e.g., "list (or select) the people you turn to most often for answers to technical questions"). These can be administered via paper or, more commonly, via specialized network survey software (e.g., Network Genie), which may generate lists of employees based on existing databases or manually entered names	High
Application programming interfaces (APIs)	Most major social Web sites such as Facebook, Twitter, and YouTube have APIs that allow programmers to request data. You may need to register first, and you will often be restricted to data they make available	Medium–high
Screen scraping	If APIs are not available for a site you can write custom screen scraping software or use existing tools (e.g., VOSON), though legal restrictions may apply (e.g., the site's privacy policy)	Medium–high
Network analysis importer tools	Some network analysis tools allow users to import data from third-party sites (e.g., Facebook, Twitter). These tools may show up as import wizards (NodeXL), plug-ins (Gephi), or stand-alone network data capture tools (NameGenWeb Facebook App)	Easy
Existing datasets	An increasing number of existing network datasets are being made freely available. Examples include the Enron e-mail network, Amazon-related items, and blog networks (see http://snap.stanford.edu/data/ for more)	Easy

Collect Data

The next step is to collect the data needed in order to achieve the desired goal or answer the designated research questions. Below is a description of the sources of network data, different types of network data, and ways of representing network data.

Sources of Network Data

Depending on the specific data needs, collecting data may take considerable effort or be as easy as checking the appropriate boxes in an import wizard of an SNA software tool such as NodeXL. Table 1 shows the key sources of data that can be used in network analysis. Those that require more effort typically allow for more flexibility in the specific types of data that are collected.

Types of Social Networks

There are many types of networks. The specific type of network will determine how to appropriately analyze, visualize, and interpret the data. The type of network is determined by the underlying phenomena it represents. For example, a network of Twitter Following relationships is different from a network of Facebook Friendships because Facebook friendships must be mutual (if you are my Friend I am necessarily your Friend), while Twitter follow relationships do not have to be mutual (I can Follow you without you Following me).

Below is a brief description of the key terminology used to characterize networks.

- *Directed Versus Undirected.* Directed networks represent phenomena where the connection between two nodes is not necessarily reciprocated. Examples include communication networks (e.g., I send you an e-mail; you reply to my forum post), exchange networks (e.g., I sell you something), and awareness or following networks (e.g., I follow your updates). Undirected networks are always mutual, for example, friendship networks (such as on Facebook where one cannot friend another person without their consent) and affiliation networks (e.g., we are connected because we are affiliated with the same organization or we both edit the same wiki page).
- *Weighted Versus Unweighted.* Some edges have values associated with them. For example, edges in an e-mail network are "weighted" based on the number of messages one person sends to another person, while a wiki coedit page network is weighted based on the number of pages two people have both edited. Other edges are binary; they either exist or they do not. For example, Facebook friendships and Twitter follow relationships do not have weights.
- *Multiplex Networks.* Multiplex networks include multiple types of edges. For example, a network that connects people together based on their communication via e-mail, phone, and face-to-face interactions would include three distinct types of edges. This could be analyzed and visualized as a single multiplex network or as three distinct networks.
- *Unimodal and Multimodal Networks.* Many social networks, called unimodal networks, include only one type of node. For example, all the nodes represent people. Or, all of the nodes represent organizations. In contrast, multimodal networks include more than one type of node. For example, a network may include people who are connected to organizations or another network may include people who are connected to wiki pages they have edited. If there are only two types of nodes we call the network bimodal, which is a subset of the more general multimodal concept. Many bimodal networks, called bipartite networks, have one type of node (i.e., people) connected to another type of node (e.g., organizations) without any edges connecting nodes of the same type (e.g., people to people). These bipartite networks can be transformed into unimodal networks. For example, the person-to-organization network can be transformed into a person-to-person network where people are connected by a weighted edge that

Edge List

Node1	Node2
Brian	Marc
Derek	Marc
Derek	Adia
Marc	Ben
Maren	Derek
Maren	Adia

Matrix

	Brian	Marc	Derek	Adia	Ben	Maren
Brian	0	1	0	0	0	0
Marc	1	0	1	0	1	0
Derek	0	1	0	1	0	1
Adia	0	0	1	0	0	1
Ben	0	1	0	0	0	0
Maren	0	0	1	1	0	0

Graph

Ben
Adia
Marc
Derek
Brian
Maren

Fig. 2 Three ways of representing network data

represents the number of organizations they are both a part of. Conversely, an organization-to-organization network could be created where a weighted edge represents the number of people who are part of both organizations.

- *Partial Networks.* In practice, it is not practical or useful to collect data on an entire network (e.g., all Facebook users). Instead, analysts create partial networks in a variety of ways. One approach is to create an "egocentric network," which includes a single node (called "ego") and all of the nodes that ego is directly connected to (called "alters"). When the connections between alters are also included, the graph is called a 1.5 degree network. Adding ego's "friends of friends" makes it a 2.0 degree network and so forth. Other techniques for creating partial networks include sampling a large network (Leskovec & Faloutsos, 2006) or finding some network boundary such as membership in an organization.

It is important to recognize that a single socio-technical system inevitably includes many types of networks. For example, Facebook includes the obvious friendship network (unimodal, unweighted, undirected), the "people tagged together" network (unimodal, weighted, undirected), the "wall post" network (unimodal, weighted, directed), and the "person-to-group" network (multimodal, unweighted, undirected) to name a few. The choice of which networks to focus on depends on the goals of the particular study.

Representing Network Data

Network data is represented in three primary ways: edge lists, matrices, and graphs (see Fig. 2). An "edge list," also called an "adjacency list," contains a row that represents each edge in the network. In directed networks the first column lists the "source" node and the second column lists the "destination" node. Additional columns can be used to describe the type of edge and/or weight of the edge. An adjacency matrix lists each node as a header for both the rows and the columns, with matrix values corresponding to the weights of the edge (or a 1 or a 0 if it is unweighted). Finally, a network graph visually shows the nodes as vertices (e.g., circles or other shapes) and the edges as lines connecting them. Visual attributes can be used to represent edge weights (line thickness or opacity), directionality (lines with arrows), and node types (different shapes).

In addition to the network data, additional attribute data that describes the nodes and/or edges is often included. For example, you may have data on each person's gender, age, organizational role, membership duration, etc. Network graphs can be customized to help understand how this attribute data maps onto the network. For example, larger nodes could represent online community members who have been around longer. An analysis may reveal that larger nodes are well connected with each other but not with smaller nodes (newer members).

In practice, there are several common network file formats that most network analysis tools can import and read. These include GraphML (.graphml), Pajek (.net), Graphlet (.gml), GraphViz (.dot), and standard text files (.txt or .csv).

Analyze and Visualize Data

A wide range of analysis techniques can be used to understand and characterize social structures. New network analysis methods, metrics, models, statistical techniques, and algorithms are developed by an ever-growing, highly prolific research community consisting of researchers from a variety of fields. In this section we introduce some of the most commonly used techniques, organized into a handful of major topics within network analysis. Readers looking for comprehensive coverage should look to the additional resources mentioned later in this chapter.

Network Analysis Tools

SNA requires the use of specialized software designed to compute network metrics and visualize network graphs. The tool landscape is in constant flux (see http://en.wikipedia.org/wiki/Social_network_analysis_software for a comprehensive list). Table 2 describes five of the most commonly used tools in order of their sophistication.

Node-Specific Metrics: Focusing on the Trees

Analysts often want to characterize how important an individual is within a particular social network. Of course, there are many different ways that a person may be important. One person may be popular, another may serve as a bridge spanner between otherwise separate groups, and yet another may be connected to popular people despite having few connections of their own. Each of these is important in a different way.

Network analysts have developed a set of quantitative measures called "centrality metrics" to represent these various types of importance. The most commonly used centrality metrics are shown in Table 3. Several of them use the idea of the "distance" between two social actors, which is measured by the number of edges on

Table 2 Commonly used network analysis and visualization tools

SNA tool	Description	Expertise required	Open source	Maximum network size
Gephi	Stand-alone network analysis designed primarily for visualization. Can be extended via plug-ins	Designed for novices	Yes	Hundreds of thousands
NodeXL	Includes sophisticated graph visualizations, social media data importers, and extensibility via formulas and macros, but relatively few metrics	Microsoft Excel plug-in designed for SNA novices	Yes	Tens of thousands
Pajek	Includes a comprehensive list of network metrics and statistical tests. Steep learning curve	Designed for sophisticated analysis of large datasets	Yes	Millions
R	Open-source statistical package with social network analysis functionality via the igraph, sna, network, and statnet packages. Includes a comprehensive list of network metrics and statistical tests	Steep learning curve	Yes	Millions
UCINet	Includes a comprehensive list of network metrics and statistical tests. Designed for knowledge-able SNA researchers, but does not require coding	Designed for researchers performing social network analyses	No	Tens of thousands

the shortest path between two nodes (i.e., the geodesic distance). Variations of these metrics, as well as specialized versions of them appropriate for weighted and/or directed networks, are also available. These core metrics are calculated by all major network packages.

These metrics, along with statistical and visualization techniques, help identify the "structural signatures" of individual participants. For example, some users, such as news agencies on Twitter with their high in-degree, function as *network hubs* able to directly reach a large audience. Others may have relatively few followers on Twitter, except within a subset of users who discuss a certain topic (e.g., use the hashtag #CSCW2012), making them *topical hubs*. Users with high betweenness centrality often serve as *bridges* connecting otherwise disparate groups together by spanning "structural holes" (Burt, 1995). Users who are re-tweeted by several hubs will have high eigenvector centrality and may reveal individuals who serve as behind-the-scene *influencers*. Users who are not connected directly to others are referred to as *isolates*. Network analysts also differentiate between those in the *core* of the network (i.e., well-connected group at the "center" of the graph) and those on the *periphery* (i.e., the fringes).

Table 3 Common centrality metrics

Centrality metric	Description	Intuitive interpretation
Degree (in-degree and out-degree)	The number of edges connected to a node. For directed networks, the number of incoming links is the in-degree, while the number of outgoing links is the out-degree	Measures popularity (i.e., the number of friends one has). In-degree may measure the number of messages one receives, while out-degree may measure the number of messages sent
Betweenness	The number of shortest paths between all other nodes that a particular node is on—i.e., how often a node lies "between" other nodes	Measures how disrupted flows through a network would be if a person was removed. Helps identify "bridges spanners"
Closeness	The inverse of the average distance to all other nodes—i.e., how "close" a node is to other nodes	Measures how long it would take to disseminate information from a person to all others in the network
Eigenvector	A measure of a node's importance that considers the importance of the node's neighbors, where importance is calculated as a weighted sum of direct connections and indirect connections of every length	Measures not only the number of immediate connections (i.e., degree) but also the importance of the people one is connected to

At times it is helpful to identify classes of people who share a similar structural signature or position in a network. Such individuals often fulfill similar social roles. For example, Welser, Gleave, Fisher, and Smith used unique structural signatures to identify key individuals they called "answer people" within technical support Usenet newsgroups (Welser et al., 2007). These individuals have high out-degree (i.e., they answer many questions), are disproportionately tied to isolates (people with only one connection), and have few intense ties (i.e., multiple exchanges with the same person). Their initial insights gained from visualization were validated using regression analysis to predict and identify those filling this role (as identified through content analysis of messages) with high accuracy ($R2 = .72$). Another technique to identify social roles is to use *equivalence methods* to identify similar individuals based on their relation to others in the network (Wasserman & Faust, 1994). For example, employees all tied to a single manager and nobody else in the company likely play a similar professional role.

Aggregate Network Metrics: Focusing on the Forest

Network analysts have developed a language and set of metrics to help characterize the entire networks, just as they have to characterize the roles of individuals within those networks. This allows for the comparison of networks with one another or

Table 4 Common aggregate network metrics

Metric	Description	Intuitive interpretation
Density	Number of edges in the network divided by the number of possible edges	The amount of interconnectivity in a network
Diameter	Maximum geodesic distance (i.e., the longest "shortest" path) of all pairs in a network	The number of hops needed to reach two individuals who are as far away socially as possible
Average geodesic distance	The average geodesic distance of all pairs in a network	The average number of hops (i.e., degree of separation) between two people in the network
Network centralization	The sum of differences between the centrality of each node and the node with the highest centrality divided by the maximum possible sum of differences	A measure of how hierarchal a network is—i.e., how centralized it is around one or a few key social actors (where 0 describes a network where everyone is connected to everyone else and 1 describes a "star" network with one key person connecting everyone else together)

over time. Visualizing entire networks is often useful, as it can reveal overall structures such as the *core* or the *periphery* of a network, network *clusters* (see next section), and other patterns. However, many graphs are too large to meaningfully visualize and some properties of a graph are difficult to visualize (e.g., the longest geodesic distance) making the calculation of aggregate network metrics essential.

A different set of metrics help characterize the properties of an entire network. Like summary statistics (e.g., mean, standard deviation) help characterize attribute data, aggregate network metrics (e.g., density, diameter) help characterize network data. Also, like summary statistics, they only tell part of the story. Just as a mean does not provide any details about the distribution that generated it, a graph density metric does not provide any details about the network that generated it. Basic metrics include the number of vertices and edges, the number of connected components (i.e., clusters of vertices that are connected to each other through some path), and their size (measured in number of vertices). Other commonly used aggregate network metrics are shown in Table 4.

In addition to aggregate network metrics, network analysts often look at the distribution of node-specific metrics such as degree. This can help identify outliers and get an overall sense of the network. For example, a network that is centralized around a few key individuals, but otherwise not densely connected, will have a very skewed degree distribution with a couple of high-degree individuals and many very-low-degree individuals. In contrast, a more densely connected network where mostly everyone is interconnected will show a relatively constant (i.e., flat) degree distribution, since everyone will have a similar degree.

Network Clusters and Motifs: Focusing on the Thickets

Networks are composed of smaller components, which are often useful to examine in their own right. Some nodes may be highly interconnected forming a *clique* or a network *cluster* (see Fig. 1 for examples). Algorithms to identify these tightly knit groups are called many things including *community detection algorithms*, *network clustering algorithms*, *n-cliques*, *n-clans*, *k-plexes*, *k-cores*, *factions*, *blocks*, and *cut-points* (Hanneman & Riddle, 2005; Newman, 2010). Other recurring structures, sometimes called network *motifs*, show unique patterns such as *fans* (one person connected to otherwise isolated nodes), *tunnels* (nodes connected in a long independent chain), and *structural holes* (places where a lack of connections offers unique opportunities for those who span them) (Burt, 1995). At an even more granular level, triads (combinations of three nodes) serve as the building blocks of networks, inspiring network analysts to perform *triad censuses* wherein they characterize the distribution of the different types of triads (Hanneman & Riddle, 2005).

Many important insights can be gained from identifying and quantifying these network structures, since network topology often reflects social divides, political opinions, and other behavior of interest. For example, studies have shown a clear divide between liberal and conservative bloggers (Adamic & Glance, 2005) as well as distinct subgroups of Twitter users interested in gubernatorial elections from a national and local perspective (Himelboim, Hansen, & Bowser, 2012). Online community administrators can use network clusters to help identify potential conflicts and/or opportunities to bridge them. And system designers can identify how different collections of people utilize various collaborative features.

Network Dynamics and Information Flow

Thus far we have examined networks as static, unchanging entities. However, social networks are constantly evolving. Furthermore, information and other items can be distributed through networks over time, as happens in viral marketing campaigns (Leskovec, Adamic, & Huberman, 2007). Techniques and metrics related to the analysis of network dynamics and information propagation are highly active areas of research, particularly in technology-mediated networks (Kleinberg, 2008).

Early techniques that examine the spread of disease through social networks have been extended to better understand the spread of other phenomena such as happiness (Fowler & Christakis, 2008), obesity (Christakis & Fowler, 2007), information (Haythornthwaite, 1996), and innovations (Rogers, 1995). Increasingly, social media systems such as Twitter and Facebook are used to facilitate the flow of information, allowing researchers to examine information diffusion at a scale never before possible (Bakshy et al., 2012; Kwak, Lee, Park, & Moon, 2010). These observations serve as the foundation of theoretical models that explain information dissemination (see Kleinberg, 2008, for an introduction and additional resources). As "Google+Ripples" and comparable Twitter visualization tools become available to more closely track the flow of content through CSCW systems, practitioners will be able to better understand how certain ideas spread and perform tests to identify what leads to increased spread of information.

In addition to information flowing through networks, network structures themselves can change: people make new friends or break up with old ones, employees get hired and fired, and users change who they communicate with. For example, findings inferred from e-mail exchanges suggest that existing network topology and organizational structures shape changes in social networks (Kossinets & Watts, 2006). Researchers examine changes in networks in many ways ranging from comparing network metrics from different snapshots in time to highlighting important critical events or "bursts" in the network (Barabasi, 2010) and using computational models to simulate network changes over time. Additionally, specialized network visualization tools allow researchers to examine changes to networks over time (Ahn, Taieb-Maimon, Sopan, Plaisant, & Shneiderman, 2011). Dynamic analysis features are increasingly being added to existing network tools as well, which often allow edges and vertices to be timestamped so that network growth can be "played back."

Network Visualization

Social networks are often best understood through visualizations, which can provide both insights and inspiration. As Fig. 1 shows, visual properties such as color, size, and positioning of the nodes can highlight important nodes, subgroups, and overall network properties. However, creating meaningful network visualizations is not trivial. It involves an iterative process of filtering out nodes and edges, mapping network metrics to appropriate visual properties such as size and color, laying out the nodes in a way that shows inherent structure and network motifs (e.g., via force-directed layouts), and labeling important nodes and edges (Hansen et al., 2009).

Ideally, networks will attain "netviz nirvana" (Bonsignore et al., 2009) wherein the following goals are achieved:

- Every vertex is visible.
- Every vertex's degree is countable.
- Every edge can be followed from source to destination.
- Clusters and outliers are identifiable.
- Unnecessary edge crossings are removed.

Tools like Gephi and NodeXL provide a range of features and built-in layout algorithms that help reach these goals for most networks with vertices in the hundreds or low thousands, though larger and/or denser networks pose significant challenges. Current research is exploring the use of network readability metrics (Dunne & Shneiderman, 2009), techniques that combine network visualization and statistical overlays (Perer & Shneiderman, 2008), and graph summarization techniques (Dunne & Shneiderman, 2012) that may help us gain insights from visualizations of much larger networks in the future.

What Constitutes Good Work

Because SNA is performed by so many different communities of practice, there is a range of different expectations and criteria for determining what constitutes acceptable work. As SNA becomes more widespread in the HCI community, it is important to report appropriate metrics and use valid statistical techniques to validate claims, as opposed to simply presenting network visualizations. Below are a few best practices that apply to most SNA projects:

- Use network metrics that are appropriate for the type of network being examined. For example, if you are analyzing a directed network then in-degree and out-degree should be reported as opposed to degree. Likewise, if the network is weighted then, when possible, versions of network metrics that take the weights into consideration should be used. Where this is not possible, authors should state the reasons for using the basic, unweighted metric and associated limitations and implications.
- Do not claim more than your data can support. Network data, particularly collected from CSCW systems, is necessarily a simplification of much more complex social relations. Do not assume that Facebook friendships or e-mail exchanges necessarily equate to real-world friendships or that Twitter users are representative of the US population.
- Customize your network visualizations to illustrate the core points you are making (see Network Visualization section above for details). Remember that different network layout algorithms will highlight different properties of a network, so network visualizations should be used in conjunction with network metrics and statistical techniques.
- Use appropriate statistical techniques when mapping network properties to outcomes of interest or comparing networks. Though beyond the scope of this chapter, it is important to recognize that unique statistical techniques must be used when working with network data. For example, networks are often compared to a baseline network model (of which there are many) to demonstrate that certain features occur more often than expected. See Butts (2008) for a nice overview and introduction.
- Look at exemplary work, such as the articles cited throughout this chapter, for examples of methods and techniques appropriate for your questions. High-quality HCI work is often found in the CSCW conference, ICWSM conference, and CHI, while SNA articles using recent methods are found in *Social Networks: An International Journal of Structural Analysis*, the *Journal of Social Structure (JOSS)*, and *Connections*.

Additional Resources

The following annotated bibliography includes some good resources for becoming more expert in SNA. Books that require no relevant background are listed first, progressing to books that are written and used by experts in the field:

- John S. (2000). *Social network analysis: A handbook* (2nd ed.). Sage Publications Ltd: This is an excellent starting point for understanding SNA theory and methods, which assumes no prior knowledge. Written from a sociology perspective.
- Hansen, D., Shneiderman, B., & Smith, M. (2011). *Analyzing social media networks with NodeXL: Insights from a connected world*. Morgan Kaufmann: This introductory text focuses on analyzing social media datasets and assumes no knowledge of network analysis. It includes a tutorial-style section that shows how to conduct network analysis using the NodeXL software package as well as case studies from leading researchers in the field. Written from an HCI and marketing perspective.
- Nooy, W. D., Mrvar, A., & Batagelj, V. *Exploratory social network analysis with Pajek (Structural analysis in the social sciences)*. Revised and expanded second edition. Cambridge University Press: This introductory text introduces readers to a range of analysis techniques that can be performed by the Pajek software. Example datasets and exercises accompany the text as well as an appendix that walks the readers through the use of Pajek itself. Written from a mathematical and sociology perspective.
- Newman, M. (2010). *Networks: An introduction*. Oxford University Press: This comprehensive reference-style textbook introduces readers to the mathematics, theory, and algorithms used to analyze, model, and describe networks. Written from a physics and computer science perspective.
- Wasserman, S., & Faust, K. (1994). *Social network analysis: Methods and applications (Structural analysis in the social sciences)*. Cambridge University Press: Long considered the "bible" of SNA, this book is a comprehensive mathematically focused reference book on SNA techniques. Written from a mathematical and sociology perspective.
- Scott, J. P., & Carrington, P. J. (2011) *The SAGE handbook of social network analysis*: This reference book includes chapters on major SNA topics (e.g., social support, cyber communities, terrorist networks) and methods (network surveys, sampling, statistical models, dynamic network analysis) written by leading authors in the field.
- Newman, M., Barabási, A., & Watts, D. J. (2006). *The structure and dynamics of networks*: This edited volume covers recent developments in SNA from leading authors in the field. Articles cover historical developments, empirical studies, modeling networks, and various application domains.

How the Authors Became Enamored with Social Network Analysis

Derek Hansen

My introduction to network theory began while browsing the shelves at the original Borders in Ann Arbor as a graduate student. I came across Duncan Watt's book "Six Degrees: The Science of a Connected Age" and read half of it that night in the store. I immediately recognized its potential for understanding interactions occurring in online communities, the focus of my research. As I began teaching at the University of Maryland's iSchool I started collaborating with Ben Shneiderman and Marc Smith on the evaluation and development of the newly created NodeXL network analysis tool. As HCI researchers, we saw our role as "democratizing" network analysis by developing a tool that would help make SNA accessible to a much wider audience. Not only could researchers use NodeXL to answer compelling social science questions, but also practitioners such as online community managers could use it to gain actionable insights into their own communities. I have been amazed at how quickly my students can now adopt "network thinking" and develop compelling network visualizations and analyses that tell important stories. No longer is the analysis of relational data relegated to a backstage role because of its obscurity. It can now take its rightful position on center stage alongside other methods that analyze more traditional qualitative and quantitative data sources. I see a bright future ahead for SNA, particularly as it is integrated with other methods as when researchers use SNA to identify the people they should interview or salient topics discussed by users within a similar network cluster. As an HCI researcher, I am particularly anxious to see collaborative system designers apply SNA as a tool to evaluate, understand, and design better systems.

Marc Smith

I have been interested in social uses of technology for many years, starting with bulletin-board systems accessed with dial-up modems. As a sociologist, I want to understand social media and be able to visualize the complex relationships, structures, and changes that are possible there. I use network analysis with a range of visualization techniques to create insights into the shape and structure of social media. I think of it as a kind of hashtag or keyword group photo. I take many pictures of many groups, and I look for patterns in the network as a whole, its subgroups, and the key people within those groups. I compare many networks of the same topic or compare topics to one another. I find that there are many different types of networks in social media and that there are different roles within those networks that are occupied by key people in strategic locations. I can now tell some stories about the size and shape and key people and subgroups within social media topics. For example, political discussions in the United States are highly polarized with highly dense but separated groups, but this pattern is less visible in other nations' political discussions. Commercial discussions are often distinct from political topics; conversations about brands are often sparse even when they attract a large population. People mentioning these brands often have no connection to one another. In contrast, some products have formed communities, with populations with dense interconnections. Within communities there are often few people occupying the position of hubs and bridges. People at the center speak more often and have many more connections. Bridges often have fewer connections than hubs but have connections that reach from their own cluster across to many other clusters.

Exercises

Many kinds of things and relationships can be represented by nodes and links in SNA. Describe a network embedded within a CSCW system (e.g., Facebook's wall post network; Twitter's Follow network, Instagram's "Like" network) by describing what the nodes and edges mean, as well as the type of network (directed/undirected; weighted/unweighted; uniplex/multiplex; unimodal/multimodal;partial/complete).

Based on the network chosen above, describe actionable insights that could be gained from:

- Calculating a node-specific metric (e.g., Betweenness Centrality),
- Calculating an aggregate network metric (e.g., Density),
- Identifying clusters (e.g., subgroups) in the network, and/or
- Measuring network dynamics and/or information flow in the network.

References

Adamic, L. A., & Glance, N. (2005). The political blogosphere and the 2004 U.S. election: divided they blog. *Proceedings of the 3rd International Workshop on Link Discovery*, LinkKDD'05 (pp. 36–43). New York, NY: ACM.

Ahn, J., Taieb-Maimon, M., Sopan, A., Plaisant, C., & Shneiderman, B. (2011). Temporal visualization of social network dynamics: Prototypes for nation of neighbors. In J. Salerno, S. Yang, D. Nau, & S.-K. Chai (Eds.), *Social computing, behavioral-cultural modeling and prediction* (Lecture notes in computer science, Vol. 6589, pp. 309–316). Berlin: Springer.

Bakshy, E., Rosenn, I., Marlow, C., & Adamic, L. (2012). The role of social networks in information diffusion. *Proceedings of the 21st International Conference on World Wide Web*, WWW'12 (pp. 519–528). New York, NY: ACM.

Barabasi, A. L. (2010). *Bursts: The hidden pattern behind everything we do* (1st ed.). New York: Dutton Adult.

Beyer, H., & Holtzblatt, K. (1997). *Contextual design: Defining customer-centered systems* (1st ed.). San Francisco, CA: Morgan Kaufmann.

Bonsignore, E. M., Dunne, C., Rotman, D., Smith, M., Capone, T., Hansen, D. L. et al. (2009). First steps to Netviz Nirvana: Evaluating social network analysis with NodeXL. *International*

Conference on Computational Science and Engineering, 2009. CSE'09. (Vol. 4, pp. 332–339).

Borgatti, S. P., & Foster, P. C. (2003). The network paradigm in organizational research: A review and typology. *Journal of Management, 29*(6), 991–1013.

Burt, R. (1995). *Structural holes: The social structure of competition.* Cambridge, MA: Harvard University Press.

Butts, C. T. (2008). Social network analysis: A methodological introduction. *Asian Journal of Social Psychology, 11*(1), 13–41.

Chau, D. H., Kittur, A., Hong, J. I., & Faloutsos, C. (2011). Apolo: Making sense of large network data by combining rich user interaction and machine learning. *Proceedings of the 2011 Annual Conference on Human Factors in Computing Systems*, CHI'11 (pp. 167–176). New York, NY: ACM.

Chen, J., Geyer, W., Dugan, C., Muller, M., & Guy, I. (2009). Make new friends, but keep the old: Recommending people on social networking sites. *Proceedings of the 27th International Conference on Human Factors in Computing Systems*, CHI'09 (pp. 201–210). New York, NY: ACM.

Christakis, N. A., & Fowler, J. H. (2007). The spread of obesity in a large social network over 32 years. *New England Journal of Medicine, 357*(4), 370–379.

Cross, R., Parker, A., & Borgatti, S. P. (2002). A bird's-eye view: Using social network analysis to improve knowledge creation and sharing. *IBM Institute for Business Value*. Retrieved Jan 1, 2006, from http://www1.ibm.com/services/us/imc/pdf/g510166900abirdseyeviewusing socialnetworkanalysis.pdf.

Cross, R., Parker, A., Prusak, L., & Borgatti, S. (2001). Knowing what we know: Supporting knowledge creation and sharing in social networks. *Organizational Dynamics, 30*(2), 100–120.

Dawson, S. (2010). "Seeing" the learning community: An exploration of the development of a resource for monitoring online student networking. *British Journal of Educational Technology, 41*(5), 736–752.

Ducheneaut, N., Yee, N., Nickell, E., & Moore, R. J. (2006). "Alone together?": Exploring the social dynamics of massively multiplayer online games. *Proceedings of the SIGCHI Conference on Human Factors in Computing Systems*, CHI'06 (pp. 407–416). New York, NY: ACM.

Ducheneaut, N., Yee, N., Nickell, E., & Moore, R. J. (2007). The life and death of online gaming communities: a look at guilds in world of warcraft. *Proceedings of the SIGCHI Conference on Human Factors in Computing Systems*, CHI'07 (pp. 839–848). New York, NY: ACM.

Dunne, C., & Shneiderman, B. (2009). Improving graph drawing readability by incorporating readability metrics: A software tool for network analysts. *University of Maryland, HCIL Tech Report HCIL-2009–13*. Retrieved April 1, 2014, from http://www-lb.cs.umd.edu/~cdunne/hcil/pubs/Dunne09Improvinggraphdrawing.pdf.

Dunne, C., & Shneiderman, B. (2012). Motif simplification: Improving network visualization readability with fan and parallel glyphs. *University of Maryland, HCIL Tech Report HCIL-2012-11*. Retrieved April 1, 2014, from http://www.cs.umd.edu/~cdunne/hcil/pubs/Dunne12Motifsimplification_Improving.pdf.

Eagle, N., Pentland, A. (. S.)., & Lazer, D. (2009). Inferring friendship network structure by using mobile phone data. *Proceedings of the National Academy of Sciences, 106*(36), 15274–15278.

Ehrlich, K., Lin, C.-Y., & Griffiths-Fisher, V. (2007). Searching for experts in the enterprise: Combining text and social network analysis. *Proceedings of the 2007 International ACM Conference on Supporting Group Work*, GROUP' 07 (pp. 117–126). New York, NY: ACM.

Eveland, J. D., Blanchard, A., Brown, W., & Mattocks, J. (1994). The role of "help networks" in facilitating use of CSCW tools. *Proceedings of the 1994 ACM Conference on Computer Supported Cooperative Work*, CSCW'94 (pp. 265–274). New York, NY: ACM.

Farooq, U., Ganoe, C. H., Carroll, J. M., & Giles, C. L. (2007). Supporting distributed scientific collaboration: Implications for designing the CiteSeer collaboratory. *System Sciences, 2007. HICSS 2007. 40th Annual Hawaii International Conference on* (pp. 26–26). http://ieeexplore.ieee.org/xpl/articleDetails.jsp?arnumber=4076423.

Fisher, D., & Dourish, P. (2004). Social and temporal structures in everyday collaboration. *Proceedings of the SIGCHI Conference on Human Factors in Computing Systems*, CHI'04 (pp. 551–558). New York, NY: ACM.

Fowler, J. H., & Christakis, N. A. (2008). Dynamic spread of happiness in a large social network: Longitudinal analysis over 20 years in the Framingham Heart Study. *British Medical Journal, 337*, a 2338.

Freeman, L. C. (2004). *The development of social network analysis: A study in the sociology of science*. Vancouver: Empirical Press.

Gilbert, E. (2012a). Predicting tie strength in a new medium. *Proceedings of the ACM 2012 Conference on Computer Supported Cooperative Work*, CSCW'12 (pp. 1047–1056). New York, NY: ACM.

Gilbert, E. (2012b). Designing social translucence over social networks. *Proceedings of the 2012 ACM Annual Conference on Human Factors in Computing Systems*, CHI'12 (pp. 2731–2740). New York, NY: ACM.

Gilbert, E., & Karahalios, K. (2009). Predicting tie strength with social media. *Proceedings of the 27th International Conference on Human Factors in Computing Systems*, CHI'09 (pp. 211–220). New York, NY: ACM.

Golbeck, J., & Hansen, D. (2011). Computing political preference among twitter followers. *Proceedings of the 2011 Annual Conference on Human Factors in Computing Systems*, CHI'11 (pp. 1105–1108). New York, NY: ACM.

Granovetter, M. S. (1973). The strength of weak ties. *American Journal of Sociology, 78*(6), 1360–1380.

Hanneman, R. A., & Riddle, M. (2005). *Introduction to social network methods*. University of California Riverside. Retrieved April 1, 2014, from http://faculty.ucr.edu/~hanneman/nettext/.

Hansen, D. L., & Johnson, C. (2012). Veiled viral marketing: disseminating information on stigmatized illnesses via social networking sites. *Proceedings of the 2nd ACM SIGHIT International Health Informatics Symposium*, IHI'12 (pp. 247–254). New York, NY: ACM.

Hansen, D. L., Rotman, D., Bonsignore, E., Milic-Frayling, N., Rodrigues, E. M., Smith, M., et al. (2009). Do you know the way to SNA?: A process model for analyzing and visualizing social media data. *Human Computer Interaction Lab Tech Report HCIL-2009–17. University of Maryland.*

Hansen, D., Shneiderman, B., & Smith, M. A. (2010). *Analyzing social media networks with NodeXL: Insights from a connected world* (1st ed.). San Francisco, CA: Morgan Kaufmann.

Haythornthwaite, C. (1996). Social network analysis: An approach and technique for the study of information exchange. *Library & Information Science Research, 18*(4), 323–342.

Haythornthwaite, C. (2001). Exploring multiplexity: Social network structures in a computer-supported distance learning class. *The Information Society, 17*(3), 211–226.

Himelboim, I., Hansen, D., & Bowser, A. (2012). Playing in the same Twitter network. *Information, Communication & Society*, 1–24. http://www.tandfonline.com/doi/abs/10.1080/1369118X.2012.706316#.UzpGVa1dUSQ.

Kairam, S. R., Wang, D. J., & Leskovec, J. (2012). The life and death of online groups: Predicting group growth and longevity. *Proceedings of the Fifth ACM International Conference on Web Search and Data Mining*, WSDM'12 (pp. 673–682). New York, NY: ACM.

Keegan, B., Gergle, D., & Contractor, N. (2012). Staying in the loop: Structure and dynamics of Wikipedia's breaking news collaborations. http://dl.acm.org/citation.cfm?id=2462934.

Kempe, D., Kleinberg, J., & Tardos, É. (2003). Maximizing the spread of influence through a social network. *Proceedings of the Ninth ACM SIGKDD International Conference on Knowledge Discovery and Data Mining*, KDD'03 (pp. 137–146). New York, NY: ACM.

Kivran-Swaine, F., Govindan, P., & Naaman, M. (2011). The impact of network structure on breaking ties in online social networks: unfollowing on twitter. *Proceedings of the 2011 Annual Conference on Human Factors in Computing Systems*, CHI'11 (pp. 1101–1104). New York, NY: ACM.

Kleinberg, J. (2008). The convergence of social and technological networks. *Communications of the ACM, 51*(11), 66–72.

Kossinets, G., & Watts, D. J. (2006). Empirical analysis of an evolving social network. *Science, 311*(5757), 88–90.

Kwak, H., Lee, C., Park, H., & Moon, S. (2010). What is Twitter, a social network or a news media? *Proceedings of the 19th International Conference on World Wide Web*, WWW'10 (pp. 591–600). New York, NY: ACM.

Leskovec, J., Adamic, L. A., & Huberman, B. A. (2007). The dynamics of viral marketing. *ACM Transactions on the Web, 1*, 1.

Leskovec, J., & Faloutsos, C. (2006). Sampling from large graphs. *Proceedings of the 12th ACM SIGKDD International Conference on Knowledge Discovery and Data Mining*, KDD'06 (pp. 631–636). New York, NY: ACM.

Leskovec, J., & Horvitz, E. (2008). Planetary-scale views on a large instant-messaging network. *Proceedings of the 17th International Conference on World Wide Web*, WWW'08 (pp. 915–924). New York, NY: ACM.

Marsden, P. V. (2005). Recent developments in network measurement. In P. J. Carrington, J. Scott, & S. Wasserman (Eds.), *Models and methods in social network analysis* (Structural analysis in the social sciences). New York, NY: Cambridge University Press.

McDonald, D. W. (2003). Recommending collaboration with social networks: A comparative evaluation. *Proceedings of the SIGCHI Conference on Human Factors in Computing Systems*, CHI'03 (pp. 593–600). New York, NY: ACM.

Milgram, S. (1967). The small world problem. *Psychology Today, 2*, 60–67.

Müller-Prothmann, T. (2006). Leveraging knowledge communication for innovation. *Framework, Methods and Applications of Social Network Analysis in Research and Development.* Frankfurt a. M. et al.: Peter Lang.

Munson, S. A., & Resnick, P. (2010). Presenting diverse political opinions: How and how much. *Proceedings of the 28th International Conference on Human Factors in Computing Systems*, CHI'10 (pp. 1457–1466). New York, NY: ACM.

Newman, M. (2010). *Networks: An introduction* (1st ed.). New York, NY: Oxford University Press.

Newman, M., Barabási, A. L., & Watts, D. J. (2006). *The structure and dynamics of networks* (1st ed.). Princeton, NJ: Princeton University Press.

Perer, A., & Guy, I. (2012). SaNDVis: Visual social network analytics for the enterprise. *Proceedings of the ACM 2012 Conference on Computer Supported Cooperative Work Companion*, CSCW'12 (pp. 275–276). New York, NY: ACM.

Perer, A., & Shneiderman, B. (2008). Integrating statistics and visualization: Case studies of gaining clarity during exploratory data analysis. *Proceedings of the Twenty-Sixth Annual SIGCHI Conference on Human Factors in Computing Systems*, CHI'08 (pp. 265–274). New York, NY: ACM.

Pickering, J. M., & King, J. L. (1992). Hardwiring weak ties: Individual and institutional issues in computer mediated communication. *Proceedings of the 1992 ACM Conference on Computer-Supported Cooperative Work*, CSCW'92 (pp. 356–361). New York, NY: ACM.

Rogers, E. M. (1995). *Diffusion of innovations*. New York: Simon and Schuster.

Steinfield, C., DiMicco, J. M., Ellison, N. B., & Lampe, C. (2009). Bowling online: Social networking and social capital within the organization. *Proceedings of the Fourth International Conference on Communities and Technologies*, C&T'09 (pp. 245–254). New York, NY: ACM.

Ugander, J., Karrer, B., Backstrom, L., & Marlow, C. (2011). The anatomy of the Facebook social graph. *arXiv:1111.4503*. Retrieved April 1, 2014, from http://arxiv.org/abs/1111.4503.

Wasserman, S., & Faust, K. (1994). *Social network analysis: Methods and applications* (1st ed.). Cambridge: Cambridge University Press.

Wellman, B. (2001). Physical place and cyberplace: The rise of personalized networking. *International Journal of Urban and Regional Research, 25*(2), 227–252.

Welser, H. T., Gleave, E., Fisher, D., & Smith, M. (2007). Visualizing the signatures of social roles in online discussion groups. *Journal of Social Structure, 8*(2), 1–32.

Welser, H. T., Underwood, P., Cosley, D., Hansen, D., & Black, L. (2010). Wiki networks: connections of creativity and collaboration. In D. L. Hansen, B. Shneiderman, & M. Smith (Eds.), *Analyzing social media networks with NodeXL* (pp. 247–271). New York: Morgan Kaufmann.

Zaphiris, P., & Sarwar, R. (2006). Trends, similarities, and differences in the usage of teen and senior public online newsgroups. *ACM Transactions on Computer-Human Interaction, 13*(3), 403–422.

Zhang, J., Ackerman, M. S., & Adamic, L. (2007). Expertise networks in online communities: Structure and algorithms. *Proceedings of the 16th International Conference on World Wide Web*, WWW'07 (pp. 221–230). New York, NY: ACM.

Research Ethics and HCI

Amy Bruckman

What is Research Ethics in HCI?

Consider these ethical challenges of human–computer interaction (HCI) research:

- May I pay homeless people to answer a survey about how they use their cell phones? Is even a small gift card potentially coercive to people with extremely limited means? Is their financial need so great that their ability to make a rational decision is diminished? (LeDantec & Edwards, 2008).
- If I want to study people's behavior on Facebook, do I have to worry about whether this is allowed by the Facebook Terms of Service? (Gilbert et al., 2008).
- If I am recording data to help make evidence-based decisions in the care of a special-needs child, how do I balance the value of that data with people's fear of surveillance? (Hayes, 2004; Hayes & Abowd, 2006).

As HCI expands its domain to every corner of the human experience, issues of research ethics keep getting more complicated. HCI has come a long way from the days of laboratory studies where volunteers completed given tasks to receive $10 gift cards, cash, or course credit. The need for careful attention to ethics in HCI research is growing. Further, research in HCI has expanded its subject and methods to such a degree that aspects of the basic paradigm of "human subjects research" need to be rethought.

In this chapter, I briefly review some core concepts in the ethical treatment of human subjects before focusing on nuances that are important or unique for HCI research. I then highlight challenges raised by research on the Internet.

Research ethics involve understanding complex relationships between law, policy, and ethics. What is codified in institutional policies and procedures is typically a higher standard than merely legal, and what is ethical is a higher standard than

A. Bruckman (✉)
Georgia Institute of Technology, 85 5th. St NW, TSRB 338, Atlanta, GA 30332-0760, USA
e-mail: amy.bruckman@cc.gatech.edu

J.S. Olson and W.A. Kellogg (eds.), *Ways of Knowing in HCI*,
DOI 10.1007/978-1-4939-0378-8_18,

what is allowed by policy or law. Laws shift at political boundaries, policies shift at institutional boundaries, and ethics shift at cultural boundaries. For this paper, I will focus primarily on research at US universities.

Researchers create three primary risks:

- First, they risk harming their subjects.
- Second, they risk disturbing the environments they are trying to study (as researchers often do; see Reading and Interpreting Ethnography, this volume). This may also reduce the group's willingness to cooperate with future researchers.
- Third, they create a risk of serious consequences for their institution if ethical violations occur.

In extreme situations, the US government has suspended all human subjects research at universities for periods of time (Manier, 1999; Weiss, 1999). If one study was not properly regulated, then all studies must be rereviewed since the review process at the institution has been declared suspect. Such extreme actions are rare, but the database of what are called determination letters, descriptions of violations of regulations, kept by the US Department of Health and Human Services Office of Human Research Protection makes interesting reading (OHRP, 2012). The stakes are high for universities.

The History of Human Subjects Research

During World War II, researchers in Nazi Germany used humans in horrific experiments that today we would not allow to be performed on animals. In response, the Nuremberg Code was drafted in the wake of The Nuremburg War Trials (1949), the trials that revealed these atrocities. This established the basic principles that subjects should provide voluntary consent, and be free to end their participation at any time. Further, the experiment must "yield fruitful results for the benefit of society," and unnecessary risk should be avoided.

However, problems still existed in the ethics of human subjects research. Most notably, in the Tuskegee syphilis experiments, 400 African-American men with syphilis were monitored for 40 years (1932–1972) without being told of their disease (UNLV, 2012). The study continued for decades after the discovery in 1947 that penicillin cures the disease (CDC, 2011). When this atrocity came to light, a commission was created which led to the creation of The Belmont Report (Belmont, 1979). The Belmont Report remains the primary statement of principles of ethical research on human subjects today. It lays out three main principles:

1. Respect for persons

    (a) Treat people as ends in themselves, not means to an end.

2. Beneficence
   (a) Do not harm, and
   (b) Maximize possible benefits and minimize possible harms.
3. Justice
   (a) Distribute the burdens and benefits of research equally across society.

The first principle draws on Immanuel Kant's categorical imperative, which says that you must "act in such a way that you treat humanity, whether in your own person or in the person of any other, never merely as a means to an end, but always at the same time as an end" (Kant, 1964). For Kant, this is the essence of what it means to be moral. In the context of research ethics, this means seeing research participants not as a means to obtain data but as morally relevant beings whose autonomy and needs matter.

The second principle is somewhat misleading in its wording, because "do not harm" is too absolute. There are risks in simply walking across a room—you might trip and fall. But if a subject trips and falls (due to no special risks created by the study), that does not mean the study was unethical. People can be harmed through their participation in ethical studies. The challenge for evaluating the ethics of a proposed study is to weigh possible benefits and harms. An Institutional Review Board (IRB) member once told me the story of a study his IRB rejected that had no risks beyond those of participating in every day life. The board judged that the research design was so weak that it had no potential benefits to research. In this case the miniscule risk outweighed the nonexistent benefit and the study proposal was rejected. The ethics of a study and the research value of a study are inseparable concepts.

Weighing the benefits and risks of a study is made more challenging by the fact that disparate things are being compared, and the benefits and costs typically fall on different individuals. The third Belmont principle, justice, asks us to try to address this problem by distributing benefits and risk equally across society. In practice, the principle of justice is the most neglected of the three in a variety of ways. For example, people of low economic means are the more likely to agree to participate in risky studies in return for monetary compensation. Pregnant women are often excluded from studies for fear of added risk to them or their fetus, sometimes in cases where no such added risk exists. Children, defined as people below the age of 18 years of age, may be excluded simply because obtaining parental consent creates extra work for investigators. To begin to address these challenges, some human subjects' review boards now require explicit justification when groups such has pregnant women and children are excluded from a study population.

In the USA, US code Title 46 contains the laws regulating federally funded human subjects research (OHRP, 2001). Most universities in the USA apply these regulations to all research they conduct, even though technically the rules apply only to federally funded work. US corporations are subject to these laws only if they accept federal research funding. Some corporations have their own human subjects review boards, and some do not. Corporate review boards are not required by law and do not need to meet government standards.

Universities in the USA that accept federal funding are required to maintain an IRB committee. IRB members must represent diverse disciplines, and must include a non-scientist and a member of the community not affiliated with the university (US §46.107). Members typically take on a relatively heavy service load, and are required to understand issues in a wide range of disciplines, predominantly outside their own areas of expertise. Investigators seeking to work with human subjects must first obtain certification in human subjects research, typically by taking a short course (which is often offered online). The Collaborative Institutional Training Initiative (CITI) is one organization that offers training (see http://www.citiprogram.org). Next, they submit a proposal to do human subjects research to their institution's IRB for review. Some IRBs will as a matter of policy request at least one change to every protocol submitted, to clearly document that the protocol was actually reviewed and not simply approved without scrutiny. While the primary job of an IRB is to protect human subjects, a secondary and legitimate function is to protect their institution from liability and official sanctions.

The Evolving IRB

IRBs were originally created with medical and psychological research in mind. Policies and procedures developed in this context do not always translate easily into new research domains, like HCI. Weighing the benefits and risks of a study is impossible without understanding the domain. In the 1980s and 1990s, many IRBs had limited knowledge of computing and the Internet. IRBs are made up of people with different backgrounds and knowledge, and each board over time becomes comfortable with certain kinds of research they have seen repeatedly. An institution that rarely reviews HCI research may have more difficulty reviewing such protocols than one (like our IRB at Georgia Tech) that reviews them on a regular basis. Each IRB has more experience and comfort in some areas than others. In other words, an IRB is a kind of knowledge-building community of practice. HCI researchers versed in theories of sociotechnical systems will find their knowledge is useful in helping them to understand their IRB.

As computers have become pervasive, most IRBs have developed experience and knowledge of HCI research. If your research area is somewhat unfamiliar to your IRB, then you can request to make a short presentation to the board to help them become better informed. If your IRB seems to be erecting unreasonable barriers to HCI research, your best recourse is to *join them*—to volunteer to be an IRB member, and change the culture from within. Numerous faculty have taken this step, for example Richard Nisbett at University of Michigan and Robert Kraut at Carnegie Mellon University. They were able to greatly advance HCI research at their universities while maintaining high ethical standards (Kraut, 2012; Nisbett, 2012). Joining the IRB is more advisable for a tenured faculty member, because political complications may occur.

You may also have the opportunity to place a thoughtful graduate student on the IRB as a student representative. This is a fantastic learning opportunity for the student, and a series of students from the College of Computing have made strong contributions to Georgia Tech's IRB over the last 15 years.

Nisbett and others later participated in a board organized by the National Research Council (NRC) that produced a draft report calling for changes in our protections for human subjects. Conversations begun by this report are ongoing. The process of official change is slow and requires extensive effort on the part of concerned stakeholders.

IRB members accept a huge service burden on behalf of the community, and struggle to keep up with the broad areas of knowledge they need to understand. They accept significant legal liability by assuming this post, and are typically offered special insurance coverage by the university[1] (Bordas, 1984). It is a challenging and often thankless job, and they generally appreciate anything that makes it easier for them to do their work well.

In approaching an IRB, an attitude of humility and sincere concern is essential. If you sincerely care about ethics and see the IRB as partners in collaboratively solving a difficult problem, they are more likely to be receptive to your proposals. Many IRB members will make themselves available to discuss nuances of a proposal before it is submitted. The IRB process is a ideally collaborative and mutually supportive among board members and applicants.

For HCI professionals considering employment at a particular university, asking questions about the institution's IRB is strategic. With patience and education, well thought-out proposals can win approval by almost any IRB. However, what is quick and easy at one institution may be challenging and time consuming at another.

Balancing benefits and costs of research can be particularly challenging in the context of emerging technologies. In 2001, MIT Media Lab researcher Pascal Chesnais requested IRB permission to do research on e-mail filtering. Though it is hard to comprehend this now, at the time unsolicited e-mail was relatively rare. Chesnais's work would have been an early contribution to the new field of spam filtering. Unfortunately, the MIT IRB denied Chesnais permission to do the research. The IRB members determined that if an individual uses Chesnais' e-mail filter, it affects whether he or she receives all mail sent. Clearly the study volunteers could consider whether this risk was acceptable and make an informed decision on whether to participate. However, the board also reasoned that this affects *everyone who could potentially send the volunteer e-mail*. If I try to mail you and my message does not arrive, then I have been negatively affected by the study; however, I did not choose to opt into the study. Therefore, the board reasoned that everyone who could potentially send a user mail also would need to give their consent. After more than a year of arguing with the IRB, Chesnais abandoned his study.

At this time, the technology was new, and Chesnais was ahead of his time. His proposed study had both potential risks and potential benefits. As we have seen,

[1] IRB members employed by state institutions may be protected by the doctrine of sovereign immunity. Exact laws vary by state. See Bordas (1984) for a detailed discussion.

research ethics is about balancing risk versus benefit. The need for e-mail filtering was not yet well understood, and hence the risk–benefit scale tipped towards unjustified risk. Anyone who has ever failed to receive an important e-mail message because of misbehavior of a spam filter understands that risk is real. The nature of the potential benefit only became clear over years that followed as we all became inundated with advertisements for pornography, malware, and scams. The IRB's decision was unfortunate but not unfounded in its reasoning. Chesnais needed both to acknowledge the risks and make a more compelling argument about the coming epidemic of unsolicited e-mail and potential benefit of his research.

Process

The first step for any researcher is to determine whether you are in fact doing human subjects research. According to the US code (§46.102), a human subject is "a living individual about whom an investigator (whether professional or student) conducting research obtains (1) data through intervention or interaction with the individual, or (2) identifiable private information." Research means "a systematic investigation, including research development, testing and evaluation, designed to develop or contribute to generalizable knowledge." (See http://www.hhs.gov/ohrp/humansubjects/guidance/45cfr46.html). Under these definitions, for example, simply analyzing publicly available information on the Internet is not human subjects research. If you are not doing human subjects research, you are not formally held to human subjects procedures and regulations; however, you still should take care to proceed in an ethical fashion.

If you are doing human subjects research, the normal process for obtaining permission to do human subjects research at a US university starts with the researchers obtaining a certification in human subjects research, obtained typically by reading material online and taking short quizzes to ensure understanding. Next a proposal to do research is submitted, to be reviewed by the university's IRB.

An IRB proposal includes a description of the research and consent forms that the subjects will read and sign. Consent forms must explain any risks of participating, any benefits, a statement that subjects can withdraw at any time, and more. It is important to check your IRB's exact requirements. The purpose of the form is to help potential research volunteers make an informed decision about whether to participate.

IRBs generally have a regular meeting schedule, and deadlines for when materials need to be received to be considered at each meeting. Checking the calendar is important so you do not end up accidentally missing the deadline and having to wait an extended period of time for your protocol to be reviewed.

There are three levels of IRB review: exempt, expedited, and full board. When you file an IRB proposal, you typically may request which kind of review you believe is appropriate; however, your IRB will make the final determination. For research at US universities, exempt categories are defined by law in

§45CFR 46.101(b)(1)-(6). They include research "conducted in established or commonly accepted educational settings, involving normal educational practices," research on existing public records or where individuals are unidentifiable, and others. If what you are doing is human subjects research, you still need to file an IRB application and your IRB determines whether your proposed research is exempt. Investigators are not allowed to determine this for themselves. However, in practice, it is often a fine line to judge between research which should be proposed and declared exempt versus work that is not even technically human subjects research. The formal definition of human subjects research discussed above should be carefully applied to any research to help make this determination.

If a proposal is not exempt, it may qualify for *expedited* review. Expedited review means an IRB staff person may quickly approve the protocol without sending it to the full board. This saves significant time, because full board meetings typically take place only once a month (or even less often) and a protocol that just misses the deadline may wait longer. Much HCI research qualifies for expedited review. Work that qualifies for expedited review is defined in §46.110 (see http://www.hhs.gov/ohrp/policy/expedited98.html). Expedited review includes "Research on individual or group characteristics or behavior (including, but not limited to, research on perception, cognition, motivation, identity, language, communication, cultural beliefs or practices, and social behavior) or research employing survey, interview, oral history, focus group, program evaluation, human factors evaluation, or quality assurance methodologies."

In designing an IRB protocol, and important factor to consider is whether vulnerable persons might be involved. Children and individuals with "diminished autonomy" (including those with cognitive disabilities) are not permitted to give consent, since they are not allowed to enter into legal contracts. Instead, they may be asked to assent, and are given an assent form written at a level appropriate for their reading level. Their parent or guardian must sign a consent form on their behalf. Checking reading level is advisable for both consent and assent forms; many word processors provide tools to evaluate the reading level of a document.. A common rule of thumb it that documents should have an eighth grade (age 14) reading level or lower (See http://en.wikipedia.org/wiki/Literacy_in_the_United_States).

To the strictest standard, a consent form must be signed by the volunteer, the subject, and a witness. The lower the risk of the study, the more these requirements may be relaxed. In many low risk studies, an IRB may waive the requirement to document consent. US §46.117 states:

> (c) An IRB may waive the requirement for the investigator to obtain a signed consent form for some or all subjects if it finds either:
>
> (1) That the only record linking the subject and the research would be the consent document and the principal risk would be potential harm resulting from a breach of confidentiality. Each subject will be asked whether the subject wants documentation linking the subject with the research, and the subject's wishes will govern; or
>
> (2) That the research presents no more than minimal risk of harm to subjects and involves no procedures for which written consent is normally required outside of the research context.

Studies which allow for a "click to accept" Web-based consent forms require a waiver of documentation of consent, since clicking on a Web form is not legal consent. Note that the above refers to waiving the requirement for *documentation* of consent. Waiving the requirement for consent entirely is more rare, but may be permitted if the study is low risk and could not be carried out without the waiver (US §46.116). For example, I obtained a waiver of consent to conduct a qualitative study of children's learning in an after-school computer clubhouse. Different children attend the drop-in program each day. For months we sent members home with forms for their parents to sign; however, few were returned. We eventually applied for and were granted the waiver. (The study was also arguably exempt from IRB review as a "normal educational setting" under §46.101).

It is important to keep careful research records. IRBs may audit a study to verify that it is in compliance with the approved research protocol.

The Nature of the Subject in HCI Research

The traditional paradigm of human subjects research presumes an expert investigator interacting with subjects who are lay people with no special knowledge of the research domain. The subjects are viewed as vulnerable individuals who will be compensated appropriately for their contribution to research, and anonymized in written accounts of that research for their protection. These assumptions are increasingly anachronistic. The nature of any vulnerability depends on the nature of the research, and many kinds of HCI research were simply not envisioned when the rules were first written.

In many research areas, our subjects are increasingly knowledgeable about our domains of inquiry. To give an easy example, if we are developing software to support a more efficient work process, then we might employ a participatory design method (Ehn, 2008), working with engineers at our study site as design partners. In this case, the professionals who helped with the design may wish to comment on drafts of any publications that result. Depending on the exact circumstances, they may prefer to be credited for their work rather than anonymized in written accounts. They may even want to be co-authors (See Knowing by Doing: Action Research as an Approach to HCI, this volume). If they have significant input, to deidentify them is to deny them credit for their work. As a result, in some circumstances, anonymizing can *cause* harm rather than protect individuals from harm.

Participatory design in partnership with professional engineers seems like a far cry from anthropologists studying a tribal culture or medical researchers trying out a new antibiotic or medical device. However, the situations are not as different as they may appear. The medical literature increasingly views patients as partners who must accept responsibility in managing their own care (Mamykina & Mynatt, 2007). Patients now often may read obsessively about their own conditions, and are increasingly encouraged to not simply blindly accept instructions from healthcare providers. The Internet has brought them a wealth of information (and misinformation), and current thinking in public health suggests that patients fare better when they are not passive in managing their care. If patients are collaborators in managing their own care, participants in studies of new medical systems in a sense become collaborators in that research. Our view of who "ordinary people" are and what their role in the research process might be has evolved.

Our fundamental view of the nature of "the research subject" has shifted significantly. A post-modern interpretation of research practice situates these shifts in a broader intellectual framework. For example, the anthropologist Renato Rosaldo writes:

> The Lone Ethnographer's guiding fiction of cultural compartments has crumbled. So-called natives do not "inhabit" a world fully separate from the one ethnographers "live in." Few people simply remain in their place these days. When people play "ethnographers and natives," it is ever more difficult to predict who will put on the loincloth and who will pick up the pencil and paper. More people are doing both, and more so-called natives are among ethnographer's readers, at times appreciative and at times vocally critical. (Rosaldo, 1993, p. 45)

A critical view of the subject brings the Belmont Report's first principle of "respect for persons" into new focus. The notion of ignorant subjects incapable of understanding or responding to our representations has always been misguided to a degree, but is even more so in the age of the Internet. For subjects who are literate and computer literate, our scholarly works are increasingly accessible, and our subjects are more and more likely to be both able to and interested in engaging with our research findings.

As a result, it is important for the ethical researcher to rethink power relations between researchers and participants, and offer participants more voice (Borning & Muller, 2012). Subjects may wish to be named in written accounts. This is particularly true, for example, in work my students and I have done on online content creators, who are justifiably proud of their creations. As a result, most of our consent forms contain this statement:

> In some cases, people we interview are proud of things they have done online (for example creative projects) and would like to have their real name listed in our published reports. If you would like to request that we use your real name if possible, please sign below. We will not be able to use your real name if we feel there is anything that might embarrass you in our report. For most people, using a fake name (that we invent) is the right choice, so you do not need to sign here but you may if you wish.

If subjects do nothing, we anonymize them as usual. They must sign to have us use their real name, and the word "optional" appears next to the signature line in case people are quickly signing without reading the content.

There are some nuances to the above wording. Respect for persons suggests that we honor the wishes of our subjects and name them if they wish to be named. However, people do not always make good choices about when to seek publicity, as we see for example in individuals who allow themselves to be filmed being arrested on reality television shows. Individuals sometimes put an undue value on public recognition and make unwise decisions about what to publicize about themselves. In this case, the value of "respect for persons" conflicts with the value to minimize harm. To suggest that we rethink knowledge and power differentials between researchers and subjects does not mean that those differences do not exist—sometimes researchers really do still know better. The question of legal liability also comes into play. Even if the individual requested their name be used, the university might be liable if the portrayal of a subject caused harm. As a result, whether to publish real names when requested is ultimately a judgment call that rests with the researcher.

If our subjects are knowledgeable individuals who understand the area of study, they may be capable of and interested in responding to drafts of research results before publication. Since they may obtain copies of research results and express opinions about those results after publication anyway, it is in everyone's interests to solicit that input before publication when it still might be taken into account. This is more relevant for qualitative studies with descriptive accounts of individuals and groups rather than simply aggregate statistics. Subjects often will not follow through on sending comments in a timely manner, so it is important to give subjects a reasonable amount of time and clear notice of when that time will be up.

In practice, this process is often unproblematic and improves the quality of the research. Subjects may often be able to correct researcher misconceptions. However, in some cases this process becomes quite complicated. Individuals may resist interpretations that they find unflattering. Resistance leaves the researcher with two dilemmas. The first dilemma is epistemological: how do I know who is right? Is the subject just being defensive, or did I really misinterpret the situation? There may not be one correct answer. As Rosaldo notes, listening to many voices often means they can not be seamlessly integrated into a coherent, single, master narrative (Rosaldo, 1993). The researcher has dual ethical obligation to produce accurate research results, and to minimize harm. Ideally obtaining more data can help resolve any misalignments. Alternatively, the existence of multiple, irreconcilable views can simply be acknowledged.

The second part of the dilemma is pragmatic: Do I want to work with these subjects again in the future? Will this account endanger an important relationship? It is unethical to publish an inaccurate account, but strategic omissions can sometimes help resolve such dilemmas. If a controversial or unflattering piece of information is not strictly relevant to the research question being addressed, then it can be safely omitted. For example, in one early draft of a paper about an online system, I described a regular user as "socially awkward" in real life. While that comment provides context to understand the person and his participation in the online community, it was not essential to our research questions about use of the system and was ultimately omitted.

Ethical Challenges of Internet Research

Doing research on the Internet introduces a host of surprisingly complicated and intriguing ethical challenges (Ess, 2002; Ess & T. A. E. W. Committee, 2002; Berry, 2004; Kraut et al., 2004). The first challenge is to even determine when you are in fact doing "human subjects research." Accessing information posted on the Internet accessible without a password is not technically human subjects research according to IRB definitions. That information is published, and may be freely used for any purpose without IRB review. US §46.102 says:

> "(f) *Human subject* means a living individual about whom an investigator (whether professional or student) conducting research obtains
>
> (1) Data through intervention or interaction with the individual, or (2) Identifiable private information."

If information is published online, you are not interacting with its author and the information is not private. However, should you decide to ask the author a question about their online posting, then you have interacted and are indeed engaged in human subjects research.

This situation is unfortunate, because it *creates a disincentive to conduct high-quality research*. The quality of research is inevitably improved by asking questions of content creators. However, doing so requires significant extra work on the part of the investigator such as writing a human subjects research protocol, waiting for approval, and obtaining consent from subjects. It is not surprising that many investigators choose not to ask any questions, and write maddeningly incomplete papers with statements like "We speculate that users might engage in this behavior pattern because…." Speculation is in fact not necessary, because the users are often available to comment if one only takes the time to get approval to ask them.

Once one does start interacting with Internet content creators as human subjects, another complication emerges: their online postings may be findable with a search engine. This creates a linkage between confidential human subjects research data and their real identities. This is not a problem if the subjects wish to be identified by name. However, if they wish to be anonymized, then we have a dilemma. For studies involving controversial subject matter and significant risk to the participants, the solution is to not quote any online postings of the participants.

One clever trick is to present the same individual as two separate people in the written account—the quote is included but not connected to the person interviewed. Alternatively, if the subject matter is uncontroversial, then a kind of "light disguise" is possible, where a determined person could unmask the subject's identity but it is not immediately obvious without some effort (Bruckman, 2002). Research protocols from my lab warn of this possibility:

> To protect your confidentiality, your name will not appear in any publications; a pseudonym (a fake name) will be used instead. However, in the case of quotes from things you have done online (such as blog entries, forum posts, etc.), this disguise could be vulnerable—a determined and skillful person could potentially break it. Since many online sites are open to the world, search engines (i.e. Google) index them. As such, a person could take a quote and use a search engine to find the actual page, thereby breaking the pseudonym disguise. We do not anticipate that this research will uncover sensitive information, but in case it does, we will omit direct quotes that could be found in a search engine.

Disguising subjects in written accounts often requires omitting or changing identifying details and names. In some studies it is possible to omit the name of the online site being used; however, this is impractical in many cases. For very large sites or sites with few competitors, the site may be obvious. It is also not possible to disguise the site in design-based studies where researchers are studying an online service they created. Disguising subjects is about striking an appropriate balance. Where disguising is necessary, the quality of the research is improved by disguising as little as possible; however, the privacy of the participants is maximized with more disguising.

Terms of Service

So far I have discussed information available online without a password. If it is necessary to log in to a service to access data, then you are arguably subject to the site's Terms of Service (TOS). The TOS may set conditions for when research is allowed, or may prohibit research altogether.

Abiding by the TOS is always a safe choice. However, some researchers argue that it is not always necessary or appropriate. First, some clauses of the TOS may not be legally valid or reasonable. Companies often include clauses in legal disclaimers that would not hold up in a court of law, and also would not hold up to ethical scrutiny. For example, changes that Google made to its privacy policy in 2012 are being disputed as not complying with European Union (EU) privacy law (http://www.guardian.co.uk/technology/2012/oct/15/google-privacy-policy). The terms of service state that "By using our Services, you agree that Google can use such data in accordance with our privacy policies." (http://www.google.com/intl/en/policies/terms/) What is the status of that agreement for EU citizens if the privacy policy is not in compliance with EU law?

More importantly, the independence and intellectual integrity of the research community is at stake. Imagine that a large Internet site that is reshaping business and culture decides to erect major barriers to research. Does the site's regulations mean this aspect of the human experience is off limits for scholarly inquiry? Further, a corporation might allow researchers access only on their own terms, and permit work (or its publication) only if it is flattering to the company. Internet researchers are increasingly dependent on the good will of corporations in providing access to data, and the intellectual independence of the academy is challenged by this state of affairs.

So if a site's TOS may not be valid and you do not accept the corporation's authority to control what is said about it, is it acceptable to ignore an Internet site's TOS? Doing so may have practical consequences (you may be denied further access) or legal ones. My cautious answer is that it may sometimes be permissible to ignore TOS, but I would only do so after careful reflection (and with sound legal advice). The rationale for doing so should also be explained in the publication.

The unresolved quandary for the research community concerns how reviewers of scholarly manuscripts should handle this issue. In some cases, work has been rejected from conferences and journals for failing to comply with TOS. The same work with different reviewers might be accepted. This inconsistency is unfortunate. The situation would be improved if conference committees and professional societies establish clear standards and apply them consistently.

More generally, reviewers often struggle with whether the ethics of a study are within their purview. At one program committee meeting I attended, a program committee member argued that the study in question had passed IRB review at the author's institution, and that was sufficient. After all, the researchers got approval that their approach was ethical and proceeded accordingly. This heuristic would greatly simplify the task before reviewers; however, it is built on the assumption that local ethics review is being done well everywhere. If reviewers are expected to double check ethical issues in research, then they need clear standards to guide them. Ethical codes by organizations like the ACM (ACM, 1992) and IEEE (IEEE, 2012) are candidates, but can sometimes be too general to help with specific challenges, and also may not be updated frequently enough to account for emerging issues. The role of program committees in double-checking ethics has not been clearly resolved in many research communities.

Recruiting Subjects Online

Recruiting subjects and sampling data for any study are critical processes that affect the quality of research results. Internet-based recruiting introduces the additional challenge that the researcher may disturb the research environment through the process of soliciting participants. In the early days of scholarly interest in users on the Internet, members on some sites joked that the researchers might outnumber the natives. In the Georgia Tech course CS 6470 Design of Online Communities, students do participant observation and interviews with members of an online site. Although we do our best to teach students how not to annoy site members through the process of requesting interviews, students are still occasionally asked to leave a site because their requests for interviews are perceived as disruptive. Requests for research participation both impose on people's time and also foster a self-awareness that may disturb normal patterns of online behavior. Students in the online communities class are given a general rule of thumb is that any online participant should see a request for study participation only once or at most twice. As a result, targeted inquiries sent to specific individuals are often more effective than broadcast messages seen by large groups.

Researcher presence and request for study participation is more likely to be perceived as disruptive if the researchers are seen as not belonging on the site. For example, two student researchers were not welcomed by a divorce support forum they attempted to study because the students were themselves happily married. Although all their communications were seemingly appropriate in tone and content, their lack of personal experience with divorce made them unwelcome. This is more likely to occur in topics that address sensitive subjects like divorce or health support. For an extended discussion of this topic, see (Bruckman 2012).

In recruiting subjects online, researchers need to be aware that responders could be anyone and anywhere. Is the volunteer a child? Do they speak fluent English and fully understand what they are volunteering for? It is important to explicitly ask people's age. IRB protocols that allow a waiver of documentation of consent with a "click to accept" Web consent form for adults often typically do not allow this procedure for minors. In our protocols, we use paper parental consent and child assent forms that must be signed and scanned or sent to us by fax.

A growing number of studies are relying on crowd sourced participants such as workers on Amazon Mechanical Turk. On such sites, people pay a small amount of money (often a few cents) for workers to complete simple tasks. Workers often receive compensation less than the legal minimum wage in the USA, and are sometimes denied compensation for work legitimately completed. The invisibility of such workers is arguably antithetical to the basic principle of respect for persons (Silberman, 2010; Bederson & Quinn, 2011). These factors need to be taken into consideration when using crowd sourced labor either as a source of research subjects or for assistance in processing research data.

Recording Ephemeral Interaction

The discussion so far has concerned content that is archived online. The issues change (and become more controversial) when we consider researchers recording otherwise ephemeral communications. Some research communities (particularly linguists) argue that it is ethical to record conversation of non-identified individuals in public places. By analogy, chat rooms and other ephemeral forms of communication may be viewed as a similar situation, particularly if user names are removed from the data (Herring, 1996). However, such recordings often make Internet users angry, violating their expectations of privacy (Hudson & Bruckman, 2004). Privacy depends on the idea of "reasonable expectations," but with new technologies expectations are emergent, may not be shared between researchers and the general public, and are evolving.

Is it ethical to record activity in a chat room or a virtual world without consent? Does this constitute activity of unidentified individuals in a public place? One compromise solution suggests that traces of online interaction may not be *recorded* without permission; however, a participant observer may take field notes on his/her own experiences without permission. Others encountered would then be anonymized in written accounts. This compromise is splitting hairs because as soon as the participant observer asks explicit questions about the environment for research purposes, then informed consent is required. There is a thin boundary between normal conversation as a participant observer and interviewing.

In some cases, the process of requesting consent for recording may be more disruptive to the online environment than the study itself. This is particularly true in synchronous communication media. For example, suppose you want to study a synchronous chatroom. Can you ask participants to either opt in or opt out of the study by typing a command? In our chatroom study (Hudson & Bruckman, 2004), we addressed this question and found that few people opted in, few opted out, and in fact the researcher was kicked out of the chatroom in most cases because the process of requesting participation annoyed members. Researcher presence was perceived as less intrusive in larger groups. For every additional 13 people present, the chance of being kicked out dropped by 50 %. In cases in which the process of requesting participation is itself disruptive, if a study is low risk, researchers may have a compelling case for a waiver of consent.

In 2012, the preponderance of online interaction is archived, in either synchronous or semi-synchronous modes. A relatively small number of researchers study real-time environments such as Massively Multiplayer Online games (MMOs) where researchers may desire to record otherwise ephemeral communication. However, going forward, the increase in data usage from mobile tracking systems like GPSs and increasingly powerful RFID technology will likely raise ethical challenges. For example, location-based data can play a role in a variety of legal cases (especially divorce lawsuits), and the creation and storage of that data creates real liability and risk. Invisibility of data collection can foster a gap between how much privacy citizens expect and how much privacy some experts judge they are entitled to expect. Just as we can ask whether someone using a pseudonym in a chatroom is really an unidentified person in a public place, we can ask the same question of an individual carrying an RFID tag in a mall. The challenges are similar and answers are not entirely clear.

Design Ethics

HCI research involves creating new artifacts that reshape the human experience. Shaowen and Jeffrey Bardzell quote Papanek, commenting that "Design is intervention, an intentional effort to create change. As design theorist Papanek defines it, design's job is 'to transform man's environment and tools and, by extension, man himself.'" The Bardzells note that "HCI is increasingly engaging with matters of social change that go beyond the immediate qualities of interaction. In doing so, HCI takes on scientific and moral concerns" (Bardzell & Bardzell, 2011). Batya Friedman proposes a model of Value-Sensitive Design (VSD) where examining ones values and the values inherent in the design is an explicit step in the design process (Friedman, 1996; Friedman & Kahn, 2006; Borning & Muller, 2012).

In truth, all design has embedded values. For example, consider something as seemingly innocuous as a word processor. Jeanette Hofmann details how the design of word processing software has embedded conceptions not only of the level of technical sophistication of the user, but what it means to be a writer and who is expected to write. Early word processors assumed a secretary was typing text written by an executive. A system like WangWriter assumed that secretary had limited ability to learn commands. The resultant system made it unlikely to make a mistake but also impossible to do anything quickly. In contrast, systems like WordPerfect and WordStar assumed that the secretary was a professional who had the time and intelligence to learn complex commands, leading to greater productivity and a much wider range of capabilities after a significant initial learning curve. It was not until the creation of the word processor on the Xerox Star that the concept emerged of a knowledge worker who would work directly on a keyboard. And paradoxically, the Xerox Star assumed the knowledge worker was a computer dilettante who had limited time to learn commands. Assumptions about the capabilities and goals of typists and writers shaped the tools which in turn significantly shaped the lives of those using the tools (Hofmann 1999). Design ethics challenges researchers to answer the question, what kind of world are we creating?

A full discussion of this topic is beyond the scope of this chapter, but see Knowing by Doing: Action Research as an Approach to HCI, this volume, for a key perspective. This topic is of growing importance—it would be difficult to overstate its significance.

Conclusion

Ethical research on human subjects in any field involves balancing the potential benefits with potential harms. New technologies can make it more challenging to determine the nature of the potential benefit (like the need for a spam filter) and the nature of the potential harm (like violating someone's privacy). Traditional human subjects research seeks to protect vulnerable research participants by anonymizing them in written accounts. However, in a growing number of circumstances, our participants may want credit for their work. To anonymize them would be to cause harm. Further, our participants are increasingly qualified to and interested in responding to our representations of them.

Simply analyzing data available on the Internet without a password does not technically constitute human subjects research. However, it often substantially improves the quality of the research findings to interview or survey or otherwise interact with creators of online content, and this *does* constitute human subjects research. It is inappropriate to speculate about what users were intending without taking the time to ask them.

A number of issues in HCI research on human subjects still pose unresolved challenges. Is it OK to sometimes ignore Terms of Service of an online site? How do we maintain the independence of the academy from corporate control of access to research data? Is it OK to record unidentified activity in a public place even if that makes people angry if they know this is taking place? Should research reviewers double-check the ethics of a study, or is local ethical review sufficient? If reviewers should check the ethics of submitted work, by what standard? Since legal and ethical standards vary across countries and even among institutions within a single country, whose standards should be applied? Leadership is needed in resolving these issues from both individuals and professional societies. And emerging technologies will inevitably create new challenges that have not yet been named.

Research ethics for HCI has evolved in a tremendously positive way over time. Regulators have become better informed about computers and HCI, and ethical norms have evolved to catch up with challenges created by new technologies. At the same time, HCI research has pervaded every day life in new ways that make the need for regulation more apparent and accepted by the research community. Finally, in the dynamic context of HCI, our basic concept of what it means to do "human subjects research" has evolved in ways that have implications for other fields.

Useful References

A good general review article on research ethics in psychological studies came out of an American Psychological Association (APA) workshop led by Robert Kraut (Kraut et al., 2004). Many qualitative methods texts have thoughtful commentaries on research ethics. I recommend the ethics chapter in *Interviewing as Qualitative Research* by Irving Seidman (Seidman, 2006).

Some of the most impressive positive examples of research ethics done well concern people successfully working with vulnerable populations in a thoughtful and respectful manner. For example, Christopher LeDantec studied use of communications technologies by the homeless, and did a noteworthy job of understanding the power relationships between homeless and researchers and accounting for them in his study design (LeDantec & Edwards, 2008). Similarly, Gillian Hayes has done impressive work in understanding the needs of autistic children and their families. Her intervention uses recording technologies to help caregivers understand the cause of behavioral outbursts, and significant effort went into careful management of the privacy implications. For recording data in school settings, other students and teachers who are not participating in the study may inadvertently appear in the recordings. In general, people's fear of surveillance may interfere with the significant benefit of having data to make evidence-based care decisions. Careful rules about data collection, retention, and use is needed to address these challenges (Hayes et al., 2004; Hayes & Abowd, 2006). These are excellent models for how to handle tricky situations.

While those are strong examples of situations where extra care is needed, at the other end of the spectrum researchers might consider that sometimes more is possible than they might think. For example, Jim Hudson and I were granted a complete waiver of consent for a deceptive study with no debriefing for our study of how chatroom participants feel about being studied. We most definitely annoyed people with our study. However, our IRB was persuaded that the benefits of the study outweighed the burden on users (Hudson & Bruckman, 2004).

Finally, one good example of how to conduct research on sites with restrictive Terms of Service is work by Eric Gilbert comparing the social networks of rural versus urban Facebook users. To comply with Facebook's terms of service, Gilbert had subjects come to the lab to log in and show their information to researchers. It is against the Facebook TOS to scrape information from Facebook, but if the users log in they voluntarily reveal information (Gilbert et al., 2008).

My Involvement in Research Ethics

I first got interested in questions of research ethics because I had to. My PhD dissertation, MOOSE Crossing, was a text-based virtual world for kids. When I started to research how people usually handle research ethics in such situations, I discovered there were no precedents. What I was doing was new. I had to collaborate with my IRB to invent a reasonable approach.

My institution's IRB at that time typically erred on the side of caution and worked from a rather expansive definition of reasonable expectations of privacy. I remember being requested to present my study to the full IRB board meeting. I was terrified. A turning point in the conversation was when I explained to the board that I would not be giving Internet access to any children. The participants had to already have Internet access, and I was providing a (somewhat) safer activity for them to do online.

That initial conversation took place in 1993. Twenty years later (at the time of this writing in 2013), I find we are still struggling with these issues. My current institution recently required me to take an IRB refresher course. The online course contained a new module on Internet research, and the content mostly said things were not yet clear to anyone! In the face of these uncertainties, researchers must understand the core ethical principles and construct a well-reasoned argument about how to apply them in any given situation.

Exercises

1. What is the difference between exempt research and expedited research, according to IRB rules?
2. What populations of people would you be studying if the IRB review has been neither exempt nor expedited?
3. In your institution, what is the minimum turn-around time for getting feedback on an IRB application? How often does the committee meet? Is there a separate board that reviews medial from behavioral applications? Is there a separate form? What kinds of questions are asked on a medical application that are not relevant to a behavioral application?

References

(1949). *Trials of war criminals before the nuremberg military tribunals under Control Council Law No. 10*. Retrieved June 13, 2012 from http://www.hhs.gov/ohrp/archive/nurcode.html

ACM (1992). *ACM code of ethics and professional conduct*. Retrieved Jue 13, 2012 from http://www.acm.org/about/code-of-ethics

Bardzell, S. & Bardzell J. (2011). Towards a feminist HCI methodology: Social science, feminism, and HCI. *Proceedings of CHI* (pp. 675–684). Vancouver, BC, Canada: ACM

Bederson, B. & Quinn A. (2011). Web workers unite! addressing challenges of online laborers. *Proceedings of alt.CHI*. Vancouver, BC, Canada: ACM

Belmont. (1979). *The belmont report*. Washington, DC: US Department of Health, Education, and Welfare.

Berry, D. M. (2004). Internet research: Privacy, ethics and alienation—An open source approach. *Internet Research, 14*(4), 323–332.

Bordas, L. (1984). Tort Liability of Institutional Review Boards. *West Virginia Law Review, 87*(1), 137–164.

Borning, A. & Muller M. (2012). Next steps for value sensitive design. *Proceedings of CHI* (pp. 1125–1134). Austin, TX: ACM

Bruckman, A. (2002). Studying the amateur artist: A perspective on disguising data collected in human subjects research on the internet. *Ethics and Information Technology, 4*(3), 217–231.

Bruckman, A. (2012). Interviewing members of online communities: A practical guide to recruiting participants. In A. Hollingshead & M. S. Poole (Eds.), *Research methods for studying groups and teams* (pp. 199–210). New York: Routledge.

CDC (2011). *The Tuskegee Timeline*. Retrieved October 31, 2012 from http://www.cdc.gov/tuskegee/timeline.htm

Ehn, P. (2008). Participation in design things. *Partcipatory Design Conference* (pp. 92–101). Indianapolis, IN: ACM

Ess, C. (2002). Internet research ethics. *Ethics and Information Technology, 4*(3), 177–188.

Ess, C. & T. A. E. W. Committee (2002). *Ethical decision-making and internet research, Association of Internet Researchers*. 33

Friedman, B. (1996). Value-Sensitive Design. *ACM Interactions, 3*, 17–23.

Friedman, B., & Kahn, P. H. (2006). Value-sensitive design and information systems. In P. Z. Dennis Galletta (Ed.), *Human-computer interaction and management information systems: Applications* (pp. 348–372). New York: ME Sharpe.

Gilbert, E., Karahalios, K., et al. (2008). The network in the garden. *Proceedings of CHI*. ACM

Hayes, G. R. & Abowd G. (2006). Tensions in designing capture technologies for an evidence-based care community. *Proceedings of CHI* (pp. 937–946). Montreal, Quebec, Canada: ACM

Hayes, G. R., Kientz, J. et al. (2004). Designing capture applications to support the education of children with autism. *Proceedings of UbiComp*. Tokyo, Japan: ACM

Herring, S. (1996). LInguistic and critical analysis of computer-mediated communication: Some ethical and scholarly considerations. *The Information Society, 12*, 153–168.

Hofmann, J. (1999). Writers, texts, and writing acts: Gendered user images in word processing software. In D. MacKenzie & J. Wajcman (Eds.), *The social shaping of technology*. New York: Open University Press.

Hudson, J. M., & Bruckman, A. (2004). "Go Away": Participant objectiions to being studied. *The Information Society, 20*(2), 127–139.

IEEE (2012). *Software engineering code of ethics and professional practice*. Retrieved June 13, 2012 from http://www.computer.org/portal/web/certification/resources/code_of_ethics

Kant, I. (1964). *Groundwork for the metaphysic of morals*. New York: Harper & Row.

Kraut, R. (2012). *Personal communication to A*. Munich, Germany: Bruckmann.

Kraut, R., Olson, J., et al. (2004). Psychological research online: Report of Board of Scientific Affairs' Advisory Group on the Conduct of Research on the Internet. *American Psychologist, 59*(4), 1–13.

LeDantec, C. A. & Edwards, W. K. (2008). Designs on dignity: Perceptions of technology among the homeless. *Proceedings of CHI* (pp. 627–636). Florence, Italy: ACM

Mamykina, L. & Mynatt, E. (2007). *Proceedings of Healthnet* (pp. 49–54). San Juan, Puerto Rico: ACM

Manier, J. (1999). *UIC suspended from doing most human research*. Chicago, IL, Tribune Company: Chicago Tribune.

Nisbett, R. (2012). *Personal communication to A*. Munich, Germany: Bruckmann.

OHRP (2001). *US Code of Federal Regulations Title 45 Part 46, Protection of Human Subjects*. United States Federal Code. Retrieved June 15, 2012 from http://ohrp.osophs.dhhs.gov/humansubjects/guidance/45cfr46.htm

OHRP (2012). *Compliance oversight, determination letters*. From http://www.hhs.gov/ohrp/compliance/letters/index.html
Rosaldo, R. (1993). *Culture and truth: The remaking of social analysis*. Boston, MA: Beacon.
Seidman, I. (2006). *Interviewing as qualitative research*. New York: Teachers College Press.
Silberman, M. S., Ross, J. et al. (2010). Sellers' problems in human computation markets. *Proceedings of KDD-HCOMP* (pp. 18–21). Washington, DC: ACM
UNLV (2012). *History of research ethics*. Retrieved June 13, 2012 from http://research.unlv.edu/ORI-HSR/history-ethics.html
Weiss, R. (1999). *US halts human research at duke*. Washington, DC: The Washington Post.

Epilogue

Wendy A. Kellogg and Judith S. Olson

The chapters in this volume attest to the diversity of our field as it has grown from its origins in applying cognitive psychology to single-user interfaces to the wonderfully eclectic set of practices and settings in which human–computer interaction (HCI) is studied today. Even so, the chapters contained herein do not represent all the ways of knowing currently in use. That is pretty amazing: 30 years from the beginning of HCI, one volume is insufficient to represent all of its important methods. That could be a really good sign of the health and vitality of the field, or a clear harbinger of intellectual chaos. We think it is a good sign.

The approaches described in this volume differ in a number of ways—for example, in the academic fields from which they originated, and the intellectual commitments each entails. These differences play out in a number of important ways that affect us as researchers—from how we frame a problem, to what we think about, how we think about it, and what we do. This is one reason that we urge readers to go through all of the chapters, not just the ones they know best or in which they already have some interest. The contrasts that emerge are fascinating and thought-provoking.

One contrast that is obvious upon reading through the set is vast differences in the stance various methods take towards the whole enterprise of knowing. What constitutes data? Is data something "out there" in the world to be uncovered by a researcher, or is it generated through the researcher's selective understanding based on sharing experiences with users in a particular context? What constitutes a good theory in HCI? Should all high-quality work aim at generating or evaluating theory,

W.A. Kellogg (✉)
IBM T. J. Watson Research Center, P.O. Box 218, Yorktown Heights, NY 10598, USA
e-mail: wkellogg@us.ibm.com

J.S. Olson
Donald Bren School of Information and Computer Sciences, University of California Irvine,
5206 Donald Bren Hall, Irvine, CA 92797-3440, USA
e-mail: jsolson@uci.edu

J.S. Olson and W.A. Kellogg (eds.), *Ways of Knowing in HCI*,
DOI 10.1007/978-1-4939-0378-8_19,

or at producing some kind of abstraction or generalization that is useful beyond the situation studied? If not, how can HCI as a field accumulate knowledge? This question goes all the way back to the beginnings of HCI and formative discussions that took place among early HCI researchers with different views—from HCI as an applied discipline that could best contribute by generating engineering approximations of human cognition and behavior (Card, Moran & Newell, 1983; Newell & Card, 1985), to those who adopted a hermeneutic stance and believed that understanding users and contexts was situated and interpretive and that nothing meaningful could be generalized from one occasion of use to another (Whiteside & Wixon, 1987), to those that sought ground in-between, emphasizing technical artifacts themselves as embodied HCI theories and focusing on task analyses and scenarios of use (Carroll, 1990; Carroll & Campbell, 1986; Carroll & Kellogg, 1989). These concerns of what a way of knowing entails—what constitutes data, theory, and the conceptualization of researchers and their relationship to users—are evident in many of the chapters.

Another noticeable contrast among ways of knowing is that the type of accountability expected of good work differs. Classical approaches to HCI pursue theory building through control of variables and/or modeling (more recently simulation). They have what Gaver ["Science and Design" this volume] calls epistemological accountability—they must carefully account for how they know what they (claim to) know. Grounded theory method seeks to synthesize a theory from data as the researcher's understanding grows through iterative analysis; it must be accountable to the data as it builds its abstractions. Ethnography and action research also have epistemological accountability, but it tends to be expressed quite differently from (say) experimental studies. Rather than substantiating claims by eliminating alternative explanations, they build credibility through characterizing and disclosing the qualities of the engagement and of the researcher as the instrument of knowing. Design research and system building have what Gaver calls aesthetic accountability—they must show that they "work" in a variety of ways specific to the problem being explored. Ultimately, how well research satisfies accountability requirements will determine its quality and whether it has met its goals. If we misunderstand the way of knowing employed by a study, or apply to it an inappropriate accountability framework, we risk missing the insights that may be contained within it. As mentioned previously, a primary goal of the tutorials on which this volume was based was to educate ourselves about methods and methodologies that were unfamiliar and in which we had little (or no) training. Of course, sometimes what is required of the researcher to understand is difficult or time consuming; for example, Dourish ["Reading and Interpreting Ethnography" this volume] points out that the conceptual claims of ethnography may reside largely in the field's corpus, rather than in single works.

Perhaps the most obvious contrast among ways of knowing is what researchers using the method actually do: the activities carried out to produce and analyze data, and how they engage with users and their contexts of use. Methods vary greatly in the type and quantity of data produced, the issues of data quality that obtain, and the scale of analysis. Ethnography immerses the researcher in a particular context as a participant with the goal of understanding and richly describing lived experience. Action research engages a community with the goal of seeking social change while engaging in systematic enquiry. Many other ways of knowing, such as social network analyses, sensor-based analysis, eye movement analysis or multi-agent simulations build understandings through acquiring and analyzing data from (real or algorithmic) users. These methods, along with log analysis and crowdsourcing, may involve extremely large data sets that require new approaches to analysis. Still other approaches involve elaborate preparation to control (lab experiments) or deploy (field experiments, system building) systems as a way of learning about people and their interactions with technology.

These differences in HCI ways of knowing—in epistemology, methodology, methods, and language—have a direct impact on the field, making it challenging to understand and evaluate all the research being published, to know what methods to teach or to employ in one's own research. Methods can complement each other, of course; indeed, it is already common practice to use multiple methods. The question we ponder is: how far can this be pushed (and should it be?!)? It is common, for example, to combine surveys, log analysis, and interviews. It is common to combine interviews and observational data with grounded theory method. In other words, it is common to combine methods that share or largely overlap in their intellectual commitments. But what about more disparate pairings? Could sensor and eye-tracking data be combined with multi-agent simulation and sophisticated analysis to create more intelligent human–machine partnerships? Could ethnography be crowd-sourced? Or perhaps more realistically, could methods from one intellectual tradition be improved or extended by borrowing from other ways of knowing? In many domains the answer might be "not really"; but in a multifaceted field like HCI we are not so sure.

Another key question to consider is the extent to which the diverse perspectives described herein need to be reconciled or brought into richer communication with each other. The subcommunities carrying out, for example, cognitive modeling, ethnomethodology, research through design, agent simulation, and social network analysis are distinct. In our view, for HCI to remain a coherent intellectual discipline with a cohesive set of practices, it is desirable to have some degree of shared understanding across ways of knowing. The higher purpose of HCI, of course, has always been understanding people and their contexts, how technology interacts with both, and how and to what extent it is possible to adapt technology to people, to their work, play, and aspirations—rather than the other way around. In our experience, it has sometimes been too easy for researchers in one camp to dismiss research based on unfamiliar ways of knowing. That is a shame, since insights about HCI from any way of knowing may provide profound and/or pragmatic inspiration for work based in another genre, if only we could speak enough of each others' languages and share enough understanding to let a variety of works speak to us.

The future of HCI looks bright from where we sit; already a third generation of extremely talented and creative researchers is shaping the field. We fully expect the repertoire of ways of knowing to expand even further. An interesting movement in this direction is a resurgence and evolution of phenomenological approaches, as conceived in works such as Harrison, Sengers, and Tatar's (2011) characterization of "third-paradigm" HCI (also Williams & Irani, 2010), and thoughtful explorations of what a feminist HCI might entail (Bardzell & Churchill, 2011; Bardzell & Bardzell, 2011; Bardzell, 2010). As new ways of knowing in HCI are developed and spread, they can create (if successful) a familiar tension that we might characterize as paradigm shift (Kuhn, 1963). The thing about HCI, however, is that the domain requires a broad repertoire of methods and approaches. The old paradigms do not go away; they continue to function well on the problems for which they are best suited. And although it no doubt does not always feel like it to methodological pioneers, HCI as a field is still young, evolving, and open to change. The challenges of gaining and maintaining a comprehensive understanding of such a rapidly expanding field affects new and seasoned researchers alike. It is our hope that the chapters contained here will serve as a substantive first step in establishing a common ground.

References

Bardzell, S. (2010). Feminist HCI: Taking stock and outlining an agenda for design. *Proceedings of CHI 2010*, pp.1301–1310

Bardzell, S. & Bardzell, J. (2011). Towards a feminist HCI methodology: Social science, feminism, and HCI. *Proceedings of CHI 2011*, pp. 675–684

Bardzell, S., & Churchill, E. (2011). Feminism and HCI: New perspectives. *Interacting with Computers, 23*(5), 385–564.

Card, S., Moran, T., & Newell, A. (1983). *The psychology of human-computer interaction*. Hillsdale, NJ: Lawrence Erlbaum Associates. Now CRC Press.

Carroll, J.M. (1990). Infinite detail and emulation in an ontologically minimized HCI. *Proceedings of CHI 90*, pp. 321–327

Carroll, J. M., & Campbell, R. (1986). Softening up hard science: Reply to Newell and Card. *Human Computer Interaction, 2*(3), 247–249.

Carroll, J.M. & Kellogg, W.A. (1989). Artifact as theory-nexus: Hermeneutics meets theory-based design. *Proceedings of CHI'89*, pp. 8–14

Harrison, S., Sengers, P., & Tatar, D. (2011). Making epistemological trouble: Third-paradigm HCI as successor science. *Interacting with Computers, 23*, 385–392.

Kuhn, T. (1963). *The structure of scientific revolutions*. Chicago, IL: The University of Chicago Press.

Newell, A., & Card, S. K. (1985). The prospects for psychological science in human-computer interaction. *Human Computer Interaction, 1*(3), 209–242.

Whiteside, J., & Wixon, D. (1987). Improving human-computer interaction: A quest for cognitive science. In J. M. Carroll (Ed.), *Interfacing thought: Cognitive aspects of human-computer interaction* (pp. 353–365). Cambridge, MA: Bradford/MIT Press.

Williams, A. & Irani, L. (2010). There's methodology in the madness: Toward critical HCI ethnography. *CHI EA 2010*, pp. 2725–2743

Printed in the United States
By Bookmasters